Birkhäuser

Earthquake Source Physics on Various Scales

Edited by
Adrien Oth
Kevin Mayeda
Luis Rivera

Previously published in *Pure and Applied Geophysics* (PAGEOPH), Volume 171, No. 10, 2014

Birkhäuser

Editors
Adrien Oth
European Center for Geodynamics
and Seismology
19 Rue Josy Welter
7256 Walferdange
Luxembourg

Luis Rivera
Institut de Physique du Globe
Université de Strasbourg
5 Rue René Descartes
67084 Strasbourg
France

Kevin Mayeda
University of California Berkeley
Seismological Laboratory
215 McCone Hall
Berkeley, CA 94720
USA

ISBN 978-3-0348-0867-5 ISBN 978-3-0348-0868-2 (eBook)
DOI 10.1007/978-3-0348-0868-2

Library of Congress Control Number: 2014955440

Cover illustration: Based on Fig. 1 in "Stress Release and Source Scaling of the 2010–2011 Canterbury, New Zealand Earthquake. Sequence from Spectral Inversion of Ground Motion Data" by Adrien Oth and Anna E. Kaiser

Cover design: deblik, Berlin.

Printed on acid-free paper

Springer Basel AG is part of Springer Science+Business Media

www.springer.com

Contents

Pure Appl. Geophys. 171 (2014), 2533–2536
© 2014 Springer Basel
DOI 10.1007/s00024-014-0942-4

Pure and Applied Geophysics

Preface to the Topical Volume *Earthquake Source Physics on Various Scales*

ADRIEN OTH,[1] KEVIN MAYEDA,[2,3] and LUIS RIVERA[4,5]

A robust understanding of the physical processes that occur during earthquake rupture is of key importance for improving our ability to reliably predict earthquake ground motions from future damaging events. The past few decades have seen intensive research in this field and scientists became aware of the remarkable complexity of earthquake rupture, which has been shown to be highly variable at all scales and to span a wide dynamic range, from slow to super-shear earthquakes.

Despite significant advances that have been made and the availability of more and more high-quality data sets, many aspects of the earthquake source process remain unclear and highly controversial. A prime example for this situation is the scaling of source parameters between small and large earthquakes, where conflicting results are still obtained by different research groups, even when considering the same data sets. Scientists around the world are working towards improving the understanding of these open questions, using data sets from earthquakes of all sizes, ranging from induced and natural micro-earthquakes to giant megathrust events. More recently, state-of-the-art laboratory rock friction experiments have also received growing attention, being able nowadays to more closely recreate the in situ conditions on a fault during rupture, such as slip velocity, normal stresses and displacements.

This Topical Volume of *Pure and Applied Geophysics* represents an attempt to bring together studies of the seismic source process on all scales in order to put better constraints on their interlinkage, give a state-of-the-art overview of current knowledge, and contribute to the conciliation of conflicting views. The idea for and objectives of this volume stem from a workshop with the same title, *Earthquake source physics on various scales*, organized by the European Center for Geodynamics and Seismology (ECGS) in Luxembourg in October 2012, where most of the included papers were presented. The 17 articles cover theoretical and observational aspects of the seismic source process, from "laboratory earthquakes" as small as magnitude (M) −6 to source complexity and radiated energy of the world's greatest earthquakes, such as the 2011 M 9 Tohoku event. They address the relationship of earthquake recurrence time with fault frictional parameters, how the results of laboratory friction experiments relate to observational source studies, whether small and large earthquakes scale self-similarly or show differences in their dynamic source characteristics, and how geometrical source complexity can be quantified.

The volume begins with contributions that deal with numerical rupture models and theoretical aspects of earthquake source physics. BIZZARRI and CRUPI investigate the relationship between earthquake recurrence time and fault frictional characteristics on the basis of a synthetic catalogue of nearly 500 sequences using a one-dimensional (1D) single-degree-of-freedom spring-slider model with three versions of rate-dependent and state-dependent friction laws. By exploring the parameter space, they show, among others, that the recurrence time is

[1] European Center for Geodynamics and Seismology (ECGS), 19 Rue Josy Welter, Walferdange 7256, Grand Duchy of Luxembourg. E-mail: adrien.oth@ecgs.lu
[2] Weston Geophysical Corporation, 181 Bedford Street, Suite 1, Lexington, MA 02420, USA. E-mail: kevin.mayeda@gmail.com
[3] University of California Berkeley, Seismological Laboratory, 283 McCone Hall, Berkeley, CA 94720, USA.
[4] Université de Strasbourg, Institut de Physique Du Globe de Strasbourg 5, Rue René Descartes, 67084 Strasbourg, France. E-mail: luis.rivera@unistra.fr
[5] Seismological Laboratory, California Institute of Technology, Pasadena, CA 91125, USA.

directly proportional to recovery stress and inversely proportional to the loading rate, but that the exact relation strongly depends on many aspects, such as the specific form of the used constitutive model. They therefore conclude that even in the most idealized case of a characteristic earthquake model, the exact calculation of recurrence intervals becomes at least highly complicated, if not virtually impossible. Huang *et al.* present two-dimensional (2D) dynamic rupture models of the 2011 Tohoku earthquake, assuming that the fault is characterized by slip-weakening friction and composed of asperities with varying frictional properties. Their dynamic models show rupture behavior (such as final slip, slip rate, and rupture velocity) consistent with observed slip distributions, finding that the subduction wedge significantly promoted the up-dip rupture propagation. Avlonitis and Papadopoulos study an analytical enrichment of a 2D spring-block model of the Olami-Feder-Christensen type. They show that low b values observed in foreshock sequences can be modeled by a process of material softening within the seismogenic volume, i.e., by a process of strain rate increasing with time under constant stress conditions, but that during aftershocks, material processes are not expected to strongly affect the macroscopic b value behavior. Gusev revisits the classical ω^{-2} earthquake source model in order to study its theoretical foundation. To this end, he introduces stochastic elements both into the final structure of the fault and into the mode of rupture propagation, which leads to the definition of a "doubly stochastic" model. The resulting source spectra behave as ω^{-2} at high frequencies, with source spectral level inversely related to the slip pulse width for a given seismic moment and rms stress drop.

Following these numerical and theoretical studies, laboratory scale and micro-seismic mining-induced earthquakes are investigated. McLaskey *et al.* discuss whether mm-scale earthquake-like seismic events generated in laboratory experiments are consistent with our understanding of the physics of large earthquakes in nature. The combined usage of a large experimental apparatus, high-fidelity sensors and rigorous treatment of wave propagation effects allow the authors to show that the M −6 events generated in their apparatus are consistent with double-couple focal mechanisms caused by left-lateral shear-slip on a mm-scale fault patch and show source properties that are consistent with those determined from large natural earthquakes, such as stress drops of the order of 1–10 MPa. Verberne *et al.* perform direct shear tests on calcite-rich fault gouges sheared at sub-seismic sliding velocities. They carry out an in-depth microstructural analysis and show that both wet and dry gauges, while depicting somewhat different levels of steady-state frictional strength, undergo a transition from stable velocity-strengthening to velocity-weakening slip above 80–100 °C. The authors suggest that frictional behavior in their experiments is controlled by a competition between crystal plastic and granular flow processes that are active in the observed shear bands, and discuss the geological implications of their findings. Niemeijer and Collettini carry out rotary shear experiments on rock samples obtained from the low-angle normal Zuccale fault exposed on the Isle of Elba in Central Italy. They vary the testing conditions on the samples, ranging from room temperature to in situ conditions of temperature, normal stress and fluid saturation, and a large range of sliding velocities in order to analyze the effect of the hydrothermal conditions on frictional stability and strength. All their samples exhibit velocity-strengthening behavior under room temperature, while for the strongest samples low in phyllosilicate content, velocity-weakening and stick–slip behavior is observed at high temperatures and sliding velocities. Naoi *et al.* investigate the frequency-magnitude distribution of micro-seismic earthquakes induced around a mining-front in a deep South African gold mine. The question whether or not the frequency-magnitude distribution follows a power law down to such extremely small events is of great importance, since it relates to fault frictional properties. The authors remove data from blast events and show that the frequency-magnitude distribution of their acoustic-emission and accelerometer data obeys a Gutenberg-Richter relationship with a b value of about 1.2 between M −3.7 and 1.

The article of Malagnini, Mayeda *et al.* provides a bridge from the laboratory and microseismic scale to seismological source scaling observations on local and regional scales. The authors study coda-derived moment-rate spectral ratios of 20 crustal earthquake sequences that occurred in various regions of the world. Their analysis indicates that apparent stress and stress drop increase with earthquake size, contrasting the typical findings of self-similar source scaling on large earthquake data sets from a range of previous studies, but that for earthquakes with moment magnitude (M_W) above about 5.5, apparent stress and stress drop seem to stop increasing, and their variability at large magnitude seems to be significantly smaller than at lower magnitudes. The authors link these observations with the results of recent laboratory friction experiments performed at seismic slip velocities, interpreting their scaling results in terms of increasing fault lubrication with increasing slip and consequent slip-weaking up to full lubrication at $M_W > 5.5$. In a second paper, MALAGNINI, MUNAFO' *et al.* extend on these concepts, analyzing in detail the 2009 L'Aquila and the 1994 Northridge seismic sequences. The authors study the scaling properties of two subsets of events belonging to these sequences and also study the scaling characteristics of seismologically estimated fracture energy, proposing a functional form for the evolution of a fault's shear stress with increasing slip that asymptotically tends to a steady-state value. ROVELLI and CALDERONI present a stress drop study of several earthquakes belonging to the 1997–1998 Colfiorito earthquake sequence, which took place in the Umbria–Marche region in Central Italy. The authors use a spectral ratio approach combined with a second step of ground motion modeling to estimate stress drop values, and discuss their results in the context of previous studies that present contradictory findings on stress drop scaling in order to better understand what the reasons for these discrepancies might be, invoking the importance of the crustal transfer function in stress drop computations. The work of SOMEI *et al.* deals with stress drop estimates from coda spectral ratios of nearly 300 earthquakes that occurred in 17 inland crustal earthquake sequences in Japan, spanning a wide magnitude range. Similar to the study of MALAGNINI, MAYEDA *et al.* mentioned above, the authors describe a break in self-similar scaling properties, in their case for events with magnitudes below and respectively above M_W 4.5. They also consider the determined stress drop values in conjunction with focal depth and mechanism, as well as the event locations with respect to the high strain-rate Niigata-Kobe tectonic zone, and conclude that crack size values determined from their corner frequency analysis correspond to the total rupture area of heterogeneous slip models for the large events in their data set. OTH and KAISER apply a generalized inversion technique to study the stress drops of more than 200 events belonging to the 2010–2011 Canterbury sequence on the South Island of New Zealand. With this approach, they isolate the source spectra from path and site contributions and calculate stress drop values based on the classical ω^{-2} model. The results show that the stress drops of these intraplate type events in a low strain-rate region are generally very high, elevated by about a factor of five compared to results from the more seismically active Japan using the same methodology. The authors find nearly self-similar scaling characteristics and discuss the lateral stress drop variations during the sequence, where the highest stress drop values tend to cluster at the fault edges of the two major events, the M_W 7.2 Darfield and M_W 6.2 Christchurch events.

The volume also includes studies that use focal mechanism investigations for gaining insights into earthquake source complexity. A case where such a description is highly valuable is, for instance, related to hydraulic stimulation of enhanced geothermal reservoirs, where the mode of fracturing (e.g., tensile vs. shear) during induced seismic events is of interest. However, for small earthquakes, the usage of full moment tensor inversion often leads to ambiguous results due to insufficient data quality. ŠÍLENÝ *et al.* systematically investigate this issue, using the case of the Soultz-sous-Forêts reservoir in France as an example. They perform a synthetic study to explore how effects such as mislocations, mismodeling, and noise contamination of the data influence the results, showing that the quality of the monitoring system and potential noise contamination are the most important factors for obtaining robust inversion results. With this synthetic study as a foundation, they investigate several observed small earthquakes in the Soultz-sous-Forêts system, finding that their source mechanisms are dominantly pure shear slips on pre-existing faults. ORTEGA *et al.* present a method for analyzing source complexity of earthquakes of different sizes using a simple formulation that relates the elastic constants obtained from independent studies with the angle between the slip and fault normal vector. With this methodology, the authors analyze earthquakes in the Gulf of California that exhibit a significant isotropic component, coming to the conclusion that small and moderate earthquakes are equally complex, and that in areas where the most relevant earthquakes occur on transform faults, full moment tensor inversion better resolves the fault orientation trend as compared to standard deviatoric inversion.

On the top end of the scale range investigated in this volume, Rivera and Kanamori use W phase inversions with teleseismic data to investigate source geometrical complexity of large earthquakes. The authors analyze large earthquakes with $M_W \geq 7.9$ that occurred since 1990 and focus on the variation of the determined scalar moment, the amount of non-double couple component and the overall focal mechanism with frequency. Their results show that these characteristics can be used to quickly diagnose geometrical complexity. Finally, the volume concludes with a study by Baltay *et al.* on the radiated energy characteristics of the six largest earthquakes worldwide that occurred during the past 10 years, where the authors expand on a teleseismic empirical Green's functions technique used in previous work on the 2011 Tohoku earthquake. They present apparent stress values consistent with self-similar source parameter scaling as obtained in a range of studies for smaller events, and find that the 2012 Off-Sumatra strike-slip event shows a higher apparent stress than the megathrust events in their data set, consistent with previous studies that argue for higher apparent stresses in strike-slip events. The technique proposed in their article might be an attractive tool to calculate radiated energies of great earthquakes in near-real time, an approach that could potentially contribute to rapid response measures.

The results presented in the above contributions provide an excellent illustration of the complexities of the earthquake rupture process and the strong potential of combining the states of knowledge from the study of these processes on different scales, but also of the variety of methodological approaches that can be used in this endeavor. In particular, in the source scaling debate, the wide breath of methodologies and their often significantly different underlying assumptions and constraints still hamper a full and clear comparison of results. Only recently, systematic and robust comparisons of the influence of different approaches have become possible through the availability of the vast databases necessary for this purpose, and further studies are therefore required to clarify how much this factor contributes to contradicting findings. As demonstrated by this volume, significant advances have been made to include all possible scales into the study of the seismic source process. However, the most challenging part of the path, i.e., the full integration of the results obtained on these various scales into a unified picture, still lies ahead of us.

Acknowledgments

We thank the authors of the papers for their contributions. Most of the papers included in this volume were presented at the 2012 ECGS workshop *Earthquake Source Physics on Various Scales*, held in Luxembourg and co-funded by ECGS, the National Research Fund of Luxembourg (FNR), the Council of Europe Major Hazards Agreement (EUR-OPA) and the National Museum for Natural History (Mnhn) of Luxembourg. We thank all the participants who made that workshop such a big success. We also thank the reviewers of each paper for their critical and constructive comments that helped to improve the scientific quality of the volume, and Renata Dmowska, Editor-in-chief for Topical Issues in PAGEOPH, as well as the editorial staff at Springer for their dedication and continuous support.

Pure Appl. Geophys. 171 (2014), 2537–2553
© 2013 Springer Basel
DOI 10.1007/s00024-013-0743-1

Pure and Applied Geophysics

Linking the Recurrence Time of Earthquakes to Source Parameters: A Dream or a Real Possibility?

Andrea Bizzarri[1] and Paola Crupi[2]

Abstract—By using a single-degree-of-freedom spring-slider analog fault model, we generate a synthetic catalog of nearly 500 different seismic sequences. We explore the parameter space by assuming different values of constitutive parameters and tectonic environment. We also consider three different versions of the rate-dependent and state-dependent friction laws [the Dieterich-Ruina (DR), the Ruina-Dieterich (RD) and the Chester-Higgs (CH) models], and different approximations of the behavior of the friction at high sliding speeds, as well as the radiation damping effects. Our results indicate that for all the considered models, the recurrence time (T_{cycle}) exhibits an inverse proportionality on the loading rate; a linear, positive dependence on the effective normal stress; and a linear, negative dependence on the characteristic distance controlling the state variable evolution. These results confirm and generalize previous studies. Remarkably, we found here that the coefficients of proportionality strongly depend on the adopted friction model, on the high speed behavior and on the reference set of parameters. Notably, we also found that the positive proportionality between T_{cycle} and the difference $b - a$, confirmed for DR and RD laws, does not hold in general for the CH law. Overall, we conclude that even in the simplest (and idealized) case of characteristic earthquakes considered here, in which the limiting cycle is reached by the system, and even in the framework of a very simplified fault model, the possibility to a priori predict, through an universal analytical relation, the inter-event time of an impending earthquake still remains only a dream. On the other hand, a numerical prediction of T_{cycle} would require the exact knowledge of the rheological model (and its parameters at all times over the entire life of the fault) and the actual state of the fault, which indeed are often unknown.

Key words: Fault mechanics, recurrence time, earthquake prediction, constitutive models, computational seismology.

[1] Istituto Nazionale di Geofisica e Vulcanologia, Sezione di Bologna, Via Donato Creti, 12, 40128 Bologna, Italy. E-mail: bizzarri@bo.ingv.it

[2] Dipartimento di Scienze della Terra e Geoambientali, Università degli Studi di Bari, Via Orabona 4, 70125 Bari, Italy. E-mail: paola.crupi@uniba.it

1. *Introduction*

The grand challenge of seismology is to predict earthquakes, which is not only a very important scientific goal, but is also a major societal objective, both in terms of human losses and economic damages.

Contrary to other ambits of physics (e.g., in quantum mechanics the Heisenberg's uncertainty principle establishes that it is theoretically impossible to exactly predict the position and the momentum of a particle, simultaneously; Heisenberg 1927), there are no theoretical reasons impeding the spatio–temporal prediction of an earthquake event (Bizzarri 2012b). Despite technical and epistemic difficulties, the prediction of earthquakes still remains a prominent objective for the entire scientific community. Indeed, seismology is a relatively young discipline, and it suffers some severe limitations, represented by (1) the epistemic, relatively incomplete understanding of the physics of earthquakes; and by (2) the ignorance about the state of stress of the faults. In other words, our knowledge of both the equations to be solved (more explicitly, the fault constitutive model, which complements the equation of motion) and of the initial conditions is still lacking, making the physics of the earthquake source a very peculiar exact science. Moreover, we do not have the possibility to plan experiments at the real-world scale, such as scientists in the fields of high energy physics or molecular biology do. On the other hand, we can perform cutting-edge numerical simulations, but do not yet have yet a general consensus on the most appropriate equations to describe the physical behavior of a seismic source (see Bizzarri 2011b for a thorough review). This is in contrast, for instance, to climatology, which still can not plan experiments, but knows the equations describing the problems to be handled.

Statistical analysis—which accepts that several properties of the faulting processes are out of range and can be described only in probability terms (Vere-Jones 2010)—relies on seismological records, and is therefore affected by the relative infrequency of large events. These events can then provide only a limited data set for the study of their impact on modern cities (e.g., Allen 2007). Moreover, some authors have shown (Wyss *et al.* 2012; Kossobokov and Nekrasova 2012; see also Zuccolo *et al.* 2011) that the seismic hazard maps generated by the Global Seismic Hazard Assessment Program (GSHAP; www.seismo.ethz.ch/static/GSHAP), do not correctly give the seismic hazard for disastrous earthquakes with $M \in (6.9, 8.6)$. Indeed, Kossobokov and Nekrasova (2012), by performing a quantitative comparison between the GSHAP maps and the factual effects under strong earthquakes, formulate a verdict useless for this kind of probabilistic product in the framework of any type of seismic hazard evaluation. The scientific debate on this topic is indeed vigorous (see for instance Panza *et al.* 2011, 2012). It is also true that during the last 50 years there have been insufficient seismological records to constrain the dynamics of a seismogenic fault over a time window of 100–1,000 years. In other words, the earthquake cycle could often be far larger than the duration of the instrumental, scientific observations.

The physical (or deterministic) description of earthquake sources—which excludes any random process—is affected by the epistemic, severe limitations discussed above. Finally, we mention here that the so-called neo-deterministic seismic hazard analysis (NDSHA)—which originated after a careful scrutiny of the standard hazard maps (Wyss *et al.* 2012)—produces a set synthetic ground motions based upon the physics of the earthquake source and wave propagation models (e.g., Panza *et al.* 2001; Zuccolo *et al.* 2011 and referenced cited therein).

By considering a single-degree-of-freedom mass-spring analog fault model (e.g., Gu *et al.* 1984), Bizzarri (2012b) emphasizes that a large number of factors can dramatically influence the recurrence time T_{cycle}; (1) fault interactions and stress triggering phenomena, (2) the analytical expression of the fault governing law, (3) the thermally-activated pressurization of pore fluids, (4) wear processes, (5) the variations of the hydraulic properties of the fault zone (such as permeability and porosity) and (6) the Arrhenius nature of the so-called direct effect in the framework of the rate-dependent and state-dependent friction laws (e.g., Ruina 1983). Moreover, (7) Crupi and Bizzarri (2013) demonstrate that the analytical formulation of the equation of motion of the spring–slider system (more explicitly, the incorporation of the contribution of the energy lost as radiating waves, ignored for a long time in previous 1-D models) has relevant consequences in the determination of the seismic cycle time. Finally, (8) Bizzarri and Crupi (2013) point out that even the initial thermal state of the fault can have significant effect on the prediction of T_{cycle} in the framework of a rate-dependent, state-dependent and temperature-dependent fault rheology (Chester and Higgs 1992). All these results can somehow endanger the concept itself of seismic cycle.

In this paper, we intentionally neglect all the above-mentioned complications, and assume a constant effective normal stress. We consider a perfectly homogeneous and isolated fault, so that the 1-D spring-slider model adopted in the numerical computation presented and discussed in the present study can be adequate to describe, at least as a first-order approximation, such a seismogenic system; the limitations of the model are discussed in more details in Sect. 4. By conducting a large number of numerical experiments (we produce an ensemble of nearly 500 different seismic sequences), we analyze whether it is possible to extract from our catalog of synthetic events a universal, analytical (empirical) equation that can fit all the data, with the aim to explore whether it is possible to analytically express a priori the cycle time as a function of the source parameters.

2. *Methodology*

As anticipated in the previous section, we consider the single-degree-of-freedom spring-slider model, so that the we deal with a 1-D fault model, in which the unique independent variable is time.

The analytical and numerical details are summarized in Appendix A. The fault system can be governed by different formulations of the rate-dependent and state-dependent friction laws, namely the Dieterich-Ruina (DIETERICH 1994), the Ruina-Dieterich (RUINA 1983) and the Chester-Higgs (CHESTER and HIGGS 1992) constitutive models (for the remainder of the present paper, we will refer to those laws with the notations DR, RD and CH, respectively). In these fault governing models, recalled for completeness in Appendix A [see Eqs. (8), (9) and (10)], there are basically three constitutive parameters; a, which accounts for the so-called direct effect and represents a velocity-hardening preparatory stage in the early stages of the rupture; b, which controls the evolutionary effects and basically describes the stress release process; and L, which represents the characteristic slip distance over which the state variable Ψ evolves. The latter physically accounts for the memory effects (i.e., previous slip instability episodes) of the sliding interface, and has a different meaning (and dimension), depending on the adopted model (see BIZZARRI 2011b; his Sect. 7).

We consider two rather different configurations, listed in Table 1. Configuration A, which is configuration A of CRUPI and BIZZARRI (2013), corresponds to a strong velocity-weakening regime, characterizing a fault with a relatively high degree of instability. Configuration B, taken from LAPUSTA and BARBOT (2012) and BARBOT *et al.* (2012), still defines a velocity-weakening regime, but with a lower degree of instability. We also emphasize that, changing from configuration A to configuration B, we do not only modify the constitutive parameters a and b, but also the hypocentral depth (and thus the resulting effective normal stress, σ_n^{eff}, and the initial temperature, T_0, which is assumed to follow a geothermal gradient), the initial shear stress and the sliding velocity, as well as the tectonic loading rate. Both the configurations fail in the unstable regime, in that the ratio κ between the critical stiffness, $k_{cr} = (b - a)\sigma_n^{eff}/L$, and the machine stiffness, k, exceeds 1, so that the limiting

Table 1

Reference parameters adopted in this paper

Parameter	Value	
	Configuration A	Configuration B
Model parameters		
Loading velocity, v_{load}	3.17×10^{-10} m/s	1.17×10^{-9} m/s
Machine stiffness, k	10 MPa/m	7.5×10^{-2} MPa/m
Tectonic loading rate, $\dot{\tau}_0 = kv_{load}$	3.17×10^{-3} Pa/s	8.775×10^{-3} Pa/s
Vibration period of the analog freely slipping system, $T_{a.f.} = 2\pi\sqrt{m/k}$	5 s[a]	5 s[a]
Radiation damping constant, c (if present)	4.5 MPa s/m	2.86 MPa s/m
Critical velocity, v_c	1×10^{-4} m/s	1×10^{-4} m/s
Threshold velocity defining a seismic instability, v_l	0.1 m/s	0.1 m/s
Fault constitutive parameters		
Effective normal stress, σ_n^{eff}	30 MPa	100 MPa
Initial sliding velocity, v_0	3.17×10^{-10} m/s ($=v_{load}$)	1.17×10^{-9} m/s ($=v_{load}$)
Initial state variable, Ψ_0	3.15×10^7 s ($=L/v_0$)	8.55×10^6 s ($=L/v_0$)
Initial shear stress, τ_0	16.8 MPa ($=\mu_* \times \sigma_n^{eff}$)	65 MPa ($=\mu_* \times \sigma_n^{eff}$)
Initial temperature, T_0	483.15 K (=210 °C)	373.15 K (=100 °C)
Logarithmic direct effect parameter, a	0.008	0.010
Evolution effect parameter, b	0.016	0.014
Characteristic scale length, L	0.01 m	0.01 m
Critical stiffness, $k_{cr} = (b - a)\ \sigma_n^{eff}/L$	24 MPa/m	40 MPa/m
Reference value of the friction coefficient, μ_*	0.56	0.65
Reference value of the sliding velocity, v_*	3.17×10^{-10} m/s ($=v_0$)	1×10^{-6} m/s
Reference value of the temperature, T_*	210 °C ($=T_0$)	100 °C ($=T_0$)

The initial conditions (denoted by the subscript 0) refer to $t = 0$

[a] We do not change the period of the analog freely slipping system, because we are not interested in performing a stability analysis of the system

cycle is reached (Gu *et al.* 1984). As mentioned above, we emphasize that the limiting cycle can not be always reached if we include in the model additional complications, such as wear mechanism and thermal pressurization (see Bizzarri 2012b and references cite therein) or if we consider the Burridge-Knopoff (e.g., Erickson *et al.* 2011) with special parameters settings.

In order to explore whether the behavior at high speeds can influence the results, we have also considered the so-called frozen approximation of the friction law. In this approximation, the steady state friction becomes independent of the sliding velocity for speeds larger than a threshold value v_T (Weeks 1993; Bizzarri 2012c). In this case, the formulation of the RD law is expressed by Eq. (12); the counterparts for the DR and the CH can be derived in a straightforward way. In this paper, we also assume that $v_T = v_*$ (see Table 1); the results pertaining to the frozen simulations are reported as open symbols in the subsequent figures.

Finally, we have also considered the introduction of the radiation damping term (RDT thereinafter; Rice 1993; Beeler *et al.* 2001), which is known to control the dynamics of the system [Crupi and Bizzarri 2013; see Eq. (13)].

Overall, we have thoroughly explored the parameter space of both Configuration A and B, and we generated a synthetic catalog. From this catalog, we extracted the resulting cycle time (T_{cycle}). The latter, which is the limiting cycle of this system (e.g., Thompson and Stewart 2002), is determined as the time separating two subsequent instabilities: $T^{[n]}_{cycle} \underset{df}{=} t^{[n]} - t^{[n-1]}$ (see, among many other examples, Fig. 3 of Rice and Tse 1986) where $t^{[n]}$ is the time occurrence of the nth instability, which in turn is defined when the sliding speed exceeds a threshold value (v_l; see Table 1).

The aim of the present paper is not to perform a detailed analysis of the stability of the system. Indeed, the single mass-spring damped harmonic oscillator is described by a system of three nonlinear ordinary differential equations (ODEs), which have universal properties summarized in a very comprehensive way by Thompson and Stewart (2002). The body of the literature about the stability analysis of the spring-slider system is immense; among many others we cite here, the early studies of Gu *et al.* (1984) and Rice and Tse (1986), and the more recent generalizations of Ranjith and Rice (1999), Erickson *et al.* (2008), and Putelat *et al.* (2008, 2010, 2012). As stated above, we want the present work to be a contribution in the stream of papers dealing with the problem of the prediction of earthquake recurrence. In particular, we aim to explore whether it is possible, from a numerical and/or analytical point of view, to predict, in a deterministic way, the recurrence time of the next earthquake.

3. Numerical Results

3.1. Variation of a and b

In Fig. 1a, we plot the simulated cycle times depending on different values of the constitutive parameter a. We can clearly see that as long as a increases, T_{cycle} decreases with a linear pattern, and this holds for all the considered friction laws in the case of Configuration A (blue symbols). Remarkably, the slope of the T_{cycle} vs. a curve (namely, the ratio dT_{cycle}/da) is the same for a given friction law, but it changes if we consider a different set of parameters; this is apparent from the slope of the dashed lines. Moreover, that slope also depends on the adopted constitutive model; this is evident by comparing, for instance, the slope pertaining to the DR law (full blue squares), that pertaining to the RD law (full blue triangles) and that of the CH law (full blue circles). On the other hand, if we modify the frictional behavior at high speeds (i.e., if we include the frozen approximation), even for the same governing equation, the ratio dT_{cycle}/da changes. The same result holds for the DR and the RD if we also add the radiation damping term (RDT; small open squares and rectangles).

The behavior is quite different if we consider the CH law. Indeed, in the case of configuration B, we observe a net increase of T_{cycle} with increasing a (full red circles). On the other hand, the same behavior is obtained for configuration A if we consider the frozen approximation (open blue circles) and if we add the RDT (small open circles). To interpret this result, we have to remember that parameter a is responsible for

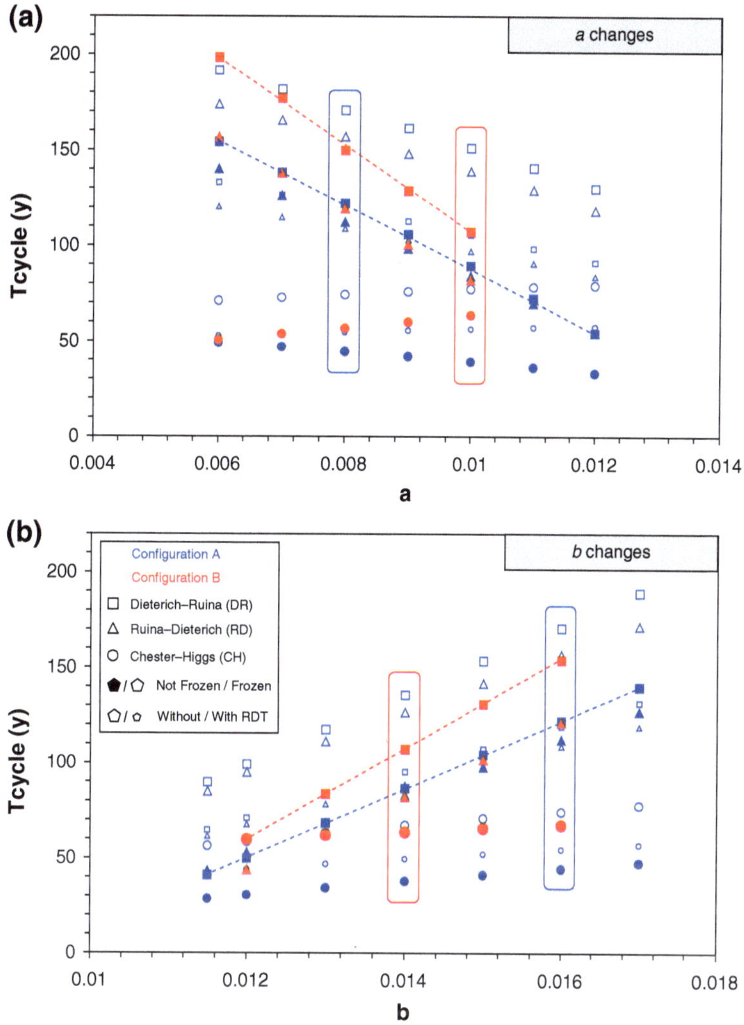

Figure 1
Cycle time (T_{cycle}) resulting by changing the constitutive parameters a [(panel (**a**)] and b [panel (**b**)]. All the other parameters are kept unchanged with respect to those listed in Table 1. *Blue symbols* refer to configuration A, while red symbols refer to configuration B. *Squares*, *triangles* and *circles* pertain to the Dieterich-Ruina [DR, Eq. (8)], Ruina-Dieterich [RD, Eq. (9)] and Chester-Higgs [CH, Eq. (10)] laws, respectively. *Empty symbols* denote numerical simulations where the frozen approximation at high speeds has been considered [see Eq. (12)]. Finally, *small symbols* indicate synthetic earthquakes in which the radiation damping term is included in the equation of motion [see Eq. (13)]. The *vertical rectangles* emphasize the reference set of parameters for both configuration A and configuration B

the so-called direct effect of friction, and it determines the amount of frictional stress that has to be exceeded in order to cause slip instability and thus the stress release (see also Bizzarri 2011b for a discussion). In the framework of the CH law, the frictional resistance depends explicitly on the temperature [see Eq. (10)], and this dependence is particularly relevant at sustained speeds. Once the slider tends to accelerate (so that the sliding speed becomes to be significant), the frictional heat starts to increase rapidly [see Eq. (11)], and its contribution to the value of the frictional resistance is not negligible. In other words, since the temperature affects the rupture threshold, the effect of the changes in a can be somehow masked by the overall dynamics of the system. A similar situation happens when we consider the parameter a varying explicitly with the temperature, so that $a = a(T)$ (Bizzarri 2012a).

This result suggests, even at this early stage of exploration of the parameter space, that it is very

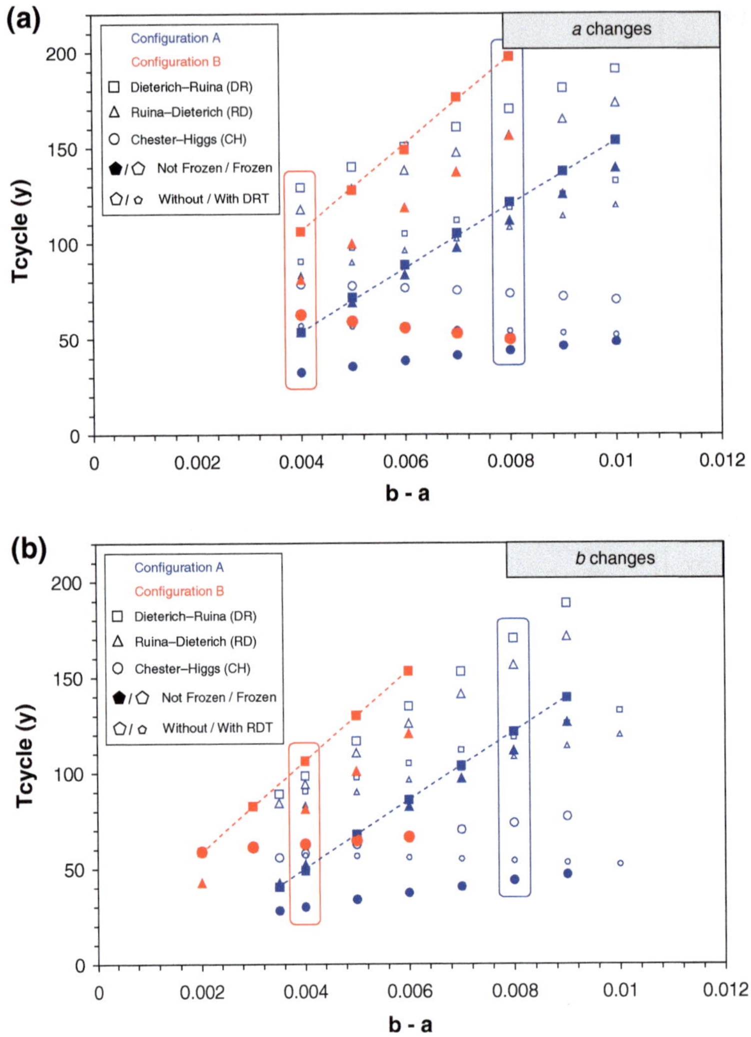

Figure 2

The same as in Fig. 1, but now we change the difference $b - a$. **a** Only the parameter a is changed. **b** Only the parameter b is changed

complicated to find an universal, analytical, empirical relationship expressing the value of the recurrence time as a function of the governing parameters, especially in the case of the CH friction model.

In Fig. 1b, the behavior of T_{cycle} for various b (and keeping all the other parameters as in Table 1) is plotted. Here we can see that, in general, T_{cycle} has a direct proportionality to b, for all the friction models (DR, RD and CH), configurations (A and B) and approximations (frozen or not, with and without the RDT). Nevertheless, the slope of the T_{cycle} vs. b curve (i.e., the ratio dT_{cycle}/db) is slightly different in the various cases, although it is constant for a given law and configuration (as emphasized by the dashed lines in Fig. 1b).

The results discussed above can be summarized by plotting T_{cycle} vs. the difference $b - a$; this is done in Fig. 2a (where only a is changed) and in Fig. 2b (where only b is changed). We know that recurrence time depends on the recovery time, i.e., on the time required to increase the traction from the friction level attained after a slip instability up to the upper yield stress, which defines the rupture point (the latter physically represents the maximum shear stress a fault can sustain by elastic deformations). The situation is described in Fig. 3 of Bizzarri and Crupi

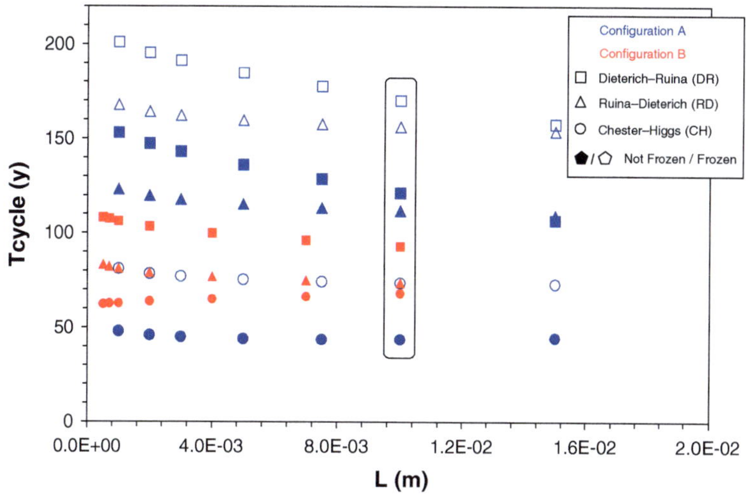

Figure 3
Behavior of T_{cycle} depending on the different values of the characteristic distance (L) controlling the evolution of the state variable. The *symbols* have the same meaning as in Fig. 1

(2013). Both the minimum and maximum traction depend on $b - a$, as also discussed by BIZZARRI and COCCO [2003; see their Eqs. (12) and (13)]. Since the difference $b - a$ is directly proportional to the degree of instability of a fault [in other words, the greater $b - a$ is, the larger amount of stress is released; see also GU and WANG (1991) and HE *et al.* (2003)], we theoretically expect that the stress to be recovered during the interseismic fault restrengthening phase, which increases too, therefore leading to longer inter-event times. This general behavior emerges from Fig. 2, with the only exception of the CH law occurring when a is changed (see open blue and full red circles in Fig. 2a). This reflects exactly what we observe in Fig. 1a.

3.2. *Variation of the Characteristic Length*

The constitutive parameter L is the characteristic length scale appearing in the time evolution of the state variable. Physically, it represents the amount of slip over which the state variable Ψ evolves from one steady state to another (e.g., RUINA 1983). Moreover, it is related to the breakdown time (BIZZARRI and COCCO 2003), which is spent during the process of the stress release (stress drop). Since the various governing models considered here have a different evolution law [see Eqs. (8)–(10)], we expect to have different values of the recurrence time even for the same value of L; this is exactly what we observe in Fig. 3, where we report the behavior of T_{cycle} for different L. With the exception of the CH for configuration B (full red circles), we can envisage from Fig. 3 a nearly linear decrease of T_{cycle} for increasing L. Contrary to what happens for variations of a and b, in the case of the variation of L, the introduction of the frozen approximation at high speeds does not change the slope of the T_{cycle} vs. L curve, but merely causes an increase of the recurrence time.

3.3. *Variation of the Effective Normal Stress*

Figure 4 reports the variations of T_{cycle} for varying σ_n^{eff}. The variation of the effective normal stress is associated with different focal depth and/or different values of the pore fluid pressure in the fault zone. It is clear that the dependence of the recurrence time on the effective normal stress is linear over a very broad range of compressive stresses. This behavior is a stable feature for both the configurations A and B and for all the considered friction laws. Also in this case, we can see that the introduction of the frozen approximation increases the value of T_{cycle}, but does not deface the linear dependence between the two quantities, although the slope of the T_{cycle} vs. σ_n^{eff} curve tends to increase if we include the frozen approximation. These results are consistent with the theoretical predictions; indeed, when σ_n^{eff} increases,

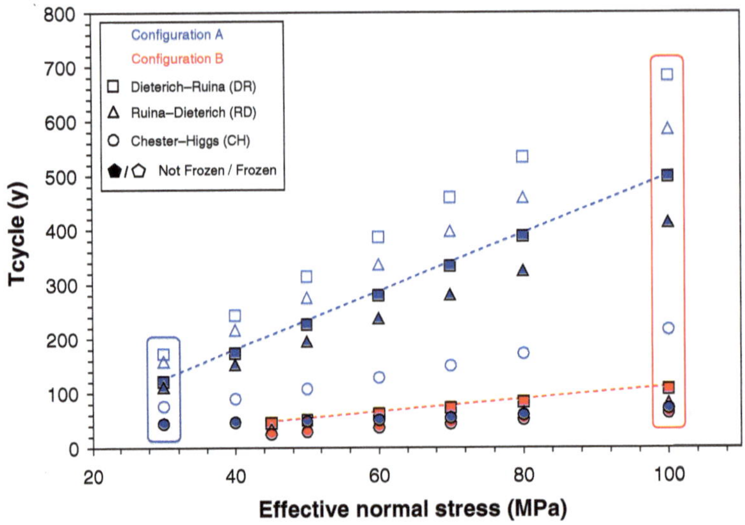

Figure 4
Behavior of T_{cycle}, depending on the different values of the effective normal stress (σ_n^{eff}). All the other parameters, including T_0 and T_*, are those tabulated in Table 1

the upper yield stress (τ_u^{eq} in the naming convention of Bizzarri and Cocco 2003) also increases. At the same time, also the dynamic stress drop increases, because the fault becomes more unstable; we recall that although $\beta \equiv B/A$ remains constant by changing σ_n^{eff} ($B = b\sigma_n^{eff}$ and $A = a\sigma_n^{eff}$), the other parameter controlling the instability of the fault, $\kappa \equiv k_{cr}/k$ (Bizzarri 2011a), increases for increasing σ_n^{eff}. As a net result, the stress to be recovered after a dynamic instability in order to again reach the rupture point increases for increasing effective normal stress, thus leading to longer inter-event times.

An interesting issue is that the slope of the T_{cycle} vs. σ_n^{eff} curve strongly depends on the adopted governing model; in particular, for both the configurations A and B, the DR law is the model that is the most sensitive to variations of σ_n^{eff}, while the CH model is the least sensitive.

3.4. *Variation of the Loading Rate*

The last dependence we explore is that of the loading rate. We change $\dot{\tau}_0$ by changing the value of the loading velocity (v_{load}) and by keeping k (and thus κ) constant; this means that we do not change the rheological properties of the fault, but only the tectonic environment (i.e., the external conditions). Physically, we expect that the larger $\dot{\tau}_0$ is, the shorter T_{cycle} is. This makes sense, because when the loading force applied to the spring increases, then the interseismic recovery phase is faster (and correspondingly T_{cycle} is smaller). This expectation is verified from the results of our synthetic catalog, as depicted in Fig. 5, from which we can clearly appreciate an inverse dependence of T_{cycle} on $\dot{\tau}_0$. Remarkably, this holds for all the adopted governing laws, although, as for the dependence on σ_n^{eff}, the DR law is the most sensitive to variations in $\dot{\tau}_0$, the CH is the least sensitive, and the RD is intermediate between the two. This is a stable feature for both configuration A and configuration B.

The dependence of T_{cycle} on $\dot{\tau}_0$ can be also understood from the analysis of the phase diagrams shown in Fig. 6. They refer to configuration B, but the results are qualitatively the same for Configuration A. We denote with the $\Delta\tau_{rec}$ the amount of stress that the fault has to recover to again reach the rupture point (and thus cause a new slip instability). It is clear that as the loading rate increases, the resulting $\Delta\tau_{rec}$ decreases, because, for the DR and the RD laws, the minimum sliding speed and the final level of friction increase, and because, at the same time, the upper yield stress decreases (Fig. 6a, b). Interestingly, in the case of the CH (Fig. 6c) the opposite holds, in that as

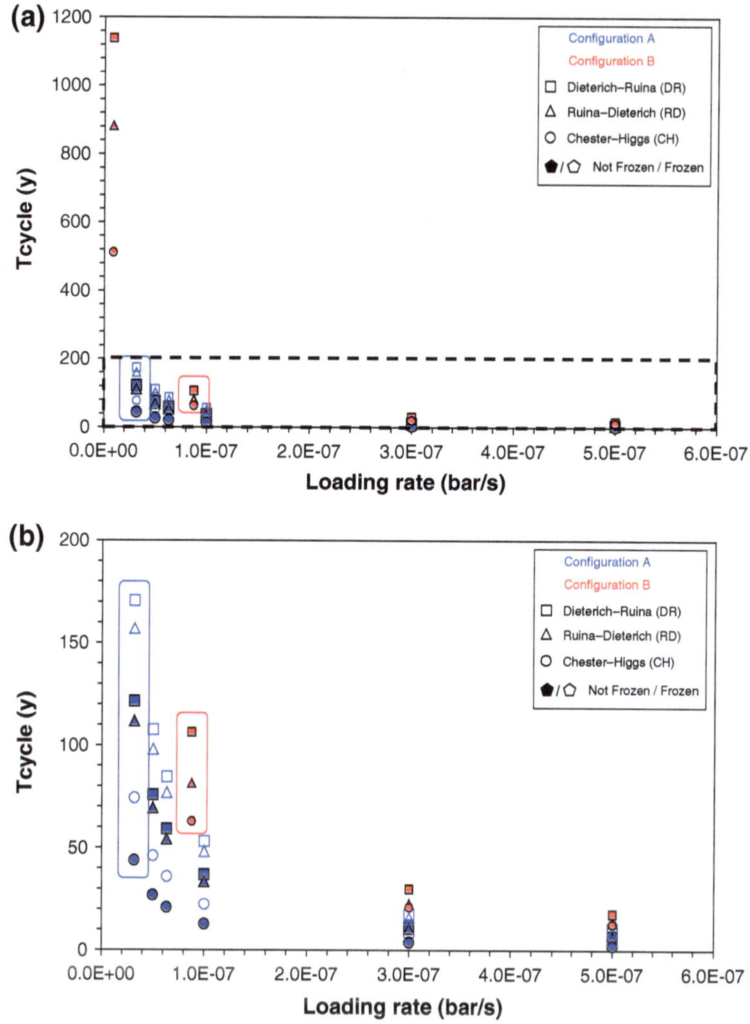

Figure 5
Behavior of T_{cycle} depending on the different values of the loading rate ($\dot{\tau}_0$). In panel (**b**), we expand the *dashed region* marked in panel (**a**)

$\dot{\tau}_0$ increases, the minimum sliding speed and the final level of friction decrease, but the upper yield stress increases. As observed previously, this "anomalous" behavior (if compared to the DR and the RD law) is imputable to the explicit dependence of friction on developed temperature. The net result for the CH law is that $\Delta\tau_{rec}$ decreases for increasing $\dot{\tau}_0$, but more slowly than in the case of DR and RD laws.

4. Discussion

In this paper, we have considered a single-degree-of-freedom mass-spring analog fault system (1-D) to investigate whether it is possible to express the recurrence time of earthquakes (T_{cycle}) through an universal, analytical equation depending on the constitutive properties of the seismogenic structure. This amenable picture requires some caveats. First of all, we emphasize that the concept itself of the cycle time (or limiting cycle) can somehow be misleading; indeed, as discussed by Bizzarri (2012b) and in Sect. 1, some different mechanisms can take place during faulting, affecting the overall dynamics of the seismogenic structure and therefore making the concept of T_{cycle} even meaningless. In idealized conditions, it is well known that a velocity-weakening 1-D system will reach its limiting cycle (e.g., Gu

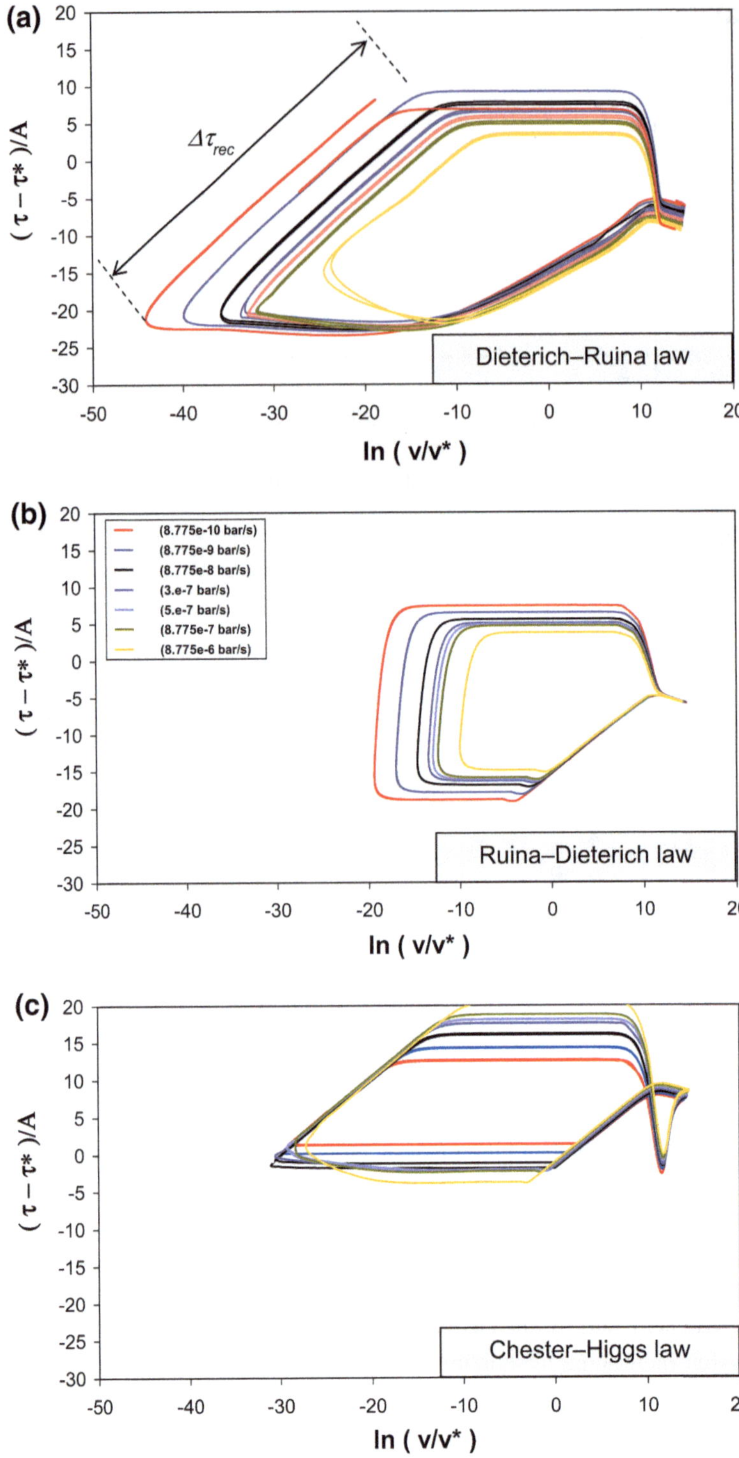

Figure 6

Phase diagrams in the case of configuration B for various values of the loading rate. **a** Orbits for the DR law, which correspond to the full *red squares* of Fig. 5a. **b** Orbits for the RD law, which correspond to the full red triangles of Fig. 5a. **c** Orbits for the CH law, which correspond to the full *red circles* of Fig. 5a. All the quantity are normalized; τ_* and v_* are references values for the shear stress and the sliding velocity, respectively *et al.* 1984, among many others), which is characterized by the same recurrence time, because the states of the dynamical system (expressed in terms of sliding speed and shear stress) are defined by a limiting orbit in the phase space (i.e., the limiting cycle; see THOMPSON and STEWART 2002). This behavior mathematically expresses the concept of the characteristic earthquake (CE; REID 1910). This simplistic (and idealized) picture is dramatically altered if we account for possible phenomena, such as stress triggering and fault interaction, wear processes on the fault core, permeability and porosity evolution, etc. (BIZZARRI 2012b). Practically, these phenomena can impede the fault from reaching its limiting cycle, so that T_{cycle} can not be determined, but this intimately depends on the actual state of the fault (which is usually unknown for natural faults).

Another important point is inherently related to the choice of the analog fault model. As pointed out earlier by RICE and TSE (1986), the single mass-spring model is certainly oversimplified, in that it predicts that an asperity fails once it reaches its critical state; indeed, we can expect that at natural conditions some patches on the fault can reach the failure point and nucleate the rupture. Moreover, the spring-slider model exhibits only a dependence on time, implying that all the properties of the fault are spatially homogeneous and that they can be represented, in an average sense, by the adopted values of the constitutive parameters. Indeed, the role of spatial heterogeneities in the fault rheology are important in the whole dynamic of a seismic fault zone (e.g., OGLESBY and DAY 2002; TINTI *et al.* 2005; AMPUERO *et al.* 2006; RIPPERGER *et al.* 2008; BIZZARRI *et al.* 2010; SONG and SOMMERVILLE 2010). Therefore, the single spring-slider model can only be regarded as a first approximation of the behavior of a real-world fault. Indeed, since it naturally represents the stick-slip mechanism, it has been successfully employed to study the traction evolution on a seismic fault and to simulate repeated earthquake events (BOATWRIGHT and COCCO 1996; BEELER *et al.* 2001; BIZZARRI 2010; MITSUI and COCCO 2010 among many others). As pointed out by BIZZARRI and BELARDINELLI (2008), the agreement between results from a single spring-slider model and from an continuum, extended (3-D) fault model is excellent during the slow nucleation phase; during the subsequent acceleration phase, in a given fault point, the 3-D model experiences the contribution of the stress redistribution on the fault (i.e., of the load due to the neighboring points that are already slipping and releasing stress). This phenomenon is obviously neglected in the single spring-slider model. We also mention here the comparison performed by WEATHERLEY and ABE (2004) between a 1-D chain of mass-block models and a lattice solid model.

Given these significant limitations, we have considered here the most simple situation in which the system reaches the limiting cycle; in such a situation, the system forgets the initial conditions ($t = 0$). In this case, which represents the CE model, we can write (see also BARBOT *et al.* 2012):

$$T_{\text{cycle}} = \frac{\Delta\tau_{\text{rec}}}{\dot{\tau}_0} \tag{1}$$

The inverse proportionality between stress drop and loading velocity (recall that the tectonic loading rate equals $k\ v_{\text{load}}$) and the direct proportionality between recurrence intervals and stress drop have been previously shown in the case of the RD law by CAO and AKI (1986) and GU and WANG (1991) and extended to the DR model by HE *et al.* (2003).

In the framework of the rate-dependent and state-dependent friction laws, both the maximum yield stress (τ_u^{eq}) and the dynamic stress drop [$\Delta\tau_d$; see Eq. (12) of BIZZARRI 2012b] are a priori unpredictable (BIZZARRI 2003). This translates into the fact that $\Delta\tau_{\text{rec}}$ can not be predicted exactly a priori.

By adopting two rather different configurations (see Table 1), and by adopting three different governing models (the Dieterich-Ruina (DR), the Ruina-Dieterich (RD) and the Chester-Higgs (CH) laws; see Appendix A), we have generated a synthetic catalog of nearly 500 different seismic sequences.

Our simulations clearly confirm for all the adopted governing equations (including the CH model, the frozen versions of governing laws and with the RDT) the inverse proportionality between T_{cycle} and $\dot{\tau}_0$ stated by Eq. (1) (see Fig. 5), which is relatively obvious from a physical point of view. Moreover, our numerical results suggest, in general, the following dependencies:

$$T_{\rm cycle} \propto b - a \quad (2)$$

(see Fig. 2),

$$T_{\rm cycle} \propto \sigma_n^{\rm eff} \quad (3)$$

(see Fig. 4), and

$$T_{\rm cycle} \propto -L \quad (4)$$

(see Fig. 3).

The dependencies we found in Eqs. (2) and (3) confirm the results that Gu and Wang (1991) and He *et al.* (2003) obtained for the canonical formulations (i.e., without frozen and without RDT) of the RD and DR models, respectively. Indeed, they found [see Eq. (4) in Gu and Wang 1991] that

$$\Delta\tau_{\rm rec} = (b - a)\,\sigma_n^{\rm eff}\left[K - \zeta \ln\left(\frac{v_{\rm load}}{v_*}\right)\right] \quad (5)$$

where K is a parameter with an inverse proportionality with $1/\kappa$ and having a weak dependence on m and $\zeta \approx 2$. Remarkably, Gu and Wang (1991) and He *et al.* (2003) do not consider the dependence of $\Delta\tau_{\rm rec}$ (and thus of $T_{\rm cycle}$) on L.

Lapusta *et al.* (2012) and Barbot *et al.* (2012), in order to fit the recurrence times of the Parkfield segment of the San Andreas fault with the DR friction law, proposed the following analytical relation:

$$T_{\rm cycle} = \frac{(b - a)\,\sigma_n^{\rm eff}}{\dot{\tau}_0} \ln\left(\frac{v_{\rm co}}{v_{\rm inter}}\right) \quad (6)$$

where $v_{\rm co}$ is a coseismic values of the fault slip velocity (they use $v_{\rm co} = 1$ m/s) and $v_{\rm inter}$ is the interseismic fault slip velocity (they use $v_{\rm inter} = 1 \times 10^{-12}$ m/s, so that the correcting, non-dimensional factor $\ln\left(\frac{v_{\rm co}}{v_{\rm inter}}\right)$ equals 27.63). The parameters of Eq. (6) have been tuned in order to reproduce a posteriori the observed recurrence times at Parkfield. We note that Eq. (6) has been derived by assuming that the stress drop between two subsequent instabilities can be determined analytically in terms of the governing parameters and of the two velocities $v_{\rm co}$ and $v_{\rm inter}$.

In general, $\sigma_n^{\rm eff}$ can somehow be estimated, based on the representative hypocentral depth of a seismogenic zone, and $\dot{\tau}_0$ can be inferred from geological and seismotectonic observations. However, from a predictive point of view, Eq. (6) has some problems, in that it is very difficult to a priori constraint the values of $v_{\rm co}$ and $v_{\rm inter}$; both of them are known to vary over several orders of magnitude, especially $v_{\rm inter}$ (see for instance Fig. 6, from which is it clear the high variability of the sliding speeds attained during the interseismic phase). Moreover, it is also known that $v_{\rm co}$ can be higher than 1 m/s; for instance, the thermal pressurization of pore fluids, the flash heating of micro-asperity contacts (and in general every mechanism causing a dramatic fault weakening) are usually associated with very high values of coseismic slip peak (see for instance Bizzarri 2012a). We emphasize that a variation of a factor of 10 in $\frac{v_{\rm co}}{v_{\rm inter}}$—very common in numerical simulations—causes a variation roughly of 8 % in $T_{\rm cycle}$. Incidentally, we mention here that if we ad hoc tune the values of $\frac{v_{\rm co}}{v_{\rm inter}}$ differently for all the set of parameters used in the DR models, we can adequately reproduce the observed cycle time with Eq. (6), but we emphasize that his tuning is done a posteriori and without any physical or observational constraints.

Moreover, Eqs. (6)— and (5) as well—do not explicitly incorporate the dependence on the characteristic distance L, which controls the evolution of the state variable, and which, although small, cause variations of $T_{\rm cycle}$ (see Fig. 3).

In the framework of the rate-dependent and state-dependent friction laws, the dynamic stress drop can be associated with the difference $b - a$ (e.g., Ruina 1983; Gu *et al.* 1984; Gu and Wang 1991; He *et al.* 2003, among many others). However, we have seen that it is extremely difficult to a priori predict the two states in the phase plane which determine $\Delta\tau_{\rm rec}$ (i.e., the minimum in (v, τ) and the upper yield stress), and therefore it is hard to exploit Eq. (1) in order to give a quantitative prediction of the recurrence time. This is particularity true in the case of the CH law, for which the explicit dependence on temperature complicates, with respect to the DR and the RD laws, the traction evolution at high speeds. More explicitly, temperature evolution controls the rupture point (i.e., the upper yield stress level), the dynamic stress release and the final level of friction (and thus the level of friction at which the long-stage interseismic phase begins).

5. Conclusions

The quantitative determination of the recurrence time of an earthquake definitively remains one of the most pursued objectives in seismology. It has been already demonstrated that the concept of recurrence time itself can be misleading or even meaningless, in that many causes can alter the idealized cyclic behavior postulated by the theory of characteristic earthquakes (CE; see the discussion on BIZZARRI 2012b).

At a fundamental level, in addition to statistical inferences (i.e., probabilistic predictions), which are disregarded here, there are analytical and numerical predictions of the recurrence of a next earthquake. The former can be ideally retrieved theoretically, or empirically inferred from data. The latter is a result of a specific numerical experiment, where the physics of the fault, as well as all the parameters of the model, are assumed to be known.

The laboratory-derived, macroscopic theory of the rate-dependent and state-dependent friction laws provides a framework in which the whole life of a seismogenic structure can be modeled, and thus the cycle time can be calculated. In this framework, by using a 1-D spring-slider dashpot model and neglecting all possible phenomena that can cause deviations from the CE model, we have generated a synthetic catalog of nearly 500 different seismic sequences, by considering a large class of fault governing models and approximations. Through a systematic exploration of the parameter space, we confirm that T_{cycle} is directly proportional to the recovery stress ($\Delta\tau_{rec}$; e.g., Fig. 6a) and inversely proportional to the loading rate ($\dot{\tau}_0$; see Fig. 5); see Eq. (1). We also found a linear, positive dependence of T_{cycle} on the effective normal stress [σ_n^{eff}; see Fig. 4 and Eq. (3)], previously found in some cases. Moreover, our numerical experiments indicate a linear, negative dependence of T_{cycle} on the characteristic distance [L; see Fig. 3 and Eq. (4)]. Remarkably, these dependences hold for all the governing models considered here, the Dieterich-Ruina (DR), the Ruina-Dieterich (RD) and the Chester-Higgs (CH) laws, and also with the frozen approximation and or with the RDT. Our study therefore generalizes previous findings by GU and WANG (1991) and HE *et al.* (2003).

One of the major outcome of the present study is that the proportionality coefficients strongly depend on the specific form of the adopted constitutive model, on its behavior at high speeds and on the possible introduction of the radiation damping term (mimicking the energy lost as radiated seismic waves).

Moreover, we also found that the positive proportionality between T_{cycle} and the difference $b - a$, confirmed in general for DR and RD laws, does not hold for the CH law in some cases (see Fig. 2). This difference is basically attributed to the effect of the temperature developed by frictional heat, which directly controls the evolution of traction in the CH case, because it is incorporated explicitly in the governing equations.

As an overall conclusion, although our catalog is qualitatively compatible with previous relations inferred for the RD by LAPUSTA and BARBOT (2012); (see also BARBOT *et al.* 2012) in the case of the DR and the RD laws, the exact, a priori determination of the recurrence time (through an universal, analytical equation) is very complicated, even in the simplest (and idealized) case of the CE model. This is because the proportionality coefficients depend on the adopted friction law and on the global set of model parameters (just as an example, from Fig. 4 we have that the slope of the T_{cycle} vs. σ_n^{eff} curve for the DR is significantly different from configuration A to configuration B). The a priori prediction, again through a unique equation, of the recurrence time is virtually impossible if we postulate that the traction evolution of the fault is described by the CH law; in such a case, we can even invert the proportionality relation observed for the other two friction models (see Fig. 2).

To conclude, we have shown that, even in the simplest case of the CE models considered here, the belief that it is possible to predict the recurrence time of a next earthquake through an universal, *analytical*, empirical equation is not supported by numerical evidence. Indeed, the results presented here can provide some general guidance about how the recurrence intervals can depend on the adopted model parameters.

As mentioned above, there is another approach to try to predict (still in a deterministic framework) the recurrence time. Numerical simulations (such as those performed here or even with more elaborated models, such a Burridge-Knopoff or continuum models) should enable us to have a *numerical* estimate of a subsequent earthquake. But the problem here is that, in spite of significant advances in the understanding of the plate tectonics and fault structures, we currently know little about the stress state in the Earth (initial and external conditions) and about the most proper friction law that can describe a given earthquake sequence (earthquake source physics). Moreover, we have to exactly know the values of the governing parameters (input data in the forward models), which are often not so well constrained by independent observation or can even change through time (i.e., during the whole life of the fault), making the problem highly nonlinear.

From another point of view, it might be emphasized that recurrence intervals are not the best way to infer the frictional properties of a given seismogenic structure, because we do not know if the available fault constitutive models are adequate to describe the whole life of that fault, or simply which is the best one among the various possibilities. However, variations in stress drop in repeating earthquakes can give some indication on the healing rate (e.g., Marone 1998; Marone *et al.* 1995), which becomes dominant once the fault has slowed significantly. At the same time, the analysis of repeating slip failures are used to constrain the loading rate (e.g., Nadeau and McEvilly 1999; Igarashi *et al.* 2003).

The present exercise falls in the stream of papers dealing with the earthquake prediction, and can be regarded as a possible, although preliminary, way to connect the source physics to the earthquake prediction.

Acknowledgments

The first author (A.B.) thanks the editors of this topical volume, A. Oth, K. Mayeda and L. Rivera, for having invited him to contribute, and G. F. Panza for some discussions on the general topic of earthquakes prediction. The guest editor, A. Oth, and three anonymous referees are acknowledged for their stimulating comments and criticism, which contributed to improvements to the paper.

Appendix: Numerical Details of the Model and Constitutive Equations

In this section, we briefly recall all the equations adopted in the present study. Readers can refer to the cited references for a more verbose discussion of the equations.

As mentioned in Sect. 2, we use here the single-degree-of-freedom (1-D) spring-slider analog fault model. The equation of motion of such a system is that of an harmonic oscillator (far of being exhaustive, we mention here Gu *et al.* 1984; Gu and Wang 1991; He *et al.* 2003; Ranjith and Rice 1999):

$$m\ddot{u} = k(u_{load} - u) - \tau \qquad (7)$$

in which the overdots indicate the time derivative, m is the mass per unit surface, k denotes the elastic constant of the spring (which mimics the elastic behavior of the medium surrounding the fault), u_{load} is the displacement of the loading point (which moves at the prescribed velocity $v_{\text{load}} \equiv \dot{u}_{\text{load}}$), u is the displacement and τ is the frictional resistance acting of the sliding surface. As described in the following, the choice of the analytical expression of τ defines the adopted governing model. The elastodynamic problem for our 1-D fault system is solved numerically by using the fourth-order Runge-Kutta algorithm with auto-adaptive time stepping (Press *et al.* 1992).

The single mass-spring system is an obvious simplification of the more elaborated Burridge-Knopoff model (see Burridge and Knopoff 1964; Carlson and Langer 1989a, b; Carlson *et al.* 1991; Erickson *et al.* 2011, among many others), where multiple masses are connected with springs. This inherently discrete fault model can account for spatial heterogeneities and it is a first step toward the continuum finite difference or finite element models.

In the present paper, we use various governing laws; the Dieterich-Ruina model (Dieterich 1978, 1994):

$$\begin{cases} \tau = \left[\mu_* + a\ln\left(\frac{v}{v_*}\right) + b\ln\left(\frac{\psi v_*}{L}\right)\right]\sigma_n^{\text{eff}} \\ \frac{\text{d}}{\text{d}t}\psi = 1 - \frac{\psi v}{L} \end{cases} \qquad (8)$$

the Ruina-Dieterich model (Ruina 1983; Marone 1998):

$$\begin{cases} \tau = \left[\mu_* + a\ln\left(\frac{v}{v_*}\right) + b\ln\left(\frac{\psi v_*}{L}\right)\right]\sigma_n^{eff} \\ \frac{d}{dt}\psi = -\frac{\psi v}{L}\ln\left(\frac{\psi v}{L}\right) \end{cases} \quad (9)$$

and the Chester-Higgs model (CHESTER and HIGGS 1992; CHESTER 1994):

$$\begin{cases} \tau = \left[\mu_* + a\ln\left(\frac{v}{v_*}\right) + b\ln\left(\frac{\Psi v_*}{L}\right) + \frac{aQ_a}{R}\left(\frac{1}{T} - \frac{1}{T_*}\right)\right]\sigma_n^{\rm eff} \\ \frac{d}{dt}\Psi = -\frac{\Psi v}{L}\left[\ln\left(\frac{\Psi v}{L}\right) + \frac{Q_b}{R}\left(\frac{1}{T} - \frac{1}{T_*}\right)\right] \end{cases} \quad (10)$$

In Eqs. (8)–(10) v is the sliding velocity, Ψ is a state variable (an empirical variable accounting for the previous sliding history of the system and usually interpreted as the average characteristic lifetime of contacting asperities; MARONE 1998; RUINA 1983), a, b and L are the constitutive parameters, μ_* and v_* are reference values of the friction coefficient and sliding velocity, respectively (namely, μ_* is the steady state value of the friction coefficient when the sliding velocity equals v_*) and $\sigma_n^{\rm eff}$ is the effective normal stress (assumed constant through time in the present modeling). In Eq. (10) Q_a and Q_b are apparent activation energies (we use $Q_a = Q_b = 1 \times 10^5$ J/mol as in CHESTER 1994), R is the universal gas constant, T is the (absolute) temperature produced by frictional heat and T_* is a (absolute) reference temperature value. T is computed as follows (MCKENZIE and BRUNE 1972; KATO 2001; BIZZARRI and CRUPI 2013):

$$T(t) = T_0 + \frac{1}{2C\sqrt{\pi\chi}}\int \frac{\tau(t')\,v(t')}{\sqrt{t - t'}}\,dt' \quad (11)$$

in which T_0 is the (absolute) initial temperature, C is the heat capacity of the bulk composite for unit volume [we use $C = 3 \times 10^6$ J/(m^3 K)], χ is the thermal diffusivity (we use $\chi = 1 \times 10^{-6}$ m^2/s) and t is time.

To perform a more comprehensive comparison of the various governing models that can be used to describe the traction evolution, we also consider a modification of the canonical rate-dependent and state-dependent friction laws, where the steady state friction (defined as the friction when the state variable is constant through time, i.e., when $\frac{d}{dt}\Psi = 0$) becomes independent on the sliding velocity at high speeds. Namely, for v greater than a threshold value v_T the Ruina-Dieterich model (10) becomes:

$$\begin{cases} \tau = \left[\mu_* + a\ln\left(\frac{v_T}{v_*}\right) + b\ln\left(\frac{\psi v_*}{L}\right)\right]\sigma_n^{\rm eff} \\ \frac{d}{dt}\psi = -\frac{\psi v}{L}\ln\left(\frac{\psi v_T}{L}\right) \end{cases} \quad (12)$$

This modification, basically due to WEEKS (1993), comes from the laboratory inferences of SCHOLZ and ENGELDER (1976) and DIETERICH (1978).

Finally, we consider a modification of the equation of motion (7), where the so-called radiation damping term (RDT) is introduced (XU and KNOPOFF 1994; BEELER 2001, 2006; BEELER *et al.* 2002; CRUPI and BIZZARRI 2013),

$$m\ddot{u} = k(u_{load} - u) - \tau - cv \quad (13)$$

where

$$c \equiv \frac{G}{2\,v_S} \quad (14)$$

with G being the rigidity of the elastic medium and v_S the S wave velocity away from the fault plane. Physically, the radiation damping approximation (RICE 1993) mimics the energy lost due to seismic wave propagation, which is inherently present continuum fault models.

REFERENCES

ALLEN, R. M. (2007), Earthquake hazard mitigation: New directions and opportunities, In: Treatise on Geophysics (G. Schubert Ed.). 4—Earthquake seismology (H. Kanamori Ed.), 607–648, Elsevier, doi:10.1016/B978-044452748-6/00083-3.

AMPUERO, J.-P., RIPPERGER, J., and P. M. MAI (2006), Properties of dynamic earthquake ruptures with heterogeneous stress drop, in *Radiated Energy and the Physics of Earthquakes*, Abercrombie, R., A. McGarr, H. Kanamori, and G. di Toro (Editors), Geophys. Monogr., *170*, American Geophysical Union, Washington, DC., 255–261.

BARBOT, S., LAPUSTA, N., and J.-P. AVOUAC (2012), Under the hood of the earthquake machine: Toward predictive modeling of the seismic cycle, Science, *336*, 707–710.

BEELER, N. M. (2001), Stress drop with constant, scale independent seismic efficiency and overshoot, Geophys. Res. Lett., *28*, No. 17, 3353–3356.

BEELER, N. M. (2006), Inferring earthquake source properties from laboratory observations and the scope of lab contributions to source physics, In Radiated Energy and the Physics of Faulting, Abercrombie, R., A. McGarr, H. Kanamori, and G. di Toro (Editors), Geophys. Monogr., *170*, American Geophysical Union, Washington, DC., 99–119.

BEELER, N. M., HICKMAN, S. H., and T.-F. WONG (2001), Earthquake stress drop and laboratory-inferred interseismic strength recovery, J. Geophys. Res., *106*, No. B12, 30,701–30,713.

Beeler, N. M., Wong, T.-F., and T. E. Tullis (2002). Slip-weakening distance and fracture energy during dynamic sliding with laboratory-inferred fault strength, Proc. 2002 SCEC Meeting, Oxnard, CA.

Bizzarri, A. (2010), On the recurrence of earthquakes: Role of wear in brittle faulting, Geophys. Res. Lett., *37*, L20315, doi:10.1029/2010GL045480.

Bizzarri, A. (2011a), Temperature variations of constitutive parameters can significantly affect the fault dynamics, Earth Plan. Sci. Lett., *306*, 272–278, doi:10.1016/j.epsl.2011.04.009.

Bizzarri, A. (2011b), On the deterministic description of earthquakes, Rev. Geophys., *49*, RG3002, doi:10.1029/2011RG000356.

Bizzarri, A. (2012a), Rupture speed and slip velocity: What can we learn from simulated earthquakes?, Earth Plan. Sci. Lett., *317–318*, 196–203, doi:10.1016/j.epsl.2011.11.023.

Bizzarri, A. (2012b), What can physical source models tell us about the recurrence time of earthquakes?, Earth–Science Reviews, *115*, 304–318, doi:10.1016/j.earscirev.2012.10.004.

Bizzarri, A. (2012c), Formulation of a fault governing law at high sliding speeds: Inferences from dynamic rupture models, Earth Plan. Sci. Lett., *355–356*, 223–230, doi:10.1016/j.epsl.2012.09.007.

Bizzarri, A., and M. E. Belardinelli (2008), Modelling instantaneous dynamic triggering in a 3–D fault system: application to the 2000 June South Iceland seismic sequence, Geophys. J. Int., *173*, 906–921, doi:10.1111/j.1365-246X.2008.03765.x.

Bizzarri, A., and M. Cocco (2003), Slip–weakening behavior during the propagation of dynamic ruptures obeying rate– and state–dependent friction laws, J. Geophys. Res., *108*, No. B8, 2373, doi:10.1029/2002JB002198.

Bizzarri, A., and Crupi, P. (2013), Is the initial thermal state of a fault relevant to its dynamic behavior?, Bull. Seism. Soc. Am., *100*, No. 3, 2062–2069, doi:10.1785/0120120279.

Bizzarri, A., Dunham, E. M., and P. Spudich (2010), Coherence of Mach fronts during heterogeneous supershear earthquake rupture propagation: Simulations and comparison with observations, J. Geophys. Res., *115*, B08301, doi:10.1029/2009JB006819.

Boatwright, J., and M. Cocco (1996), Frictional constraints on crustal faulting, J. Geophys. Res., *101*, No. B6, 13,895–13,909, doi:10.1029/96JB00405.

Burridge, R., and L. Knopoff (1964), Body force equivalents for seismic dislocations, Bull. Seismol. Soc. Am., *54*, 1875–1888.

Cao, T., and K. Aki (1986), *Seismicity simulation with a rate- and state-dependent fricion law*, Pure Appl. Geophys., *124*, No. 3, 487–513.

Carlson, J. M., and J. S. Langer (1989a), Properties of earthquakes generated by fault dynamics, Phys. Rev. Lett., *62*, 2632–2635, doi:10.1103/PhysRevLett.62.2632.

Carlson, J. M., and J. S. Langer (1989b), A mechanical model of an earthquake fault, Phys. Rev. A, *40*, 6470–6484, doi:10.1103/PhysRevA.40.6470.

Carlson, J. M., Langer, J. S., Shaw, B. E. and C. Tang (1991), Intrinsic properties of a Burridge–Knopoff model of an earthquake fault, Phys. Rev. A, *44*, No. 2, 884–897.

Chester, F. M. (1994), Effects of temperature on friction: Constitutive equations and experiments with quartz gouge, J. Geophys. Res., *99*, 7247–7261, doi:10.1029/93JB03110.

Chester, F. M. and H. G. Higgs (1992). Multimechanism friction constitutive model for ultrafine quartz gouge at hypocentral conditions, J. Geophys. Res., *97*, B2, 1859–1870.

Crupi, P., and Bizzarri, A. (2013), The role of radiation damping in the modeling of repeated earthquake events, Ann. Geophys., *56*, No. 1, R0111, doi:10.4401/ag-6200.

Dieterich, J. H. (1978), Time–dependent friction and the mechanics of stick slip, Pure Appl. Geophys., *116*, 790–806, doi:10.1007/BF00876539.

Dieterich, J. H. (1994), A constitutive law for rate of earthquake production and its application to earthquake clustering, J. Geophys. Res., *99*, No. B2, 2601–2618.

Erickson, B., Birnir, B., and D. Lavallée (2008), *A model for aperiodicity in earthquakes*, Nonlin. Proc. Geophys., *15*, 1–12.

Erickson, B. A., Birnir, B., and D. Lavallée (2011), Periodicity, chaos and localization in a Burridge–Knopoff model of an earthquake with rate–and–state friction, Geophys. J. Int., *187*, 178–198, doi:10.1111/j.1365-246X.2011.05123.x.

Gu, J. C., Rice, J. R., Ruina, A. L., and S. T. Tse (1984), Slip motion and stability of a single degree of freedom elastic system with rate and state dependent friction, J. Mech. Phys. Sol., *32*, 167–196, doi:10.1016/0022-5096(84)90007-3.

Gu, Y., and T. -f. Wong (1991), Effects of loading velocity, stiffness, and inertia on the dynamics of a single degree of freedom spring–slider system, J. Geophys. Res., *96*, No. B13, 21,677–21,691.

He, C., Wong, T. -f., and N. M. Beeler (2003), Scaling of stress drop with recurrence interval and loading velocity for laboratory-derived fault strength relations, J. Geophys. Res., *108*, 2037, doi:10.1029/2002JB001890.

Heisenberg, W. (1927), Über den anschaulichen Inhalt der quantentheoretischen kinematik und mechanik, Zeitschrift für Physik, *43*, Nos. 3–4, 172–198.

Igarashi, T., Matsuzawa, T., and A. Hasegawa (2003), Repeating earthquakes and interplate aseismic slip in the northeastern Japan subduction zone, J. Geophys. Res., *108*, No. B5, doi:10.1029/2002JB001920.

Kossobokov, V. G., and A. K. Nekrasova (2012), Global seismic hazard assessment program maps are erroneous, Seismic Instruments, *48*, No. 2, 162–170.

Kato, N. (2001). Effect of frictional heating on pre-seismic sliding: a numerical simulation using a rate-, state- and temperature-dependent friction law, Geophys. J. Int., *147*, 183–188.

Lapusta, N., and S. Barbot (2012), Models of earthquakes and aseismic slip based on laboratory–derived rate and state friction laws, In: The Mechanics of Faulting: From Laboratory to Real Earthquakes (A. Bizzarri and H. S. Bhat Eds.), Research Signpost, ISBN 978-81-308-0502-3.

Marone, C. J. (1998), The effect of loading rate on static friction and the rate of fault healing during the earthquake cycle, Nature, *391*, 6662, 69–72.

Marone, C. J., Vidale, J. E., and W. L. Ellsworth (1995), Fault healing inferred from time dependent variations in source properties of repeating earthquakes, Geophys. Res. Lett., *22*, No. 22, 3095–3098, doi:10.1029/95GL03076, 1995.

McKenzie, D., and J. N. Brune (1972). Melting on fault planes during large earthquakes, Geophys. J. R. Astr. Soc., *29*, 65–78.

Mitsui, Y., and M. Cocco (2010), The Role of porosity evolution and fluid flow in frictional instabilities: A parametric study using a spring–slider dynamic system, Geophys. Res. Lett., *37*, L23305, doi:10.1029/2010GL045672.

Nadeau, R. M., and T. V. McEvilly (1999), Fault slip rates at depth from recurrence intervals of repeating microearthquakes, Science, *285*, No. 5428, 718–721, doi:10.1126/science.285.5428.718.

OGLESBY, D. D., and S. M. DAY (2002), Stochastic fault stress: Implications for fault dynamics and ground motion, Bull. Seismol. Soc. Am., *92*, 3006–3021.

PANZA, G. F., ROMANELLI, F., and F. VACCARI (2001), Seismic wave propagation in laterally heterogeneous anelastic media: theory and applications to seismic zonation, Adv. Geophys., *43*, 1–95.

PANZA, G. F., IRIKURA, K., KOUTEVA, M., PERESAN, A., WANG, Z., and R. PARAGONI (2011), Advanced seismic hazard assessment, Pure App. Geophys., *169*, 1–9, doi:10.1007/s00024-010-0179-9.

PANZA, G. F., LA MURA, C., PERESAN, A., ROMANELLI, F., and F. VACCAI (2012), Seismic hazard scenarios as preventive tools for a disaster resilient society, Advances in Geophysics, *53*, 93–165.

PRESS, W. H., TEUKOLSKY, B. P., and W. T. VETTERLING (1992), *Numerical Recipes*, 2nd ed., Cambridge University Press, New York.

PUTELAT, T., DAWES, J. H. P., and J. R. WILLIS (2008), On the seismic cycle seen as a relaxation oscillation, Phil. Mag., *88*, Nos. 28–29, 3219–3243, doi:10.1080/14786430802216374.

PUTELAT, T., DAWES, J. H. P., and J. R. WILLIS (2010), Regimes of frictional sliding of a spring–block system, J. Mech. Phys. Solids, *58*, 27–53.

PUTELAT, T., WILLIS, J. R., and J. H. P. DAWES, (2012), Wave-modulated orbits in rate-and-state-friction, Int. J. Non-Lin. Mech., *47*, 258–267, doi:10.1016/j.ijnonlinmec.2011.05.016.

RANJITH, K., and J. R. RICE (1999), Stability of quasi–static slip in a single degree of freedom elastic system with rate and state dependent friction, J. Mech. Phys. Sol., *47*, No. 6, 1207–1218.

RICE, J. R. (1993), Spatio-temporal complexity of slip on a fault, J. Geophys. Res., *98*, No. B6, 9885–9907.

REID, H. F. (1910), The California earthquake of April 18, 1906. Report of the State Investigation Commission, The Mechanics of the Earthquake, vol. 2. Carnegie Inst., Washington, DC.

RIPPERGER, J., MAI, P. M., and J.-P. AMPUERO (2008), Variability of near–field ground motion from dynamic earthquake rupture simulations, Bull. Seismol. Soc. Am., *98*, 1207–1228.

RUINA, A. L. (1983), Slip instability and state variable friction laws, J. Geophys. Res., *88*, 10,359–10,370.

RICE, J. R. and S. T TSE (1986), Dynamic motion of a single degree of freedom system following a rate and state dependent friction law, J. Geophys. Res., *91*, 521–530.

SCHOLZ, C. H., and J. T. ENGELDER (1976), The role of asperity indentation and ploughing in rock friction: I. Asperity creep and stick–slip, Int. J. Rock Mech. Min. Sci. Geomech. Abstr., *13*, 149–154.

SONG, S. G., and P. SOMERVILLE (2010), Physics–based earthquake source characterization and modeling with geostatistics, Bull. Seismol. Soc. Am., *100*, No. 2, 482–496.

THOMPSON, J. M. T., and H. B. STEWART (2002), *Nonlinear Dynamics and Chaos*, 2nd ed., John Whiley and Sons, New York.

TINTI, E., BIZZARRI, A., and M. COCCO (2005), Modeling the dynamic rupture propagation on heterogeneous faults with rate– and state–dependent friction, Ann. Geophys., *48*, No. 2, 327–345.

VERE-JONES, D. (2010), Foundations of statistical seismology, Pure Appl. Geophys., *167*, 645–653, doi:10.1007/s00024-010-0079-z.

WEATHERLEY, D., and S. ABE (2004), Earthquake statistics in a block slider model and a dully dynamic fault model, Nonlin. Proc. Geophys., *11*, 553–560.

WEEKS, J. D. (1993), Constitutive laws for high–velocity frictional sliding and their influence on stress drop during unstable slip, J. Geophys. Res., *98*, No. B10, 17,637–17,648.

WYSS, M., NEKRASOVA, A. K., and V. G. KOSSOBOKOV (2012), Errors in expected human losses due to incorrect seismic hazard estimates, Natural Hazards, doi:10.1007/s11069-012-0125-5.

XU, H.-J. and L. KNOPOFF (1994), Periodicity and chaos in a one-dimensional model of earthquakes, Phys. Rev. E, *50*, 5, 3577–3581.

ZUCCOLO, E., VACARI, F., PERESAN, A., and G. F. PANZA (2011), Neo-deterministic and probabilistic hazard assessments: a comparison over the Italian territory, Pure Appl. Geophys., *168*, 69–83, doi:10.1007/s.00024-010-0151-8.

(Received March 18, 2013, revised November 11, 2013, accepted November 14, 2013, Published online December 14, 2013)

Pure Appl. Geophys. 171 (2014), 2555–2568

DOI 10.1007/s00024-013-0718-2

Pure and Applied Geophysics

Slip-Weakening Models of the 2011 Tohoku-Oki Earthquake and Constraints on Stress Drop and Fracture Energy

Yihe Huang,[1] Jean-Paul Ampuero,[1] and Hiroo Kanamori[1]

Abstract—We present 2D dynamic rupture models of the 2011 Tohoku-Oki earthquake based on linear slip-weakening friction. We use different types of available observations to constrain our model parameters. The distribution of stress drop is determined by the final slip distribution from slip inversions. As three groups of along-dip slip distribution are suggested by different slip inversions, we present three slip-weakening models. In each model, we assume uniform critical slip distance eastward from the hypocenter, but several asperities with smaller critical slip distance westward from the hypocenter. The values of critical slip distance are constrained by the ratio of deep to shallow high-frequency slip-rate power inferred from back projection source imaging. Our slip-weakening models are consistent with the final slip, slip rate, rupture velocity and high-frequency power ratio inferred for this earthquake. The average static stress drop calculated from the models is in the range of 4.5–7 MPa, though large spatial variations of static stress drop exist. To prevent high-frequency radiation in the region eastward from the hypocenter, the fracture energy needed there is in the order of 10 MJ/m^2, and the average up-dip rupture speed cannot exceed 2 km/s. The radiation efficiency calculated from our models is higher than that inferred from seismic data, suggesting the role of additional dissipation processes. We find that the structure of the subduction wedge contributes significantly to the up-dip rupture propagation and the resulting large slip at shallow depth.

Key words: Tohoku-Oki earthquake, dynamic rupture model, stress drop, fracture energy, energy partitioning, subduction wedge.

1. *Introduction*

Analyses of a wealth of data generated by the 2011 Tohoku-Oki earthquake have unveiled several unique features: (1) Rupture propagated through the shallow region (defined here as the region up-dip of the hypocenter), and it resulted in a large slip. This is supported by slip inversions (Simons 2011; Ide *et al.* 2011; Yue and Lay 2011; Yagi and Fukahata 2011; Lee *et al.* 2011; Wei *et al.* 2012; Iinuma *et al.* 2012) and static measurements by ocean-bottom pressure gauges and bathymetric data (Fujiwara *et al.* 2011; Ito *et al.* 2011; Sato *et al.* 2011; Kido *et al.* 2011; Kodaira *et al.* 2012). Several slip models are schematically shown in Fig. 1a. (2) High-frequency (~1 Hz) energy radiation was mostly concentrated down-dip from the hypocenter. Under the assumption that the advancing front of high-frequency radiation coincides with the rupture front, the down-dip rupture velocity was estimated to be as low as ~1 km/s (Meng *et al.* 2011; Kiser and Ishii 2012).

The extensive observations available for this event warrant efforts to understand the basic physics responsible for these unique observations. To this end, we carry out dynamic rupture simulations for this earthquake. Previous studies suggest that heterogeneities of fault friction or stress are needed to explain the spatial variations of rupture behavior during this earthquake (Kato and Yoshida 2011; Aochi and Ide 2011; Duan 2012; Goto *et al.* 2012; Huang *et al.* 2012) as well as the complex temporal characteristics of historical earthquakes (Igarashi *et al.* 2003; Tajima *et al.* 2013). Furthermore, a key question is what causes the rupture to propagate through the shallow region. Numerical simulations suggest that waves reflected inside the subduction wedge induce large transient stress changes on the fault, which promote the up-dip rupture propagation (Huang *et al.* 2012) despite the stable, velocity-strengthening frictional behavior expected in fault zones at shallow depth (Kozdon and Dunham 2013). Other models invoke fault weakening mechanisms in the shallow region, such as shear heating of pore fluids, to promote unstable slip (Yoshida and Kato 2011; Noda and Lapusta 2013).

[1] Division of Geological and Planetary Sciences, California Institute of Technology, 1200 E. California Blvd. MS 252-21, Pasadena, CA 91125, USA. E-mail: yihe@gps.caltech.edu

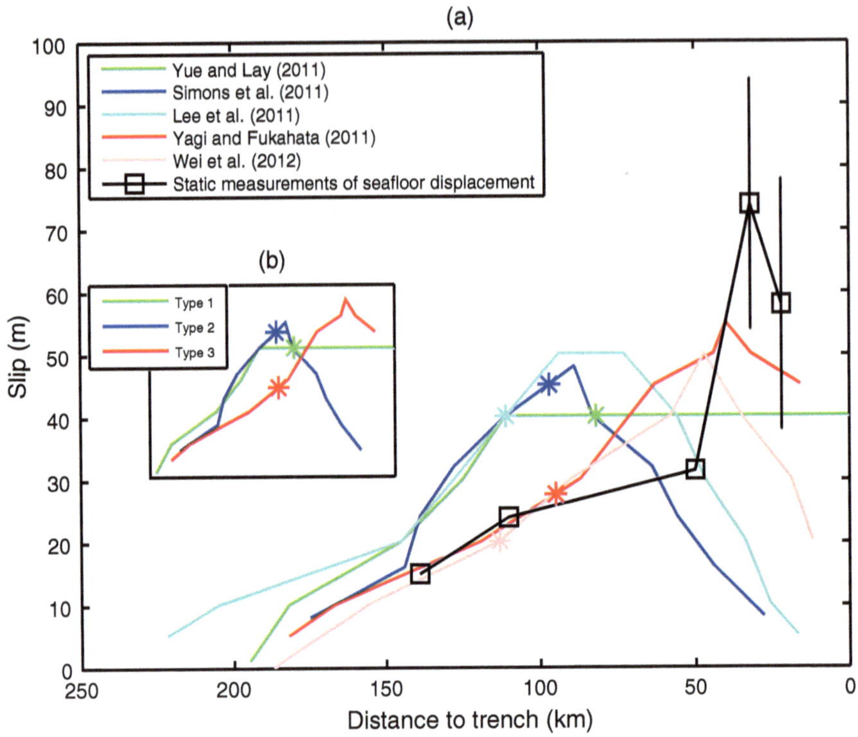

Figure 1

a The along-dip slip distributions across the hypocenter inferred from five slip inversions. The hypocenter used in each inversion is shown as a *star*. The displacement of sea floor is shown in *black*, including five measurements at or near the latitude of the hypocenter from Sato *et al.* (2011), Ito *et al.* (2011) and Kido *et al.* (2011). **b** The three types of along-dip slip distributions

In this paper, we attempt to find numerical models that can provide some useful constraints on the overall physical properties of the earthquake, such as stress drop and fracture energy. To keep the number of assumptions as few as possible, we use a simple elastic model with slip-weakening friction. We will focus on the along-dip rupture process near the latitude of the hypocenter and will try to explain the various observations in the shallow and deep regions, such as variations of slip, radiation frequency spectrum and rupture velocity.

2. *Model Setup and Observational Constraints on Model Parameters*

We consider a shallow-dipping fault with a dip angle of 14° embedded in a 2D elastic half space. The fault is 200 km long in the along-dip direction. The hypocenter is located in the middle. Material properties such as density ρ (3,000 kg/m^3), Poisson's ratio v (0.25) and shear modulus μ (30 GPa) are uniform throughout the medium. We solve the problem using a 2D spectral element code (Ampuero 2009) and the unstructured mesh shown in Fig. 2a (Huang *et al.* 2012). We prescribe an artificial nucleation procedure in the hypocentral region. The friction coefficient is forced to drop over a certain time scale from static to dynamic levels inside a region with time-dependent size (Andrews 1985). After reaching a critical nucleation size which is much shorter than the total rupture length, the rupture propagates spontaneously to both up-dip and down-dip directions. The linear slip-weakening friction law governs the remaining part of the fault, and the model contains five free parameters: initial shear stress τ_0, normal stress σ_n, static friction coefficient μ_s, dynamic friction coefficient μ_d and critical slip distance D_c (Fig. 2b). We constrain these model parameters using several observations, as described next.

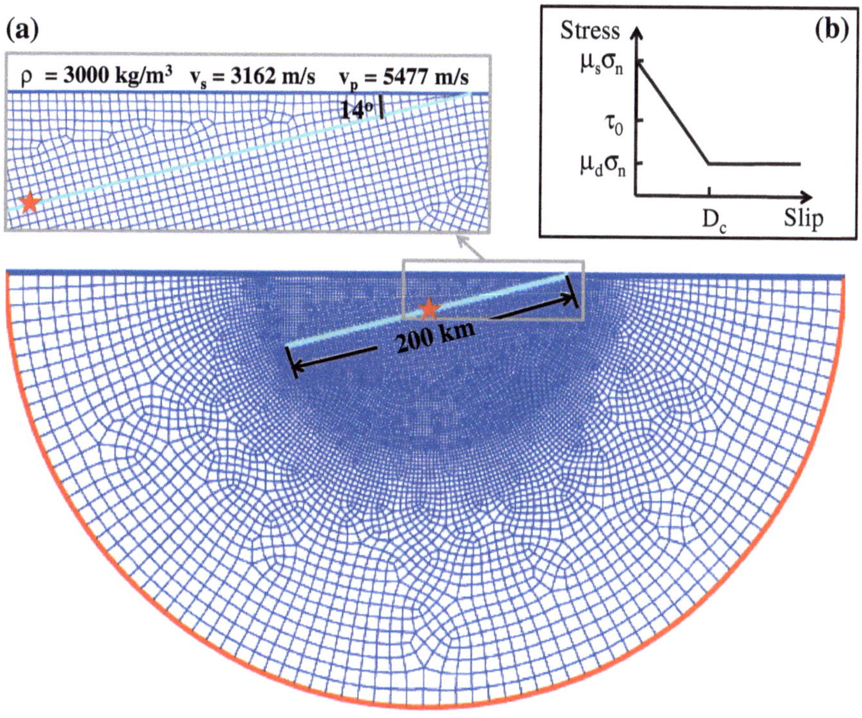

Figure 2

a The unstructured mesh with a free boundary on the top (*blue line*) and an absorbing boundary in a semicircular (*red line*). The hypocenter is in the middle of the 200-km-long fault (*turquoise line*). The zoom-in picture shows the mesh around the fault and the dip angle of the fault. The density, S velocity and P velocity are indicated on the top of the zoom-in picture. **b** Linear slip-weakening law. Stress increases from initial shear stress τ_0 to static strength $\mu_s\sigma_n$ first, and then decreases linearly to dynamic strength $\mu_d\sigma_n$ when slip reaches the critical slip distance D_c. The shear stress then remains at the dynamic strength level

2.1. *Normal Stress σ_n and Friction Coefficients μ_s and μ_d*

We adopt a normal stress profile from the Nankai region, which has an effective normal stress of about 10 MPa up to a horizontal distance of 20 km from the trench (Tobin and Saffer 2009). Further away from the trench, the normal stress is increased to 100 MPa by a vertical gradient of 6 MPa/km, and kept constant in deep regions (Fig. 3). We initially assume a constant static friction coefficient $\mu_s = 0.6$ and dynamic friction coefficient $\mu_d = 0.2$. As we will illustrate in Sect. 3, the friction coefficients have to be modified in some regions to reproduce the observations.

2.2. *Stress Drop $\tau_0 - \mu_d\sigma_n$*

The distributions of stress drop $\tau_0 - \mu_d\sigma_n$ are inferred from the coseismic final slip distributions shown in Fig. 1a. The static stress drop that results from our calculations is different from $\tau_0 - \mu_d\sigma_n$ due to overshoot, and we will discuss the distribution of the static stress drop later in Sect. 4. For each slip inversion, we measured roughly the slip at several locations in the along-dip direction across the hypocenter. The figure aims to show the overall differences of the inferred slip profiles, rather than reproduce the details of each model. We find that, although in all models large slip is concentrated in the shallow region, the slip profiles are highly variable up-dip from the hypocenters, which are shown by stars. Overall, they fall into three types (Fig. 1b): (1) almost constant slip in the shallow region (green line in Fig. 1a), (2) peak slip near the hypocenter (blue and turquoise lines in Fig. 1a) and (3) peak slip between the hypocenter and the trench (red and pink lines in Fig. 1a). Static measurements of seafloor displacements (black line in Fig. 1a) seem to favor

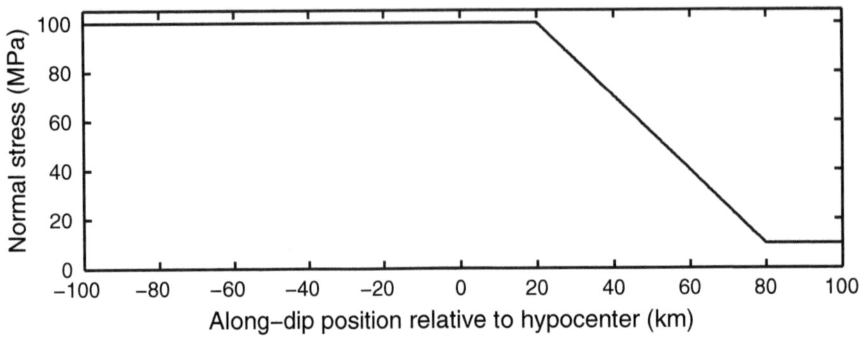

Figure 3
The along-dip distribution of normal stress in all three models

large slip near the trench, but have large uncertainties (Ito *et al.* 2011) and possibly involve post-seismic deformations. Hence, we consider the three possible slip profiles in our numerical models.

2.3. Critical Slip Distance D_c

The results from back-projection source imaging constrain the high-frequency slip-rate power in the deep region. Huang *et al.* (2012) found that the ratio between deep and shallow high-frequency slip-rate power is at least 10. In order to generate high-frequency radiation in dynamic rupture models, heterogeneities of either fracture energy or initial stress are needed (Madariaga 1983). However, only stress concentrations such as the residual stresses at the edge of a previous slip event can be as efficient as an abrupt change of fracture energy. For computational convenience, we choose to set heterogeneities of fracture energy by varying the value of D_c in our slip-weakening model. To reproduce the spatial contrast of high-frequency radiation, we set a uniform D_c in the shallow region, but several small asperities with much smaller D_c in the deep region. In reality the deep region may have variations of both fault strength and stress, which in combination can give rise to strong high-frequency radiation. A deep region with small asperities also agrees with the fact that earthquakes have repeatedly occurred there in the past (e.g., Igarashi *et al.* 2003; Tajima *et al.* 2013). The spacing of small asperities in our model is conceptual rather than corresponding directly to earthquakes of specific magnitude. However, given the same stress conditions, the spacing needs to be large enough to prevent the down-dip rupture from propagating faster than 1 km/s.

It is noteworthy that in such an asperity model the deep/shallow D_c ratio is determined by the high-frequency slip-rate power ratio, as illustrated in Fig. 4. The amplitude spectrum of slip rate at a certain location on the fault tends to the final slip D at very low frequency. Figure 4 shows two amplitude spectra, one for a deep region with final slip D^d and the other for a shallow region with final slip D^s. Each amplitude spectrum has two corner frequencies: the lower one is related to the time required for slip to reach its final value, or rise time t_{ris}; and the higher one is related to the time required for slip to reach D_c, or process zone time t_{pz} (see also Fig. 5c in Kaneko *et al.* 2008). The latter can be approximated as $t_{pz} \sim \frac{\mu D_c}{(1-\nu)\Delta\tau_s A_{II}(v_R)v_R}$, where ν is Poisson's ratio, v_R is rupture velocity, $\Delta\tau_s$ is the strength drop, i.e., the difference between static and dynamic strength $(\mu_s - \mu_d)\sigma_n$, and A_{II} is a function of rupture speed in mode II given by equation 5.3.11 in Freund, L. B., Dynamic fracture mechanics (Cambridge Univ. Press, Cambridge (1990). As D^d, D^s, t^d_{ris} and t^s_{ris} can be inferred from the slip inversions, and the high-frequency slip-rate power ratio from the back projection, the ratio t^d_{pz}/t^s_{pz} can be determined, so is the deep/shallow D_c ratio D^d_c/D^s_c. Besides, the back projection is carried out in a certain frequency band (e.g., 0.5–1 Hz in Meng *et al.* 2011). The center of the back-projection frequency band, f_{bp}, should be larger than the second corner frequency in the shallow region, $f_{bp} > \frac{1}{t^s_{pz}}$. This provides a lower bound for D^s_c. We show a detailed mathematical derivation of deep/

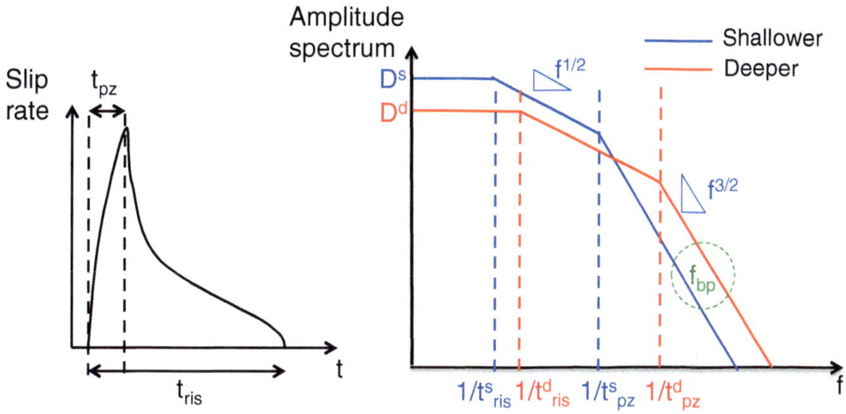

Figure 4
Slip rate function with two time scales: rise time t_{ris} and process zone time t_{pz} (*left*). Amplitude spectrum of the slip rate functions for shallow and deep regions, respectively, denoted by subscript 's' and 'd' (*right*). The frequency of back projection f_{bp} is indicated by a *dot circle*

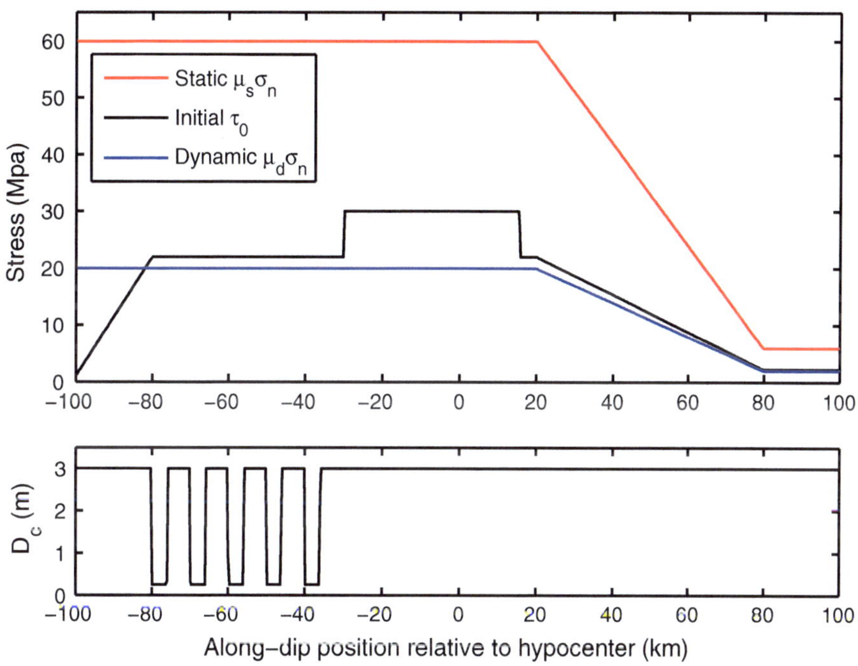

Figure 5
The along-dip distributions of initial shear stress, static strength, dynamic strength (*top*) and critical slip distance (*bottom*) in the first model

shallow D_c ratio in terms of high-frequency slip-rate power ratio in the "Appendix".

3. Results from the Three Models

In this section, we will present dynamic rupture simulations for the three different models that reproduce the three types of along-dip slip profiles (Fig. 1b).

3.1. First Model (Constant Slip in Shallow Region)

To reproduce the first slip profile that has a constant slip in the shallow region, we keep the static and dynamic friction coefficients constant. The

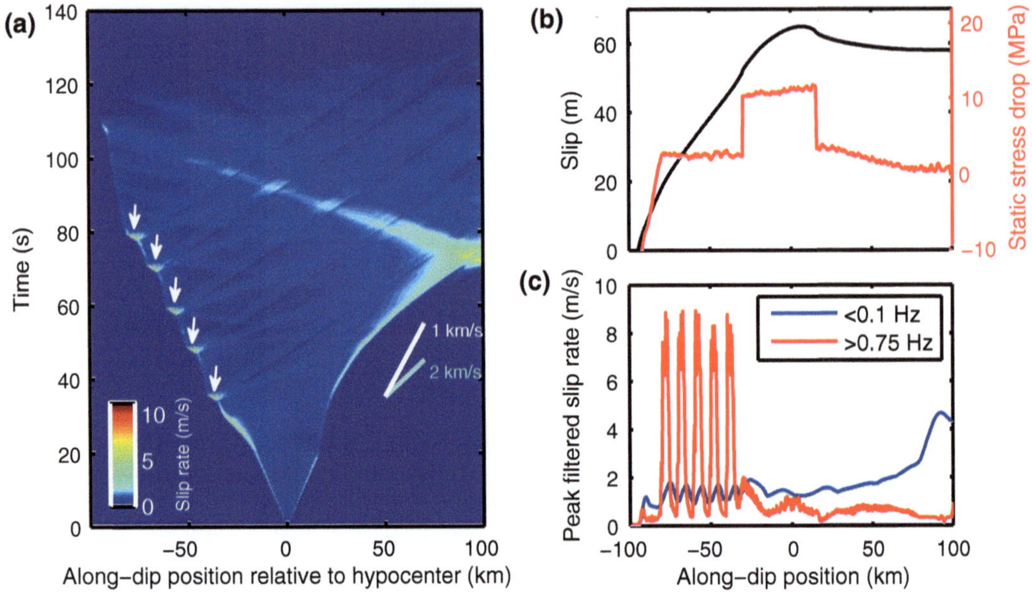

Figure 6
The spatial temporal distribution of slip rate (*left*), the along-dip distribution of final slip and static stress drop (*top right*), and the along-dip distribution of low-pass filtered (<0.1 Hz) and high-pass filtered peak slip rates (>0.75 Hz) (*bottom right*) in the first model. The *white arrows* in the left figure denote the regions of high-frequency bursts

distributions of model parameters are shown in Fig. 5. Rupture is forced to propagate bilaterally at 800 m/s inside the nucleation region (Fig. 6a). After about 20 s, the down-dip rupture starts to propagate spontaneously and accelerates until it reaches the low stress-drop region at 30 km from the hypocenter. It then propagates at a speed of about 1 km/s and generates high-frequency bursts when it propagates through the small asperities (Fig. 5). Due to the uniform frictional properties assumed eastward of the hypocenter, the up-dip rupture propagates smoothly. It reaches an average speed of about 2 km/s and produces an almost constant slip in the shallow region (Fig. 6b). To quantify the distribution of high-frequency radiation we compute at each fault location the peak value of the slip rate high-passed filtered above 0.75 Hz. The resulting high-frequency peak slip rate is much larger in the deep region than in the shallow region (Fig. 6c). In contrast, the peak values of the slip rate low-passed filtered below 0.1 Hz are more uniform. Their values are in the range of 1–2 m/s, consistent with the average slip rate from slip inversions (Lee *et al.* 2011; Wei *et al.* 2012), except in the region near the trench. To compare with the high-frequency power ratio in the back projection, we compute the power of the high-passed slip rate over a 10 s sliding window and apply a spatial Gaussian smoothing of half width 50 km, a conservative estimate of the spatial smearing in the back-projection source imaging. This leads to a deep/shallow power ratio of about 10.

3.2. *Second Model (Peak Slip in Hypocentral Region)*

In the second model, peak slip near the hypocenter indicates a larger stress drop there (Fig. 7), which can promote rupture acceleration. Thus, we reduce the initial shear stress in the deep region to keep the rupture velocity as low as 1 km/s. We found that the steep decrease of slip from the hypocenter to the trench (Fig. 1b) can only be achieved by a negative stress drop. This suggests that either the initial shear stress is lower or the dynamic friction coefficient is higher than in our first model. However, since the normal stress near the trench is very low and so is the dynamic strength, it is not possible to reduce the initial shear stress enough while keeping its sign consistent with thrust faulting. We, hence, increase the dynamic friction coefficient linearly in the

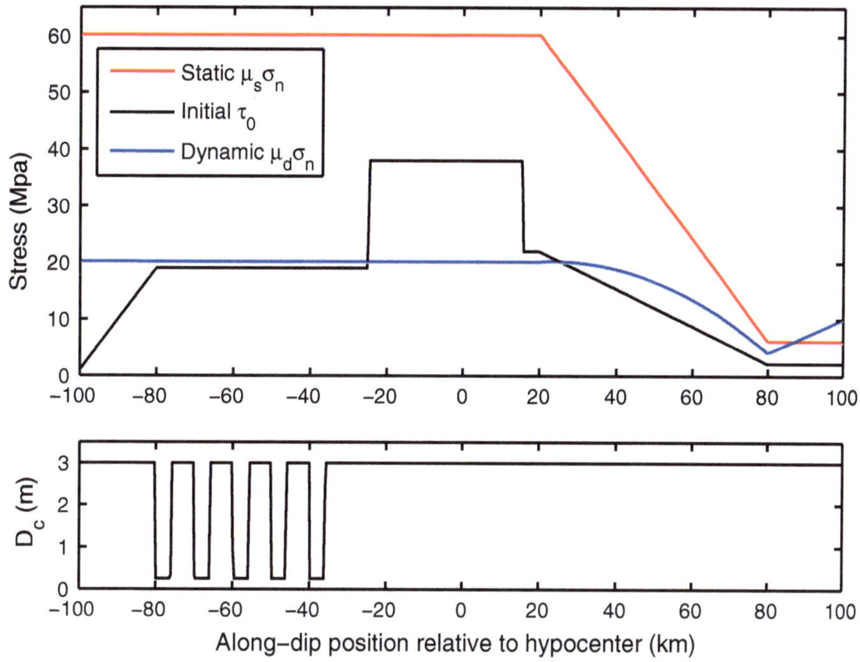

Figure 7
The along-dip distributions of initial shear stress, static strength, dynamic strength (*top*) and critical slip distance (*bottom*) in the second model

shallow region, which produces a curved profile of dynamic strength (Fig. 7). The resulting dynamic rupture (Fig. 8a) is similar to our first rupture model, except for the shorter nucleation stage. The final slip distribution (Fig. 8b) and high-frequency power ratio (Fig. 8c) are also consistent with the observations.

3.3. *Third Model (Peak Slip in Shallow Region)*

The third model features peak slip between the hypocenter and the trench, as suggested by many slip inversion studies. In this model, the largest stress drop is located up-dip from the hypocenter and decreases in both directions along-dip. Because of the lower stress drop in the nucleation region compared to our previous two models, the static friction coefficient and D_c are reduced there to achieve rupture nucleation (Fig. 9). Successful down-dip rupture requires the region with reduced D_c, which is as low as within the small asperities, to extend 40 km down-dip from the hypocenter. The small asperity located from 36 to 40 km down-dip from the hypocenter in the previous two models is, hence, not present in the third model. To avoid significant slowing down of the down-dip rupture, the value of D_c in between the small asperities is smaller than in the previous two cases. As large stress drop promotes high-frequency radiation, we also increase D_c in the shallow region. The details of the resulting rupture are shown in Fig. 10a. The rupture reaches a down-dip speed of about 1 km/s and an up-dip speed of about 2 km/s. Again, the down-dip rupture generates much stronger high-frequency radiation than the up-dip rupture (Fig. 10c).

4. *Constraints on Static Stress Drop, Fracture Energy and Energy Partitioning*

4.1. *Static Stress Drop*

The average static stress drop inferred from different slip inversions of the Tohoku-Oki earthquake is 4.8 MPa (Koketsu *et al.* 2011), 6 MPa (Yagi and Fukahata 2011) and 7 MPa (Lee *et al.* 2011), respectively. We compare these values to $\Delta\tau_E = \frac{\int \Delta\tau D ds}{\int D ds}$, the slip-weighted average of the static stress drop distributions obtained in our rupture models. This averaging procedure is appropriate for energy estimates (Noda and Lapusta 2012). The

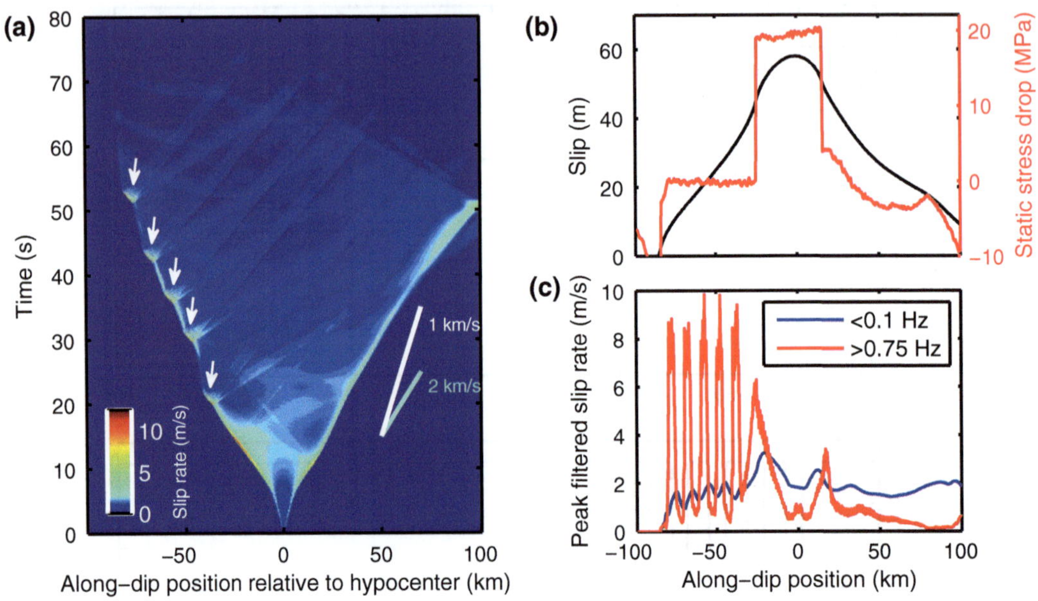

Figure 8

The spatial temporal distribution of slip rate (*left*), the along-dip distribution of final slip and static stress drop (*top right*), and the along-dip distribution of low-pass filtered (<0.1 Hz) and high-pass filtered peak slip rates (>0.75 Hz) (*bottom right*) in the second model. The white arrows in the left figure denote the regions of high-frequency bursts

values of $\Delta\tau_E$ thus calculated for the three models are 4.5, 7 and 4.6 MPa, similar to the average static stress drop inferred from slip inversions. Note that the static stress drop can vary in space by almost two orders of magnitude, e.g., from 12 MPa in the peak-slip region to 0.2 MPa near trench in the first model (Fig. 6b). In the other two models the stress drop near the trench is negative but the overall stress drop is still positive (Figs. 8b and 10b). The first and second models provide the two end members of the distributions of static stress drop in the Tohoku-Oki earthquake. In the first model, the static stress drop in the shallow region is rather small and almost zero near the trench. The large stress drop around the hypocenter pushed the whole shallow region eastwards and the resulting slip is large but almost constant. This behavior is somewhat similar to block sliding. In contrast, the second model requires a large negative stress drop in the shallow region. The large stress drop around the hypocenter still pushes the shallow region eastwards, but the final slip decreases due to the resistance caused by the larger dynamic friction prescribed there, resulting in the negative stress drop. The third model is similar except that the region with large stress drop is located eastward from the hypocenter.

The negative stress drops in the second and third models can result from velocity-strengthening materials in nature. The shallow velocity-strengthening region is usually considered as the upper limit of the seismogenic zone, defined as the zone where earthquakes can nucleate. Our numerical models show that ruptures can break through this region and result in a large slip there. Our models constrained by observed slip profiles (Fig. 1) show that the velocity-strengthening region needs to reach at least 40 km down-dip from the trench, or even more than 70 km as suggested by the slip profiles from Lee *et al.* (2011) and Simons (2011). Usually the upper limit of the seismogenic zone on subduction megathrusts is at a depth of 5–15 km (Hyndman *et al.* 1997), or at an along-dip distance of 20–60 km given the subduction geometry in our models. However, the different in situ pressures, temperatures and minerals make it hard to determine the upper limit of the seismogenic zone in specific subduction zones. Resolving this question in the Tohoku region requires a reliable identification of the interplate seismicity and determination of the coseismic slip profile of the Tohoku-Oki earthquake at shallow depth. While multiple observations point to large slip close to trench,

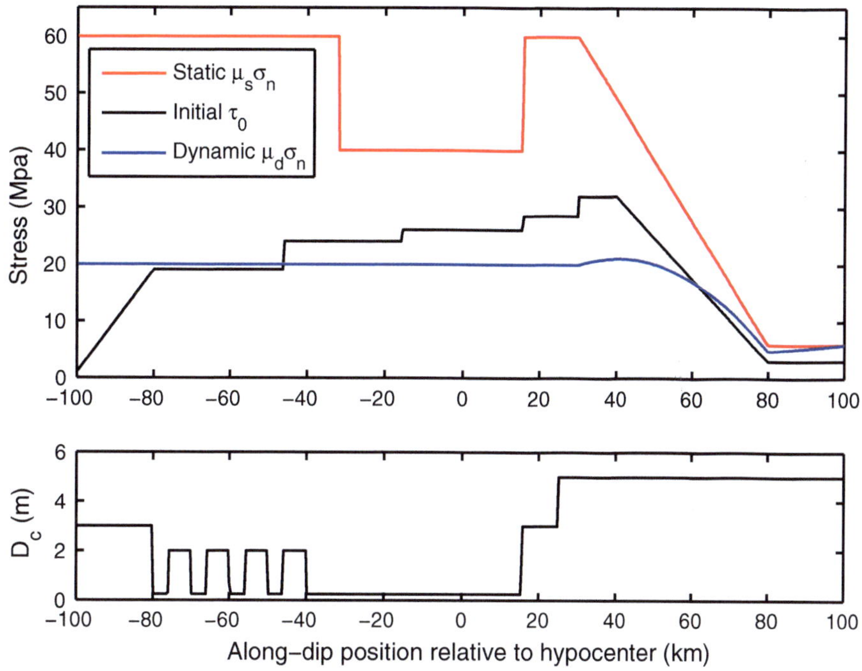

Figure 9
The along-dip distributions of initial shear stress, static strength, dynamic strength (*top*) and critical slip distance (*bottom*) in the third model

whether they involve significant postseismic deformation is still an open question. Studies of early postseismic deformation are necessary to understand this better.

4.2. Fracture Energy

A significant product of our models is the constraint on fracture energy or D_c in the shallow region. As discussed in Sect. 2, the back-projection frequency f_{bp} should be larger than the second corner frequency in the shallow region, which suggests:

$$D_c^s > \frac{(1-\nu)\Delta\tau_s^s v_R^s A(v_R)}{\mu f_{bp}}. \quad (1)$$

For example, when $f_{bp} = 0.75\,\text{Hz}$, $\Delta\tau_s^s = 22.5\,\text{MPa}$ and $v_R^s = 2\,\text{km/s}$, we find $D_c^s > 1.9\,\text{m}$ in the first model. Thus, for a given D_c^d, we can find a D_c^s that satisfies the deep/shallow high-frequency power ratio ("Appendix") and the lower bound given by f_{bp} (Eq. 1). Figure 11 summarizes the deep/shallow high-frequency slip-rate power ratio (>0.75 Hz) obtained in the first model for a range of values of fracture energy and D_c^s in a 80 km wide region, 20 km eastward from the hypocenter and beyond. This result shows that if $D_c^d = 0.25\,\text{m}$, D_c^s needs to be larger than ~2.7 m, or fracture energy larger than ~60 MJ/m^2, in order to satisfy a high-frequency power ratio of at least 10. The large fracture energy in the shallow region also prohibits the acceleration of up-dip rupture, limiting the rupture speed to about 2 km/s except in the region near the trench where supershear rupture tends to occur.

4.3. Energy Partitioning

The nature of an earthquake is controlled by partitioning of energy between the radiated energy, E_R, and the fracture energy, E_G, which is the energy used for advancing the fracture against resistance at the fault tip. Depending on whether E_R/E_G is large or small, we expect rapid or slow earthquakes, respectively. (Here, rapid earthquake is an earthquake with strong seismic radiation, and slow earthquake means an earthquake deficient in high frequency energy, like tsunami earthquakes.) We define the total available energy by $E_{T0} = E_R + E_G$, and call the ratio $\eta_R = E_R/E_{T0}$ the radiation efficiency (e.g., Kanamori and

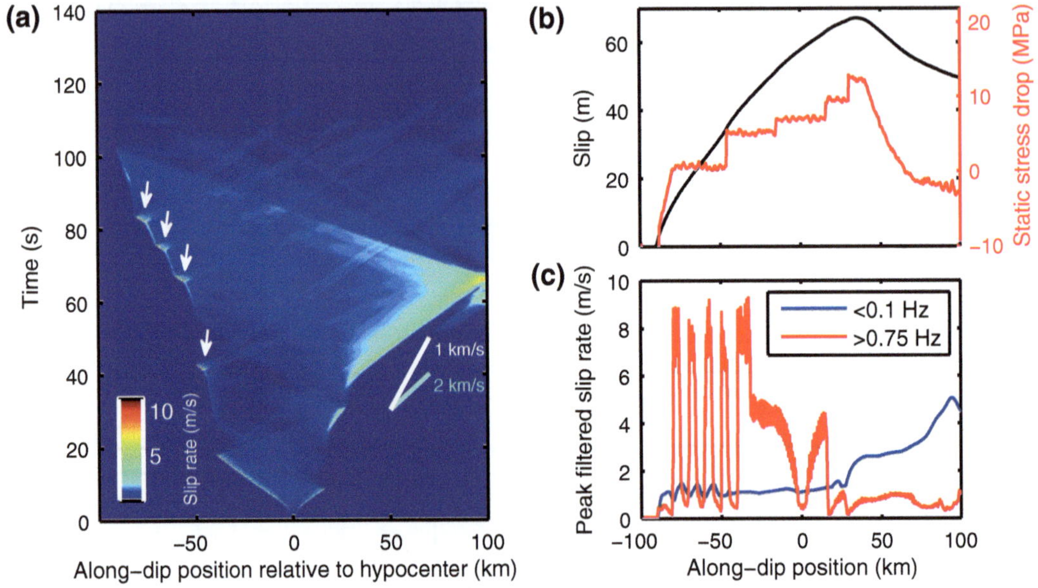

Figure 10
The spatial temporal distribution of slip rate (*left*), the along-dip distribution of final slip and static stress drop (*top right*), and the along-dip distribution of low-pass filtered (<0.1 Hz) and high-pass filtered peak slip rates (>0.75 Hz) (*bottom right*) in the third model. The *white arrows* in the left figure denote the regions of high-frequency bursts

Rivera 2006). In seismology, we can directly determine E_R from observations. We cannot estimate the total available energy directly from seismic observations, but if the friction follows the simple slip-weakening curve and if overshoot or undershoot is not very large, we can approximate E_{T0} by $\Delta\tau DS/2$, where $\Delta\tau$ is stress drop, D is the average slip and S is the fault area.

We compare η_R estimated from seismological observations with that estimated from the dynamic models studied here. Using the relation for the total available energy, the radiation efficiency can be written as $\eta_R = \left(\frac{2\mu}{\Delta\tau}\right)\left(\frac{E_R}{M_0}\right)$ which can be estimated from the shear modulus μ, the stress drop $\Delta\tau$, the radiated energy E_R, and the moment M_0. We use the CMT moment $M_0 = 5.31 \times 10^{22}$Nm, which is determined at a depth of 20 km, and $\mu = 44.1$ GPa at this depth in PREM. The estimation of radiated energy by different investigators ranges from 3 to 9 $\times$ 10^{17} J (Ide *et al.* 2011; Newman 2011; Lay *et al.* 2012). Then as the estimated stress drop varies from 4.8–10 MPa (Koketsu *et al.* 2011; Lay *et al.* 2011; Lee *et al.* 2011; Yagi and Fukahata 2011), the radiation efficiency η_R ranges from 0.05 to 0.31.

In the dynamic models, we can compute E_R and E_{T0} directly using the stress parameters and slip in the models:

$$E_{T0} = \frac{1}{2}\int_0^L (\tau_0(x) - \tau_1(x))D(x)\mathrm{d}x \quad (2)$$

$$E_R = E_{T0} - \int_0^L \left\{ \int_0^D \left[\tau(\tilde{D}, x) - \tau_1(x)\right]\mathrm{d}\tilde{D}(x) \right\}\mathrm{d}x \quad (3)$$

where $\tilde{D}$, D and $\tau_1$are the slip, final slip and final stress, respectively, at a given location x. The radiation efficiencies of the first, second and third models thus calculated are 0.33, 0.39, and 0.5, respectively. These values are higher than current seismological estimates of η_R. Including energy dissipation mechanisms such as off-fault plasticity may help reduce the radiation efficiency (Ma and Hirakawa 2013), but it can also diminish the final slip near the trench. Considering the many assumptions we made for our dynamic models (2-dimensionality and the specific dissipation mechanism) and the uncertainties in the

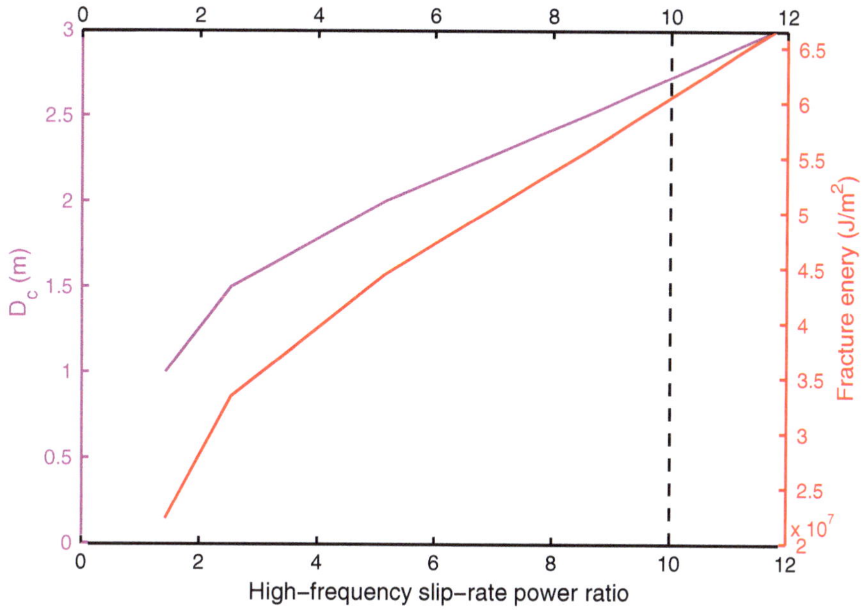

Figure 11
The high-frequency slip-rate power ratio for different values of D_c and fracture energy in an 80-km-wide region 20 km eastward from the hypocenter, when D_c in the deep region is kept at 0.25 m

seismological parameters, we consider the values of η_R to be in agreement only approximately. Our objective here is to illustrate how dynamic models can be compared with the real earthquake through energy partitioning. As seismological methodology improves, we expect the uncertainties in seismological parameters to decrease significantly. Also, incorporating more realistic 3-dimensional structure and fault zone constitutive laws in dynamic modeling may eventually enable more meaningful comparisons between the physical models and real earthquakes. What is presented here is an illustration of how to make such comparisons.

4.4. *Effects of the Subduction Wedge*

One unexpected feature of the Tohoku-Oki earthquake is its propagation to the shallow region and the resulting large slip. To find how the subduction wedge (the structure between free surface and plate interface) with shallow dip angle can affect the rupture propagation, we first compare our results to a rupture simulation on a fault with a dip angle of 90°. The hypocenter is 100 km deep, and model parameters are the same as those prescribed in the first model (Fig. 5). We find that on the vertical fault the rupture stops when it reaches the region with a small stress drop at ~27.5 km upward from the hypocenter. In contrast, the up-dip rupture reaches the trench in our models of a shallow dipping fault. This indicates that without the subduction wedge the rupture is unable to propagate to the shallow region, unless stress drop is increased or fracture energy is reduced there.

We also run a rupture simulation on a horizontal fault at a depth of 25 km (the hypocentral depth in our models of the Tohoku-Oki earthquake) from the free surface. Given the same model parameters, we find that the rupture can propagate as far as where the trench would be (100 km eastward from the hypocenter). This shows that the effect of the free surface enables rupture to the trench despite the small stress drop in the shallow region. The structure of the subduction wedge is even more favorable for rupture propagation as the shallow region is closer to the free surface than is the hypocenter. Waves reflected by the free surface induce transient reduction of normal stress and increase of shear stress on the fault that lead to large transient stress drop (Kozdon and Dunham 2013), which promotes the rupture propagation.

Our assumed fault geometry is simplified with a constant dip angle of 14°, while the real plate interface starts at a much smaller angle 4.6° (Kimura *et al.* 2012) and steepens with depth. As the subduction wedge can amplify the final slip, the real stress drop in the region near the trench needs to be much smaller than in our models. The subduction wedge also leads to rupture acceleration, which produces high-frequency radiation. This effect is expected to be stronger in a subduction wedge with a smaller dipping angle. Thus, fracture energy (or D_c) may also need to be larger in order to inhibit the high-frequency radiation in the shallow region. Our models provide an upper bound for the average stress drop and a lower bound for the fracture energy in the shallow region.

5. Conclusions

We presented dynamic rupture models of the Tohoku-Oki earthquake by integrating key observational constraints. We assumed a fault governed by slip-weakening friction, with asperities of different frictional properties, and used the final slip distribution and the high-frequency radiation to constrain the stress drop and D_c. We used three models to reproduce the three typical along-dip slip profiles obtained by finite fault source inversions. Rupture properties estimated by our dynamic modeling such as rupture velocity and slip rate are consistent with the observations, though variations do exist in different models. Overall, in our models the average static stress drop of the event is in the range of 4.5–7 MPa, and fracture energy in the shallow region is in the order of 10 MJ/m^2. Stress drop reaches values of order 10 MPa in the regions of maximum slip, as also suggested by slip inversions (Yagi and Fukahata 2011). The coseismic slip distributions constrain the size of the shallow region of negative stress drop, which can be associated with strengthening materials that delineate the upper limit of the seismogenic zone. We find that the radiation efficiency computed for our models is larger than 0.3, somewhat larger than that inferred from seismic data. This may indicate that additional forms of energy dissipation are needed.

Acknowledgments

This work was supported by NSF grants EAR-0944288 and EAR-1015704, the Gordon and Betty Moore Foundation, the Japan Nuclear Energy Safety Organization (JNES) and SCEC (funded by NSF EAR-0106924 and USGS 02HQAG0008 cooperative agreements).

Appendix

The amplitude spectrum of slip rate in a rupture model with process zone (Fig. 4) can be expressed as a function of final slip D, the first corner frequency $f_{\mathrm{ris}} = 1/t_{\mathrm{ris}}$ and the second corner frequency $f_{\mathrm{pz}} = 1/t_{\mathrm{pz}}$. The amplitude spectrum is flat until the first corner frequency, and then decreases as a function of $f^{-1/2}$. After the second corner frequency, the amplitude spectrum decreases as a function of $f^{-3/2}$ in the case of the linear slip-weakening friction law (Fig. 5c in Kaneko *et al.* 2008). Thus, the amplitude spectrum can be expressed as:

$$v(f) = \begin{cases} D \text{ if} f < f_{\mathrm{ris}} \\ D\left(\frac{f_{\mathrm{ris}}}{f}\right)^{1/2} \text{if} \quad f_{\mathrm{ris}} < f < f_{\mathrm{pz}} \\ D\left(\frac{f_{\mathrm{ris}}}{f_{\mathrm{pz}}}\right)^{1/2}\left(\frac{f_{\mathrm{pz}}}{f}\right)^{3/2} \text{if} \quad f > f_{\mathrm{pz}} \end{cases} \qquad (\mathrm{A}-1)$$

As can be seen from Fig. 4, the largest deep/shallow ratio of slip rate α happens when $f \geq f_{\mathrm{pz}}$:

$$\alpha = \frac{D^{\mathrm{d}}\sqrt{f^{\mathrm{d}}_{\mathrm{ris}} f^{\mathrm{d}}_{\mathrm{pz}}}}{D^{\mathrm{s}}\sqrt{f^{\mathrm{s}}_{\mathrm{ris}} f^{\mathrm{s}}_{\mathrm{pz}}}}. \qquad (\mathrm{A}-2)$$

The formula for second corner frequency f_{pz} in Mode II is:

$$f_{\mathrm{pz}} = \frac{(1-\upsilon)\Delta\tau_{\mathrm{s}} v_{\mathrm{R}}}{\mu D_{\mathrm{c}}} A_{\mathrm{II}}(v_{\mathrm{R}}), \qquad (\mathrm{A}-3)$$

where v is Poisson's ratio, μ is shear modulus, $\Delta\tau_{\mathrm{s}}$ is strength drop, v_{R} is the rupture velocity, and A_{II} is a function of v_{R} (Equation (5.3.11) in Freund 1990). By combining (A-2) and (A-3) and assuming a uniform Poisson's ratio and shear modulus, the shallow/deep D_c ratio can be expressed as a function of α:

$$\frac{D_c^s}{D_c^d} = \alpha \frac{D^s}{D^d} \left(\frac{f_{ris}^s}{f_{ris}^d}\right)^{\frac{1}{2}} \frac{\Delta\tau_s^s v_R^s}{\Delta\tau_s^d v_R^d} \frac{A_{II}^s(v_R^s)}{A_{II}^d(v_R^d)}. \qquad (A-4)$$

The final slip, rise time and rupture velocity can be inferred from slip inversions. The strength drop is our model parameter. For example in our first model, as $\left(\frac{f_{ris}^s}{f_{ris}^d}\right)^{\frac{1}{2}} \sim 1$, $\frac{\Delta\tau_s^s v_R^s}{\Delta\tau_s^d v_R^d} \sim 1$, and $\frac{A_{II}^s(v_R^s)}{A_{II}^d(v_R^d)} \sim 1$, we can simplify (A-4) to:

$$\frac{D_c^s}{D_c^d} \approx \alpha \left(\frac{D^s}{D^d}\right). \qquad (A-5)$$

As the observed high-frequency slip rate power ratio is at least 10, α has to be larger than $\sqrt{10}$. Given $\frac{D^s}{D^d} \sim 2$–3 the shallow/deep D_c ratio is larger than 6 to 9. Note that the needed D_c ratio (>10) in our model is larger than this range, as the observed power ratio involves smoothing in time and space. Thus, (A-4) and (A-5) only give a lower bound of the shallow/deep D_c ratio.

References

Ampuero, J. P. (2009), *SEM2DPACK, A spectral element method tools for 2D wave propagation and earthquake source dynamics*, User's Guide, version 2.3.6. Available online at http://www.sourceforge.net/projects/sem2d/.

Andrews, D. J. (1985), *Dynamic plane-strain shear rupture with a slip-weakening friction law calculated by a boundary integral method*, Bull. Seismol. Soc. Am., *75*, 1–21.

Aochi, H., and Ide, S. (2011), *Conceptual multi-scale dynamic rupture model for the 2011 off the Pacific coast of Tohoku Earthquake*, Earth Planets Space, *63*, 761–765.

Duan, B. (2012), *Dynamic rupture of the 2011 Mw 9.0 Tohoku Oki earthquake: Roles of a possible subducting seamount*, J. Geophys. Res., *117*, B05311.

Freund, L. B., Dynamic fracture mechanics (Cambridge Univ. Press, Cambridge 1990).

Fujiwara, T., Kodaira, S., No, T., Kaiho, Y., Takahashi, N., and Kaneda, Y. (2011), *The 2011 Tohoku-Oki earthquake: Displacement reaching the trench axis*, Science, *334*, 1240.

Goto, H., Yamamoto, Y., and Kita, S. (2012), *Dynamic rupture simulation of the 2011 off the Pacific coast of Tohoku Earthquake: Multi-event generation within dozens of seconds*, Earth Planets Space, *64*, 1167–1175.

Huang, Y., Meng, L., and Ampuero, J.-P. (2012), *A dynamic model of the frequency-dependent rupture process of the 2011 Tohoku-Oki earthquake*, Earth Planets Space, *64*, 1061–1066.

Hyndman, R. D., Yamano, M., and Oleskevich, D. A. (1997), *The seismogenic zone of subduction thrust faults*, Island Arc, *6*, 244–260.

Ide, S., Baltay, A., and Beroza, G. C. (2011), *Shallow dynamic overshoot and energetic deep rupture in the 2011 Mw 9.0 Tohoku-Oki earthquake*, Science, *332*, 1426–1429, 2011.

Igarashi, T., Matsuzawa, T., and Hasegawa, A. (2003), *Repeating earthquakes and interplate aseismic slip in the northeastern Japan subduction zone*, J. Geophys. Res., *108*, 2249.

Iinuma, T., Hino, R., Kido, M., Inazu, D., Osada, Y., Ito, Y., Ohzono, M., Tsushima, H., Suzuki, S., Fujimoto, H., and Miura, S. (2012), *Coseismic slip distribution of the 2011 off the Pacific Coast of Tohoku Earthquake (M9.0) refined by means of seafloor geodetic data*, J. Geophys. Res., *117*, B07409.

Ito, Y., Tsuji, T., Osada, Y., Kido, M., Inazu, D., Hayashi, Y., Tsushima, H., Hino, R., and Fujimoto, H. (2011), *Frontal wedge deformation near the source region of the 2011 Tohoku-Oki earthquake*, Geophys. Res. Lett., *38*, L00G05.

Kanamori, H., and Rivera, L., Energy partitioning during an earthquake, Earthquakes: Radiated energy and the physics of faulting (AGU Monograph Series 170, Washington D. C., 2006) pp. 3–14.

Kaneko, Y., Lapusta, N., and Ampuero, J.-P. (2008), *Spectral element modeling of spontaneous earthquake rupture on rate and state faults: Effects of velocity-strengthening friction at shallow depth*, J. Geophys. Res., *113*, B09317.

Kato, N., and Yoshida, S. (2011), *A shallow strong patch model for the 2011 great Tohoku-oki earthquake: a numerical simulation*, Geophys. Res. Lett., *38*, L00G04,

Kido, M., Osada, Y., Fujimoto, H., Hino, R., and Ito, Y. (2011), *Trench-normal variation in observed seafloor displacements associated with the 2011 Tohoku-Oki earthquake*, Geophys. Res. Lett., *38*, L24303.

Kimura, G., Hina S., Hamada, Y., Kameda, J., Tsuji, T., Kinoshita, M., and Yamaguchi, A. (2012), *Runaway slip to the trench due to rupture of highly pressurized megathrust beneath the middle trench slope: The tsunamigenesis of the 2011 Tohoku earthquake off the east coast of northern Japan*, Earth Planet. Sci. Lett., *339–340*, 32–45.

Kiser, E., and Ishii, M. (2012), *The March 11, 2011 Tohoku-oki earthquake and cascading failure of the plate interface*, Geophys. Res. Lett., *39*, L00G25.

Kodaira, S., No, T., Nakamura, Y., Fujiwara, T., Kaiho, Y., Miura, S., Takahashi, N., Kaneda, Y., and Taira, A. (2012), *Coseismic fault rupture at the trench axis during the 2011 Tohoku-oki earthquake*, Nature Geosci., *5*, 646–650.

Koketsu, K., Yokota, Y., Nishimura, N., Yagi, Y., Miyazaki, S., Satake, K., Fujii, Y., Miyake, H., Sakai, S., Yamanaka, Y., and Okada, T. (2011), *A unified source model for the 2011 Tohoku earthquake*, Earth Planet. Sci. Lett., *310*, 480–487.

Kozdon, J. E., and Dunham, E. M. (2013), *Rupture to the trench: Dynamic rupture simulations of the 11 March 2011 Tohoku earthquake*, Bull. Seismic. Soc. Am., *103*, 1275–1289.

Lay, T., Ammon, C. J., Kanamori, H., Xue, L., and Kim, M. J. (2011), *Possible large near-trench slip during the 2011 Mw 9.0 off the Pacific coast of Tohoku earthquake*, Earth, Planets Space, *63*, 687–692.

Lay, T., Kanamori, H., Ammon, C. J., Koper, K. D., Hutko, A. R., Ye, L., Yue, H., and Rushing, T. M. (2012), *Depth-varying rupture properties of subduction zone megathrust faults*, J. Geophys. Res., *117*, B04311.

Lee, S.-J., Huang, B.-S., Ando, M., Chiu, H.-C., and Wang, J.-H. (2011), *Evidence of large scale repeating slip during the 2011 Tohoku-Oki earthquake*, Geophys., Res. Lett., *38*, L19306.

Ma, S., and Hirakawa, E. T. (2013), *Dynamic wedge failure reveals anomalous energy radiation of shallow subduction earthquakes*, Earth Planet. Sci. Lett., *375*, 113–122.

Madariaga, R. (1983), *High-frequency radiation from dynamic earthquake fault models*, Ann. Geophys., *1*, 17–23.

Meng, L., Inbal, A., and Ampuero, J.-P. (2011), *A window into the complexity of the dynamic rupture of the 2011 Mw 9 Tohoku-Oki earthquake*, Geophys. Res. Lett., *38*, L00G07.

Newman, A. (2011), Written Communication, May 3, 2011.

Noda, H., and Lapusta, N. (2012), *On averaging interface response during dynamic rupture and energy partitioning diagrams for earthquakes*, J. Appl. Mech., *79*, 031026.

Noda, H., and Lapusta, N. (2013), *Stable creeping fault segments can become destructive as a result of dynamic weakening*, Nature, *493*, 518–523.

Sato, H., Ishikawa, T., Ujihara, N., Yoshida, S., Fujita, M., Mochizuki, M., and Asada, A. (2011), *Displacement above the hypocenter of the 2011 Tohoku-Oki earthquake*, Science, *332*, 1395.

Simons, M., *et al.* (2011), *The 2011 magnitude 9.0 Tohoku-Oki earthquake: Mosaicking the megathrust from seconds to centuries*, Science, *332*, 1421–1425.

Tajima, F., Mori, J., and Kennett, B. L. N. (2013), *A review of the 2011 Tohoku-Oki earthquake (Mw 9.0): Large-scale rupture across heterogeneous plate coupling*, Tectonophysics, *586*, 15–34.

Tobin, H. J., and Saffer, D. M. (2009), *Elevated fluid pressure and extreme mechanical weakness of a plate boundary thrust, Nankai Trough subduction zone*, Geology, *37*, 679-682.

Wei, S., Graves, R., Helmberger, D., Avouac, J.-P., and Jiang, J. (2012), *Sources of shaking and flooding during the Tohoku-Oki earthquake: A mixture of rupture styles*, Earth Planet. Sci. Lett., *333–334*, 91–100.

Yagi, Y., and Fukahata, Y. (2011), *Rupture process of the 2011 Tohoku-oki earthquake and absolute elastic strain release*, Geophys. Res. Lett., *38*, L19307.

Yoshida, S., and Kato, N. (2011), *Pore pressure distribution along plate interface that causes a shallow asperity of the 2011 great Tohoku-oki earthquake*, Geophys. Res. Lett., *38*, L00G13.

Yue, H., and Lay, T. (2011), *Inversion of high-rate (1sps) GPS data for rupture process of the 11 March 2011 Tohoku earthquake (Mw 9.1)*, Geophys. Res. Lett., *38*, L00G09.

(Received May 2, 2013, revised September 8, 2013, accepted September 17, 2013, Published online October 13, 2013)

Pure Appl. Geophys. 171 (2014), 2569–2580

DOI 10.1007/s00024-014-0799-6

Pure and Applied Geophysics

Foreshocks and *b* Value: Bridging Macroscopic Observations to Source Mechanical Considerations

M. Avlonitis[1] and G. A. Papadopoulos[2]

Abstract—Spring-block models, such as the Olami-Feder-Christensen (OFC) model, were introduced several years ago to describe earthquake dynamics in the context of self-organized criticality. With the aim to address the dependency of the seismicity style on source's material properties, we present an analytical enrichment of a 2D OFC model. We conclude with an analytical expression which introduces, through an appropriate constitutive equation, an effective dissipation parameter a^{eff} related analytically not only with the elastic properties of the fault plane, but also with stochastic structural heterogeneities and structural processes of the source through a gradient coefficient. Moreover, within the proposed formulation, the low *b* values experimentally observed in foreshock sequences can be modeled by a process of material softening in the seismogenic volume. To check our analytical findings, a cellular automaton was built-up whereas simulation results have verified the model's predictions for the evolution of *b* in macroscopic records.

Key words: Spring-block models, *b* values, stochastic heterogeneity, softening.

1. Introduction

The forecasting or prediction of earthquakes remains an ever-green topic in the area of seismology in the last decades. In the aftermath of the lethal earthquake (Mw = 6.3) of 6th April 2009 in L' Aquila, Italy, it was shown that a strong foreshock signal preceded the mainshock in the domains of space, time and size (Papadopoulos *et al.*, 2010), and a discussion re-opened regarding the operational implementation of the forecasting of tectonic earthquakes (Jordan *et al.*, 2011). Time independent forecasts are useful for long-term seismic hazard analysis. However, in a time-dependent forecast, the probabilities depend on the information available at the time when the forecast is made. In the same work, it was found that the most useful information for operational forecasting has come from seismic catalogs and the geologic history of surface ruptures.

In the short-term sense the incidence of foreshocks as a precursory pattern useful for the forecasting of the mainshock attracted the interest of the seismological community since the 1960's (Mogi, 1963a, b). Since foreshocks may provide information which may lead to a more robust estimation of the probability for the occurrence of a future strong mainshock (e.g. Agnew and Jones, 1991), they are generally considered as one of the most promising precursory phenomena (Wyss, 1997; Vidale *et al.*, 2001). However, some mainshocks are preceded by foreshocks while others are not, which underlines the need to investigate the physical conditions that favour or disfavour the foreshock occurence.

The methodological approach introduced in this paper is to bridge mechanical considerations of the earthquake source to macroscopic observations resulting from seismicity analysis. In particular, we investigate links between foreshock activity and one of the most important parameters of seismicity, the "*b* value" of the magnitude-frequency or G–R relation (Ishimoto and Iida, 1939; Gutenberg and Richter, 1944),

$$\log N = d - bM, \tag{1}$$

where N is the cumulative number of events of magnitude equal to or larger than M and d, b are parameters determined by the data. It is shown that, by relaxing the assumption of constant shear strength within the earthquake source it is possible to

[1] Department of Informatics, Ionian University, Platia Tsirigoti 7, Corfu, Greece. E-mail: avlon@ionio.gr
[2] Institute of Geodynamics, National Observatory of Athens, Athens, Greece. E-mail: papadop@noa.gr

analytically model the effect of both the material heterogeneity and structural evolution of the source to the macroscopically observed *b* values. More specifically, it is shown how the introduction of a softening mechanism before mainshocks can adequately model the macroscopically observed drop of *b* value in foreshock sequences.

In the next two sections we review briefly macroscopic observations as well as the OFC spring-block model. Then, our contribution in understanding the variability of *b* value via analytical derivation is presented, while the predicted behavior of *b* variability is validated via simulation results.

2. *Macroscopic Observations*

Seismicity space–time clusters have been examined around the world long ago. Omori (1894) was the first to introduce the concept of aftershock sequence, that is of a space–time cluster of shocks following a larger mainshock with power-law time decay of the number of aftershocks known as Omori's law. Later on, three patterns of earthquake sequences were described by Mogi (1963b): mainshock-aftershocks, foreshocks-mainshock-aftershocks, and swarms. The mainshock-aftershocks pattern implies that foreshocks precede only some mainshocks and not others, although perhaps it is only an apparent result given that, in the routine seismic analysis performed in seismograph centers, low-magnitude shocks including short-term foreshocks usually escape recognition and are not listed in standard earthquake catalogues due to limitations in monitoring capabilities. A detailed case study in the east Aegean Sea was published by Papadopoulos *et al.* (2006). It is of relevance to note that the completeness magnitude cut-off in the earthquake catalogues varies from one region to another again due to different monitoring capabilities. Spatio-temporal seismicity clusters that exhibit a gradual rise and fall in seismic moment release, lacking a mainshock-aftershocks pattern, are termed earthquake swarms (Yamashita, 1998). It is quite common that in a swarm the magnitude range is too short which results in high or even very high values of *b*.

Foreshock sequences are characterized by some distinct features. Laboratory material fracture experiments (e.g. Mogi, 1963a; Scholz, 1968) along with numerical modeling in spring-block models (e.g., Hainzl *et al.*, 1999) and analytical damage mechanics modeling (e.g. Main, 2000) showed a clear acceleration of the fracturing process before the main fracture. Studies regarding seismicity in Japan, western United States, Greece, Italy and elsewhere verified this in nature showing that foreshock activity increases approximately as the inverse of time before mainshock (e.g., Papazachos, 1975; Kagan and Knopoff, 1978; Jones and Molnar, 1979; Papadopoulos *et al.*, 2000, 2010, 2011). In the G–R relation and in global scale seismicity, *b* was found around unity (Frohlich and Davis, 1993). This parameter, however, is a variable dependent on local seismotectonic conditions, such as the material heterogeneity, the degree of symmetry in stress distribution, and on the existence or not of asperities in the fault zone. Such conditions are reflected in the mode of seismic activity, e.g., background seismicity, swarms, foreshocks, and aftershocks.

Observations on seismic sequences have shown that the parameter *b* usually drops and becomes significantly lower in foreshocks than in aftershocks or in background seismicity (Papazachos, 1975; Jones and Molnar, 1979; Main *et al.*, 1989; Molchan et al., 1999; Enescu *et al.*, 2001; Papadopoulos *et al.*, 2010, 2011; Nanjo *et al.*, 2012). On the other hand, in laboratory experiments, analysis of seismicity in mines, pore pressure records and simulations in spring-block models indicated that locally high stress may drop the normal $b = 1$ to lower values (Mogi, 1963b, Scholz, 1968, Wyss, 1973, Main *et al.*, 1989, Urbancic *et al.*, 1992, Hainzl *et al.*, 1999). On the contrary, creeping segments of the fault are found to be characterized by high *b* values (Amelung and King, 1997), which is consistent with the high *b* value usually found in aftershock sequences as compared to the *b* value in foreshocks and in the background seismicity. Therefore, for seismically active parts of the fault zone, it was proposed that *b* can be used as an indicator where asperities may be located, that is where the creeping and locked patches are situated along the fault zone, and where mainshocks are most likely (Wyss *et al.*, 2000, 2001; Zhao and Wu, 2008). Swarm-type activity is characterized by high to very high values of the parameter *b* at around 1.5 or more.

The strong variations of b across different stress regimes implies that this parameter acts as a stress meter that depends inversely on differential stress (SCHORLEMMER *et al.*, 2005; NARTEAU *et al.*, 2009).

From the point of view of mechanical properties of the earthquake source, there is evidence that the incidence of foreshocks is disfavored by increasing focal depth, and that it likely depends on the faulting type and orientation (JONES, 1984; OHNAKA, 1992; ABERCROMBIE and MORI, 1996; MAEDA, 1996; REASENBERG, 1999). Given that normal stress in the crust, that is the regional tectonic stress plus the loading stress, increases with depth, supports the hypothesis that increasing normal stress inhibits foreshock occurrence and that at high normal stress foreshock sequences tend to be short or non-existent (ABERCROMBIE and MORI, 1996). Since heterogeneity decreases with depth and normal stress, foreshock activity should increase with the degree of small-scale crustal heterogeneity. The stronger the heterogeneity the larger the percentage of mainshocks preceded by foreshock sequences and the longer the duration of each individual foreshock sequence.

3. *Source Considerations and the OFC Model*

Most models of earthquake dynamics have in common that if the applied stress between the two fault segments exceeds a certain threshold value a displacement occurs instantaneously and an earthquake is generated. In order to model this kind of elastic behavior, the so called spring block models were introduced such as the BK (BURRIDGE and KNOPOFF 1967) and OFC (OLAMI *et al.* 1992) models. Although a wide variety of spring-block models were proposed and studied by several authors (see review in RUNDLE *et al.*, 2003), so far they are only weakly related to distributed seismicity (TURCOTTE *et al.*, 2009). A modification of the OFC model was studied by JAGLA (2010) who introduced a structural relaxation mechanism, thus, generating earthquake sequences that contain features observed in real seismicity. The relaxation was introduced in a rather phenomenological way by means of an arbitrary energy measure E stored locally in the system and which can not directly mapped to some known source mechanical parameters.

In this paper we address the mapping of source mechanical parameters to macroscopic observations by presenting a theoretical framework which models analytically the dependence of the macroscopic b values on the structural heterogeneity as well as on material processes, such as structural relaxation, through appropriate constitutive equations. A specific case of the theoretical framework is examined where a model for structural softening is proposed by relaxing the assumption of constant shear strength in order to reproduce realistic b values in foreshocks sequences.

Following the lines of the OFC model, we note that at the basis of spring-block models, which exhibit Self-Organized Criticality (SOC), we find the assumption of two rigid plates, moving relative each other, where their interface is modeled by means of a number of blocks (nodes) interconnected with springs of a certain elastic constants, K, for each direction. If a one-dimension scheme is adopted, the well-known BK model is obtained. Here we consider the two-dimension scheme of the problem as it was presented by the OFC model depicted in Fig. 1.

A dynamic variable, f_{ij}, is assigned to each node of a two-dimension grid which represents the fault plane of an earthquake source. The variable, f_{ij}, being the total force exerted at the node n_{ij}, grows through time, t, at a constant quantity, $r = K_L \upsilon t$ where υ is the constant velocity of the upper moving plate and K_L is the elastic constant at the vertical direction of the fault plane. Assuming, without loss of generality, the isotropic case $K_1 \equiv K_2 \equiv K$ at the horizontal plane (K_1, K_2 are the elastic constants along the two axes), then the dynamic variable, in the continuum limit, is expressed by (u_{ij} being the node's displacement at x_{ij})

$$f_{ij} = Kl^2\nabla^2 u_{ij} + K_L \delta u_{ij}, \qquad (2)$$

where $\delta u_{ij} = \upsilon t - u_{ij}$ and the second order gradient term in the right hand-side models as usual in the continuum limit the expression $2u_{ij} - u_{i-1j} - u_{i+1j} + 2u_{ij} - u_{ij-1} - u_{ij+1}$. The l is a characteristic scale associated with space discretization (hereafter, without loss of generality, l is assumed unity). When the force on an arbitrary node is larger than a threshold value F_{th}, which is the maximal static friction, the

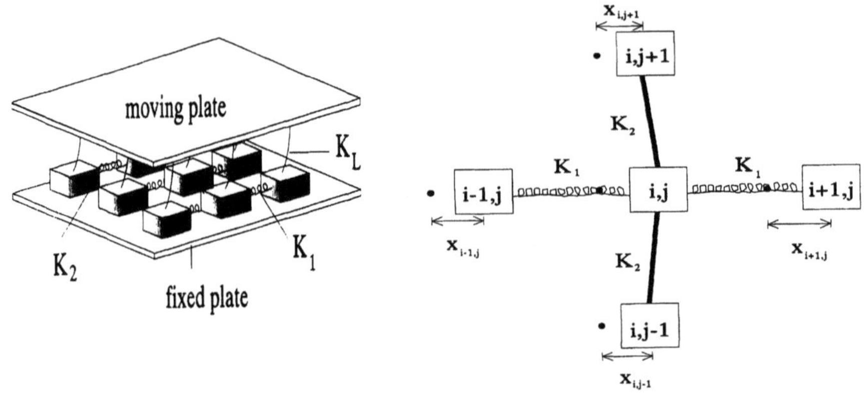

Figure 1
The geometry of the OFC model for earthquake source (after OLAMI *et al.* 1992)

node slips. For the node n_{ij} of the fault plane to become unstable, the following inequality must be fulfilled,

$$K\nabla^2 u_{ij} + K_L \delta u_{ij} - F_{th} \geq 0. \qquad (3)$$

Then, redistribution of forces to the neighbor sites is taking place as follows,

$$f_{ij} \to 0, \quad f_{\mathrm{mm}} \to a f_{ij}, \qquad (4)$$

where f_{mm} are the forces for the four nearest neighbors and a is the ratio at which force is transferred to the neighbor sites. For the assumed isotropic case, the parameter a takes the form $a = 1/(4 + K_L/K)$ (OLAMI *et al.* 1992) and can be interpreted as a parameter of the source's material depending on the corresponding elastic constants of the fault plane. Note that since always $K_L/K > 0$ then always $a < 0.25$ and as a result the OFC model describes a non-conservative system. Decreasing a reflects increase in dissipation, while when this parameter equals 0.25 the case is the conservative one.

Within this context, simulations of the original OFC model by means of a continuous automaton qualitatively predicted the emergence of power-law for the earthquake magnitude distribution. On the other hand, initially the presence of foreshocks and aftershocks was not recognized in the OFC model as pointed out, for example, by HAINZL *et al.* (1999). However, HERGARTEN and NEUGEBAUER (2000) and HERGARTEN (2002) showed with the help of numerical simulations that the OFC model exhibits foreshocks and aftershocks with results consistent with Omori's empirical law for the time decay of events, although the exponents predicted by the model are lower than observed in nature. Then, foreshocks and aftershocks can be attributed to the non-conservative character of OFC model. On the other hand, the experimentally observed variations of b value before and after the occurrence of an earthquake have not been modeled so far in terms of SOC.

From simulation results, OLAMI *et al.* (1992) showed that in their model of the G–R power-law, and, consequently, the b value, depends reversely on the conservation level. The measured values of b around 1 are obtained for $a \approx 0.20$. With the decrease of a, the critical exponent b increases. Furthermore, it was found that the exponent varies continuously when changing the boundary conditions smoothly (CHRISTENSEN and OLAMI, 1992). These results indicated that there is no universality of the critical exponent. Introducing anisotropy changes the scaling of the distribution function, but not the power-law exponent. The dependence of the exponent b on the conservation level, that is on α, was also shown for simulated foreshock and aftershock sequences (HERGARTEN and NEUGEBAUER, 2002).

At this point the contradiction of the OFC model with experimental findings arises: while the OFC model assumes a constant material parameter a and consequently a constant power-law exponent b, the last varies significantly with the style of seismicity, e.g., before and after a mainshock or during seismic swarms. As a consequence, the OFC model is not capable of correctly modeling the material processes taking place within the source, which seems to be more complex than described by the simplified picture of the spring-block model suggested earlier. In the next lines, we present an analytical enrichment of the spring-block model, and support it by simulation results, with the aim to address the dependency of the seismicity style on a source's mechanical properties.

4. *Analytical OFC Modeling of the b Value Variability*

It has been proposed that the weaknesses of the existing models of SOC for seismicity lie in the fact that internal mesoscopic material processes, such as viscous relaxation, were not taken into account (e.g., HAINZL *et al.*, 1999; PELLETIER, 2000). Consequently, interactions between network sites are by no means the only dynamic process taking place but also material parameters, that were assumed constant so far, are also involved in the process of seismogenesis in the time and space domains. More recently, it was hypothesized that the threshold value F_{th} of the maximal static friction is not a constant material parameter but a dynamic internal variable which varies with the evolution of seismicity (AVLONITIS and TASSOS, 2010). As will be shown in the next paragraphs, this leads to a drastic change in the dynamic behavior of the spring-block model for earthquakes, resulting in robust model predictions for the variability of the b value.

Let us assume a shear displacement induced hardening–softening law for the shear strength of the fault plane as proposed by PALMER and RICE (1973) and MCCLUNG (1979) (Fig. 2). In the simplest case, adapting a linear idealization of the hardening–softening curve, we may write,

$$F_{th} = F_{\text{th}}(u_{ij}) = F^{*} + \lambda u_{ij}, \tag{5}$$

where λ is a proportionality parameter modeling the hardening or softening process of the Earth's material and F^{*} a constant parameter. The sign of the proportionality parameter (as well as the parameter F^{*}) depends at which branch of the hardening–softening law of the fault plane the material deforms. Within the framework of the so-called gradient theory, the

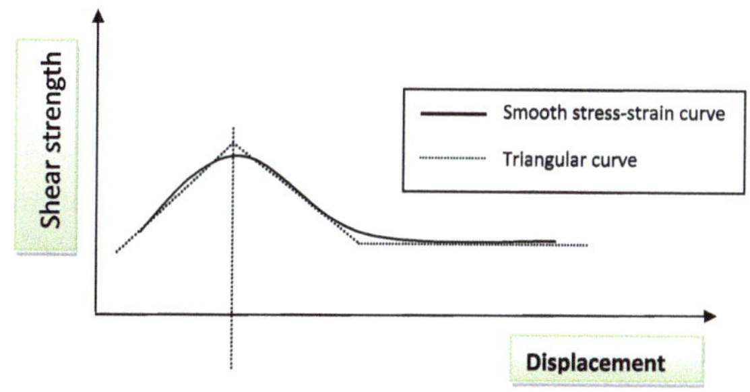

Figure 2
A non-convex shear strength hardening–softening law of the fault plane

non-local behavior of the displacement manifold can be modeled by gradient terms (AIFANTIS 2003, and references therein), i.e.,

$$u_{ij} = u_{ij}^{0} + c\nabla^{2}u_{ij}, \tag{6}$$

where u_{ij}^{0} is uniform over the entire space displacement field, and c is the gradient coefficient. The second order gradient term in the right hand-side of Eq. (6) was introduced in order to model the local material non-uniformities into the macroscopic stress–strain relation of Eq. (3). Substituting in (5), we get

$$F_{\text{th}} = F^{*} + \lambda(u_{ij}^{0} + c\nabla^{2}u_{ij}) \quad \text{or} \quad F_{th} = F_{th}^{0} + K'\nabla^{2}u_{ij} \tag{7}$$

with $F_{\text{th}}^{0} = F^{*} + \lambda u_{ij}^{0}$ and

$$K' = \lambda c. \tag{8}$$

After relaxing the assumption of constant homogeneous shear strength and interpreting it as an internal spatially evolving variable, the final inequality for the node n_{ij} of the fault plane to become unstable, reads

$$(K - K')\nabla^{2}u_{ij} + K_{L}\delta u_{ij} - F_{th}^{0} \geq 0. \tag{9}$$

As a result, an effective dissipation parameter a^{eff}, which controls the macroscopically observed values of b, is introduced as follows,

$$a^{\text{eff}} = \frac{1}{4 + \frac{K_{L}}{K-K'}}. \tag{10}$$

Taking into account Eq. (8), we end up with the final expression

$$a^{\mathrm{eff}}(c,\lambda) = \frac{1}{4 + \frac{K_L}{K - K'(c,\lambda)}}. \qquad (11)$$

Equation (11) is of crucial importance since it redresses the generic weakness of the OFC model. In fact, it relates the effective structural parameter a^{eff} not only with K, K_L, that is the elastic properties of the fault plane, but also with structural heterogeneity through the gradient coefficient c. More importantly, Eq. (11) relates a^{eff} with material processes of the earthquake source through an appropriate constitutive equation for source evolution as expressed in Eq. (5), i.e., whether the source material deforms under a softening or hardening mode expressed by the appropriate sign—minus or plus correspondingly—of the proportionality parameter λ.

Within the proposed physical framework, a one-to-one relation between an effective structural parameter a^{eff} and the macroscopically observed b value is provided either theoretically or synthetically through simulation experiments on the enriched OFC model. As a result, within this context it is possible to bridge the macroscopically observed b values with specific styles of seismicity and, therefore, to specific processes of seismogenesis. Then, the low b values experimentally observed in foreshock sequences can be modeled or mapped by a process of material softening in the seismogenic volume. Indeed, as the source material enters to the descending branch in Fig. (2), Eq. (5) takes the following form (for convenience the term F^* was omitted),

$$F_{\mathrm{th}} = F_{\mathrm{th}}(u_{ij}) = -|\lambda| u_{ij}. \qquad (12)$$

The negative value $-|\lambda|$, results to negative values of $K'(c, \lambda)$ and, via Eq. (11), as $|\lambda|$ increases the effective structural parameter $a^{\mathrm{eff}}(c, \lambda)$ increases too. However, from simulation experiments, it comes out that a^{eff} and b evolve always reversely as noted before. Then, as the source material enters into the softening regime, macroscopic lower values of b are expected, which is in perfect accordance with the macroscopic observations in foreshock sequences. As a consequence, within the proposed analytical formulation, a bridge between macroscopic observations and specific material processes in the source was established, providing a robust tool in the effort to understand better the behavior of seismicity not only in the presence of foreshocks at the final acceleration stage before mainshock generation, but also in aftershocks, in swarms and in the background seismicity as well.

5. Simulation Results

To check the analytical findings of our formulation presented before a cellular automaton was build-up in order to simulate the enriched OFC model. Practically, if the simulation of the evolution of the source material when it enters into the softening regime is the goal, then a variation of the classic OFC simulator is needed. Then, we can reconstruct the macroscopic magnitude-frequency relation and check whether lower b values are emerging in accordance with observations in foreshock sequences. To this end, in order to mimic the evolution of source material when it enters into the softening regime, we built the classical OFC simulator as explained in Sect. 3 and introduced a linear reduction of the threshold value F_{th} of each site according to Eq. (5). The results presented below refer to the values $a^{\mathrm{OFC}} = 0.245$ and $a^{\mathrm{OFC}} = 0.235$ of the OFC classical structural parameter, but it is noted that similar behavior was found to lower values of a^{OFC} for which more realistic values of b are reproduced.

Our main finding consists of the variation of the b value with parameter λ of the introduced material process, that is hardening or softening. Indeed, in Fig. 3 the cases of no softening and of gradually increasing softening are depicted for different system sizes indicating that both the qualitative and quantitative behavior of the system dynamics are similar for different system sizes. Moreover, it is evident that the parameter b changes inversely proportionally with the softening coefficient, in accordance with our findings in the previous sections. The procedure is analytically presented in Fig. 4. For different values of the softening parameter, λ, mainly when $\lambda = 0, -2, -4$, the estimation of the emerged b value is given.

It can be seen that a clear linear reduction of the b value versus the absolute value of the softening parameter λ emerged (Fig. 4d).

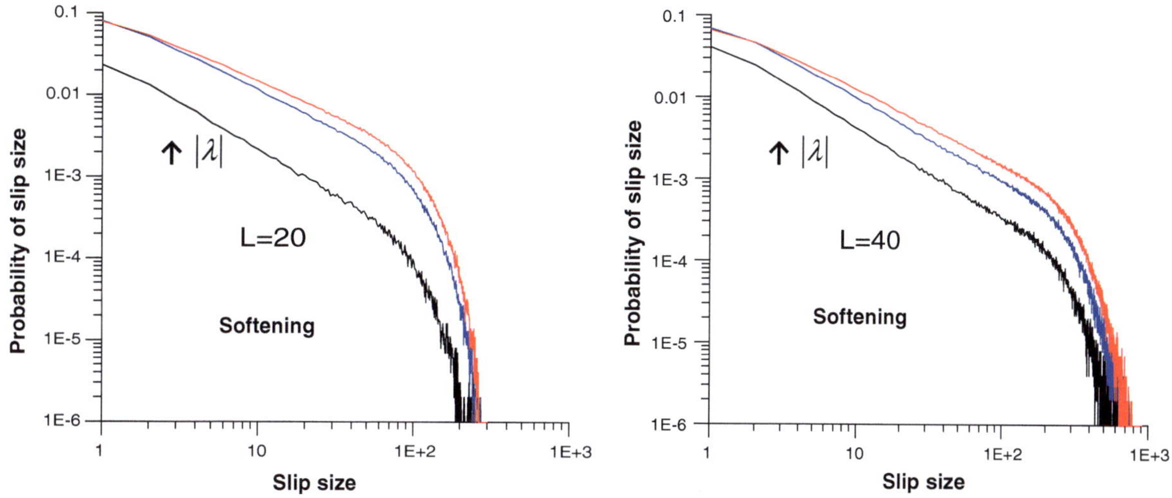

Figure 3
Simulation results for the macroscopic magnitude-frequency relation within the softening regime for different system sizes, $L = 20$, 40 and for $a^{OFC} = 0.245$. Cases of no softening (*first curve* at the *bottom*) and gradually increasing softening are shown

Although the linear dependency of the macroscopic *b* value on the softening coefficient within the softening regime is pronounced, this is not the case within the hardening regime, that is for positive λ. In fact, while an initially significant increase for the corresponding *b*-exponent is found, an almost no dependency regime is emerging for greater hardening parameters (Fig. 5). Notably the dynamics breaks down for hardening parameters greater than unity.

In Fig. 6a a comparison of the estimated linear relation of the *b* values versus the softening parameter $|\lambda|$ is depicted for different values of the structural parameter a^{OFC}. It can be seen that, while the linearity is preserved, a slight decrease of the corresponding slope is observed. Finally, in Fig. 6b the dependency of the *b* value from the structural heterogeneity is also depicted. Here, we mimic source heterogeneities by introducing a distribution of the shear strength values. Indeed, the different curves correspond to different Weibull distributions of the shear strength, F_{th}. More specifically, different simulations were performed for different shape parameters w of the Weibull distribution ($p(x; m, w) = (w/m)(x/m)^{w-1} \exp[-(x/m)^w]$, $x \geq 0$) of F_{th}, thus, imposing gradually stronger heterogeneity: higher w corresponds to higher variability in the corresponding distribution, and as a result, it corresponds to higher heterogeneity. It can be seen that the stronger the heterogeneity, the higher the *b* value, in accordance with our derivation in Eq. (10) and because, as we noted before, the dissipation parameter a^{eff} and the *b* value always evolve reversely.

6. Discussion

The simulation results presented in the previous section refer to an artificial situation where the earthquake source evolves always under softening. This simplified simulation setup should be elaborated by introducing a material mechanism that reproduces the non-convex hardening–softening law of the fault plane, as illustrated in Fig. (2). As an instance, it could be the mechanism of transient creep as was proposed by Hainzl *et al.* (1999). In any case, the adaptation of the specific mechanism taking place at the source will be incorporated into the equation that introduces the effective parameter a^{eff} as expressed here by Eq. (11). Moreover, the dependency of the *b* value from the structural heterogeneity through the gradient coefficient c, as expressed in Eq. (11), implies that different structures at the source may result in different macroscopic *b* values. This became evident from our simulation results (Fig. 6b). Then, in an ideal scenario, the parameter λ (modeling the structural properties of the source) and the gradient

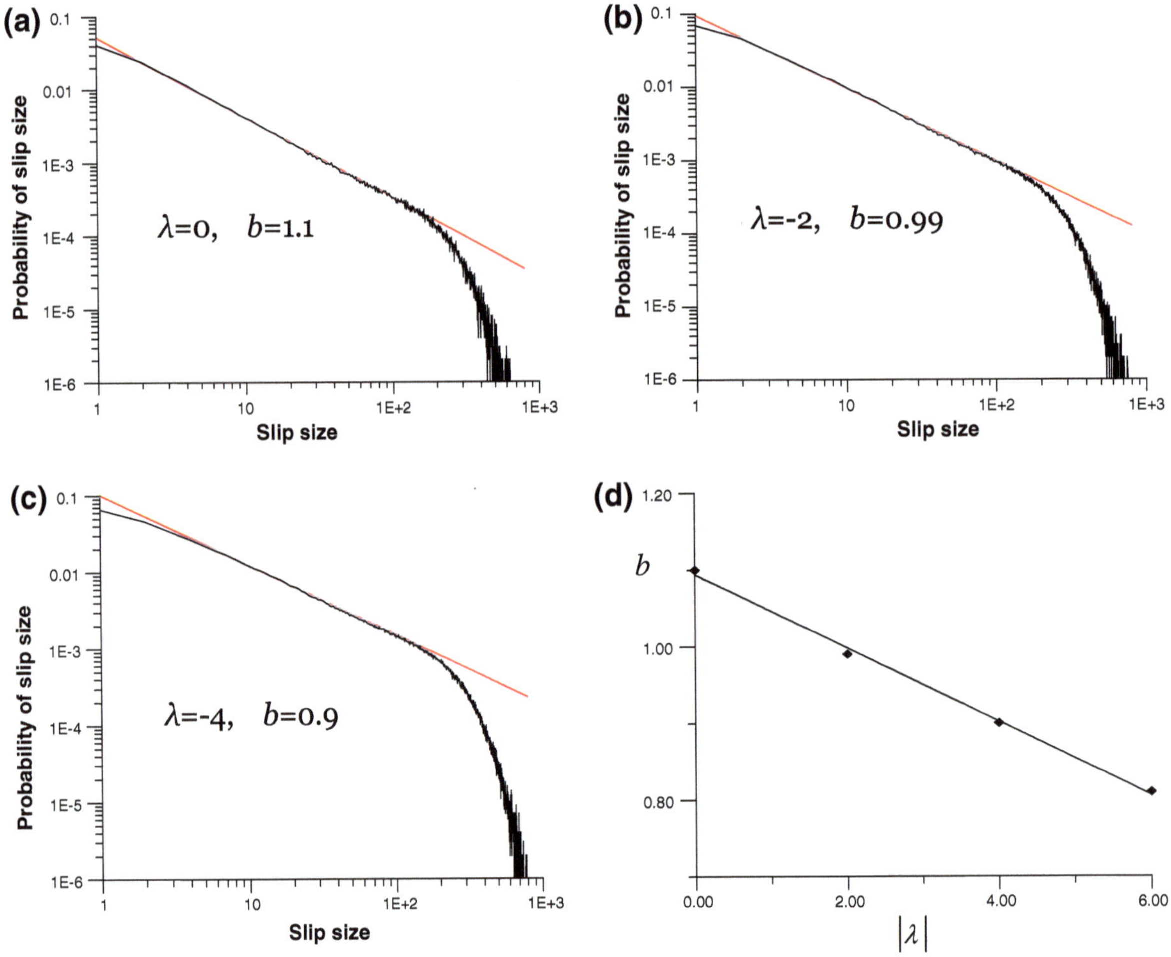

Figure 4
Explicit estimation of the b values within the softening regime (**a**), (**b**), (**c**). In **d** a clear linear relation between b and $|\lambda|$ is depicted. The simulations referred to $L = 40$ and $a^{\mathrm{OFC}} = 0.245$

coefficient c (modeling crustal heterogeneity properties) should be determined in a different context both from theoretical considerations and/or experiments in laboratories. At this point a well known limitation arises, since there is an ongoing debate in the literature about the robustness of the source's material parameters estimated from different theories and experimental configurations. Unknown parameters such as the exact stress and strain fields within the source, the exact geometry of the asperities as well as size effects that may alter the values estimated in the scale of an laboratory experiment, are some of the unanswered questions concerning the robust estimation of the parameters λ and c. It is our aim to show, in a future work, that one can overcome this generic difficulty in a more constructive way. For example, it is possible to use a learning dataset from past seismicity records of a certain fault in order to calibrate the two basic simulation parameters λ and c. Indeed, within the context of the present work, it is possible to calibrate λ and c by forcing the simulation outcomes to reproduce the exact evolution of the b value. In that way, we can estimate effective values of the source parameters without knowing the exact values of stress and strain fields, the asperity roughness, the elastic properties or even the specific mechanism of the fault.

According to Aki (1981), if there is a basic generic difference between foreshocks and normal earthquakes (background seismicity), there remains a

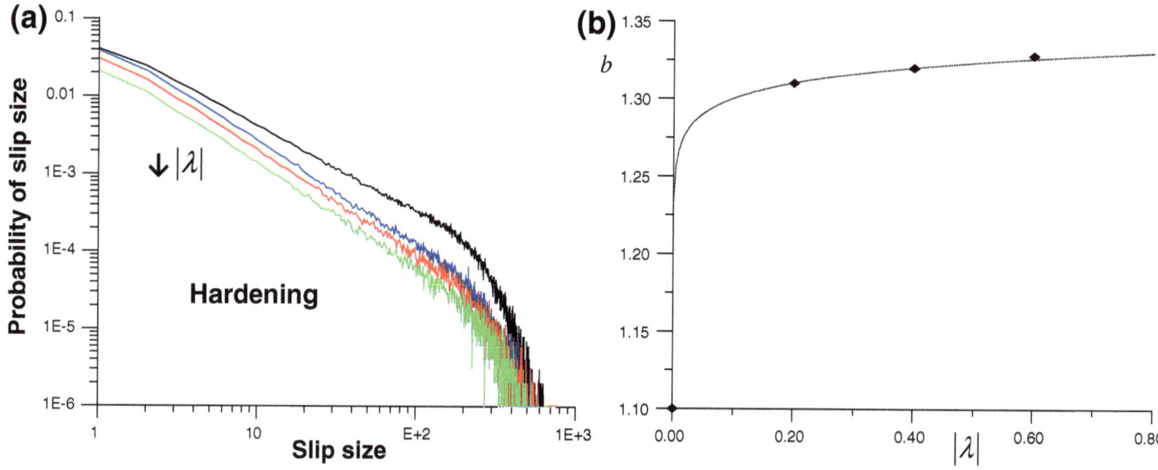

Figure 5
In **a** simulation results for the macroscopic magnitude-frequency relation within the hardening regime is shown for system size $L = 40$ and for $a^{OFC} = 0.245$. In **b** the almost no dependency regime between b and $|\lambda|$ is depicted

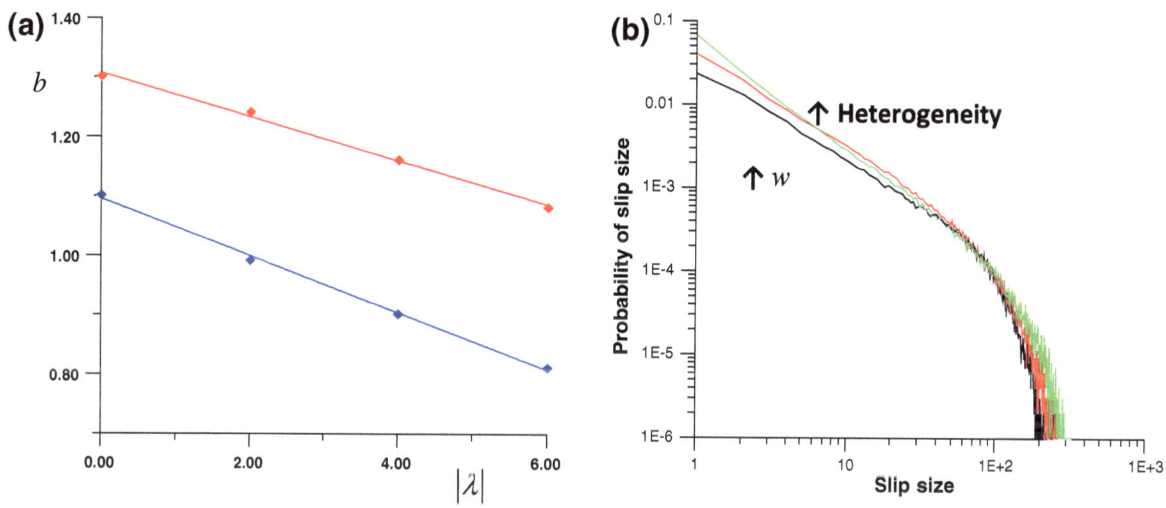

Figure 6
a Comparison of the estimated linear relation of b vs $|\lambda|$ for different values of the structural parameter, $a^{OFC} = 0.245$ and $a^{OFC} = 0.235$, and for $L = 40$. **b** Probability distributions for different levels of heterogeneity of F_{th} by means of different shape parameters w of the Weibull distribution

hope to discriminate them. Smaller b values observed for some foreshocks than for aftershocks certainly agree with the idea that the percolation model applies to the former and the barrier model to the latter (Aki, 1981). However, what certainly remains unexplained is why some mainshocks are preceded by foreshocks and others are not. This behaviour may be related to mainshocks that take place under conditions of plasticity. It has been proposed that when a mainshock nucleates within the brittle seismogenic layer and near the base of the seismogenic layer, immediate foreshocks for this mainshock are necessarily restricted to lie within a localized region shallower than the hypocentral depth of the mainshock (Ohnaka, 1992). By contrast, the nucleation process below the base of the seismogenic layer is aseismic in nature, as it happens with interplate earthquakes that nucleate below the base of the brittle seismogenic layer, thus, explaining also why strong intermediate-depth and deep earthquakes are not associated by foreshocks while the aftershock activity is much lower than in shallow earthquakes (see review in FROHLICH, 2006). ROBERTS and TURCOTTE (2000) considered the plastic instability hypothesis against mechanical friction, provided numerical solution using plastic rheology and concluded that, because the yield (failure) stress decreases with increasing temperature, earthquake nucleation at the base of the seismogenic zone follows naturally. Plastic rheology includes work hardening and thermal softening. In this context the proposed formulation could be a valuable tool in order to model such material processes.

On the other hand, NODA *et al.* (2013) were based on hierarchical asperity models and showed that there is a variety of nucleation scenarios for large mainshocks; thus, it may not be surprising that foreshocks are missing for some earthquakes. As a conclusion, the dependency of the seismicity style, e.g., foreshocks or aftershocks, on source's mechanical properties was studied by an analytical enrichment of a 2D spring-block model of the OFC type. The analytical expression reached relates, through an appropriate gradient term, the newly introduced generalized effective dissipation parameter a^{eff} not only with the elastic properties of the fault plane, but also with structural heterogeneity and specific structural processes of the source. Simulation results with a modified OFC cellular automaton have verified the predicted evolution of b in macroscopic records, i.e., low b values are experimentally observed in foreshock sequences that can be modeled by a process of material softening in the seismogenic volume that is by a process of strain rate increasing with time under constant stress. During aftershocks and beyond, however, it is not expected that the material processes affect drastically the macroscopic behavior as expressed by the magnitude-frequency relation. Within this context the value of the exponent b in the magnitude-frequency relation is introduced mainly as a structural dependent measure or parameter of the earthquake source, thus, generalizing previous considerations where the b value was assumed as a "strainmeter" (e.g., SCHORLEMMER *et al.*, 2005; SPADA *et al.*, 2013). This generalization results in a more robust formalization in modeling the mechanical behavior of the source since the same constitutive values of stress or/and strain may refer to a completely different type of source structure e.g., in the case of a non-convex constitutive stress–strain relation of the source.

REFERENCES

ABERCROMBIE, R.E., and MORI, J. (1996), *Occurrence patterns of foreshocks to large earthquakes in the western United States*, Nature, *381*, 303–307.

AGNEW, D.C., and JONES, L. (1991), *Prediction probabilities from foreshocks*, J. Geophys. Res., *96*(B7), 11959–11971.

AIFANTIS, E.C. (2003), *Update on a class of gradient theories*, Mech. Mater., *35*, 259–280.

AKI, K. (1981), A probabilistic synthesis of precursory phenomena, In Earthquake prediction-An international review (eds Washington. D.C., Simpson. D.W. and Richards. P.G.) (Maurice Ewing series 4, Am. Geophys. Union, 1981), pp. 566–574.

AMELUNG, F., and KING, G. (1997), *Earthquake scaling laws fro creeping and non-creeping faults*, Geophys. Res. Lett., *24*, 507–510.

AVLONITIS, M., and TASSOS, S. (2010). Towards realistic self-organized critical models for earthquakes, ESC 2010, Montpellier, France, (abstr.).

BURRIDGE, R., and KNOPOFF, L. (1967), *Model and theoretical seismicity*, Bull Seismol. Soc. Am., *57*, 341–371.

CHRISTENSEN, K., and OLAMI, Z. (1992), *Scaling phase transitions, and nonuniversality on a self-organized critical cellular-automaton model*, Phys. Review A, *46*, 1829–1838.

ENESCU, B., and ITO K. (2001), *Some premonitory phenomena of the 1995 Hyogo-Ken Nanbu (Kobe) earthquake: seismicity, b-value and fractal dimension*, Tectonophysics, *338*(3–4), 297–314.

FROHLICH, C., Deep Earthquakes, (Cambridge University Press, UK, 2006).

FROHLICH, C., and DAVIS, S.D. (1993), *Teleseismic b values; Or, much ado about 1.0*. J. Geophys. Res., *98*(B1), 631–644.

GHOSH, A., NEWMAN, A.V., THOMAS, A.M., and FARMER, G.T. (2008), *Interface locking along the subduction megathrust from b-value mapping near Nicoya Peninsula, Costa Rica, Geophys.* Res. Lett., *35*, L01301, doi:10.1029/2007GL031617.

GUTENBERG, B., and RICHTER, C. (1944), *Frequency of earthquakes in Caifornia*, B. Seismol. Soc. Am., *34*, 185–188.

HAINZL, S., ZOLLER, G. and KURTHS, J. (1999), *Similar power laws for foreshock and aftershock sequences in a spring-block model for earthquakes*, J. Geophys. Res., *104*, 7243–7253.

HERGARTEN, S., Self-organized criticality in earth systems (Springer, Berlin, Heidelberg, 2002).

HERGARTEN, S., and NEUGEBAUER, HJ. (2000), *Self-organized criticality in two-variable models*, Phys. Rev. E, *61*, 2382–2385.

HERGARTEN, S., and NEUGEBAUER, HJ. (2002), *Foreshocks and aftershocks in the Olami-Feder-Christensen model*, Phys. Review Lett., 88, 238501/1–238501/4.

JAGLA, E.A. (2010), *Realistic spatial and temporal earthquake distributions in a modified Olami-Feder-Christensen model*, Phys. Rev. E *81*, 046117/1–046117/8.

JONES, L.M. (1984), *Foreshocks (1966–1980) in the San Andreas System, California*, B. Seismol. Soc. Am., *74*, 1361–1380.

JONES, L.M., and MOLNAR, P. (1979), *Some characteristics of foreshocks and their possible relationship to earthquake prediction and premonitory slip on faults*, J. Geophys. Res., *84*, 3596–3608.

JORDAN, TH.H., CHEN, Y.T., GASPARINI, P., MADARIAGA, R., MAIN, I., MARZOCCHI, W., PAPADOPOULOS, G.A., SOBOLEV, G, YAMAOKA, K. And ZSCHAU, J. (2011), *Operational Earthquake Forecasting: State of Knowledge and Guidelines for Utilization. Annals of Geophysics, 54*(4), 316–391, doi:10.4401/ag-5350.

ISHIMOTO, M., and IIDA, K. (1939), *Observations of earthquakes registered with the microseismograph constructed recently*, Bull. Earthq. Res. Inst. Tokyo Univ., *17*, 443–478.

KAGAN, Y., and KNOPOFF, L. (1978), *Statistical study of the occurrence of shallow earthquakes*, Geophys. J. Roy. Astr. Soc., *55*, 67–86.

MAEDA, K. (1996), *The use of foreshock in probabilistic prediction along the Japan and Kuril trenches*, Bull. Seismol. Soc. Am., *86*(1A), 242–254.

MAIN, I. (2000), *Apparent breaks in scaling in the earthquake cumulative frequency-magnitude distribution: fact or artifact?* Bull. Seism. Soc. Am., *90*(1), 86-97.

MAIN, I., MEREDITH, PH.G., and JONES, C. (1989), *A reinterpretation of the precursory seismic b-value anomaly from fracture mechanics*, Geophys. J. Internat., *96*, 131–138.

MCCLUNG, D.M. (1979), *Shear fracture precipitated by strain softening as a mechanism of dry slab avalanche release*, J. Geophys. Res. *84*, 3519–3526.

MOGI, K. (1963a), *The fracture of a semi-infinite body caused by an inner stress origin and its relation to the earthquake phenomena*, Bull. Earthq. Res. Inst., Univ. Tokyo, *41*, 595–614.

MOGI, K. (1963b), *Some discussion on aftershocks, foreshocks and earthquake swarms – the fracture of a semi-infinite body caused by an inner stress origin and its relation to the earthquake phenomena*, Bull. Earthq. Res. Inst., Univ. Tokyo, *41*, 615–658.

MOLCHAN, G.M., KRONROD, T.L., and NEKRASONA, A.K. (1999), *Immediate foreshocks: time variation of the b-value*, Phys. Earth Planet. Int., *111*, 229–240.

NANJO, K. Z., HIRATA, N., OBARA, K., and KASAHARA, K. (2012), *Decade-scale decrease in b value prior to the M9-class 2011 Tohoku and 2004 Sumatra quakes*, Geophys. Res. Lett., *39*, L20304, doi:10.1029/2012GL052997.

NARTEAU, C., BYRDINA, S., SHEBALIN, P., and SCHORLEMMER, D. (2009), *Common dependence on stress for the two fundamental laws of statistical seismology*, Nature, *462*, 642–645.

NODA, H., NAKATANI, M., and HORI, T. (2013), *Large nucleation before large earthquakes is sometimes skipped due to cascade-up—Implications from a rate and state simulation of faults with hierarchical asperities*, J. Geophys. Res., *118*, 2924–2952, doi:10.1002/jgrb.50211.

OLAMI, Z., FEDER, H., and CHRISTENSEN, K. (1992), *Self-organized criticality in a continuous, nonconservative cellular automaton modeling earthquakes*, Phys. Rev. Lett. *68*, 1244–1247.

OMORI, F. (1894), *On the aftershocks of earthquakes*, J. Coll. Sci Imp., Univ. Tokyo, *7*, 111–200.

OHNAKA, M. (1992), *Earthquake source nucleation: a physical model for short-term precursors*, Tectonophysics, *211*, 149–178.

PALMER, A.C., and RICE, J.R. (1973), *The growth of slip surfaces in the progressive failure of over-consolidated clay*, Proc. R. Soc. London, Ser. A, *332*, 527–548.

PAPADOPOULOS, G.A., DRAKATOS, G., and PLESSA, A. (2000), *Foreshock activity as a precursor of strong earthquakes in Corinthos Gulf, Central Greece*, Phys. Chem. Earth, *25*, 239–245.

PAPADOPOULOS, G.A., LATOUSSAKIS, I., DASKALAKI, E., DIAKOGIANNI, G., FOKAEFS, A., KOLLIGRI, M., LIADOPOULOU, K., ORFANOGIANNAKI, K., and PIRENTIS, A. (2006), *The East Aegean Sea strong earthquake sequence of October–November 2005: lessons learned for earthquake prediction from foreshocks*, Nat. Hazards Earth Syst. Sci., *6*, 895–901.

PAPADOPOULOS, G.A., CHARALAMPAKIS, M., FOKAEFS, A., and MINADAKIS, G. (2010), *Strong foreshock signal preceding the L'Aquila (Italy) earthquake (Mw 6.3) of 6 April 2009*, Natural Hazards & Earth System Science, *10*, 19–24.

PAPADOPOULOS, G.A., MINADAKIS, G., AVLONITIS, M., and DI FIORE, B. (2011), Foreshock precursory patterns before strong mainshocks. AGU Final Meeting, Session NH22A, San Francsisco, 5–9 December 2011 (abstr.).

PAPAZACHOS, B.C. (1975), *Foreshocks and earthquake prediction*, Tectonophysics, *28*, 213–226.

PELLETIER, J.D., Spring-block models of seismicity: review and analysis of a structurally heterogeneous model coupled to a viscous asthenosphere, In GeoComplexity and the Physics of Earthquakes (eds Rundle, JB., Turcotte, DL., and Klein, W.) (Am. Geophys. Union, Washington, DC, 2000)

REASENBERG, P.A. (1999), *Foreshock occurrence before large earthquakes*, J. Geophys. Res., *104*, 4755–4768.

ROBERTS, D.C., and TURCOTTE, D.L., Earthquakes: friction or a plastic instability?, In GeoComplexity and the Physics of Earthquakes (eds Rundle, JB., Turcotte, DL., and Klein, W.) (Am. Geophys. Union, Washington, DC, 2000).

RUNDLE, J.B., TURCOTTE, D.L., SCHERBAKOV, R., KLEIN, W., and SAMMIS, C. (2003), *Statistical physics approach to understanding the multiscale dynamics of earthquake fault systems*, Reviews of Geophysics, *41*, 1019, doi:10.1029/2003RG000135.

SCHOLZ, C. H. (1968), *Microfractures, Aftershocks, and Seismicity*, Bull. Seismol. Soc. Am., *58*, 1117–1130.

SCHORLEMMER, D., and WIEMER, S. (2005), *Microseismicity data forecast rupture area*, Nature, *434*, 1086.

SCHORLEMMER, D., WIEMER, S., and WYSS, M. (2005), *Variations in earthquake-size distribution across different stress regimes*, Nature, *437*, 539–542.

SPADA, M., TORMANN, T., WIEMER, S., and ENESCU, B. (2013), *Generic dependence of the frequency-size distribution of earthquakes on depth and its relation to the strength profile of the crust*, Geophys. Res. Lett., *40*, 709–714, doi.10.1029/2012GL054198.

SUYEHIRO, S., ASADA, T., and OHTAKE, M. (1964), *Foreshocks and aftershocks accompanying a perceptible earthquake*, Meteorol. Geophys., *15*, 71–88.

TURCOTTE, D.L., RUNDLE, J. B., and SCHERBAKOV, R., Complexity and earthquakes, In Earthquake Seismology (eds Kanamori, H., and Schubert, G.) (Treatise on Geophysics, Elsevier, 2009).

URBANCIC, T.I., TRIFU, C-I., LONG, J.M., and YOUNG, R.P. (1992), *Space-time correlations of b values with stress release*, Pure Appl. Geophys., *139*, 449–462.

VIDALE, J., MORI, J., and HOUSTON, H. (2001), *Something Wicked This Way Comes: Clues from Foreshocks and Earthquake Nucleation*, EOS Trans. AGU, *82*, 68.

WYSS, M. (1973), *Towards a physical understanding of the earthquake frequency distribution.* J. R. Astron. Soc., *31*, 341–359.

WYSS, M. (1997), *Second round of evaluations of proposed earthquake precursors*, Pure Appl. Geophys., *149*, 3–16.

Wyss, M. (2001), *Locked and creeping patches along the Hayward fault, California*, Geophys. Res. Lett., *28*, 3537–3540.

Wyss, M., Schorlemmer, D., and Wiemer, S. (2000), *Mapping asperities by minima of local recurrence time: The San Jacinto-Elsinore fault zones*, J. Geophys. Res., *105*, 7829–7844.

Yamashita, T. (1998), *Simulation of seismicity due to fluid migration in a fault zone*, Geophys. J. Int., *132*, 674–686.

Zhao, Y.Z., and Wu Z.L. (2008), *Mapping the b-values along the Longmenshan fault zone before and after the 12 May 2008, Wenchuan, China, Ms 8.0, earthquake*, Nat. Hazards Earth Syst. Sci., *8*, 1375–1385.

(Received May 2, 2013, revised January 20, 2014, accepted February 12, 2014, Published online March 8, 2014)

Pure Appl. Geophys. 171 (2014), 2581–2599

DOI 10.1007/s00024-013-0764-9

Pure and Applied Geophysics

Doubly Stochastic Earthquake Source Model: "Omega-Square" Spectrum and Low High-Frequency Directivity Revealed by Numerical Experiments

A. A. GUSEV[1,2]

Abstract—Since its formulation in 1967–1970, the classical ω^{-2} model of earthquake source spectrum awaits a consistent theoretical foundation. To obtain one, stochastic elements are incorporated both into the final structure of the fault and into the mode of rupture propagation. The main components of the proposed "doubly stochastic" model are: (1) the Andrews's concept, that local stress drop over a fault is a random self-similar field; (2) the concept of rupture with running slip pulse, after Heaton; (3) the hypothesis that a rupture front is a tortuous, multiply connected ("lacy") fractal polyline that occupies a strip of finite width close to the slip-pulse width; and (4) the assumption that the propagation distance of fault-guided, mostly Rayleigh waves from a failing spot on a fault is determined by the slip-pulse width. Waveforms produced by this model are determined based on the fault asperity failure model after Das and Kostrov. Properties of the model are studied by numerical experiments. At high frequency, simulated source spectra behave as ω^{-2}, and acceleration spectra are flat. Their level, at a given seismic moment and rms stress drop, is inversely related to the relative width of the slip pulse. When this width is relatively low, a well-defined second corner frequency (lower cutoff of acceleration spectrum) is seen. The model shows clear dependence of propagation-related directivity on frequency. Between the first and the second corner frequency, amplitude spectra are strongly enhanced for the forward direction; whereas, above the second corner frequency, directivity is significantly reduced. Still, it is not inhibited totally, suggesting incomplete incoherence of the simulated radiator at high frequencies.

Key words: Self-similar, random, stress drop field, fractal, acceleration, slip pulse, random rupture front, frequency-dependent directivity.

1. *List of denotations and abbreviations*

The following three earthquake source models will be discussed: Case F1: asperity-source on an infinite planar fault P with zero cohesion (DAS and KOSTROV 1983, henceforth referred to as DK83); Case F2: fault-source P of finite area that contains single smaller-size asperity and whose cohesion equals zero over the rest of its surface (DAS and KOSTROV 1986; henceforth referred to as DK86); and Case F3: finite fault-source P whose entire area is a composition of spots/asperities of finite or infinitesimally small area (DAS and KOSTROV 1988, henceforth referred to as DK88; BOATWRIGHT 1988, henceforth referred to as B88; GUSEV 1988, henceforth referred to as G88; GUSEV 1989). The superscript "∞" is used for Case F1 where appropriate.

$\mathbf{x} = \{x, y, z\}$	Cartesian coordinates of the receiver point; selected so that $z = 0$ on a planar fault P
$\xi = \{\xi_1, \xi_2, 0\}$	Cartesian coordinates of a radiating point on P
Σ_a	Asperity area in Cases F1 and F2; located on P
$2R_a$	Its characteristic size (e.g. diameter)
Σ	Entire source area in Cases F2 and F3, located on P
$2R_c$	Its characteristic size. For Case F2, $R_c \gg R_a$. It is assumed that fault length $L \approx 2R_c$
dS	Infinitesimally small element of Σ_a or of Σ
ΔS	Small element of Σ in Case F3, of size $\Delta S^{0.5}$. A generalization of a single asperity treated in Cases F1 and F2, of size $2R_a$

[1] Institute of Volcanology and Seismology, Russian Ac. Sci, 9 Piip Blvd, Petropavlovsk-Kamchatsky 683006, Russia. E-mail: gusev@emsd.ru
[2] Kamchatka Branch, Geophysical Service, Russian Ac. Sci, 9 Piip Blvd, Petropavlovsk-Kamchatsky 683006, Russia.

R_s	Characteristic propagation distance of inhomogeneous waves (Rayleigh and other) guided by the free fault surface. These waves are generated by asperity failure in Cases F1 and F2, and by each ΔS in Case F3. Called "slip radius" in B88. In Case F2, $R_s = R_c$. At distances close to R_s, inhomogeneous fault-guided waves die off and do not propagate farther
c_S	*S*-wave velocity
v	Mean velocity of rupture front; it is assumed to be close to c_S
$T_a \approx 2R_a/c_S \approx 2R_a/v$	Characteristic duration of rupture propagation over in Cases F1 and F2
$T_c \approx (1-2)R_c/c_S \approx (1-2)R_c/v$	Characteristic duration of rupture process in Cases F2 and F3
$T_s \approx R_s/c_S$	Characteristic time associated with R_s
$\tau(\xi)$	Local dynamic stress drop on d*S;* it is assumed that $\tau(\xi) > 0$ In time, this stress drop is assumed to occur abruptly: its time history $\tau(\xi, t) \equiv \tau(\xi)H(t)$ where *H*(*t*) is the Heaviside step
T_r	Rise time, defined as twice the centroid of slip rate time history at ξ. It is assumed that $T_s \approx T_r$
L	Width of running slip pulse $l = vT_r$
C_H	Haskell–Heaton constant; $C_H = l/L$. In the case when T_r and l vary over a fault, C_H is understood as the average value over fault area
w	Characteristic width of the "thick" lacy random rupture front or "front strip". It is assumed here to coincide with l
T_w	Temporal width of the "front strip"; $T_w = w/v$
R	Hypocenter-to-receiver distance, it is assumed that $R \gg R_a$ and $R \gg R_c$
$F_0 = \int_{\Sigma_a} \tau(\xi)\mathrm{d}S$	Seismic force of an asperity source; introduced in DK83 for Case F1; still meaningful for Case F2. Valid for the Case F3 when the area of integration is replaced by Σ
f_{c1}	Common (leftmost) corner frequency, crossover point where theoretical or empirical source spectrum trend turns from f^0 to f^{-2}, f^{-1} or, generally f^{γ}. A model-independent definition of f_{c1} is used here based on log-spectrum expansion into MacLaurin series (Silver 1983); f_{c1} is defined as $1/2\pi T_{rms}$, where T^2_{rms} is the second normalized central power moment for a displacement waveform
f_{c2}	Second from the left corner frequency, crossover point where source spectrum trend turns from (accurately or approximately) f^{-1} to f^{-2}; simultaneously: left cutoff of approximately flat source acceleration spectrum. May coincide with f_{c1}; this case is one of the "ω^{-2}" spectral model
f_{ucoh}	Frequency of transition between ranges of high and low directivity; these cases are understood as the manifestations of, respectively, effectively coherent and effectively incoherent space–time organization of the radiating fault; the denotation means "upper bound of coherent behavior". Introduced in (Gusev 2013a). Is close to $1/T_0$ as defined by Bernard and Herrero (1994)

$f_{c1,90}$, $f_{c2,90}$	The values of f_{c1} and f_{c2} for the ray normal to the fault
$t_{fr}(\xi)$	Is the time of arrival of rupture front to ξ; it is simulated numerically as the sum of three terms: deterministic low-wavenumber term $Q_{\mathrm{det}}(\xi)$ that describes systematic propagation of the front, random high-wavenumber term $Q_{\mathrm{rnd}}(\xi)$ that describes fine details of the front geometry, and random low-wavenumber term $Q_{\mathrm{lk}}(\xi)$ that perturbs $Q_{\mathrm{det}}(\xi)$, of secondary importance
CCM	Composite-crack source model: a source consists of a number of isolated slip patches or cracks abutting one another; proposed by Boatwright (1982) and Papageorgiou and Aki (1983)
SPM	Slip-pulse source model; proposed and substantiated by Heaton (1990) who developed the concept of Haskell (1964)

2. *Introduction*

The first stochastic model of an earthquake fault is one proposed by Haskell (1966), who modified his running dislocation model (Haskell, 1964) by introducing a random local dislocation (slip) rate. This space–time function was specified through its correlation functions: in time (1D) and in space (2D). Aki (1967) modified Haskell's correlation function to make displacement spectra of radiated body waves to behave as ω^{-2} at high frequencies (HF), instead of the less realistic ω^{-3} of Haskell's (1966) model. Brune (1970) reduced this concept to a simpler, deterministic, one. The best-known part of his theory treats the simple one-corner ω^{-2} spectral model, with a single corner frequency denoted here as f_{c1}. He also noted, however, that a more realistic spectral model of the general ω^{-2} kind may need an intermediate ω^{-1} branch. This behavior of "ω^{0}–ω^{-1}–ω^{-2}" needs two corner frequencies: f_{c1}, and $f_{c2} > f_{c1}$. Brune (1970) associated the emergence of the second corner with the fractional stress drop over a fault. Gusev (1983) noted that spectral shapes with the second characteristic frequency clearly above f_{c1} are ubiquitous for moderate-to-large magnitudes; in essence, he provided empirical support for the idea of a two-corner spectrum. Since then, this guess was reliably confirmed further by analyses of many studied data sets; see Gusev (2013a) for a fresh review on f_{c2}. Hanks and McGuire (1981) found that the flat acceleration spectrum of Brune's (1970) ω^{-2} model fits well the smoothed observed spectra of strong motion. In essence, they treated the deterministic spectra of Brune (1970) as rms spectra of a certain, implicitly introduced, stochastic model, and with great success. Andrews (1980) proposed to treat the final slip over a fault as a random self-affine function, whose 2D Fourier spectrum is asymptotically of the $1/k^{\alpha}$ type. He also has shown that the field of local stress drop has a spectrum of a similar ($1/k^{\beta}$) kind. Andrews (1980) also suggested that real faults are not only self-affine, but self-similar; in this case, $\alpha = 2$ and $\beta = 1$. Inverted slip distributions over real faults generally support this idea (Tsai 1997, Somerville *et al.* 1999) despite certain deviations. Bernard and Herrero (1994; see also Herrero and Bernard 1994) noticed that the instant dislocation jump that propagates over Andrews' fault with $\alpha = 2$ generates random signal with ω^{2} spectrum. In general, the idea to explain ω^{2} spectra through spatial spectra of fault heterogeneity looks attractive.

Two-corner spectra represent a characteristic property of multiple-subsource, composite-crack source models (CCM) (Papageorgiou and Aki 1983). Both in general, and in CCM in particular, the lower corner f_{c1} is close to inverse rupture duration, whereas the second, upper corner f_{c2} of CCM is related to the duration of a subsource. Although the assumption of ubiquitous isolated subsources of CCM raises doubts from a tectonophysical viewpoint (Gusev 1983, 2013a), CCM constitutes a rather successful scheme for practical numerical modeling of HF radiation (Beresnev and Atkinson 2002; Halldorsson and Papageorgiou 2005). Thus, any competing source model should operate not worse than CCM. In CCM, local stress drop (i.e. average over a subsource) is many times above the average/global stress drop over an entire composite fault. Taking local stress drop as a reference value, the CCM can be classified as one with fractional stress drop as introduced by Brune (1970).

Another concept of rupture evolution that also features fractional stress drop, is the slip-pulse model (SPM) (Heaton 1990), which inherited important features of Haskell (1964, 1966). Different from CCM, this model does not assume any unbreakable barriers: slip pulse sweeps the entire fault. Such a behavior is much more plausible tectonophysically. Each of the two models—CCM and SPM—contains a dimensionless parameter of scale ratio, namely, for SPM, the running strip width l to source length L ratio, l/L, in the 0.025–0.15 range (Heaton 1990), and, for CCM, the subsource/composite source size ratio of comparable magnitude. In both cases, such a ratio gives a rough estimate of the ratio of mean global stress drop to local stress drop, thus describing numerically the fractional character of stress drop. The value of l is proportional to rise time T_r of dislocation.

Let us consider HF fault radiation, i.e. radiation in the frequency range well above f_{c1}. Its important property is low directivity. Weak HF directivity is a well-expressed feature, clearly manifest as the typical lack of expressed asymmetry for isoseismal maps (and for peak acceleration maps) around a finite surface-focus earthquake source of moderate-to-large magnitude, even for clearly unilateral ruptures; see examples in Gusev (2013a). This feature is seen best for the two highest isoseismals. There are other causes for asymmetric isoseismals: radiation pattern of a point source, non-uniform attenuation, as well as uneven site effects. All these make various imprints on isoseismal pattern, but none among them might be capable of suppressing a specific pattern expected for deterministic unilateral ruptures, with an expressed unilateral lobe, specific for the discussed kind of directivity (see e.g. Boatwright 1982). As for frequency-dependent directivity of HF amplitudes, see Tsai (1997), Somerville (1997) for deeper analysis. For deterministic fault rupture models, both of a dislocation or of crack kind, one can expect expressed directivity; thus, the property that needs theoretical explanation is, rather, significant diminution of directivity at HF. The probable cause of low HF directivity is the incoherence of HF radiation (Kostrov 1975; Boatwright 1982; Gusev 1983). From the observational viewpoint, incoherence is suggested by a noise-like appearance of HF records, and conceptually it can be associated with random phases of waveforms contributed by elemental spots of a fault. Randomness of phases is a key factor that underpins the application of rms amplitudes (and, essentially, energy spectra) in strong motion analyses at HF, e.g. in Hanks and McGuire (1981). For lower frequencies (typically below 0.5–1 Hz), however, directivity becomes quite noticeable; near the source it shows itself particularly clearly as the emergence of "forward-directivity pulses" (Somerville 1997). For smaller magnitudes, directivity is clearly manifest at high frequencies (Boatwright 2007) if one defines "high frequency" in absolute terms, as this is commonly done; but for smaller events, this part of the spectrum seems to be associated with coherent radiation, and thus shows clear directivity. Thus, the presence of directivity at low magnitude seems to be an apparent problem that calls for a physically consistent, non-dimensional definition of "high frequency". See more on this point in (Gusev 2013a). Bernard and Herrero (1994) associated the transition point from higher low-frequency directivity to diminished HF directivity with characteristic frequency $1/T_0$ associated with rise time: $T_0 \approx T_r$. This guess is generally confirmed below.

As a different cause of diminution of directivity effects (as well as radiation pattern ones), one may imagine wave scattering, usually expressed just at HF. Simulation suggests, however (Gusev and Abybakirov 1996), that significant randomization of radiation pattern and/or directivity by scattering may be expected only at the "diffusion mode" scattering. This mode appears at lapse times on the order of 3× (transport mean free path TMFP)/(wave speed c_S) or later. At values TMFP = 50–100 km and c_S = 3.5 km/s, typical for regional distances, this means delays like 150–300/3.5 = 40–90 s. This is much above typical direct wave travel time for waveforms that produce strong motion/isoseismal data. At more relevant travel times like 10–30 s, scattering is mostly of the "forward" kind (preferably at small angle relative to the ray direction). Directivity is only slightly suppressed in this case (whereas nodal-plane holes in the radiation pattern may be significantly smoothed over).

The three listed features—ω^{-2} HF spectral asymptote, two-corner spectral shape of, roughly, the ω^{0}–ω^{-1}–ω^{-2} kind, and expressed frequency dependence of directivity—all lack systematic theoretical explanation. Textbooks (e.g. SHEARER 1999), illustrate spectral features by source time function (STF) that is a convolution of two boxcars, whose widths are defined by (1) rupture propagation duration, and (2) rise time of local slip. In general terms, the imprint of these two characteristic times on spectral shapes in the form of two corresponding corner frequencies is a quite reasonable idea, and it is completely supported by the present study. However, in the time domain, convolution of the two boxcars creates the deterministic displacement STF of trapezoidal shape, and, therefore, unrealistic far-field accelerogram that consists of exactly four delta-functions and nothing else. An attempt to iron out this problem by smoothing of any of the two unrealistically sharp boxcar shapes immediately leads to significantly faster spectral decay as compared to the ω^{-2} kind. Thus, such a description of STF is oversimplified. More realistic model of STF is needed, such that might preserve two-corner spectra with ω^{-2} overall behavior, and, simultaneously, show realistic, white-noise style, acceleration waveform. Note that CCMs generate two-corner spectra, but the rise time concept is alien to these models. As for low HF directivity, although its connection with the incoherency property of a HF source has been suggested in general terms, there is no model that would reproduce this property in a conceptually consistent way. The CCM does show incoherency (formed by randomly perturbed timing of subsources) but these "subsources" seem to be an artificial construct. At any rate, an appropriate model that might explain the mentioned properties must probably be of a stochastic kind. To propose a model of this type is the key intention of the present paper.

Fault description developed here is based on local stress drop, and differs from the common approach based on local slip rate. The general idea to represent radiated earthquake waves as a sum of contributions of fault spots characterized by their stress drops was introduced in parallel, but from non-identical viewpoints in DK88, in B88 ("composite asperity model") and in G88. (The last paper is in Russian; the paper in English, GUSEV (1989), covers similar points and is easier to access; only the earlier paper will be referred to in the following). DK88 and G88 considered only strong spots (small asperities) dotted over background of negligible strength; whereas B88 considered all fault spots. In DK88, a number of multiple-asperity example faults are accurately simulated deterministically, whereas both G88 and B88 analyzed a multitude of asperities in less detail and in a stochastic manner. In particular, they assumed pulses from asperities to add incoherently at high frequency. G88 also discussed the structure of the rupture front and treated it as a "wave of breaking of asperities", i.e. a loosely organized feature, resulting in randomly phased contribution of asperity-generated pulses at a receiver. The present work elaborates on these concepts.

The main idea here is to combine two separate stochastic models. One of them is purely spatial and describes the fields of final slip and of local stress drop over a fault as random functions, following ANDREWS (1980). Another model describes fault evolution in space–time: this evolution is represented by a propagating rupture front whose instant geometry is a wiggling, multiply connected ("lacy") random polyline of fractal type (GUSEV 2013a), instead of a commonly assumed smooth curve. This random polyline is thought to be "thick": at any time instant, it occupies a strip with loosely defined margins, but of finite width w. This polylinear geometrical object will be called "front strip." The combination of these two models will be called "doubly stochastic source model". Other significant components of the present approach are: the notion of running slip-pulse rupture (HEATON 1990), whose width l is thought to be close to w; and application of the fault asperity failure model of DK83 and DK86 that permits one to relate formation of the complete earthquake wavetrain to fault failure and to propagation of fault-guided (mostly Rayleigh) waves over the non-locked part of a fault. The last critical assumption made here following B88 and G88 is that an arbitrary failing spot on a fault can be treated as an asperity that generates fault-guided waves that travel a certain distance along the non-locked part of the fault. However, in contrast to B88 and G88, this value is thought here to be close to l and w. (The fault-guided waves discussed here are strictly confined within the slipping part of the source-fault, and should not be confused with regular guided waves that may propagate in a fault-related low-velocity waveguide.)

The set of described assumptions was implemented as a simulation code based on a simplified kinematic description of rupture evolution. The properties of the doubly stochastic source model are studied through numerical experiments. In these, all the three above-mentioned empirical properties of fault-generated radiation are manifest. The results may provide a solution for an old puzzle, and are also important for applications.

A preliminary, highly compressed version of a part of the results presented here was recently published in Gusev (2013b).

3. General Background

As an elastodynamical basis of the present approach, the fault asperity concept after Das and Kostrov is employed. They analyzed, in DK83, radiation generated by failure of a strong patch—an asperity—on an infinite fault which is weak, i.e. with negligible cohesion (Case F1). They considered this analysis as a first step in understanding seismic waves radiated by real faults with their widely varying strength. In the next step in DK86, they modified their results for the case of a similar single asperity on a *finite* weak fault (Case F2). Lastly, in DK88 they analyzed a weak fault with multiple asperities. To understand better their way of treatment of fault radiation one should realize that it is somewhat unusual. Generally, there are two equivalent ways to represent the source of elastic radiation field of a fault: via double-couple density—the common one, and via stress change/drop. Despite similarity of appearance of corresponding integrals, there are significant differences. In the double-couple description, no fault dislocation exists: a fault with dislocations is *represented* by double-couples inserted into homogeneous medium whose cut is welded. This representation, in addition to its many positive points, has some inconveniences. One cannot determine dislocations and their rates while stepping in time (because of point-to-point interaction). On the contrary, the use of stress change permits direct integration in time along the rupture front, and thus presents some advantage. But only in the case of infinite fault is it a relatively easy procedure. In this case, there are two contacting halfspaces, and in each halfspace, the asperity failure is equivalent to the switching on of a force spread over the asperity area. The forces in the two halfspaces are equal and of opposite sign. Body wave displacement time history from such a failure represents a smoothed step at any distance, including far field. The case of finite fault was analyzed numerically in DK86 and DK88; of course, no static step for the far field appears in these cases (although numerical integration in DK88 cannot show this with full clarity). Use of stress-change description of finite fault evolution is the key point of the present paper. However, the presentation below begins with the introductory Case F1 of infinite fault.

A remarkable feature of the DK86 model (F2) is the emergence of two balanced forces applied to the opposite walls of the finite weak fault, i.e. of the cut in the elastic space. Just these forces generate initial parts of body waveforms that are essentially static far field displacements. These displacements persist until fault-guided, mostly Rayleigh, waves reach the boundary of the fault and in their diffraction, create backward static displacement that precisely compensates the first one; consequently, the summary static offset equals zero. Fault-guided waves are a strictly transient phenomenon; it accompanies formation of the free surface of the cut that arises temporarily during rupture evolution.

The approach of DK83 and DK86 was developed in DK88, and, on a less strict level, in B88 and G88, who tried to apply it to observable entities. Analysis in B88 is concentrated on spectral description of the radiator. It is noted in B88 that one can treat the fault radiation as a combination of contributions of fault spots/asperities covering the entire fault surface. Therefore, the usage of the term "asperity" in B88 is somewhat loose and a large fraction of these "asperities" could only be relatively weak. For the present paper, it is important to note that the analysis of B88 is applicable for any decomposition of the entire fault into "asperities", in particular into the great number of small ones.

In G88, both spectral and time-domain features are analyzed. It is supposed that significant contribution into strong motion comes from strongest fault patches, and the role of the remaining part of fault area (taken as 90–95 %) can be neglected. Rupture evolution was thought to be a "wave" of the switching on of asperities, with some average tendency that creates a loosely defined front, but with random timing of precise moments of failure of individual asperities. This assumption permits one to explain incoherence of HF radiation in the time domain, introduced earlier in (GUSEV 1983) on an abstract level. The model presented here elaborates on this scheme.

In the frequency domain, both in G88 and in B88, the incoherent ("energy") summation of spectra generated by individual asperities was assumed, resulting in a spectral model with two characteristic corner frequencies and an interval of ω^{-1} trend between them. From an observational viewpoint, this spectral pattern explained the emergence of two characteristic frequencies of an empirical source spectrum of GUSEV (1983) (whose f_2 is essentially f_{c2} of the present paper); it was also supported by the limited analysis of observational data in G88. Recently, GUSEV (2013a) reviewed many empirical studies of strong motion Fourier acceleration spectra and source acceleration spectra. The vast majority of these data indicate that f_{c2} exists at least in the form of low-frequency cutoff of acceleration spectrum. This cutoff practically never matches f_{c1} that is close to inverse rupture duration; very often $f_{c1} << f_{c2}$. These facts reliably indicate the reality of the intermediate spectral branch with its slope between 0 and 2; however, this slope may deviate from the accurate value 1.0 (cf. BOATWRIGHT and CHOY 1989).

Both in B88 and G88, the distance R_s over which fault-guided waves propagate was discussed. In B88, this point was treated in more detail, and the special notion of "slip radius" was introduced for R_s. In terms of an estimate, both in B88 and G88 it is assumed that R_s is close to fault radius R_c. Therefore, as seen retrospectively from the spectral patterns constructed below, they could not construct even qualitatively realistic theoretical spectra. To correct this deficiency of the early work is one of the main aims of the present paper. It will be shown that the value of R_s has a prominent effect on spectral shapes. And specifically when it approaches the fault radius, two-corner spectra and the intermediate ω^{-1} spectral slope tend to *disappear*. For spectral shapes to be realistically two-cornered, one needs R_s much below R_c. Slip-pulse width l is an evident candidate to provide an appropriate distance parameter, and equating $2R_s$ to l is the main idea of the present paper. Note that this assumption is physically plausible: slip pulse is just the place where fault walls are not engaged; in this way, slippage becomes possible, and fault-guided waves are able to propagate.

It should be added for clarity that there are two processes that are deeply interwoven in Case F3: (1) development of dislocation/slip within the slip-pulse strip (that follows from its definition as *slip* pulse) is associated with the *static* component of the fault-guided wavetrain; and (2) formation of *radiated body waves* is caused by the *wave* component of the same wavetrain, when it dies off. Thus, the two phenomena, slip and radiation, are two aspects of the same process.

4. *Fault Asperity Theory (Cases F1 and F2) as the Basis of the Present Model*

In DK83 (Case F1), an infinite fault is considered formed by frictionless contact of two elastic half-spaces. There is a limited welded patch or "asperity" loaded with shear, and it fails. In the process of failure, the rupture front propagates over the asperity surface; at the moment of arrival of this front to an element of the asperity, local stress drops to its residual value on this element, and stress drop arises. The rupture front generates body waves P and S in each halfspace, and also a fault-guided (Rayleigh, inhomogeneous S and static) wavetrain that propagates along surfaces of the fault. For the SH wave, the displacement and velocity at a far-field receiver, at a point $\mathbf{x} = \{x, y, z\}$ and time t can be written as

$$u^{SH,\infty}(\mathbf{x},t) = A\int_{\Sigma_a} \tau(\xi)H(t-(R-\xi\cdot\gamma)/c_S - t_{fr}(\xi))\mathrm{d}S;$$

$$A = \frac{\Re_F^{SH}}{4\pi\rho\, c_S^2 R} \tag{1a}$$

$$\dot{u}^{SH,\infty}(\mathbf{x},t) = A\int\limits_{\Sigma_a} \tau(\xi)\delta(t-(R-\xi\cdot\gamma)/c_S - t_{fr}(\xi))\mathrm{d}S \tag{1b}$$

where Σ_a is the asperity patch, located on a planar fault with its element d*S;* the hypocenter is at $\xi = 0$; $R = |\mathbf{x}|$; $\gamma = \mathbf{x}/R$; ρ is density; $\Re_F^{SH}$ is the radiation pattern of *SH* waves radiated by a point *force*; $H(\cdot)$ is the unit step and $\delta(\cdot)$ is a delta-function. Similar formulas exist for *P* and *SV* waves, with the *SV* case technically more sophisticated. Note that no waves are generated by parts of the fault outside Σ_a (fault-guided waves generate no far-field body waves). To be accurate, the model permits non-zero, finite friction outside Σ_a; the crucial property is not the level of cohesion but the lack of displacement-jump-related weakening (sliding friction must be equal to static friction). Therefore, the stress drop equals zero over these parts; henceforth they will be referred to somewhat loosely as "low-cohesion" parts. In this case, qualitatively, the waveform defined by (1a) is a step $H(\cdot)$, convolved with a smoothing window with its width close to T_a. Correspondingly, (1b) is a one-sided, say, positive, pulse with similar duration. See sketch on Fig. 1a, d for illustration. The area of the pulse (1a) equals the final step-like static displacement whose value, as can be derived from (1a), is

$$u^{SH,\infty}(\mathbf{x},\infty) = \int\limits_{\Sigma_a} \dot{u}^{SH,\infty}(\mathbf{x},t)\mathrm{d}t = A\int\limits_{\Sigma_a}\tau(\xi)\mathrm{d}S \equiv \mathrm{AF}_0 \tag{2}$$

For the average amplitude of positive velocity signal of duration $\approx T_a$, based on (1, 2), one can derive an estimate

$$\dot{u}^{SH,\infty}(\mathbf{x},t) \approx \mathrm{AF}_0/T_a \tag{3}$$

The $\tau(\xi)$ function that enters the above discussion is

$$\tau(\xi) = \sigma_{dyn}(\xi,t)|_{t=t-\mathrm{failure}} - \sigma_f(\xi) \tag{4}$$

where $\sigma_f(\xi)$ is residual friction and $\sigma_{dyn}(\xi,t)$ is the time-dependent stress level, taken at the moment of failure; thus, it can be equated to yield stress $\sigma_y(\xi)$. It is often assumed (e.g. B88; Oglesby and Day 2002) that pre-earthquake initial stress $\sigma_0(\xi)$ is closely correlated with $\sigma_y(\xi)$, and local values of $\sigma_0(\xi)$, being below $\sigma_y(\xi)$ everywhere but at the rupture nucleation point, are still comparable to $\sigma_y(\xi)$. To justify such an assumption, it may be noted that at points ξ where $\sigma_0(\xi)$ is only infinitesimally below $\sigma_y(\xi)$, rupture propagation velocity would be near to or above its critical value (c_S or c_R). Conversely, in the case of a barrier, where $\sigma_0(\xi)$ is significantly below $\sigma_y(\xi)$, the rupture must significantly decelerate (or even stop in the case of an impenetrable barrier). The actual range of observed average rupture velocities, typically 0.65–$0.85c_S$, directly indicates that $\sigma_0(\xi)$ is only marginally below $\sigma_y(\xi)$ over most of the fault area. Following Oglesby and Day (2002), one can believe that $\sigma_0(\xi)$ values are 1.2–2 times below $\sigma_y(\xi)$. Another argument in support of close correlation of fields of $\sigma_0(\xi)$ and $\sigma_y(\xi)$ follows from the argument of Lay and Kanamori (1981) who suggested the existence of large-size stress peaks or asperities on the fault plane, created in the course of interseismic fault evolution; their reasoning applies both to $\sigma_0(\xi)$ and $\sigma_y(\xi)$, and may well be valid for asperities of various sizes, not only for large ones. Based on these considerations, random field models, which are used henceforth, are thought to be applicable similarly to $\Delta\sigma(\xi) = \sigma_0(\xi) - \sigma_f(\xi)$ and $\tau(\xi)$.

In the present paper, like (implicitly) in DK83, DK86, $\tau(\xi, t)$ is assumed to drop instantly at the rupture front. This is simplification; generally, it might be more realistic to assume a less abrupt dying-out of cohesion (i.e. its evolution from $\sigma_y(\xi)$ to the "stationary", low slip rate $\sigma_f(\xi)$ at a point, related to slip-weakening). However, the numerical estimates of the cohesion zone width derived from laboratory, geological and accurate seismological data (see e.g. Ben-Zion and Sammis 2003) are very low, much smaller than the cell size of the grid used here, and related stress evolution during failure may be assumed abrupt. There is, however, another possible mechanism for stress drop to develop gradually, with time-dependence of $\sigma_f(\xi)$ related to possible expressed velocity weakening (see e.g. Di Toro *et al.* 2011). In this way, σ_f values may drop down to values significantly below those characteristic for low slip rates; and formation of these low σ_f values, related to accelerated local slip, may take appreciable time. This possibility will be ignored in the following.

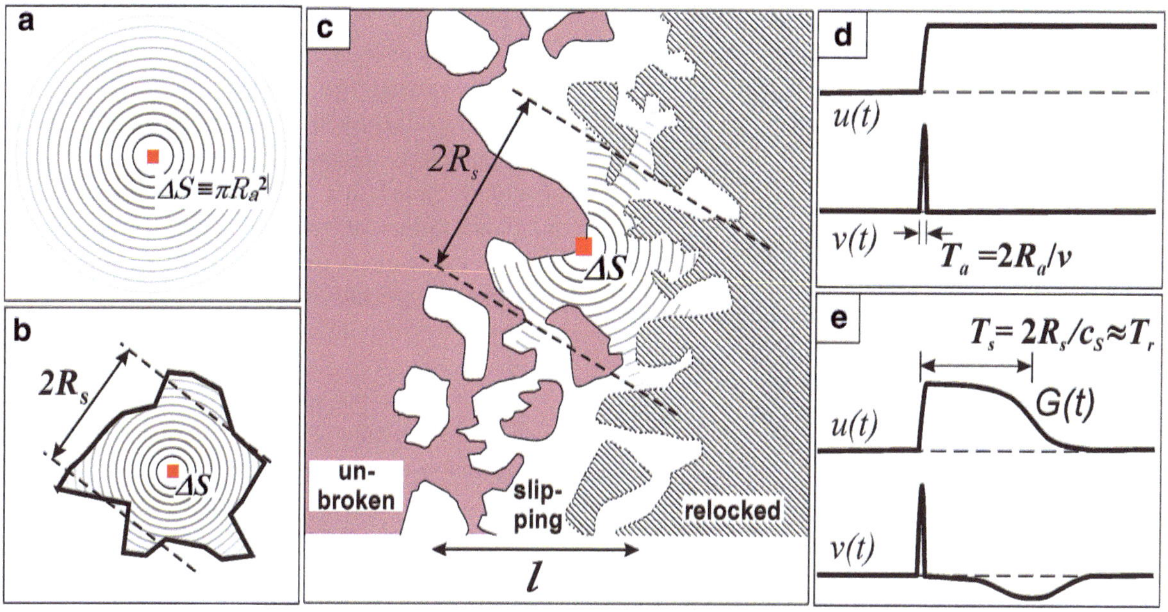

Figure 1
Cartoons of fault-guided (mostly Rayleigh) waves propagating from a single failing fault element ΔS, and of related far-field waveforms. **a** (Case F1)—Rayleigh waves from a fault element ΔS that constitutes a small failing asperity on an unbounded frictionless fault; **b** (Case F2)—same for an asperity located within a limited cohesionless area of size $2R_s$. **c** (Case F3)—as the source of Rayleigh waves, an individual element ΔS of earthquake rupture front is supposed (among many such elements). In **c**, instant positions are shown: of random multiply connected rupture fronts, represented as a boundary (*solid tortuous line*) between unbroken (*grey or pink in color*) and slipping (*white*) areas; and of a healing front with similar geometry, represented as a boundary (*dashes*) between slipping and locked-in (*hatched*) areas. The general direction of rupture propagation (i.e. of the front strip or of "macroscopic front") is from left to the right. The area where Rayleigh waves from ΔS propagate coincides with the "slip patch", of size $2R_s$, associated with this ΔS. For each particular ΔS, an individual patch of this kind exists. Rayleigh waves can propagate only within the limits of the slipping (*white*) area; this area is an overlap of slip patches associated with all individual ΔS that constitute the instant position of a rupture front. Surface wave diffraction and tunneling are ignored; the abrupt screening of the wavefield by obstacles shown on the sketch is a simplification only. Looking at the sketch *c*, one should keep in mind that it depicts a snapshot; actually, both fronts are travelling and change their shape; a more adequate picture must be movie-like. **d**—far-field time histories of displacement and velocity for the Case **a** (F1); **e**—same for cases **b** (F2) and **c** (F3)

The model DK83 has been extended in DK86 (Case F2) for an asperity surrounded by a larger low-friction fault region of *finite* size $2R_c$. In this case, fault-guided waves do not run to infinity, but die away at the boundary of this region, where they are diffracted and converted to body waves. Having arrived at the receiver, these constitute an additional term to the signal (1a). This term is lagged behind by, approximately, T_c, and its sign is opposite to that of (1a), say, negative. At $t \gg T_c$ the contribution of this term to residual body wave displacement precisely compensates (1a). The displacement signal can now be written as:

$$u^{SH}(\mathbf{x},t) = A \int_{\Sigma_a} \tau(\xi) G(\mathbf{x}, t - [(R - \xi \cdot \gamma)/c + t_{fr}(\xi)]) \mathrm{d}S \approx A(F_0/T_a)T_c \qquad (5)$$

where $G(\mathbf{x}, t)$ is an elementary signal generated by radiator $\mathrm{d}S$ at ξ. It is assumed for simplicity that $G(\cdot)$ are the same for all ξ within Σ_a. $G(\mathbf{x}, t)$ is a one-sided, e.g. positive, trapezoid-like pulse, asymmetric, with step-like leading edge and gradually decaying trailing edge:

$$G(\mathbf{x}, t) = H(t) - \int \Lambda(\mathbf{x}, s) H(t - s) \mathrm{d}S \qquad (6)$$

where $\Lambda(\mathbf{x}, t)$ is an appropriate unit-area window function, with duration on the order of T_c; in the case of far field, $\Lambda(\mathbf{x}, t)$ can be considered as independent of ξ. In integration (5) the property of positivity, inherent for $G(\mathbf{x}, t)$, is preserved. Thus, displacement pulse $u^{SH}(x, t)$ is one-sided, as usual in seismology. Its characteristic duration equals (1–2) T_c.

5. Earthquake Source Model (Case F3)

5.1. Running Slip Pulse and "Slip Patch"

Case F2, considered in DK86, is one when fault-guided waves propagate away from a small single asperity. On the basis of DK86 and DK88, and following B88 and G88, one can generalize this case and believe that radiation from an earthquake source of a general kind can be represented again by the same Eq. (5), modified as:

$$u^{SH}(\mathbf{x},t) = A\int_{\Sigma} \tau(\xi)G(\mathbf{x},t-[(R-\xi\cdot\gamma)/c+t_{\mathrm{fr}}(\xi)])\mathrm{d}S \approx A(F_0/T_{\mathrm{a}})T_{\mathrm{c}} \qquad (5*)$$

Here, integration is spread from a small patch Σ_a onto the complete fault area Σ, of size $2R_c$. Now let us replace the integral in (5*) by the integral sum over small elements ΔS. Each ΔS, be it strong (an element of a true asperity or an entire such asperity) or weak, now plays the role of a single asperity of DK86 (of size $2R_a$) and is now included in summation. Elements ΔS are assumed to be of comparable size and isometric, and their characteristic size can be written as $\Delta S^{0.5}$.

In a partial analogy with Case F2, let us consider a patch on the fault plane, around ΔS ("slip patch" of B88), that temporarily loses cohesion after rupture passes over ΔS. Within this patch, fault slip is localized associated with stress drop at $\Delta S(\xi)$, and fault-guided waves generated by ΔS have non-zero amplitude. The characteristic size of the slip patch is denoted as $2R_s \gg \Delta S^{0.5}$. Different from with G88 and B88, it is assumed that this size is significantly smaller than the source size: $R_s \ll R_c$. Overall, the following double inequality is assumed true: $\Delta S^{0.5} \ll R_s \ll R_c$. It should be emphasized that for each individual ΔS (each ξ), there is an individual, specific slip patch. Still, the notion of a characteristic or average slip radius is considered meaningful.

A critical further assumption is based on the observation that the conditions of low cohesion around a point ξ are realized just when the running slip pulse crosses its neighborhood. Thus, one can believe that the slip patch size $2R_s$ and the slip-pulse width l of SPM (Fig. 1c) are close to one another. The assumption $2R_s \approx l$ is the key one here. The relative value of R_s as compared to R_c is a significant model parameter. The denotation $C_H = l/L$ will be used to recall Haskell and Heaton, originators of SPM, and it is assumed here that $R_s/R_c = C_H$. It should be noted that slip patches associated with individual failing fault elements ΔS that constitute the entire instant rupture front overlap to a large degree, and their aggregate (union) is assumed here to constitute the strip instantly occupied by propagating slip pulse.

The shape of the specific $G(\mathbf{x}, t)$ function associated with particular ξ depends on the rupture front configuration and evolution in the vicinity of ξ, on details of propagation and decay of fault-guided waves radiated by $\Delta S = \Delta S(\xi)$, and also on the ray direction γ. To obtain approximate numerical estimates, the dependence of $G(\mathbf{x}, t)$ on ξ and γ is ignored in what follows, and for any ξ, the tentative pulse shape is taken after B88 as:

$$G(\mathbf{x},t) = G_0(t) = \begin{cases} H(t) - 0.5(1+\cos \pi t/T_B); & t < T_B \\ 0 & t > T_B \end{cases} \qquad (7)$$

Here, T_B is pulse duration, set as $1.667T_r$ to provide the value of the centroid of the pulse (7) to be equal to $0.5T_s = 0.5l/v$. Thus, the distance of complete decay of guided-wave amplitude is set as $1.667l$.

The complete body wave displacement pulse (5, 5*) with its amplitude of the order AF_0T_r/T_c and duration of the order of T_c has its integral close to $AF_0T_r = (A/c_S)F_0 \cdot 2R_s$. As usual, such an integral is related to the seismic moment M_0 of the source. Generalizing the results of DK86, B88 one can write:

$$M_0 \approx F_0 \cdot 2R_s = F_0 l. \qquad (8)$$

where F_0 is the total seismic force of an earthquake source (see Eq. 2). [Note that F_0 as introduced here is a dynamic entity defined by $\tau(\xi)$, not a static one defined by $\Delta\sigma(\xi)$] For more accurate calculations, one should account for the position of the ΔS patch within source area. In DK86, asperity was positioned at the centre of a circular fault. For asperity (or ΔS patch) located arbitrarily, factor A in (5) depends on the position of asperity within the large source, and includes an additional coefficient, equal to 2/3 on the average for the case of circular source (B88). The variation of factor A over the fault area is neglected, and the average coefficient 2/3 is assumed to be incorporated into the value of factor A.

In DK88, a fault is simulated through multiple local asperities, and waveforms were analyzed that result from their breakdown. It was found that at some points, negative stress drop may appear, but with relatively small, negligible amplitude. Also, for natural earthquakes, there is a possibility of overshoot during the previous event on the same spot of a fault; this again may create limited negative stress drop. Still, the results of DK88 suggest that such effects must be secondary, not so strong as to violate the assumption of non-negativity of the wave displacement pulse (5, 5*). For this reason, the simple assumption of positive $\tau(\xi)$ combined with positive $G_0(t)$ was considered acceptable.

As one can see from DK83, the theory derived above will be valid also for P waves with evident modification of using single-force P radiation pattern. The case of SV waves (polarized in a plane orthogonal to Σ) is more complicated as complex coefficients may arise in this case. This does not significantly affect noise-like HF waveforms. Thus, one can consider the developed approach to be mostly valid for body waves and especially for their amplitude spectra.

5.2. Stress Drop Field

In his stochastic fault model, ANDREWS (1980) assumed that final slip $D(\xi)$ is a self-similar (in a broad sense, or "self-affine") random function with 2D Fourier spectrum close to $1/k^{\alpha}$. This behavior can be accurately valid only over an infinite plane. For the case of a source of finite area, low-wavenumber cutoff must be applied; also, for the case of tectonic earthquake, $D(\xi)$ must have positive average. Moreover, both seismogeology and inversions indicate that $D(\xi)$ can be assumed to be positive over entire Σ despite minor deviations found by some inversions. The assumption of self-similarity approximately agrees with the results of inversions of real sources (TSAI 1997; SOMERVILLE *et al.* 1999; MAI and BEROZA 2002). ANDREWS (1980) notes that fields of final slip and of static stress drop $\Delta\sigma(\xi)$ are rigidly connected. In the wavenumber domain, for an infinite fault, their relationship is close to multiplication by k^1; therefore, $\Delta\sigma(k) \propto 1/k^{\beta} = 1/k^{\alpha-1}$, and similar behavior is further assumed for $\tau(k)$. As a refinement of self-affinity, strict-sense self-similarity [Hurst exponent $H = 1$ for $D(\xi)$] was also proposed and substantiated for real faults by ANDREWS (1980); in such a case $\beta = 1$ (and $\alpha = 2$). Although the analysis of inverted final slip maps does not accurately match the hypothesis that $\alpha = 2$ (and thus $\beta = 1$), the simple assumption that $\beta = 1$ will be used henceforth as the most plausible from general physical considerations. The selection of amplitude spectral shape $\tau(k) \propto 1/k^{\beta}$ requires power spectrum of $\tau(\xi)$ to behave as $1/k^{2\beta}$; in this way, correlation properties of random field $\tau(\xi)$ are specified.

To perform the simulation, one must also specify the probability law for (positive) local random values of $\tau(\xi)$. The relative scatter of these values (relative with respect to their own mean) is set through coefficient of variation $CV_{\tau} = (\mathrm{Var}(\tau(\xi))^{0.5}/\mathrm{E}(\tau(\xi))$; as for the distribution law $p(\tau)$, it is assumed lognormal. Note that the particular choice of the lognormal law has no solid basis: one might use, e.g., Weibull or gamma law. What is really significant is the behavior of the tails of $p(\tau)$. Because of (1b) and (5*), these tails can be expected to be related to similar tails of velocity and acceleration amplitude peaks of HF radiation. This important connection should not be thought of as a general one; it is valid in the limited proportion of cases when the distorting effects of path and site are not too strong, and the duration of an accelerogram is not much expanded because of scattering and multipathing (cf. GUSEV 1988, 1996). It also should be kept in mind that the relationship between tails of $p(\tau)$ and those of peak amplitudes is not immediate, as discussed in GUSEV (1992). In the case when tails of the $p(\tau)$ distribution law are sufficiently heavy, they are almost directly reflected in the tails of amplitude peaks. With less expressed tails, this relationship is less close. The question of acceleration peak statistics was recently discussed in GUSEV (2011a, 2013a) in some detail; that discussion implies that the upper tail of the distribution of the observed peak acceleration is only moderately heavy. The use of lognormal law with $CV = 0.7$–1.0 for acceleration peaks seems to be a reasonable starting approximation. On a different line, in Gusev (2011b), the statistics of inverted slip distributions was analyzed, and it was found that the distribution of local final slip can be approximated by the lognormal law with $CV \approx 1.0$ (or by exponential law). Unfortunately, these fragments of information do not match well: one can expect tails of the distribution of acceleration peaks to be relatively enhanced (and corresponding CV value tangibly larger) as compared to peaks of slip distribution. Still, for want of anything better, similar lognormal law, with CV_τ in the range 0.7–1.0 was used for simulation of $\tau(\xi)$. Simulated values of $\tau(\xi)$ are positive, as assumed above. Note that the operator that converts $\tau(\xi)$ to $D(\xi)$ is convolution with positive kernel; therefore, $D(\xi)$ is also a positive function.

5.3. Random Rupture Front and a Technique for its Simulation

The shape of the earthquake rupture front is traditionally thought to be a smooth line. Such a line has a well-defined local normal, and it defines the local direction of rupture propagation. This concept needs to be generalized for the case of a "lacy" random rupture front as introduced here. (Generally, one can imagine random fronts whose realizations are kept smooth, but this degree of randomness is too weak for our purpose). Let us first consider mean (ensemble-average) evolution of a rupture front, and assume the mean front to be again a smooth line with a well-defined normal. In the vicinity of a certain point ξ let us consider the coordinate x' along this normal, with $x' = 0$ at the crossing of the normal with mean front, and let us introduce "front arrival time function" $t_{\rm fr}(x')$. Now consider points of the normal at a certain time moment. The *mean* front forms a crossing at a single point $x' = 0$ as defined; sample *random* fronts will cross the normal at various points around $x' = 0$ forming a probability distribution $p(x')$. It should be emphasized that each single sample "lacy" front will, as a rule, form *many* crossings with the normal. Thus, even a single "lacy" front will occupy a finite interval on the x' axis. When normals are considered for many points of the mean front, these intervals form a finite band along a mean front, or "front strip," of characteristic width w. Instead of fixing time moment, one can broaden a viewpoint and consider front arrival time $t_{\rm fr}(x')$ as a function of x'. In the deterministic case, $t_{\rm fr}(x')$ is a one-to-one and monotonous function. In the stochastic case, $t_{\rm fr}(x')$ is, normally, not monotonous, and the inverse function $x'(t_{\rm fr})$ becomes, typically, multiple-valued. As usual, assuming ergodicity, one can believe that mean (ensemble-average) fronts can be estimated (and visualized) as smoothed or low-k-passed versions of sample fronts. Such mean or low-resolution fronts of smooth shape, can be called "macroscopic fronts", whereas the actual high-resolution lacy shape can be called "microscopic front".

Complex non-monotonous evolution of fronts is a necessary condition for formation of usually observed incoherence of high-frequency radiation from a source. This consideration is mentioned only very loosely by Boore and Joyner (1978) but clearly in Day *et al.* (2008). Dynamical rupture models with marked contrasts of stress drop show jumps of the front (Day 1982), meaning multiple-valued x' $(t_{\rm fr})$. Observations (Spudich and Cranswick 1984; Archuleta 1984) reveal qualitatively the same phenomenon. Following Gusev (2013a), it is assumed further that the rupture front is "lacy". It is a wiggling line accompanied by "lakes" and "islands", or a "polyline"; it is multiply connected. After the works of Mandelbrot, it is natural to assume this polyline to have fractal geometry; this assumption also greatly simplifies its simulation. The polyline occupies the "front strip" of width w. It is assumed that w is close to l, and the relationship $w \approx l$ is followed in simulation. At the same time, simulation is organized so that w is close to $2R_{\rm s}$, this is attained by the proper selection of $T_{\rm B} = (5/3)\,T_{\rm s}$ in (7). To provide these properties, and make the contours of the mean or smoothed front look plausible, the front arrival time (failure time) at ξ is represented as the sum of three terms:

$$t_{\rm fr}(\xi) = Q_{\rm det}(\xi) + Q_{\rm rnd}(\xi) + Q_{\rm lk}(\xi) \qquad (9)$$

Here

(i) $Q_{\rm det}(\xi)$ is deterministic term that describes the systematic behavior of rupture front, simulated as $Q(\xi) = |\xi - \xi_{\rm h}|/v$ where $\xi_{\rm h}$ is the hypocenter that represents the vertex of the cone $t = Q_{\rm det}(\xi_1, \xi_2)$. In itself it forms accurately circular mean fronts.

(ii) $Q_{rnd}(\xi)$) is a random, stochastic term that provides the geometry of the rupture front that is "lacy" at any time instant, with wiggling shape and fragmented structure. It is a sample of a self-similar random function, is positive, and has a preset spectrum $\propto 1/k^{\delta}$. The distribution law for local values of $Q_{rnd}(\xi)$ is selected as Rayleigh law, with twice the mean value equal to $T_w = l/v = T_s = T_r$. Therefore, the microscopic front wiggles within the front strip, of width $w = l$, that trails after the macroscopic front. The trailing edge of this strip constitutes the "macroscopic healing front". The condition $T_w \approx T_s$ provides the required match between the propagation distance of fault-guided waves and the width of slip-pulse.

(iii) $Q_{lk}(\xi)$ is another random, stochastic term of limited amplitude and secondary importance. It has smooth relief, ensured by imposed low-k spectrum. It perturbs unrealistic, primitive circular shapes of mean fronts as generated by $Q_{det}(\xi)$. It is simulated by the same procedure as used for $Q_{rnd}(\xi)$, with low-k filter added and a somewhat modified set of other parameters.

Overall, a smoothly propagating "macroscopic" front is formed by the sum of $Q_{det}(\xi)$ and $Q_{lk}(\xi)$, whereas $Q_{rnd}(\xi)$ is a perturbation term that provides fine "microscopic" details.

The described procedure has certain weakness. The value of l (and thus T_w) was implicitly assumed as fixed during front propagation. The assumption of a fixed l value makes relative front perturbation overly strong at the early stages of rupture growth, when instant rupture size is comparable or even lower than w and/or l. To amend this deficiency, T_w (and implicitly T_r) is made time dependent: $T_w(\xi) \propto Q_{det}(\xi)^{\eta}$. The preset T_w value discussed above is considered as an average value over entire fault, and used to properly scale the $T_w(\xi)$ function.

6. Simulation and Its Results

The developed numerical procedure includes the following steps (the accepted parameter values are given in parentheses):

(a) selecting: the size of a rectangular source ($L \times W = 38 \times 19$ km, corresponds approximately to the magnitude range $M = 6.8$–6.9), time step dt (0.013 s), distance step dx (0.037 km); v (3.0 km/s); c_S (3.5 km/s); hypocenter position $\xi_h(\xi_1 = 0.12L, \xi_2 = 0.24W)$;

(b) setting control parameters: $\beta(1.0)$, C_H, ($0.0075 \div 0.24$), CV_τ (0.8), $\eta(0.5)$ and $\delta(1.2)$; the shortest wavelength for $Q_{lk}(\xi)$) is set close to 1.5 km.

(c) generation of random fields $t_{fr}(\xi)$ and $\tau(\xi)$ (Fig. 2);

(d) calculation of $u^{SH}(\xi, t) \equiv u(t)$ through (5*, 6, 7) (Fig. 3a) and of its amplitude spectrum $u(f)$;

(e) determination of normalized displacement amplitude spectrum $u_n(f) \equiv u(f)/u(f)|_{f=0}$ and of associated acceleration spectrum $\ddot{u}_n(f)$ (Fig. 3b).

The discussed modeling procedure is entirely kinematic, and no connection is assumed between $t_{fr}(\xi)$ and $\tau(\xi)$. For plotting, all simulated spectra are smoothed at moderate-to-high frequencies using log-equal bins (three bins per octave). Spectra are normalized by displacement pulse area $u(f)|_{f=0}$ (i.e.

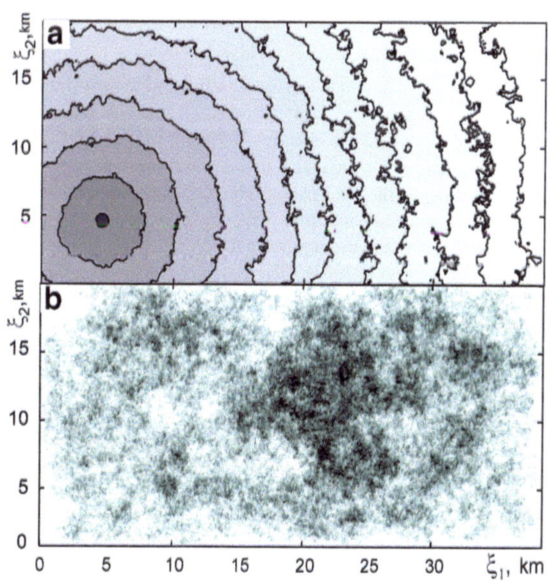

Figure 2
a Propagating rupture front of a sample simulated source. Positions of the front, i.e. isolines of the sample t_{fr} (ξ) function, are plotted each 0.89 s. Shades of *gray* code time: the later, the lighter. *Black dot* is the nucleation point. **b** Sample random field $\tau(x, y)$ with $1/k$ mean amplitude spectrum; shading reflects amplitude; maxima are darker

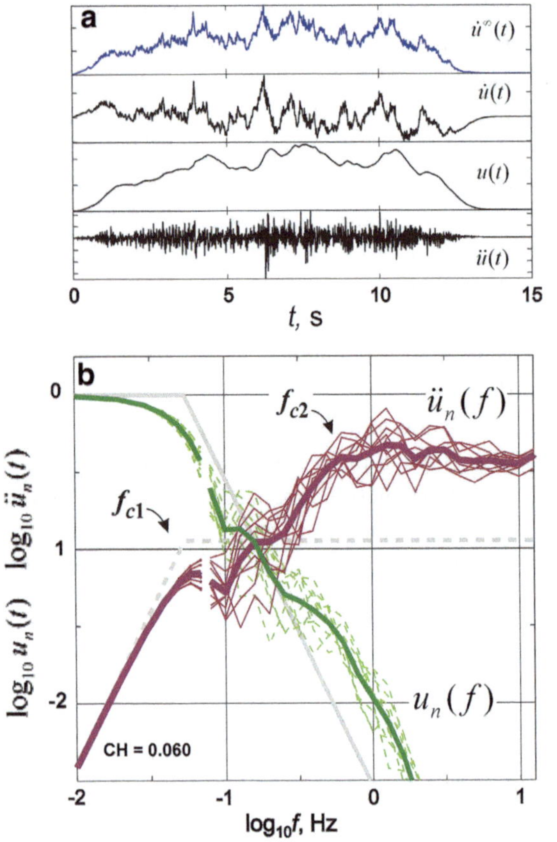

Figure 3
Typical results of simulation. **a**—Signals $\dot{u}^{\infty}(f), \dot{u}(t), u(t)$ and $\ddot{u}(t)$ at the receiver. Amplitude scale is arbitrary. **b**—Normalized spectra $\ddot{u}_n(f)$ *(dashes)* and $u_n(f)$ *(solid line)*, raw (on the left, at lower *f*) and smoothed (on the right). *Thin lines*: ten individual spectra; *thick curve*: their average. In this and further pictures, gray angular shapes on the background are schematic spectra of the common ω^{-2} model; their corner frequency f_{c1} is set equal to $1/2\pi T_{rms}$, where T^2_{rms} is the average second normalized central power moment for $u(t)$ (SILVER 1983). Gap in a spectral curve on this and next figures indicates switching from raw to smoothed spectra

always setting M_0 value to unity). This allows one to cancel all parameters involved in A, and to concentrate on shapes and relative levels of spectra that are of main interest in the present study. The lower (common) corner frequency f_{c1} of simulated spectra is essentially constant as it is mostly defined by L, ξ_h, and v, all of them fixed. This means that not only magnitude, but also global stress drop of simulated sources is fixed.

Simulated signals $u(t), \dot{u}(t)$ and $\ddot{u}(t)$ (Fig. 3a) qualitatively agree with those observed in real earthquakes at moderate distances from a fault. (No attempt is made here to imitate records obtained close to a fault.) The average smoothed acceleration spectrum (Fig. 3b) is flat, and clearly shows the second corner-frequency f_{c2}. Both these features also qualitatively agree well with observations, and moreover with the ω^{-2} model, in its two-corner ($\varepsilon < 1$) version after BRUNE (1970).

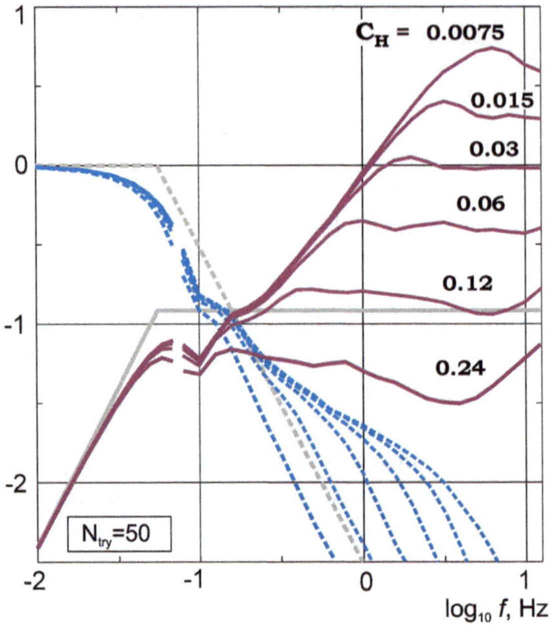

Figure 4
Normalized simulated spectra $\ddot{u}_n(f)$ (*solid lines, purple online*), and $u_n(f)$ (*solid blue online, dashes on paper*), averaged over 50 realizations, for a set of values of relative slip-pulse width C_H, at fixed values of other parameters

The parameter that controls the shape of the acceleration spectrum most significantly is C_H; it affects both the relative level of acceleration spectrum and the position and degree of expression of the f_{c2} corner. Its role is seen in Fig. 4: the f_{c2} corner is manifest clearly when $C_H \leq 0.12$. At larger C_H it disappears, and a simple ω^{-2} shape with no distinct f_{c2} corner can be expected at C_H about 0.15–0.20. Let us denote actual flat acceleration spectral level as A_{HF}, and its value for the primitive ω^{-2} shape of the case $f_{c1} = f_{c2}$ (grey dashes on Fig. 3) as A_{1HF}. The following approximate relationships hold

$$f_{c2}/f_{c1} \approx 0.67/C_H \tag{10}$$

$$A_{HF}/A_{1HF} \approx 0.17/C_H \tag{11}$$

The most significant fact here is that of approximate inverse proportionality. The numerical factors are interesting; however, they are not of general validity, and may depend, e.g., on the aspect ratio of the fault, or on the selection of the CV_τ parameter. Between f_{c1} and f_{c2} the acceleration spectral shape can have a $f^{1.0}$ segment but on the whole deviates from a simple $f^{1.0}$ line connecting two corners; it shows certain concavity in the range (1–1.4) f_{c1}.

The simulation results were tentatively checked against real spectra using the well-established empirical spectral model of Halldorsson and Papageorgiou (2005) for the interplate data set. From Fourier spectral shapes expected from their spectral family at $M_w = 6.8$ one can crudely estimate the parameters f_{c2}/f_{c1} and A_{HF}/A_{1HF}; their values are about 7 and 4, correspondingly. From (10) and (11) one can obtain from these numbers the estimates for C_H equal to 0.1 and 0.04, correspondingly. Their average value is 0.07, quite comparable with original Heaton's (1990) estimate of 0.1.

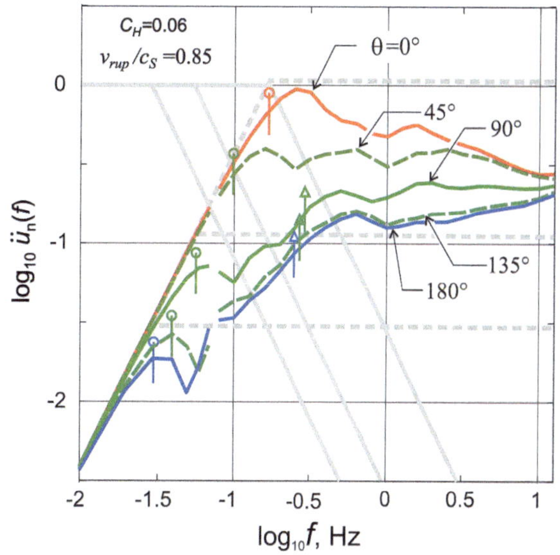

Figure 5

Angular dependence of normalized acceleration spectrum $\ddot{u}_n(f)$ obtained by averaging over 50 simulated sample sources with size, geometry and hypocenter as shown in Fig. 2, with approximate magnitude $M_w = 6.8$–6.9. Five spectra are plotted for angles $\theta = 0$ (forward), 45, 90, 135 and 180° (backward) between the ξ_1 axis and the ray to the receiver. Fault parameters: $C_H = 0.06$, $CV_\tau = 0.8$, $\delta = 1.2$. The gap in the curve indicates switching from raw to smoothed spectra. *Spoon-like symbols* indicate f_{c1} positions. They were determined via T_{rms} as explained in the caption of Fig. 3. *Arrow-like symbols* indicate approximate f_{c2} positions, where discernible

7. *Frequency-Dependent Directivity and Incoherence*

It was interesting to analyze directivity features of the described model and, particularly, frequency dependence of directivity. Two aspects of directivity are angular dependence of spectral amplitudes, and that of characteristic frequencies (Doppler effect). The numerical procedure that followed Eqs. 5–7 was performed for the fixed $C_H = 0.06$, $v/c_S = 0.85$ and for a set of angles θ between (1) the rupture propagation direction, taken as the positive direction of the ξ_1 axis (abscissa on Fig. 2), and (2) the direction of the receiver ray; these two angles define the γ vector in (5). The y component of the ray vector equals zero. The following set of θ was used: $\theta = 0$, 45, 90, 135, 180°. Average spectra were calculated over 50 simulation runs (Fig. 5); they show the following features (numerical estimates are specific for the selected v/c_S and C_H but qualitative conclusions seem to be of wider applicability):

1. In the frequency band around $f_{c1,90}$ and up to approximately $f_{c2,90}$, both mentioned angular dependencies—for spectral levels as well as for the f_{c1} value—are clearly seen, in a close match to the textbook behavior of deterministic unilateral fault models. The value of f_{c1} varies from 0.027 to 0.18 Hz, broadly deviating up and down from $f_{c1,90} = 0.055$ Hz.
2. In the discussed band, the contrast of spectral amplitudes is at its maximum (up to ten times) at $f \approx 3f_{c1,90}$; above $f = f_{c2} \approx 5\, f_{c1,90} \approx 0.27$ Hz, the contrast diminishes, down to about three times at 3 Hz, and almost disappears at 10–15 Hz.
3. No clear variation of f_{c2} with θ can be noticed.
4. Taking $f_{c2,90}$ as the reference, the critical frequency f_{ucoh} where the transition from LF enhanced to HF diminished directivity takes place is $f_{ucoh} \approx (1–2)f_{c2}$. This critical frequency can be understood as a change from mostly coherent to mostly incoherent behavior of the radiator. The general correspondence between f_{ucoh}, f_{c2} and $1/T_r$ has already been proposed in Gusev (2013a), see also Bernard and Herrero (1994).

Somerville *et al.* (1997) analyzed frequency dependence of directivity for vicinity of a fault, but their qualitative results can be expected to be approximately valid also for larger distances. Their analysis sets the just defined transition frequency f_{ucoh} at about 1.5–2 Hz. This value is in approximate agreement with the transition from high to low amplitude contrast that, according to simulation (Fig. 5), takes place within the frequency band 0.4–3 Hz. Although the analysis of Somerville *et al.* (1997) was performed in bulk for many earthquakes with various magnitudes, our approximate magnitude range $M = 6.7$–6.9 (based on the value of the fault area) generally agrees with the dominant range of magnitudes studied in Somerville (1997). Still, the degree of HF reduction of directivity attained in the present model seems to be somewhat weaker as compared to empirical directivity studies.

8. Discussion

Boore and Joyner (1978) and Joyner (1991) tried to analyze HF radiation on the basis of the notion that one can describe incoherency through non-uniform, random 2D slip rate, and/or non-uniform, random local rupture velocity. However, directivity produced by such a model is unrealistically high; the reason is that phase shifts created by such randomization are insufficient to create mutual cancellation of amplitudes which is required to provide "energy-style" summation of contributions of fault spots, in place of "amplitude-style" summation. See Gusev (2013a) for further discussion. Random phasing of pulses from fault spots (Gusev 1983) is a critical condition to provide realistically low directivity. The present model seemingly needs adjustment in order to provide a greater, more realistic degree of reduction of directivity at HF; this means, simultaneously, less degree of coherency.

Some researchers proposed locally random orientation of an instant, local rupture velocity vector within small fault patches (Koyama and Izutani 1990; Bernard and Herrero 1994). Similarly, random omnidirectional orientation of an instant, local rupture front is assumed in Day *et al.* (2008). This view contradicts in no way the model presented here. Just the opposite; it is easy to see that in a "microscopic", small-scale view, the "lacy" rupture front assumed here is indeed, to a large degree, omnidirectional. A close examination of the example case in Fig. 2a shows quite irregular orientation of local normal to the instant rupture front, created by its various wiggles, "islands" and "lakes". It should be noted that in the present model, there is a definite mean ("macroscopic") direction of propagation of rupture front/front strip; this systematic behavior must result in certain correlation between arrival times of the mentioned elementary pulses. The obtained numerical results show that the postulated "lacy" structure significantly reduces possible directivity related to such a correlation. Experiments (not shown here) show that in the case of the rupture front of zero width, but with $G_0(t)$ pulse of finite duration, no directivity reduction takes place at arbitrary high frequency.

In addition to significant general reduction of directivity, Bernard and Herrero (1994) found the secondary effect of a residual $\cos 2\theta$ directivity with enhanced 0° and 180° directions, and reduced 90° direction (see their Fig. 7a, b). The present model does not support this prediction.

Gusev (2011a) proposed a strong-motion simulation procedure that features lack of directivity for sufficiently high frequencies; the analog of f_{ucoh} was set (ad hoc) at $7.7f_{c1}$. At higher frequencies, point subsources of the source model of Gusev (2011a) have no intrinsic directivity. Time histories of these subsources are simulated as uncorrelated, thus their relative phases are random and coherence in this frequency range is suppressed. Alternatively, frequency-dependent coherence can be imposed, with correlation radius proportional to wavelength (Gusev 1983). Generally, partial coherence amplifies a signal at a receiver as compared to complete incoherence. This model may be used to clarify the question whether frequency-dependent coherence may affect directivity by itself, without any effects related to details of rupture front evolution. This research may be interesting.

There is a significant inter-event scatter among strong-motion parameters, estimated as an inter-event component of scatter of observed strong-motion amplitudes with respect to their mean trends described by GMPE for a given M_0 and distance. From the viewpoint of the present model, such a scatter of amplitudes has multiple causes, including the variability of at least the following parameters: (1) average rupture velocity v; (2) global stress drop τ_{gl}, not discussed here, on the order of M_0/LW^2; (3) rms local stress drop $\sigma_\tau = CV_\tau\tau_{gl}$, (4) relative slip-pulse width C_H. All these parameters must contribute to inter-event scatter, as well as to also very interesting inter-regional scatter, but the presented results do not permit us to isolate the most relevant parameter or parameters. My opinion is that parameters 2 and 4 are the most significant, but this statement has no solid basis.

From a tectonophysical point of view, the hypothesis of lacy rupture front is the only natural one. As noted in G88, corrugated or rough geometry of fault walls dictates spotty contact between them (or at least heavy-tailed distribution law for fault strength). With an increasing amount of data regarding complex geometry of faults and of instant rupture fronts, the idea of a smooth, simple, "Euclidean" shape of rupture front becomes less and less plausible.

9. Conclusion

The developed "doubly stochastic source model" provides a workable broadband description of earthquake source radiation. It is based on transparent concepts, most of which are not new. A fresh idea is one of a "lacy" fractal rupture front (Gusev 2013a), an almost inevitable assumption when considering the impossibility to suppress effectively HF directivity in another way. The only really new but critical idea is to relate the size of a slipping patch around an element of the fault plane with the width of propagating slip pulse.

The proposed model permitted us to emulate two well-established features of source spectra: the almost ubiquitous "omega-square" HF branch, and often-observed two-corner spectra. Formation of the second spectral corner, as proposed by a few researchers, is associated, on a conceptual level, with existence of a narrow slip pulse. The particular mechanism for this connection is proposed, and a particular tentative relationship between the slip-pulse width and spectral corner is found.

The simulation suggests that slip-pulse width controls not only the position of the second spectral corner, but also the amplitude level of acceleration spectrum. This relationship, revealed by numerical simulation, may provide at least a partial explanation of the scatter of HF ground motion amplitudes at a certain particular set of values of magnitude, global stress drop and v/c_S (Mach number w.r.t S waves). This conclusion may be important for earthquake hazard applications. In addition, it might be interesting to apply an empirical check to the model prediction (that directly follows from Eqs. 10 and 11) that the values of second corner frequency and acceleration spectral level must be positively correlated.

The range of values of the slip-pulse width to fault length ratio C_H needed to generate realistic spectral shapes, of approximately 3–15 %, matches well with the original estimates of Heaton (1990) based on radically different information. At high values of C_H, the model generates single-corner spectra. Therefore, observations of single-corner, clearly double-corner or intermediately shaped spectra in real earthquakes can be associated with the natural variations of the C_H parameter.

The developed model manifests frequency-dependent directivity, with high low-frequency directivity in accordance with common deterministic fault models, and diminished high-frequency directivity that reflects partial incoherence of the source for this frequency range. An estimate is obtained for the critical frequency f_{ucoh} where the transition between these regimes takes place; this estimate matches observations reasonably.

Acknowledgments

Discussions with V.I. Osaulenko and G.M. Molchan were highly valuable. Comments of anonymous reviewers and of the Invited Editor helped to improve the manuscript.

References

Aki, K. (1967). *Scaling law of seismic spectrum.* J. Geophys. Res., *72*, 1217–1231.

ANDREWS, D. J. (1980). *A stochastic fault model. 1. Static Case.* J. Geophys. Res., *78*, p. 3867–3877.

ARCHULETA, R. J. (1984) *A faulting model for the 1979 Imperial Valley earthquake*. J. Geophys. Res. *89*, 559–4585.

BEN-ZION, Y., and SAMMIS, C.G. (2003) *Characterization of Fault Zones* Pure Appl. Geophys. *160*, 677–715.

BERESNEV, I., and G. ATKINSON (2002). *Source parameters of earthquakes in eastern and western North America based on finite-fault modeling*, Bull. Seism. Soc. Amer. *92*, 695–710.

BERNARD, P. and A. HERRERO (1994): *Slip heterogeneity, body-wave spectra, and directivity of earthquake ruptures*, Ann. Geofis., *XXXVII* (6), 1679–1690.

BOATWRIGHT, J. (1982) *A dynamic model for far-field acceleration.* Bull. Seismol. Soc. Amer. *72* 1049–1068.

BOATWRIGHT, J. (1988) *The seismic radiation from composite models of faulting*, Bull. Seism. Soc. Am, *78*, 489–508.

BOATWRIGHT, J. (2007) *The persistence of directivity in small earthquakes*. Bull. Seism. Soc. Amer. *97*, 1850–1861, doi:10.1785/0120050228.

BOATWRIGHT, J. and CHOY, G.L. (1989) *Acceleration spectra for subduction-zone earthquakes*. J. Geophys. Res, *94*, 15541–15553.

BOORE, D. M. and W. B. JOYNER (1978). *The influence of rupture incoherence on seismic directivity*, Bull. Seism. Soc. Am. *68*, 283–300.

BRUNE, J. (1970). *Tectonic stress and the spectra of seismic shear waves from earthquakes*, J. Geophys. Res. *75*, 4997–5009.

DAS, S., and B.V. KOSTROV. (1983). *Breaking of a single asperity: rupture process and seismic radiation*, J. Geophys. Res, *88*, 4277–4288.

DAS, S., and B.V. KOSTROV. (1986). *Fracture of a single asperity on a finite fault: a model for weak earthquakes?* In: Earthquake Source Mechanics. Washington, Am. Geophys. Union. 91–96.

DAS, S., and KOSTROV, B. V. (1988), *An investigation of the complexity of the earthquake source time function using dynamic faulting models*, J. Geophys Res, *93*(B7), 8035–8050.

DAY, S. (1982). *Three dimensional simulation of spontaneous rupture: The effect of non-uniform pre-stress*. Bull. Seismol. Soc. Am., *72*, 1881–1902.

DAY, S. M., GONZALEZ, S. H., ANOOSHEHPOOR, R., and BRUNE, J. N. (2008), *Scale-model and numerical simulations of near-fault seismic directivity*, Bull Seismol Soc Am, *98*, 1186–1206. doi:10.1785/0120070190.

DI TORO, G., R. HAN, T. HIROSE, N. DE PAOLA, S. NIELSEN, K. MIZOGUCHI, F. FERRI, M. COCCO, & T. SHIMAMOTO (2011) *Fault lubrication during earthquakes*, Nature, *471*, 494–98, doi:10.1038/nature09838.

GUSEV, A. A. (1983). *Descriptive statistical model of earthquake source radiation and its application to an estimation of short-period strong motion*, Geophys J R Astr Soc, *74*, 787–808.

GUSEV, A. A. (1988). *A model of earthquake source with multitude of asperities*. Vulkanol. Seismol, *#1*, 41–55 (in Russian).

GUSEV, A.A. (1989). *Multiasperity fault model and the nature of short-period subsources*, Pure Appl. Geophys, *130*, 635–660.

GUSEV, A.A. (1992) *On relations between asperity population and earthquake population on a fault*. Tectonophysics, *211*, 85–98.

GUSEV, A.A. (1996) *Peak factors of Mexican accelerograms: evidence of non-Gaussian amplitude distribution*. J. Geophys. Res. *101*, 20083–20090.

GUSEV, A.A. (2011a) *Broadband kinematic stochastic simulation of an earthquake source: a refined procedure for application in seismic hazard studies*. Pure Appl. Geophys. *168*, 155–200. doi:10.1007/s00024-010-0156-3.

GUSEV, A.A. (2011b) *Statistics of the Values of a Normalized Slip in the Points of an Earthquake Fault. Izvestiya*, Physics of the Solid Earth, *47*, 176–185. [Original Russian Text: Fizika Zemli, 2011, No. 3, 24–33].

GUSEV, A.A. (2013a) *High-frequency radiation from an earthquake fault: a review and a hypothesis of fractal rupture front geometry*. Pure Appl. Geophys. *170*, 65–93. doi:10.1007/s00024-012-0455-y.

GUSEV, A.A. (2013b) *Fractal earthquake source with slip zone generates acceleration time histories with flat spectra*. Doklady Earth Sciences, Vol. *448*, Part 2, pp. 211–213. [original in Russian: DAN, 2013, 448, 465–467].

GUSEV, A.A., and I.R. ABYBAKIROV. (1996). *Simulated envelopes of non-isotropically scattered body waves as compared to observed ones: another manifestation of fractal heterogeneity*, Geophys. J. Int. *127*: 49–60, 1996.

HALLDORSSON, B., and PAPAGEORGIOU, A. S. (2005) *Calibration of the specific barrier model to earthquakes of different tectonic regions*, Bull Seism Soc Am, *95*, 1276–1300. doi:10.1785/0120040157.

HANKS, T. C., and R. K. MCGUIRE. (1981). *The character of high frequency strong ground motion*, Bull Seism. Soc Am, *71*, 2071–2095.

HASKELL, N. A. (1964). *Total energy and energy spectral density of elastic wave radiation from propagating faults*, Bull Seismol Soc Amer, *54*, 1811–1841.

HASKELL, N. A., (1966), *Total energy and energy spectral density of elastic wave radiation from propagating faults. II. A stochastic fault model.* Bull Seism Soc Am, *56*, 125–140.

HEATON, T. H., (1990). *Evidence for and implications of self-healing pulses of slip in earthquake rupture*, Phys. Earth Planet. Inter., *64*, 1–20.

HERRERO, A., and P. BERNARD, (1994). *A kinematic self-similar rupture process for earthquakes*, Bull Seism. Soc Am, *84*, 1216–1228.

KOSTROV, B. V. (1975). *Mechanics of the source of a tectonic earthquake*. (Nauka Moscow) (in Russian).

KOYAMA, J., IZUTANI, Y. (1990) *Seismic excitation and directivity of short-period body waves from a stochastic fault model*. Tectonophysics, *175*, 67–79.

LAY, T. and H. KANAMORI (1981). *An asperity model of large earthquake sequences, in: Earthquake Prediction–An International Review*, Maurice Ewing Series, vol. 4, D. Simpson and P.G. Richards, Editors; AGU, Washington D.C, pp 579–592.

MAI, P.M., and G.C. BEROZA (2002). *A spatial random-field model to characterize complexity in earthquake slip*, J. Geophys. Res., Vol. *107* (B11), 2308, doi:10.1029/2001JB000588.

OGLESBY, D.D. and S. M. DAY (2002) *Stochastic fault stress: implications for fault dynamics and ground motion*. Bull. Seismol. Soc. Amer *92*, 3006–3021.

PAPAGEORGIOU, A. S., and K. AKI, (1983). *A specific barrier model for the quantitative description of inhomogeneous faulting and the prediction of the strong ground motion, I: Description of the model*, Bull Seism. Soc Am, *73*, 693–722.

SHEARER, P.M. (1999) *Introduction to Seismology*, Cambridge University Press, 260 pp.

SILVER, P. (1983) *Retrieval of source-extent parameters and the interpretation of corner frequency*. Bull. Seismol. Soc. Amer., *73*,#6A, 1499–1511.

SOMERVILLE, P. G., N. F. SMITH, R.W. GRAVES, and N. A. ABRAHAMSON (1997). *Modification of empirical strong ground motion attenuation relations to include the amplitude and duration effects of rupture directivity*, Seismol. Res. Lett. *68*, 199–222.

SOMERVILLE, P., K. IRIKURA, R. GRAVES, S. SAWADA, D. WALD, N. ABRAHAMSON, Y. IWASAKI, T. KAGAWA, N. SMITH, and A. KOWADA (1999). *Characterizing crustal earthquake slip models for the prediction of strong motion*, Seism. Res. Lett. *70*, 59–80.

SPUDICH, P. and E. CRANSWICK, (1984) *Direct observation of rupture propagation during the 1979 Imperial Valley earthquake using a short baseline accelerometer array*. Bull. Seismol. Soc. Amer. *74*, 2083–2114.

TSAI, C. -C. P. (1997). *Ground motion modeling for seismic hazard analysis in the near-source regime: an asperity model*, Pure Appl. Geophys, *149*, 265–297.

(Received April 13, 2013, revised December 12, 2013, accepted December 19, 2013, Published online February 11, 2014)

Pure Appl. Geophys. 171 (2014), 2601–2615

DOI 10.1007/s00024-013-0772-9

Pure and Applied Geophysics

Laboratory Generated M -6 Earthquakes

GREGORY C. MCLASKEY,[1] BRIAN D. KILGORE,[1] DAVID A. LOCKNER,[1] and NICHOLAS M. BEELER[1]

Abstract—We consider whether mm-scale earthquake-like seismic events generated in laboratory experiments are consistent with our understanding of the physics of larger earthquakes. This work focuses on a population of 48 very small shocks that are foreshocks and aftershocks of stick–slip events occurring on a 2.0 m by 0.4 m simulated strike-slip fault cut through a large granite sample. Unlike the larger stick–slip events that rupture the entirety of the simulated fault, the small foreshocks and aftershocks are contained events whose properties are controlled by the rigidity of the surrounding granite blocks rather than characteristics of the experimental apparatus. The large size of the experimental apparatus, high fidelity sensors, rigorous treatment of wave propagation effects, and in situ system calibration separates this study from traditional acoustic emission analyses and allows these sources to be studied with as much rigor as larger natural earthquakes. The tiny events have short (3–6 μs) rise times and are well modeled by simple double couple focal mechanisms that are consistent with left-lateral slip occurring on a mm-scale patch of the precut fault surface. The repeatability of the experiments indicates that they are the result of frictional processes on the simulated fault surface rather than grain crushing or fracture of fresh rock. Our waveform analysis shows no significant differences (other than size) between the **M** -7 to **M** -5.5 earthquakes reported here and larger natural earthquakes. Their source characteristics such as stress drop (1–10 MPa) appear to be entirely consistent with earthquake scaling laws derived for larger earthquakes.

Key words: Earthquake scaling, acoustic emission, stick–slip.

1. Introduction

Extremely small earthquakes are used to monitor mine stability (e.g. URBANCIC and TRIFU 2000) and to map hydraulic fractures and fault networks (e.g. MAXWELL *et al.* 2010). In addition, small earthquakes are believed to have similar physics to large earthquakes, and their properties are often extrapolated to estimate the properties of larger but less frequent ones (e.g. SPOTTISWOODE and MCGARR 1975). The analysis of the smallest earthquakes requires the consideration of small wavelength radiated waves that attenuate rapidly with distance and require high speed recording equipment. Many researchers have employed sensors in deep boreholes (ABERCROMBIE 1995; JOST *et al.* 1998; PREJEAN and ELLSWORTH 2001; IMANISHI and ELLSWORTH 2006), or in mines (MCGARR 1984; GIBOWICZ *et al.* 1991; BOETTCHER *et al.* 2009; KWIATEK *et al.* 2011) in order to determine whether the properties of these small earthquakes (**M** -4 to **M** -2) are consistent with scaling behavior observed for larger ones. KWIATEK *et al.* (2010) studied mining induced earthquakes as small as **M** -4.4, while ELLSWORTH *et al.* (2007) report aftershocks of a naturally occurring **M** -1.8 earthquake that were as small as **M** -3.5. These small magnitudes imply source dimensions on the order of 100 mm.

A key parameter described in earthquake scaling studies is a stress parameter, typically termed stress drop $\Delta\sigma$, that describes how the amplitude of high frequency ground accelerations relates to the seismic moment M_0 (e.g. WALTER *et al.* 2006). If rupture properties are similar among small and large earthquakes, then we would expect to find that $\Delta\sigma$ is independent of M_0. Site-specific studies often find that $\Delta\sigma$ appears to increase with M_0 over a limited magnitude range which might imply a breakdown in earthquake self-similarity (GIBOWICZ *et al.* 1991). However, studies that compile many datasets from a wide range of magnitudes have found no consistent scaling of $\Delta\sigma$ with moment (e.g. MCGARR 1999; IDE and BEROZA 2001) and argue that apparent scaling is due to bandwidth limitations, data selection, or inadequate treatment of path and site effects (IDE and BEROZA 2001; IDE *et al.* 2003).

[1] United States Geological Survey, 345 Middlefield Rd. MS 977, Menlo Park, CA 94025, USA. E-mail: gmclaskey@usgs.gov

In laboratory experiments, abrupt (dynamic) slip episodes on simulated faults, often referred to as 'stick–slips', were first suggested by BRACE and BYERLEE (1966) to be analogues of natural earthquakes. The tiny, earthquake-like tremors radiated from stick–slips and other damage processes such as microcracking are often termed acoustic emission (AE). Recordings of AE signals are used to monitor damage in laboratory experiments (e.g. LOCKNER 1993) and civil infrastructure (e.g. GROSSE and OHTSU 2008). Laboratory studies have shown many similarities between AE and earthquakes: AE signals often appear to be similar to earthquake seismograms with identifiable P and S arrivals and codas, AE magnitudes obey Gutenberg-Richter power law distributions, and they exhibit Omori-type aftershock decay sequences (LOCKNER 1993). Consequently, there is now a long history of interpreting both AE and stick–slip phenomena observed in laboratory experiments as analogues to larger earth processes (e.g. SELLERS *et al.* 2003; THOMPSON *et al.* 2009; GOEBEL *et al.* 2012; MCGARR 2012).

Unfortunately, the precise source mechanisms of AE are not typically known. AE are typically recorded with resonant piezoelectric sensors for which the precise sensor response (including coupling) is typically unknown, and this makes quantitative analysis of AE waveforms challenging (PLENKERS *et al.* 2013). As a result, even sophisticated AE analyses use simplifying assumptions regarding wave propagation, (e.g., OHTSU 1991), or employ a relative source characterization method (DAHM 1996; GROSSE *et al.* 1997; SELLERS *et al.* 2003). Studies that are able to place constraints on AE sources suggest a variety of mechanisms ranging from predominantly tensile fracture (MANTHEI 2005) to a mixture of shear, tensile, and implosive sources (ZANG *et al.* 1998; STANCHITS *et al.* 2006) or the implosive destruction of surface topography (MCLASKEY and GLASER 2011). Consequently, it is not known to what extent these tiny seismic events are consistent with our understanding of the physics of larger earthquakes. Better understanding of the physical and mechanical properties of AE and stick–slip phenomena will improve our ability to relate them to earthquake source physics and determine whether or not they are consistent with common earthquake scaling relationships.

In this paper, we quantitatively describe the source characteristics of a particular set of tiny earthquakes produced in large-scale laboratory experiments and the time-domain techniques used to analyze their radiated waves. The tiny seismic events described in this paper are associated with shear slip on a well-defined 0.4 m by 2 m planar simulated fault cut diagonally through a 1.5 m square and 0.4 m deep granite sample (Fig. 1). We believe they are the result of nondestructive frictional processes on the simulated fault surface rather than fracture of fresh rock or brittle destruction of surface topography. In addition to seismicity associated with the complete rupture of the 2 m long fault, we can identify many smaller seismic events, termed foreshocks and aftershocks (MCLASKEY and KILGORE 2013). These small shocks are the subject of the current paper, and we expect them to be similar to natural earthquakes because their source areas are contained within the interior of the simulated fault. Study of these small events allows us to extend our understanding of earthquake physics to the mm scale.

The tiny seismic events described in this paper might be termed AE because of the high frequency content of their radiation, but because the current experiments and analysis techniques differ in a number of ways from more conventional AE tests, the events reported here are termed laboratory earthquakes (LabEQs). Conventional AE detected in rock fracture experiments likely originate from grain scale (<100 μm) damage processes (i.e. LOCKNER 1993; SHAH and LABUZ 1995; ZANG *et al.* 1998; SELLERS *et al.* 2003). We roughly estimate such events to be **M** < -8, assuming a few μm of slip and a 100 μm source dimension, so it is likely that the events described here are somewhat larger (**M** -7 to **M** -5.5), and may bridge the gap between conventional AE sources and the smallest observed natural earthquakes (**M** -3 to **M** -4). As described in this paper, the large sample size, high fidelity sensors, and in situ system calibration permit a more complete consideration of wave propagation and instrument response effects, which allows the LabEQs to be studied with as much rigor as larger natural earthquakes. We present a description of the recorded waveforms and the LabEQ source locations, focal mechanisms, seismic moments, and stress drops, and the variability

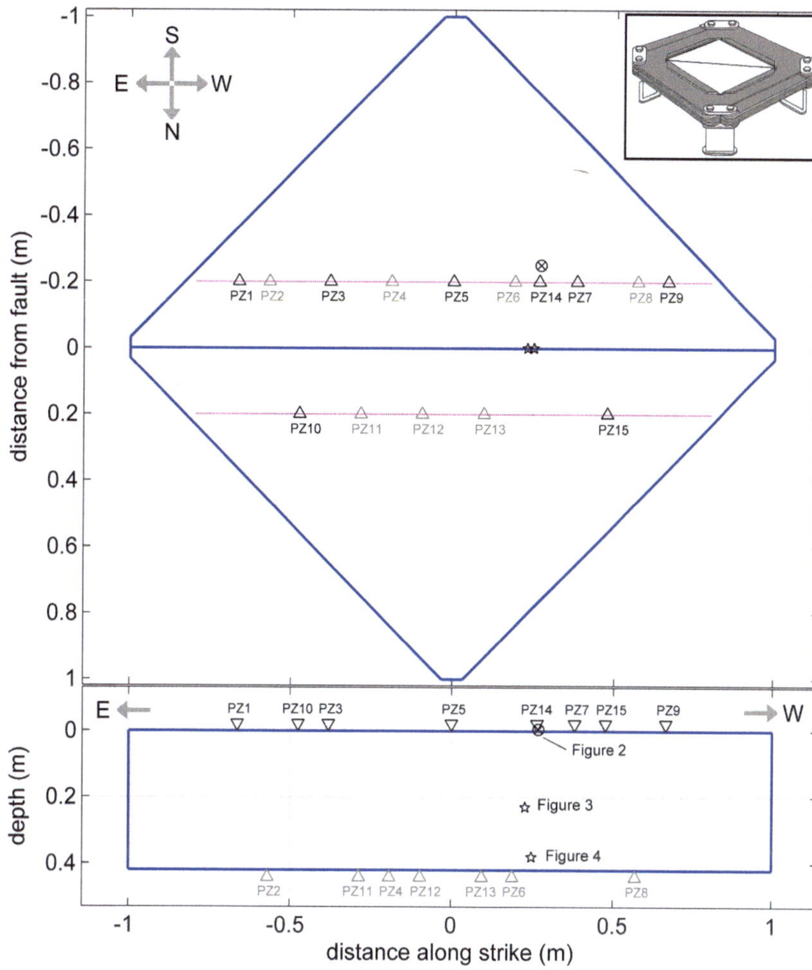

Figure 1
Schematic diagram of a top view (*upper*) and fault cross section (*lower*) of the granite sample and the locations of 15 piezoelectic sensors (PZ1–PZ15) used to record radiated ground motions. The inset in the upper right corner shows an isometric view of the sample (*white*) in the steel loading frame (*shaded*). The locations the ball impact (an '*x' within a circle*) and LabEQs (*stars*) described in Figs. 2, 3, 4 are also shown

of these parameters over a population of 48 LabEQs generated on the same fault under similar experimental conditions.

Previous work indicated that these tiny events are smaller than the expected minimum length scale for unstable fault slip, and that they are triggered by nearby aseismic slip and cannot spontaneously nucleate from slow loading conditions (McLaskey and Kilgore 2013). The current study shows that these features are not evident from an analysis of their seismograms. The stress drops of these tiny earthquakes generally fall in the 1–10 MPa range, indicating that their properties are consistent with those of larger natural earthquakes.

2. General Description of the Experiments

Experiments were conducted on the large-scale biaxial apparatus at the USGS in Menlo Park, California (Dieterich 1981) which accommodates a 1.5 m square and 0.4 m thick specimen composed of Sierra white granite from Raymond, California, USA, as depicted schematically in Fig. 1. A 2 m long simulated fault is cut diagonally through the sample. The fault was saw cut at the quarry and then initially roughened (in the laboratory) by lapping the two surfaces together with 30 grit silicon carbide abrasive, producing a peak-to-trough surface roughness of 80 μm, as measured with a profilometer over a traverse of 20 mm (Okubo and Dieterich 1984). This sample pair (with a "rough" fault, in the terminology of Okubo and Dieterich 1984) has been in use for more than 25 years and has sustained hundreds of stick–slip events and probably many cm of cumulative slip without additional surface preparation, so we expect that the surface is somewhat smoother than initially reported. Operating stresses (<10 MPa) in the current experiments are well below the failure strength of the granite host rock (unconfined strength ~200 MPa). Consequently, there is no apparent off-fault damage and experiments can be repeated over and over again on the same samples.

Each triangular half of the sample is supported at three locations by stands (~30 cm in diameter) topped with Teflon pads to provide low friction coupling with the floor. A steel loading frame surrounds the four vertical sides of the specimen (upper right inset of Fig. 1), and the sample is squeezed by increasing the hydraulic fluid pressure inside four flatjacks positioned between the sample and the loading frame. The flatjacks have roughly the same area as each of the four sample sides. In a typical experiment, pressure is increased uniformly in all four flatjacks to apply the desired normal stress σ_n of 4 or 6 MPa on the fault surface. Then, under computer control, pressure is increased in one opposing pair of flatjacks and decreased in the other pair, in order to increase fault average shear stress slowly at a rate of 0.001 MPa/s while keeping fault average normal stress constant. Under this test procedure, the fault typically slips abruptly in stick–slip fashion when the average shear stress attains a value of 0.76–0.80 of the average normal stress. We refer to these global stick–slip instabilities, in which the entire fault surface undergoes dynamic failure, as dynamic slip events (DSEs). An accelerometer on the sample is used as a trigger signal such that each DSE activates the closure of a ball valve so that the volume of hydraulic fluid in the flatjacks is held constant for some time period t_{hold} after which slow shear loading resumes and the load cycle is repeated.

Fifteen Panametrics V103 piezoelectric sensors (PZ1-PZ15) are mounted 200 mm from the simulated fault on either side and on both the top and bottom surfaces of the sample (see Fig. 1). These sensors detect surface normal motions in the frequency range of ~100 Hz to ~1 MHz and act as vertical-component seismic stations. Signals from the sensors are recorded with a high speed recording system (14 bit 5 MHz) in triggered mode for 420 ms intervals that include each DSE. In addition, capacitive slip sensors straddle the fault and measure local fault slip at 16 locations along the top surface of the sample both continuously at 100 Hz and in triggered mode for 512 ms at 1 MHz surrounding each DSE. More experimental details about the apparatus and instrumentation can be found elsewhere (Okubo and Dieterich 1984; Beeler *et al.* 2012; McLaskey and Kilgore 2013).

When a DSE occurs, the entire fault of length 2 m slips 50–150 microns. Typically, the majority of the fault slip occurs in about 3–5 ms, but some premonitory fault slip and slow afterslip usually occurs in the surrounding hundreds of ms. The premonitory slip and afterslip are predominantly aseismic, but do produce small, localized LabEQs, termed foreshocks and aftershocks, which are the subject of the current study. Unlike DSEs, these tiny LabEQs have abrupt onsets that radiate clearly defined P and S waves that can be used to calculate event locations and mechanisms. For cataloging purposes each DSE is numbered sequentially and labeled by experiment date (i.e. $SE4_{Nov2012}$) while foreshocks (FS) and aftershocks (AS) are labeled by their timing relative to the initiation of the DSE such as $SE12_{Nov2012}FS\text{-}17$ (a foreshock occurring 17 ms before the beginning of dynamic rupture of SE12) and $SE12_{Nov2012}AS + 49$ (an aftershock occurring 49 ms after initiation of SE12).

3. *Contained Foreshocks and Aftershocks Versus DSEs*

While contained stick–slip events have been produced when the stress state on the simulated fault is specifically controlled by fluid injection (Lockner *et al.* 1982), the vast majority of experiments on the current apparatus produce DSEs that rupture the entire simulated fault. For these uncontained DSEs, the free ends of the fault and the stiffness of the loading frame contribute to the energetics of the event. The stiffness of the laboratory apparatus is less than that of a crack of equivalent area within granite, so some adjustments are required for the comparison of DSEs to natural earthquakes (e.g. McGarr 1994, 2012; McGarr and Fletcher 2003). However, the small foreshocks and aftershocks described here were chosen for analysis specifically because they are contained events where nucleation, rupture, and arrest are controlled by the rigidity of the surrounding granite blocks and are not affected by characteristics of the loading frame. Indeed, with observed source durations of <6 μs, as discussed below, these events have ended before dynamic stress perturbations have traveled 25 mm from the source. We expect these contained events to be directly comparable to natural earthquakes.

We describe the characteristics of foreshocks and aftershocks produced during a set of experiments similar to those previously reported (McLaskey and Kilgore 2013) except that the hold time t_{hold} between DSEs was varied from 2 to 80 min, and sample average fault normal stress was held constant at either 4 MPa (SE1-12) or 6 MPa (SE13-27). From this study as well as previous studies, we note that important characteristics of the nucleation of the DSEs are sensitive to both hold time and normal stress. Some DSEs have a significant amount (>20 μm) of premonitory slip as well as afterslip. Other DSEs are more similar to those reported in McLaskey and Kilgore (2013) and have <10 μm of premonitory slip and no measurable afterslip or detectable aftershocks.

The tiny LabEQs discussed here (linear dimension <10 mm) are termed foreshocks and aftershocks because they always occur in close temporal proximity (typically within 500 ms) to larger DSEs that rupture the entire fault. The current dataset is consistent with previous work (McLaskey and Kilgore 2013) that showed that these events are driven by rapid stress changes due to nearby larger scale aseismic slip, and do not occur on their own as spontaneous slip events embedded in an otherwise-locked fault. We have searched for small events throughout the entire load cycle, but we only detect them when and where measured premonitory slip or afterslip rates exceed approximately 50 μm/s. McLaskey and Kilgore (2013) proposed that the source area of these tiny events is smaller than a minimum length scale for unstable slip under the conditions of homogenous fault strength and tectonic loading rates, and that the increased stressing rate caused by aseismic slip of surrounding areas is responsible for their existence. Here, we more rigorously address whether any of the unique circumstances under which these LabEQs occur might be discernible in an analysis of their radiated seismic waves.

4. *Methods*

For the accurate characterization of a seismic source, we must be able to isolate source effects from instrument effects and wave propagation effects, typically described by a Green's function. In conventional AE tests, thousands of individual AE are recorded, so the goal is to estimate the source locations (and possibly source mechanisms) in an automated procedure of data extraction and inversion. Even in AE tests that record full waveforms and include detailed post-processing of the data, a few incorrect wave arrival estimates, or inexact Green's functions, can produce inaccurate source location estimates (even physically unrealistic location estimates such as outside the sample), and this will cause severe biases in the focal mechanism results (Linzer 2005; Manthei 2005). Additionally, small sample sizes and high pressure environments limit the types of sensors that can be used and make it difficult to apply quantitative artificial sources of seismic waves for in situ verification of the transducer, transducer coupling, and recording system characteristics (Manthei 2005). As a consequence, most analyses rely on relative amplitudes rather than absolute measurements.

In the current experiments we record only a few (2–10) foreshocks or aftershocks before and after each SE. Instead of using automated algorithms, we carefully analyze the waveforms recorded from each LabEQ and compare them to full waveform synthetic seismograms generated from full elastodynamic Green's functions and instrument response functions obtained from calibration experiments. Signals recorded in the current experiments are amenable to rigorous analysis for four reasons: (1) There is essentially no off-fault damage so source-to-sensor ray paths traverse intact, homogeneous, fine-grained granite with directly measurable elastic properties. (2) The large source-to-sensor path length allows different wave arrivals to separate in time by 20–50 μs, which is greater than the source duration of the small LabEQs. Consequently, P and S waves (and a few other later arriving phases such as reflections) can be clearly isolated in time and analyzed separately. The S wave phase is particularly useful for constraining focal mechanisms, and reflections are sometimes helpful for constraining source locations. (3) Both sources and sensors are located far (0.5–1 m) from the side edges of the sample so that recorded waveforms are not complicated by side reflections or interactions with the loading frame for a relatively long period of time (hundreds of μs). (4) Because the sensors used in this study are sensitive to displacement rather than velocity or acceleration, coherent pulse-shaped farfield P and S wave arrivals are readily distinguishable from higher frequency and incoherent scattered waves. These features make waveform analysis more straightforward and allow for a more robust estimation of source characteristics described below. The main challenge of this study is that seismicity is clustered in time, so small aftershocks often occur within the coda of previous events, leading to an unavoidable reduction of signal-to-noise ratio.

4.1. Wave Propagation

For frequencies below a few MHz, wavelengths are significantly larger than the grain size, and the granite samples can be well approximated as linear elastic, isotropic, and homogeneous. Consequently, wave propagation effects can be estimated theoretically by solving elastodynamic equations. We use a generalized ray theory code to calculate Green's functions (Hsu 1985). The code assumes an infinite slab geometry; therefore, Green's functions presented here include reflections off the top and bottom free surfaces of the sample (as well as direct waves, surface waves, and nearfield terms) but do not include reflections from the four vertical sides of the sample, and the fault is assumed to be transparent. The direct arrivals (P, S, Rayleigh, and nearfield components) were checked against 3D finite element models, so we are confident in the accuracy of the synthetic seismograms for the early part of the signal. A comparison between the generalized ray theory solutions and finite element solutions is described in the appendix. We restrict our analysis to only the first few hundred microseconds of recorded signals before side reflections reach the sensors. This time of usable signal is an order of magnitude longer than typical AE recorded from cm-sized samples where reflections off the sides of the sample arrive 5–20 μs behind direct P wave arrivals and produce a 'coda' that is difficult to analyze deterministically.

4.2. Sensor Calibration and System Verification

Since theoretical Green's functions are readily obtainable for the laboratory samples, the sensors can be absolutely calibrated from theory by employing a known calibration source. The piezoelectric sensor model used in these experiments (Panametrics V103) was first calibrated when coupled to a 50 mm thick steel plate, following the methods of McLaskey and Glaser (2012). The calibration indicated that, in the ~30 kHz–1 MHz frequency band, sensor output (voltage) is very nearly proportional to ground displacement. At high frequencies, the finite aperture of the sensor causes a decrease in sensitivity to incoming waves with low incidence angle (McLaskey and Glaser 2012). For this sensor (16 mm diameter aperture), the aperture effect becomes significant for frequencies above about 200 kHz and increases with increasing frequency such that it can cause up to a 20 dB reduction in sensitivity at 700 kHz. Based on the results of the calibration experiments, we constructed a time domain model of the impulse response function of the sensor which is used to model the modest sensor distortions included in the synthetic seismograms described below. Since this model does not include aperture effects, for this study, our reliable frequency band is limited to ~30–300 kHz. The consideration of lower frequencies requires longer time windows, and the Green's functions described above are not valid for windows longer than a few hundred μs. An analysis of the LabEQ characteristics in the frequency domain, which includes the consideration of lower frequencies

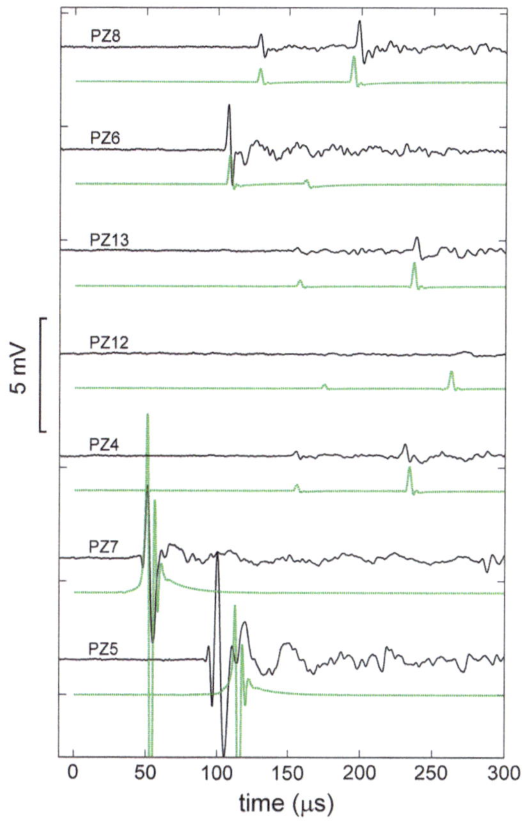

Figure 2
Unfiltered, unamplified recorded signals from the impact of a 1.00 mm glass ball dropped from a height of 1 m onto the top surface of the sample (location shown in Fig. 1). Signals from seven sensors are shown alongside synthetic seismograms (*green dotted lines*) that take into account wave propagation effects, minor sensor distortions, and a source that is represented as a 5 N force pulse with 4.5 μs duration. Sensor output is roughly proportional to vertical surface displacement with sensitivity 20 mV/nm. Positive displacement is in the direction outward from the sample

and employs an empirical Green's function, is the subject of further study (McLaskey *et al.* in preparation).

In addition to the initial transducer calibration on a steel plate, we used the seismic waves radiated from a ball dropped onto the granite samples as a way to verify the properties of the sensors and recording system when under actual test conditions. Figure 2 shows recorded waveforms and synthetic seismograms from a 1.00 mm diameter glass ball dropped 1 m onto the surface of the granite sample under 5 MPa fault normal stress, which was applied to simulate the stress conditions during testing. The comparison of recorded signals and synthetic seismograms from ball drop calibration sources at known locations is used to verify sensor coupling and sensitivity which was found to be roughly 20 mV/nm, and to calibrate our velocity model used in the Green's function code (v_p = 4,080 m/s and v_s = 2,700 m/s). The synthetic seismograms were calculated by convolving the appropriate Green's function with the sensor's impulse response function and a source function that is a single-sided pulse. This pulse is the force time history that the ball imposes on the sample. Since the ball typically bounces back to >50 % of its initial drop height, the force pulse can be estimated with Hertzian theory (McLaskey and Glaser 2010). To estimate attenuation, we constructed a few sets of synthetic seismograms with various degrees of attenuation explicitly assumed. In this way, we roughly estimate a frequency independent quality factor Q ~150, which is generally consistent with previous estimates for Sierra white granite in our ~30–300 kHz frequency band (Ulrich *et al.* 2002). We did not include the effects of attenuation in any further analyses. Future work will include corrections for attenuation and the aperture effect.

The ball impact imposes a downward force on the top surface of the specimen, so the P wave first motion is compressional (at all azimuths) rather than dilatational. Sensors located on the bottom surface of the sample should, therefore, all have down first motions. In Figs. 2, 3, 4 and Fig. 9 positive vertical displacement is in the direction outward from the sample. For the sensors on the top surface of the sample, positive displacement is up, but for sensors located on the bottom of the sample (which are oriented upside down), positive displacement is down. The two lower traces in Fig. 2 are from sensors located on the top surface of the sample. These sensors see large amplitude Rayleigh waves and very small amplitude direct P waves. We find that these waves that travel along the free surface are poorly modeled with the synthetic seismograms, so they are excluded from our analysis.

4.3. *Source Location, Focal Mechanisms, and Source Parameters*

The source properties of the individual LabEQs were estimated from forward modeling by comparing

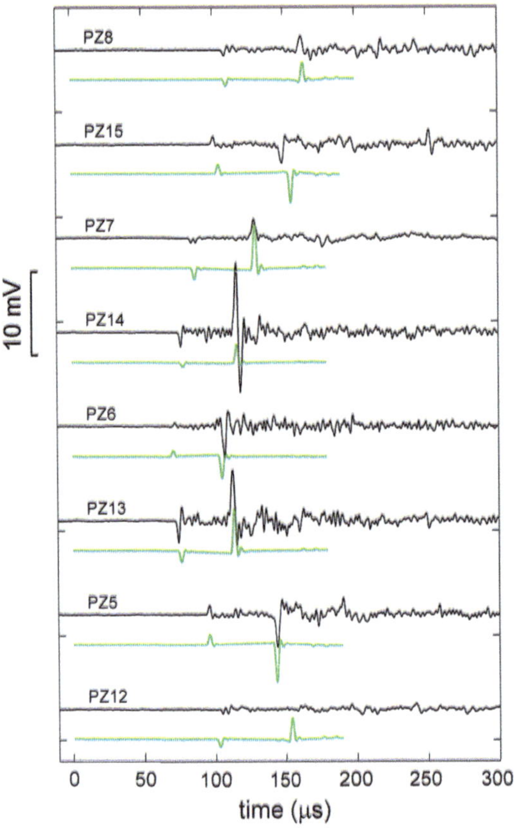

Figure 3
Unfiltered, unamplified recorded signals from **M** -6.5 LabEQ ($SE25_{Nov2012}$ FS-13). The source location relative to sensor locations is shown in Fig. 1. Signals from eight sensors are shown alongside synthetic seismograms (*green dotted lines*) which take into account wave propagation effects, minor sensor distortions, and a source which has a left-lateral strike slip double couple focal mechanism. The current geometry and source mechanism produces P and S waves which always have opposite polarity. Sensor output is roughly proportional to vertical surface displacement with sensitivity 20 mV/nm. Positive displacement is in the direction outward from the sample

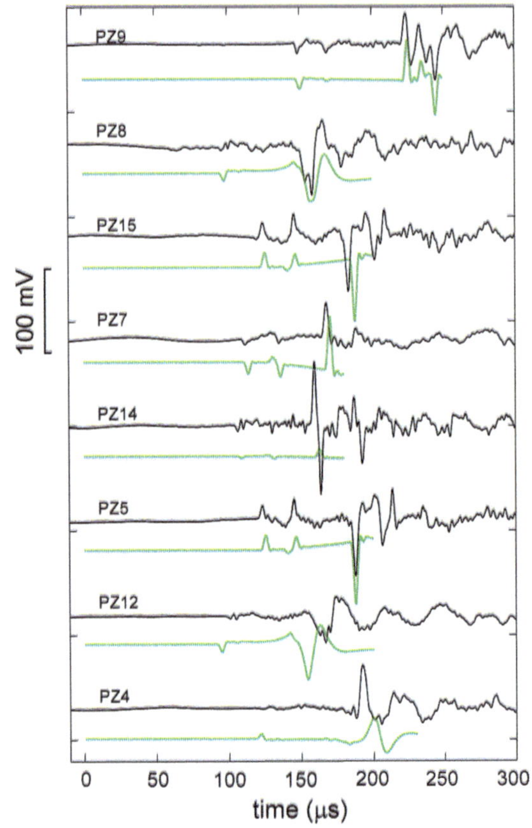

Figure 4
Same as Fig. 3, but a **M** -5.5 LabEQ ($SE16_{Nov2012}$ FS-18). Since this LabEQ is located close to the bottom surface of the sample, initially down-going rays that reflect off the bottom free surface of the sample arrive in between the direct P and S waves on the recordings from sensors located on the top surface of the sample (PZ9, PZ15, PZ7, and PZ5)

full-waveform synthetic seismograms to recorded signals. Figures 3 and 4 show raw recorded signals alongside synthetic seismograms for two well-constrained LabEQs. In our source characterization procedure, the source location is first estimated from the relative timing of arrivals for P, S, and, in some cases, reflected waves calculated from our calibrated velocity model. We estimate a ±10 mm source location accuracy in both along-strike and depth dimensions. Accuracy is likely limited by our velocity model that assumes the granite blocks are isotropic and homogeneous. Source locations are assumed to lie on the fault plane, and the good fit to the synthetics justifies this assumption at least to ±10 mm accuracy. The synthetic seismograms are calculated in a similar fashion to those shown in Fig. 2 (radiation pattern and minor sensor distortions are accounted for but not attenuation, Q, or sensor aperture effects) except the source is represented as a full moment tensor $M_{ij}(t)$, and we use the first spatial derivative of the Green's function. We assume that the time history of each component of $M_{ij}(t)$ is identical such that $M_{ij}(t) = m(t)M_{ij}$, where $m(t)$ is scalar seismic moment as a function of time. We estimate the focal mechanism M_{ij} from the polarities and relative amplitudes of direct P and S wave arrivals recorded by many nearby stations (typically 4–8). Finally, we estimate the width t_0 and height

$\dot{m}_{max}$ of a pulse-shaped moment rate function $\dot{m}(t)$ (the overdot represents a time derivative). There is some variability in source properties with angle of observation (observable in the waveforms shown in Fig. 3 and 4), but instead of estimating t_0 and $\dot{m}_{max}$ separately from recordings from individual sensors, these parameters (as well as M_{ij}) are estimated, in an average sense, for each event using constraints from direct P and S wave arrivals of many sensors. For these estimates, we place the most confidence in recordings from sensors whose source-to-sensor ray paths are far from nodes in the radiation pattern and predominantly travel through the interior of the sample rather than close to the free surfaces. For example, in Figs. 3 and 4, the signal recorded by sensor PZ14 is not well matched by the synthetic seismogram, but this station is close to a node in the P wave radiation pattern for these events.

5. *Results and Discussion*

5.1. *Locations and Catalog Considerations*

We analyze a catalog of 48 LabEQs, shown in Fig. 5, that could be accurately located and were separated enough in time from previous events that characteristics of pulse-shaped direct P and S wave arrivals could be estimated from the recorded signals. The LabEQs are categorized as either foreshocks (occurring before complete rupture of the 2 m long simulated fault) or aftershocks (occurring after complete rupture), denoted by circles and diamonds, respectively, in Fig. 5. We did not analyze the many additional foreshocks and aftershocks that were detected but had extremely small amplitudes or were buried in the coda of previous events. While LabEQs did occur close to the ends of the fault, our selection criterion removes from the catalog any LabEQs located outside of the sensor array, as shown in Fig. 5. A few larger (**M** > -5.0) foreshocks with more complicated source functions (their moment rate function $\dot{m}(t)$ could not be adequately represented with a single pulse), were also excluded. These larger events are studied with frequency domain techniques reported elsewhere (McLaskey *et al.* manuscript in preparation).

In general, both foreshocks and aftershocks are broadly distributed across the simulated fault, and there are no obvious differences between foreshocks and aftershocks, in either focal mechanisms, magnitudes, or stress drops, as described below. When the nucleation processes of successive DSEs are very similar, the foreshock locations cluster on specific fault patches. In most cases, clustered foreshocks also have similar source characteristics, such as SE25$_{Nov2012}$FS-148 and SE26$_{Nov2012}$FS-84, two foreshocks located at 0.36 m along strike and 0.38 m depth that are both relatively high stress drop events (see Table 1). The increased variability in nucleation patterns observed in the current experiments, due to more variable value of t_{hold}, is the likely reason why less clustering of seismicity is observed in the current sets of DSEs than in a previously recorded dataset (McLaskey and Kilgore 2013).

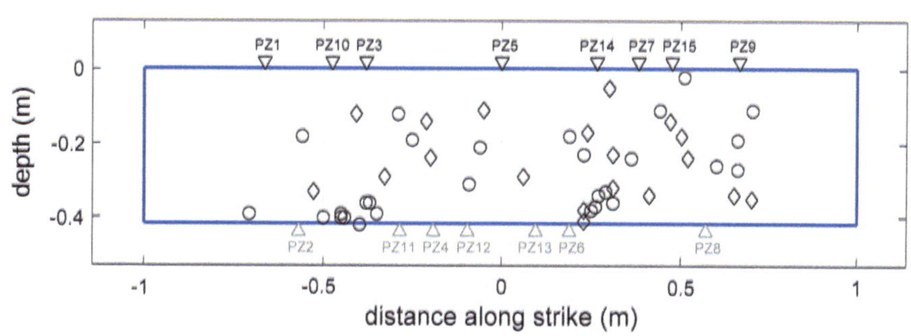

Figure 5

Fault cross section showing locations of 48 LabEQs listed in Table 1 that are foreshocks (*circles*) or aftershocks (*diamonds*) of complete rupture events on a the 2 m-long simulated fault. Locations have ±10 mm accuracy both along strike and with depth. Events are also detected near the fault ends, but those events are excluded from the current catalogue because their location uncertainties are larger

Table 1

Forty-eight LabEQs

LabEQ name	σ_n (MPa)	Dist. along strike (m)	Depth (m)	t_0 (μs)	$\dot{m}_{max}$ (kNm/s)	M_0 (Nm)	**M**	f_0 (kHz)	r_0 (mm)	$\Delta\sigma$ (Mpa)
SE6Nov2012 AS+12	4	−0.05	0.11	3.5	75	0.146	−6.6	286	3.5	1.47
SE7Nov2012 AS+29	4	0.52	0.24	3.5	55	0.107	−6.7	286	3.5	1.07
SE8Nov2012 AS+30	4	−0.21	0.14	4.0	55	0.122	−6.7	250	4.0	0.82
SE8Nov2012 FS-41	4	−0.06	0.21	3.5	15	0.029	−7.1	286	3.5	0.29
SE8Nov2012 FS-35	4	0.31	0.36	5.5	15	0.046	−7.0	182	5.5	0.12
SE10Nov2012 AS+19	4	0.65	0.34	5.0	350	0.974	−6.1	200	5.0	3.35
SE10Nov2012 AS+24	4	0.06	0.29	4.0	72	0.16	−6.6	250	4.0	1.08
SE10Nov2012 AS+25	4	0.41	0.34	3.0	150	0.25	−6.5	333	3.0	3.99
SE10Nov2012 AS+30	4	0.23	0.38	4.5	125	0.313	−6.4	222	4.5	1.48
SE10Nov2012 AS+47	4	0.30	0.05	2.5	30	0.042	−7.0	400	2.5	1.15
SE12Nov2012 FS-46	4	−0.50	0.40	3.5	430	0.837	−6.1	286	3.5	8.4
SE12Nov2012 FS-44	4	−0.56	0.18	3.5	40	0.078	−6.8	286	3.5	0.78
SE12Nov2012 FS-17	4	−0.29	0.12	4.0	200	0.445	−6.3	250	4.0	2.99
SE12Nov2012 FS-2	4	−0.35	0.39	4.0	60	0.134	−6.6	250	4.0	0.9
SE12Nov2012 AS+49	4	0.24	0.17	5.0	200	0.556	−6.2	200	5.0	1.92
SE12Nov2012 AS+17	4	0.47	0.14	3.0	20	0.033	−7.1	333	3.0	0.53
SE13Nov2012 FS-197	6	−0.45	0.39	4.0	100	0.223	−6.5	250	4.0	1.5
SE13Nov2012 AS+12	6	0.31	0.32	4.0	300	0.668	−6.2	250	4.0	4.49
SE13Nov2012 AS+1100	6	0.70	0.35	4.0	150	0.334	−6.4	250	4.0	2.24
SE14Nov2012 FS-13	6	−0.45	0.40	2.5	270	0.376	−6.4	400	2.5	10.3
SE14Nov2012 FS-11	6	0.29	0.33	5.0	300	0.835	−6.1	200	5.0	2.87
SE15Nov2012 FS-17	6	0.19	0.18	3.0	250	0.417	−6.3	333	3.0	6.65
SE15Nov2012 FS-14	6	−0.44	0.40	5.0	300	0.835	−6.1	200	5.0	2.87
SE16Nov2012 FS-32	6	−0.45	0.39	4.0	1,000	2.226	−5.8	250	4.0	15
SE16Nov2012 FS-23	6	−0.25	0.19	3.0	20	0.033	−7.1	333	3.0	0.53
SE16Nov2012 FS-18	6	0.25	0.38	5.5	2,160	6.61	−5.5	182	5.5	17.1
SE17Nov2012 FS-30	6	0.36	0.24	3.5	15	0.029	−7.1	286	3.5	0.29
SE17Nov2012 FS-22	6	−0.40	0.42	5.0	25	0.07	−6.8	200	5.0	0.24
SE18Nov2012 FS-183	6	−0.71	0.39	4.0	65	0.145	−6.6	250	4.0	0.97
SE18Nov2012 FS-28	6	0.70	0.11	3.5	400	0.779	−6.1	286	3.5	7.82
SE18Nov2012 FS-3	6	0.26	0.37	4.5	65	0.163	−6.6	222	4.5	0.77
SE18Nov2012 AS+11	6	−0.53	0.33	4.5	600	1.502	−5.9	222	4.5	7.09
SE18Nov2012 AS+15	6	−0.33	0.29	3.5	250	0.487	−6.3	286	3.5	4.89
SE18Nov2012 AS+17	6	−0.41	0.12	3.0	160	0.267	−6.4	333	3.0	4.26
SE18Nov2012 AS+19	6	0.50	0.18	5.0	400	1.113	−6.0	200	5.0	3.83
SE19Nov2012 AS+33	6	−0.20	0.24	3.5	14	0.027	−7.1	286	3.5	0.27
SE25Nov2012 FS-53	6	0.60	0.26	4.0	180	0.401	−6.3	250	4.0	2.69
SE25Nov2012 FS-36	6	0.51	0.02	4.0	420	0.935	−6.1	250	4.0	6.28
SE25Nov2012 FS-13	6	0.23	0.23	4.0	85	0.189	−6.5	250	4.0	1.27
SE25Nov2012 FS-148	6	−0.38	0.36	4.5	3,600	9.014	−5.4	222	4.5	42.6
SE25Nov2012 FS-104	6	−0.09	0.31	3.0	20	0.033	−7.1	333	3.0	0.53
SE26Nov2012 FS-84	6	−0.37	0.36	5.5	3,700	11.323	−5.4	182	5.5	29.3
SE26Nov2012 FS-39	6	0.27	0.34	3.0	84	0.14	−6.6	333	3.0	2.23
SE26Nov2012 FS-27	6	0.44	0.11	2.5	30	0.042	−7.0	400	2.5	1.15
SE26Nov2012 FS-8	6	0.66	0.27	5.0	40	0.111	−6.7	200	5.0	0.38
SE26Nov2012 FS-5	6	0.66	0.19	5.5	250	0.765	−6.1	182	5.5	1.98
SE26Nov2012 AS+48	6	0.31	0.23	4.5	25	0.063	−6.9	222	4.5	0.3
SE26Nov2012 AS+97	6	0.23	0.41	3.5	25	0.049	−6.9	286	3.5	0.49

5.2. *Focal Mechanisms*

The synthetic seismograms shown in Figs. 3 and 4 are each computed from a source that has a left-lateral double couple focal mechanism with one of the two nodal planes oriented parallel to the fault. Sensors located on the south side of the fault and east of the LabEQ hypocenter (PZ1-PZ6, see Fig. 1) see compressive P wave first motions, while sensors west of the hypocenter (PZ7–PZ9) see dilatational P wave first motions. Sensors located on the north side of the fault see opposite polarities. In the current geometry, the sign of the S wave first motions is always opposite that of the P wave for a double couple mechanism (Figs. 3, 4) while the ball drop source produces S waves that are the same polarity as P waves (Fig. 2). The good match between synthetic seismograms and recorded signals indicates that the LabEQs are well explained by the simple double couple mechanism. We cannot rule out the possibility of some percentage of non-double-couple components, but it cannot exceed about 20 %, otherwise the polarities and relative amplitudes of the P and S waves would be altered enough to be easily identified in the waveforms.

The vast majority of the LabEQs were well modeled by identical double couple focal mechanisms indicative of a shear dislocation on the fault. Only two of the 48 LabEQs studied had signals with P and S wave arrivals that were somewhat inconsistent with this model, but a more suitable source could not be identified.

5.3. Moment, Corner Frequencies, and Stress Drop

Estimates of source parameters t_0 and $\dot{m}_{max}$ derived from waveform modeling, as described in Sect. 4.3, are shown in Fig. 6 and listed in Table 1. (Note that we only estimated t_0 to the nearest 0.5 μs.) We calculate $M_0 = t_0\dot{m}_{max}/2$, and for simplicity, we assume the Brune (1970) relationship between corner frequency and source dimension $r_0 = 2.34*\beta/(2\pi f_0)$ and calculate stress drop $\Delta\sigma = 7/16 M_0 r_0^{-3}$. Here, corner frequency f_0 is approximated as $1/t_0$, and β is the shear wave velocity in the granite (2,700 m/s). Estimates of M_0 and $\Delta\sigma$, calculated from t_0 and $\dot{m}_{max}$, are shown in Fig. 7, and reported in Table 1.

For the 48 LabEQs catalogued here, t_0 estimates range from 2.5 to 5.5 μs, which correspond to f_0 of 400–180 kHz and Brune source radii of 3–6 mm. $\dot{m}_{max}$ ranges from 10 to over 1,000 kNm/s and shows little correlation with t_0. Most of the variation in M_0, from 0.05 to 10 Nm, is due to variations in the amplitude of the LabEQs ($\dot{m}_{max}$) rather than the frequency content (t_0). At a given corner frequency, M_0 varies by more than an order of magnitude.

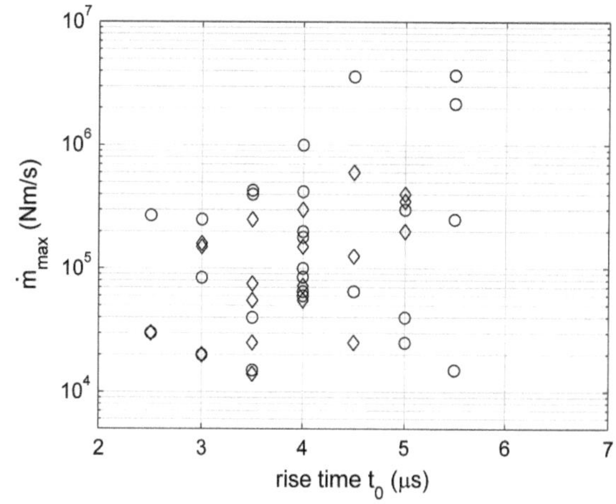

Figure 6
Width t_0 and amplitude $\dot{m}_{max}$ of the moment rate functions of 48 LabEQs listed in Table 1 that are foreshocks (*circles*) or aftershocks (*diamonds*) of complete rupture events on a the 2 m-long simulated fault. These parameters are estimated in the time domain from the width and amplitude of pulse-shaped direct P and S wave arrivals

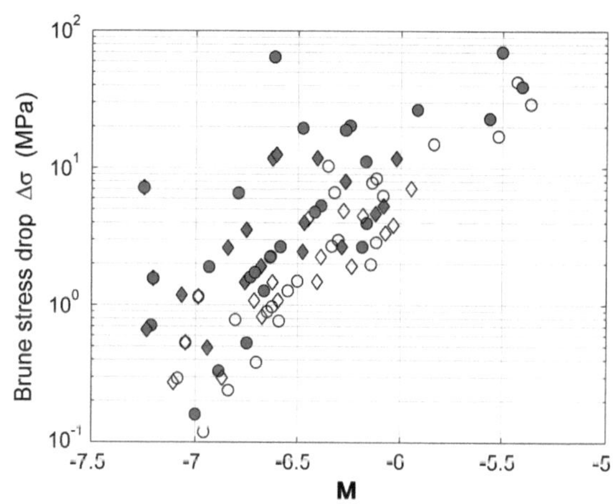

Figure 7
Estimates of moment magnitude *M* and Brune stress drop Δσ derived from parameters t_0 and $\dot{m}_{max}$ obtained from waveform fitting for 48 LabEQs listed in Table 1 that are foreshocks (*circles*) or aftershocks (*diamonds*) of complete rupture events on a the 2 m-long simulated fault. *Open symbols* are original estimates and *shaded symbols* have been adjusted to account for attenuation and finite sensor aperture

Many of the estimated corner frequencies are close to or above the upper bound of the reliable frequency band of our recording system. At these high frequencies, errors are introduced from the finite aperture of the sensor and attenuation in the granite, which would bias t_0 estimates to be somewhat larger than reality (a widening of farfield displacement pulses). We empirically tested the combined effects of sensor aperture and attenuation by comparing pulse widths estimated from ball drop calibration experiments to those expected from theory. A 1.00 mm glass ball dropped 1 m (Fig. 2) should theoretically have a pulse width of 2.9 μs, but the observed P and S wave farfield pulses were fit by a 4.5 μs pulse model. Similarly, a 1.58 mm steel ball should have 6.4 μs wide pulse width, but recorded signals were fit by a 7 μs pulse. Based on these observations, we apply an approximate correction factor to our t_0 estimates, which decreases our calculated M_0 and increases $\Delta\sigma$. These adjusted estimates (shaded symbols) are shown alongside unadjusted estimates (open symbols) in Fig. 7. This exercise indicates that at least some of the apparent scaling of stress drop with M_0 is an artifact of attenuation and sensor aperture effects described above.

Previous earthquake studies showed a similar apparent scaling of stress drop, (or, more commonly, apparent stress) with M_0 (e.g. Gibowicz *et al.* 1991), and this effect has been attributed to either attenuation, band limits of the recording system, or data selection bias (Ide and Beroza 2001; Ide *et al.* 2003). We believe that the apparent scaling of stress drop with M_0 seen in the current study is also an artifact caused by unmodeled sensor aperture and attenuation effects which attenuate events with short rise time (high corner frequencies), but have only a small effect on events with lower corner frequencies which are in our reliable frequency band. The persistent lack of events with low $\Delta\sigma$ and large M_0 may be the result of our catalogue selection criterion that throws out events without clearly identifiable pulse-shaped direct P and S wave arrivals.

Figure 8 shows how the seismic moments and corner frequencies of the current dataset compare to natural and mining-induced earthquakes from five other studies. Taking into account the uncertainty noted above, and considering only events with corner frequencies that lie within our reliable frequency

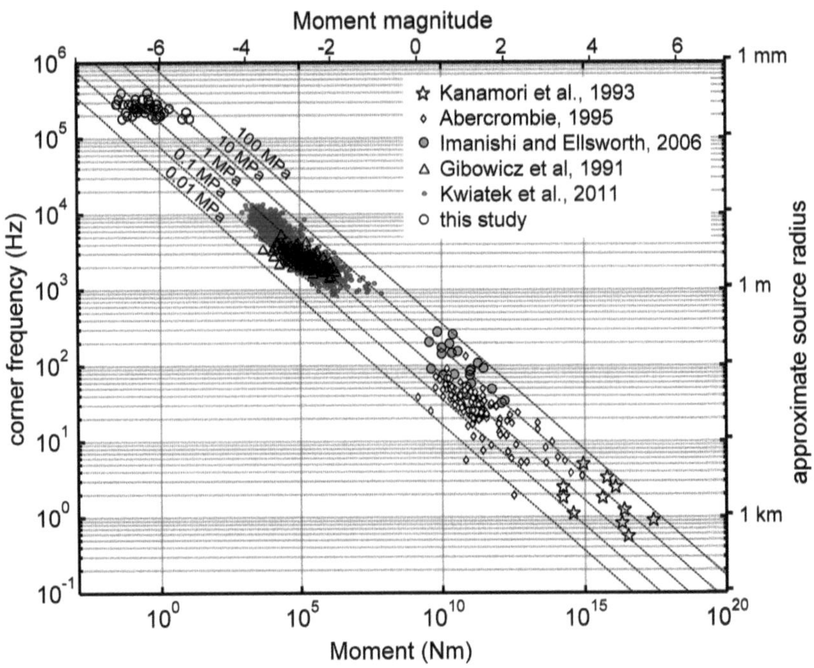

Figure 8

Corner frequencies and seismic moments of the LabEQS compared to events from five other studies of natural and mining-induced earthquakes. Approximate source radii and lines of constant stress drop (*dotted lines*) are based on the Brune model (see text) with $\beta = 3{,}300$ m/s. These parameters are model dependent and are only approximate, since β varies somewhat between the laboratory samples (2,700 m/s) and field estimates (3,300–3,700 m/s) band, our estimates of Brune stress drop still lie in the 1–10 MPa range and show at least one order of magnitude spread in stress drops. As indicated by Fig. 8, even if a very modest scaling of stress drop with magnitude is fit to results from studies of larger earthquakes, when extrapolated down to the **M** -7 to **M** -5.5 events of the current experiments, such a scaling would predict stress drops that are many orders of magnitude higher or lower than the range estimated here. We, therefore, argue that the current data set is consistent with an earthquake model where stress drop does not vary systematically with seismic moment (Aki 1967; Brune 1970).

The stress drops estimated here are in some cases up to five times higher than the fault average normal stress. This observation coupled with the large spread in stress drops may indicate that considerable stress heterogeneity is present even on the relatively flat and smooth simulated fault.

6. Conclusions

The large scale laboratory apparatus employed in the current work provides observation conditions that are particularly conducive to the seismic analysis of tiny LabEQs radiated from the simulated fault. Utilizing high fidelity sensors and a rigorous treatment of wave propagation effects, we find that these events are well modeled by double couple focal mechanisms that are consistent with left lateral shear slip occurring on a mm-scale fault patch enclosed within the 2 m long simulated strike-slip fault cut through the granite sample.

Previous work on the same sample and apparatus indicated that these tiny LabEQs only occur when larger sections (>100 mm) of the fault undergo premonitory creep or afterslip at slip rates exceeding approximately 50 μm/s, and that high local stressing rates (~20 MPa/s) due to nearby aseismic slip were required for their existence (McLaskey and Kilgore 2013).

The current work is consistent with the above observations, but also shows that the seismic source characteristics of the **M** -7 to **M** -5.5 LabEQs are consistent with larger natural earthquakes, despite the unique circumstances for their existence. The short (3–6 μs) rise times indicate corner frequencies in the hundreds of kHz range and source dimensions of a few mm. We find that the stress drops of these tiny LabEQs range from 1 to 10 MPa, which is entirely consistent with natural earthquakes of all sizes. Additionally, even on the relatively homogeneous laboratory fault, we observe more than an order of magnitude variability in LabEQ amplitude at a given corner frequency, which suggests more than an order of magnitude spread in stress drops at a given magnitude. The current experiments show that it is possible to create **M** -7 to **M** -5.5 seismic events in a laboratory that appear to be scaled versions of larger earthquakes. Careful study of such events could allow us to relate input parameters such as fault average stress state, loading conditions, and surface roughness to seismically observable source properties such as stress drop and radiated energy that are relevant to the physics of larger earthquakes.

Acknowledgments

This work benefitted from careful reviews by Art McGarr, Annemarie Baltay, and Grzegorz Kwiatek. The authors would also like to thank Brad Aagaard for assistance with the finite element modeling.

Appendix

Our estimates of seismic moment and stress drop are dependent on the absolute accuracy of the Green's functions, so we deemed it worthwhile to verify them by means of finite element modeling. We computed solutions to a 3D finite element model using the software PyLith (Aagaard *et al.* 2013). Explicit dynamic finite element models useful for the computation of elastodynamic Green's functions become unstable when the number of elements per wavelength decreases to below about 10. The element size in our model is 3 mm, so the maximum frequency for S waves ($v_s = 2{,}700$ m/s) that our model can accommodate is about 90 kHz. Figure 9 shows synthetic seismograms obtained from the finite element model (thick grey lines) compared to synthetic seismograms obtained from the generalized ray

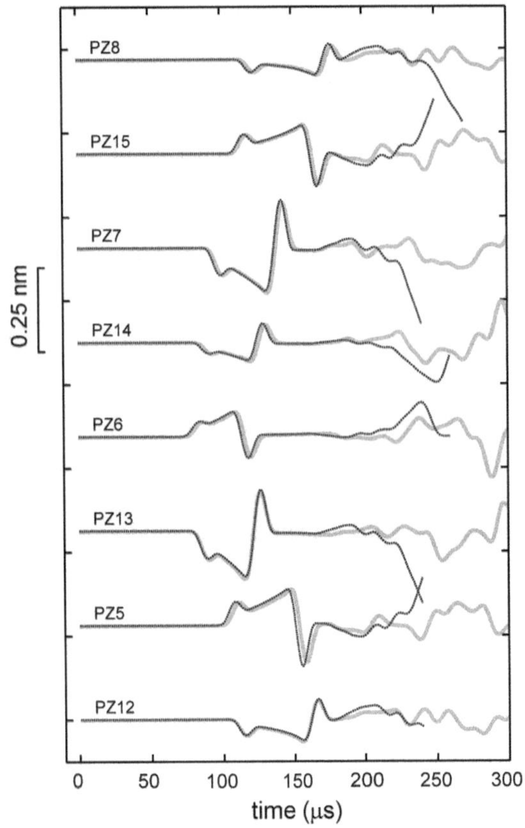

Figure 9
Comparison between synthetic seismograms obtained from our finite element model (*thick grey lines*) and synthetic seismograms obtained from the generalized ray theory code (*black dashed lines*), for the same source location as the LabEQ shown in Fig. 3 but different source duration. These synthetics include source and wave propagation components but do not include sensor distortions. The source term is a left-lateral strike-slip double couple focal mechanism with pulse-shaped moment rate function $\dot{m}(t)$ that is 15 μs wide and 85 kNm/s tall. Positive displacement is in the direction outward from the sample

theory code (black dashed lines) for the source and sensor geometries and orientations shown in Fig. 3. These synthetic seismograms are calculated exactly the same way as those shown in Fig. 3 except the instrument response function is not included and the width t_0 of a pulse-shaped moment rate function $\dot{m}(t)$ is set to 15 μs. This relatively wide pulse width (15 μs, corner frequency ~67 kHz) is required to keep the majority of the wave energy below our 90 kHz band limit, as set by the stability of the finite element model. The two solutions agree during the first part of the signals when only direct P and S waves and near field terms are present, and they diverge at later times due to (1) errors associated with later arriving reflections in the generalized ray theory code, and (2) reflections off the sides of finite element model which are not accounted for in the infinite slab geometry of the generalized ray solution.

References

Aagaard, B. T., Knepley, M. G., and Williams, C., A. (2013) *A domain decomposition approach to implementing fault slip in finite-element models of quasi-static and dynamic crustal deformation*. J. Geophys. Res., *118*, 3059–3079.

Abercrombie, R. E., (1995) *Earthquake source scaling relationships from −1 to 5 ML using seismograms recorded at 2.5-km depth*, J. Geophys. Res., *100*, 24015–24036.

Aki, K., (1967) *Scaling law of seismic spectrum,* J. Geophys. Res., *72*, 1217–1231.

Beeler, N. M., Kilgore, B., McGarr, A., Fletcher, J., Evans, J., and Baker, S. R. (2012) *Observed source parameters for dynamic rupture with non-uniform initial stress and relatively high fracture energy*, J. Struct. Geol., *38*, 77–89.

Boettcher, M. S., A. McGarr, and M. Johnston (2009), *Extension of Gutenberg-Richter distribution to M_w–1.3, no lower limit in sight*, Geophys. Res. Lett., *36*, L10307, doi:10.1029/2009GL038080.

Brace, W. F. and J. D. Byerlee (1966) *Stick slip as a mechanism for earthquakes*, Science, *153*, 990–992.

Brune, J. N., (1970) *Tectonic stress and spectra of seismic shear waves from earthquakes*, J. Geophys. Res. *75*, 4997–5009.

Dieterich, J. H., (1981), *Potential for geophysical experiments in large scale tests*, Geophys. Res. Lett., *8*, 653–656.

Dahm, T. (1996) *Relative moment tensor inversion based on ray-theory: theory and synthetic tests*, Geophys. J. Int. *124*, 245–257.

Ellsworth, W.L., Hickman, S.H., Zoback, M.D., Imanishi, K., Thurber, C.H., Roecker, S.W. (2007) Micro- Nano- and Pico-earthquakes at SAFOD: Implications for Earthquake Rupture and Fault Mechanics: Eos (Transactions, American Geophysical Union), v. 88, no. 52, supplement, abstract S12B-05.

Grosse, C., Reinhardt, H, and Dahm, T. (1997) *Localization and classification of fracture types in concrete with quantitative acoustic emission measurement techniques*, NDT & E International *30*, 223–230.

Grosse, C., and Ohtsu, M. *Acoustic Emission Testing* (Springer-Verlag: Berlin 2008).

Gibowicz, S. J., Young, R.P., Talebi, S., and Rawlence, D. J. (1991) *Source parameters of seismic events at the Underground Research Laboratory in Manitoba, Canada: Scaling relations for events with moment magnitude smaller than −2*, Bull. Seismol. Soc. Am., *81*, 1157–1182.

Goebel, T., Becker, T., Schorlemmer, D., Stanchits, S., Sammis, C. Rybacki, E., and Dresen. G. (2012) *Identifying fault heterogeneity through mapping spatial anomalies in acoustic emission statistics*, J. Geophys. Res., *117,* B03310.

Hsu, N. (1985), Dynamic Green's functions of an infinite plate—A computer program, Tech. Rep. NBSIR 85-3234, Cent. for Manuf. Eng., Natl. Bur. of Stand., Gaithersburg, Md.

Imanishi, K., and W. L. Ellsworth (2006), Source scaling relationships of microearthquakes at Parkfield, CA, determined using the SAFOD Pilot Hole Seismic Array, in Earthquakes: Radiated Energy and the Physics of Earthquake Faulting, Geophys. Monogr. Ser., vol. 170, (ed Abercrombie, R. et al.), pp. 81–90, AGU, Washington, D. C.

Ide, S. and Beroza, G. (2001) *Does apparent stress vary with earthquake size?* Geophys. Res. Lett. *28*, 3349–3352.

Ide, S., Beroza, G., Prejean, S., Ellsworth, W. (2003) *Apparent break in earthquake scaling due to path and site effects on deep borehole recordings,* J. Geophys. Res. *108*, B5, 2271.

Jost, M. L., Büßelberg, T., Jost, Ö., and Harjes, H.-P. (1998) *Source parameters of injection-induced microearthquakes at 9 km depth at the KTB deep drilling site, Germany*, Bull. Seismol. Soc. Am., *88*, 815–832.

Kwiatek, G., Plenkers, K., Nakatani, M., Yabe, Y., Dresen, G., and JAGUARS-Group (2010) *Frequency-magnitude characteristics down to magnitude -4.4 for induced seismicity recorded at Mponeng gold mine, South Africa*, Bull. Seismol. Soc. Am. *100*, 1165–1173.

Kwiatek, G., Plenkers, K., Dresen, G. and JAGUARS Research Group (2011) *Source parameters of picoseismicity recorded at Mponeng deep gold mine, South Africa: implications for scaling relations*, Bull. Seismol. Soc. Am. *101*, 2592–2608.

Linzer, L. (2005) *A relative moment tensor inversion technique applied to seismicity induced by mining*, Rock Mech. Rock Engng. *38* (2), 81–104.

Lockner, D. A., P. G. Okubo, and J. H. Dieterich (1982) *Containment of stick slip failures on a simulated fault by pore fluid injection,* Geophys. Res. Lett., *9*, 801–804.

Lockner, D. A., (1993) *Role of acoustic emission in the study of rock fracture*, Int. J. Rock Mech. Min. Soc. Geomech. Abstr., *30*, 884–899.

Manthei, G. (2005). *Characterization of acoustic emission sources in a rock salt specimen under triaxial compression*, Bull. Seismol. Soc. Am. *95*, 1674–1700.

Maxwell, S. C., J. T. Rutledge, R. Jones, and M. Fehler (2010) *Petroleum reservoir characterization using downhole microseismic monitoring*, Geophysics, *75*, 75A129–75A137.

McGarr, A. (1984). *Scaling of ground motion parameters, state of stress, and focal depth*, J. Geophys. Res. *89*, 6969–6979.

McGarr, A. (1994). *Some comparisons between mining-induced and laboratory earthquakes*, Pure Appl. Geophys. *142*, 467–489.

McGarr, A., (1999) *On relating apparent stress to the stress causing earthquake fault slip*, J. Geophys. Res., *104*, 3003–3011, 1999.

McGarr. A., (2012) *Relating stick–slip friction experiments to earthquake source parameters*, Geophys. Res. Lett., *39*, L05303.

McGarr, A., and Fletcher, J. B. (2003) *Maximum Slip in Earthquake Fault Zones, Apparent Stress, and Stick–Slip Friction.* Bull. Seismol. Soc. Am. *93*, 2355–2362.

McLaskey, G., and Glaser, S. (2010) *Hertzian impact: experimental study of the force pulse and resulting stress waves.* J. Acoust. Soc. Am. *128*, 1087–1096.

McLaskey, G., and Glaser, S. (2011) *Micromechanics of asperity rupture during laboratory stick slip experiments* Geophys. Res. Lett., *38*, L12302.

McLaskey, G. and Glaser, S. (2012) *Acoustic Emission Sensor Calibration for Absolute Source Measurements*, J. Nondest. Eval. *31*, 157–168.

McLaskey. G. and Kilgore, B. (2013) *Foreshocks during the nucleation of stick–slip instability*, J. Geophys. Res. vol 118, 1–16.

Ohtsu, M. (1991), *Simplified moment tensor analysis and unified decomposition of acoustic emission: application to hydrofracturing test*, J. Geophys. Res. *96*, 6211–6221.

Okubo, P., and Dieterich, J. (1984), *Effects of physical fault properties on frictional instabilities produced on simulated faults,* J. Geophys. Res., *89*, 5817–5827.

Plenkers, K., G. Kwiatek., and F. Krüger (2013) "Monitoring induced seismicity with AE sensors : the influence of unknown calibration functions," in EGU General Assembly 7–12 April, 2013, Vienna, Austria. Abstract Number EGU2013-3849.

Prejean, S. G. and Ellsworth, W. L. (2001) *Observations of earthquake source parameters and attenuation at 2 km depth in the Long Valley Caldera, Eastern California*, Bull. Seismol. Soc. Am., *91*, 165–177.

Sellers, E. J., Kataka, M. O. and Linzer L. M. (2003). *Source parameters of acoustic emission events and scaling with mining-induced seismicity,* J. Geophys. Res. *108*, B9 2418.

Shah, K., and Labuz, J. (1995) *Damage mechanisms in stressed rock from acoustic emission.* J. Geophys. Res., *100*, 15,527–15,539.

Spottiswoode, S. M. and McGarr, A. (1975) *Source parameters of tremors in a deep-level gold mine*, Bull. Seismol. Soc. Am. *65*, 93–112.

Stanchits, S., Vinciguerra, S., and Dresen, G. (2006) *Ultrasonic velocities, acoustic emission characteristics, and crack damage of basalt and granite*, Pure Appl. Geophys. *163*, 974–993.

Thompson, B. D., Young, R. P. and Lockner, D. A. (2009), *Premonitory acoustic emissions and stick–slip in natural and smooth-faulted Westerly granite*, J. Geophys. Res., *114*, B02205.

Ulrich, T. J., McCall, K. R., and Guyer, R. A., (2002) *Determination of elastic moduli of rock samples using resonant ultrasound spectroscopy*, J. Acoust. Soc. Am. *111*, 1667–1674.

Urbancic, T. Y., and Trifu, C. (2000) *Recent advances in seismic monitoring technology at Canadian mines*, J. Appl. Geophys. *45*, 225–237.

Walter, B. R., Mayeda, K., and Gok, R. (2006) in *Earthquakes: Radiated Energy and the Physics of Earthquake Faulting*, Geophys. Monogr. Ser., vol. 170, (ed Abercrombie, R. et al.), pp. 25–41, AGU, Washington, D. C.

Zang, A., Wagner, C., Stanchits, S., Dresen, G., Andresen, R., and Haidekker, M. (1998) *Source analysis of acoustic emissions in Aue granite cores under symmetric and asymmetric compressive loads*. Geophys. J. Int. *135*, 1113–1130.

(Received June 12, 2013, revised November 22, 2013, accepted December 30, 2013, Published online February 17, 2014)

Pure Appl. Geophys. 171 (2014), 2617–2640
© 2013 Springer Basel
DOI 10.1007/s00024-013-0760-0

Pure and Applied Geophysics

Frictional Properties and Microstructure of Calcite-Rich Fault Gouges Sheared at Sub-Seismic Sliding Velocities

B. A. Verberne,[1] C. J. Spiers,[1] A. R. Niemeijer,[1] J. H. P. De Bresser,[1] D. A. M. De Winter,[2] and O. Plümper[3]

Abstract—We report an experimental and microstructural study of the frictional properties of simulated fault gouges prepared from natural limestone (96 % $CaCO_3$) and pure calcite. Our experiments consisted of direct shear tests performed, under dry and wet conditions, at an effective normal stress of 50 MPa, at 18–150 °C and sliding velocities of 0.1–10 μm/s. Wet experiments used a pore water pressure of 10 MPa. Wet gouges typically showed a lower steady-state frictional strength ($\mu = 0.6$) than dry gouges ($\mu = 0.7$–0.8), particularly in the case of the pure calcite samples. All runs showed a transition from stable velocity strengthening to (potentially) unstable velocity weakening slip above 80–100 °C. All recovered samples showed patchy, mirror-like surfaces marking boundary shear planes. Optical study of sections cut normal to the shear plane and parallel to the shear direction showed both boundary and inclined shear bands, characterized by extreme grain comminution and a crystallographic preferred orientation. Cross-sections of boundary shears, cut normal to the shear direction using focused ion beam—SEM, from pure calcite gouges sheared at 18 and 150 °C, revealed dense arrays of rounded, ~0.3 μm-sized particles in the shear band core. Transmission electron microscopy showed that these particles consist of 5–20 nm sized calcite nanocrystals. All samples showed evidence for cataclasis and crystal plasticity. Comparing our results with previous models for gouge friction, we suggest that frictional behaviour was controlled by competition between crystal plastic and granular flow processes active in the shear bands, with water facilitating pressure solution, subcritical cracking and intergranular lubrication. Our data have important implications for the depth of the seismogenic zone in tectonically active limestone terrains. Contrary to recent claims, our data also demonstrate that nanocrystalline mirror-like slip surfaces in calcite(-rich) faults are not necessarily indicative of seismic slip rates.

Electronic supplementary material The online version of this article (doi:10.1007/s00024-013-0760-0) contains supplementary material, which is available to authorized users.

[1] High Pressure and Temperature Laboratory, Department of Geosciences, Utrecht University, Budapestlaan 4, 3584 CD Utrecht, The Netherlands. E-mail: b.a.verberne@uu.nl

[2] Biomolecular Imaging, Department of Biology, Utrecht University, Padualaan 8, 3584 CH Utrecht, The Netherlands.

[3] Structural Geology and Tectonics, Department of Earth Sciences, Utrecht University, Budapestlaan 4, 3584 CD Utrecht, The Netherlands.

1. Introduction

Natural faults are often marked by the presence of a narrow principal slip zone (PSZ) at the core, which is thought to accommodate the bulk of the slip imposed on the fault (Sibson 1986). The PSZ consists of the fine-grained granular wear material, or fault gouge, that is the product of cumulative slip along the fault (e.g. Chester and Chester 1998). Understanding the frictional properties of fault gouges and their role in seismic rupture nucleation and propagation, as well as in interseismic strength recovery, is crucial for improving our understanding of the seismic cycle. Laboratory studies focusing on the frictional properties of crustal fault rocks, in particular through experiments simulating the conditions pertaining to natural faults, have a key role to play here (e.g. Blanpied *et al.* 1995; He *et al.* 2007; Carpenter *et al.* 2011; Den Hartog *et al.* 2012; Ikari *et al.* 2013; Zhang and He 2013).

Since Brace and Byerlee (1966) identified unstable frictional slip, i.e. stick–slips, as an analogue for earthquakes, much experimental and theoretical work has focused on the processes leading to frictional instability (see Marone 1998 and Scholz 1998 for reviews). It was recognized that regular stick–slip may occur when a sliding medium decreases in strength with increasing sliding rate, i.e. when it exhibits velocity weakening behaviour (Dieterich 1979; Ruina 1983). For this reason, much laboratory work aimed to characterize the velocity dependence of (fault) rock frictional sliding, and in particular the phenomenology and constitutive modelling of transitions from stable to (potentially) unstable slip (e.g. Blanpied *et al.* 1995; He *et al.* 2007) and on the microphysical mechanisms controlling them (e.g. Bos and SPIERS 2002a; NIEMEIJER and SPIERS 2007; HE *et al.* 2013). In this context, VERBERNE et al. (2010) reported on friction experiments conducted on (simulated) fault gouges from the Longmenshan fault zone (LFZ), Sichuan, China, with the aim of characterizing the frictional strength and stability of crushed rocks sampled from surface outcrops cut by a natural, earthquake-prone fault. Their samples consisted of clay-rich, quartz-rich, and calcite-rich gouges, which were sheared at low sliding velocities (0.1–1 μm/s), under water-saturated conditions at an effective normal stress of ~53 MPa, at temperatures ranging from room temperature (~20 °C) to 150 °C. In the range of conditions investigated, the clay-rich and quartz-rich samples invariably exhibited stable velocity strengthening behaviour, while the calcite-rich gouge showed a transition from stable slip at 20–50 °C to unstable velocity weakening slip at 100–150 °C (VERBERNE *et al.* 2010). These results offer potential constraints on the depth to the upper limit of the seismogenic zone in the LFZ, implying that calcite-rich lithologies may be particularly susceptible to shallow-focus seismicity.

Indeed, recent experimental studies focusing on upper-crustal fault-frictional behaviour show growing evidence for an intrinsic tendency towards frictional instability of calcite-rich fault rocks (COLLETTINI *et al.* 2011; IKARI *et al.* 2013; ZHANG and HE 2013; NIEMEIJER and COLLETTINI, this issue). In experiment and nature, calcite(-rich) rocks are well known to exhibit ductile deformation by mechanisms such as pressure solution and crystal plasticity, at lower pressures and temperatures than generally required for these mechanisms to operate in other upper crustal rocks, such as granite (e.g. BURKHARD 1990, 1993; SPIERS *et al.* 2004). These time-dependent deformation mechanisms offer important fault gouge healing and compaction or re-strengthening mechanisms (BOS *et al.* 2000; BOS and SPIERS 2002a; NAKATANI and SCHOLZ 2004; NIEMEIJER and SPIERS 2006) and are known, from microphysical modelling and analogue experiments, to play a key role in causing potentially unstable velocity weakening behaviour in gouge-bearing faults at low slip rates (NAKATANI 2001; BOS and SPIERS 2002b; NIEMEIJER and SPIERS 2007; DEN HARTOG and SPIERS 2013; HE *et al.* 2013). Crystal plasticity or pressure solution processes may therefore offer the key to explaining the unusual frictional behaviour of calcite-rich fault rocks compared with other sedimentary rocks.

To further investigate this hypothesis, we conducted dry and wet friction experiments on simulated gouges prepared from natural limestone and pure calcite, at upper crustal effective normal stresses (50 MPa) at temperatures ranging from 18 to 150 °C, imposing sliding velocities ranging from 0.1 to 10 μm/s. Preliminary findings and microstructures from the dry experiments were published by VERBERNE *et al.* (2013). Here, we provide an in-depth comparison of the frictional behaviour of natural limestone and calcite gouge, tested dry and in the presence of water, at temperatures of 18, 50, 100, and 150 °C. Our specific research questions are:

1. What is the effect of water on the frictional strength and slip stability of simulated calcite (-rich) fault gouges?
2. Are there differences in frictional strength and/or slip stability related to the rock composition, i.e. between gouges prepared from natural limestone and pure calcite (crushed Iceland Spar), and, if so, why?
3. On the basis of the microstructural characteristics, can we identify what deformation mechanisms control the frictional strength and slip stability of calcite(-rich) fault gouges, and, if so, can we establish microstructural indicators for unstable frictional slip?

Answering these questions will provide more insight into the microphysical processes controlling frictional sliding of calcite fault gouges, and on the role of limestone faults in the seismic cycle.

2. Methodology

2.1. Sample Materials

We used simulated fault gouges prepared from natural limestone from the Xujiahe (XJ) formation, Wenchuan, China (as used by VERBERNE *et al.* 2010) and from crushed single crystals of calcite (Iceland Spar). The XJ Formation is cut by a branch of the Longmenshan fault zone that experienced surface

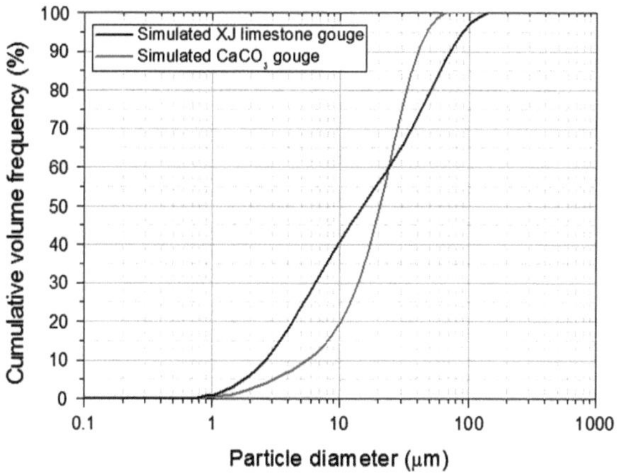

Figure 1
Initial grain size distributions of simulated Xujiahe (XJ) limestone and pure $CaCO_3$ gouge (crushed Iceland Spar) used in our experiments (taken from Verberne *et al.* 2013). Measured using a Malvern laser particle sizer

rupture during the 2008 M_w 7.9 Wenchuan earthquake (Xu *et al.* 2009) in Sichuan, China. X-ray diffraction (XRD) analysis of the crushed XJ limestone showed that it consists of 95–96 % $CaCO_3$, ~2–3 % illite, and ~1–2 % (Verberne *et al.* 2010) quartz. XRD analysis of the crushed calcite revealed a mineralogy comprising ~98 % calcite, with <2 % dolomite. Grain size distributions (GSDs) of the crushed and sieved samples, measured using a Malvern laser particle sizer, showed that the simulated XJ limestone gouge used has a wider GSD ($1 < d \leq 100$ μm) compared with that of the crushed calcite ($1 < d \leq 65$ μm), but that the median particle size (d_{50}) for both gouges is similar (~15 to 20 μm) (Fig. 1).

2.2. Experimental Apparatus

All experiments were performed using a conventional triaxial deformation apparatus, comprising a self-compensated, constant volume pressure vessel plus an Instron electro-servo-controlled ram for axial loading. The confining medium was silicone-oil. Figure 2a shows a schematic cross-section of the pressure vessel.

When performing individual experiments, axial force is transmitted to the sample using a Ti-grade 5 piston, furnished with an internal differential variable reluctance transformer (DVRT), to allow for measurement of piston distortion and hence axial force, independent of seal friction (Fig. 2a). This internal force gauge has a full-scale of 100 kN, with a resolution of ~33 N. Displacement of the Instron ram was measured externally using a linear variable differential transformer (LVDT) with 100 mm full-scale and 5 μm resolution. Confining pressure was applied using a compressed-air–driven diaphragm pump. During testing, pressure was manually regulated with a 20 mL hand pump, which enabled control to within ~0.1 MPa of the desired value. In wet experiments, we used demineralized water to apply pore pressure, which was held constant to within ~0.05 MPa of the desired value (10 MPa typically) using a three-term (P.-I.-D.) electro-servo-mechanical pump. Heating was achieved using a Thermocoax internal furnace, and temperature controlled with a three-term CAL2300 industrial controller that enables control to within 0.1 °C of set-point temperature. The sample temperature was measured by a thermocouple penetrating the upper direct shear piston to a position close to the sample layer (Fig. 2a, b).

2.3. Direct Shear Testing Assembly

We ran our experiments using a so-called 69 piston assembly designed for direct shear testing in a triaxial deformation apparatus (Fig. 2b), as described by Logan *et al.* (1992), Samuelson and Spiers (2012) and Verberne *et al.* (2013). This testing geometry has the advantage over the conventional "saw-cut" assembly (e.g. Logan *et al.* 1979; Blanpied *et al.* 1995) that the normal stress acting on the sample layer is independent of shear displacement, and is at all times equal to the confining pressure (Samuelson and Spiers 2012). Moreover, in the set-up used here, sample extrusion at the gouge layer-ends is minimized.

Each test used a set of two direct shears or 69 pistons with individual dimensions of 35 mm in diameter by 70 mm total length, comprising a 35 × 49 mm, grooved piston face (in contact with the sample), a 35 × 9 mm smooth surface, plus a cylindrical 35 × 12 mm end-piece (Fig. 2b). In wet experiments, the piston face consisted of a sintered stainless steel porous frit, which ensured even

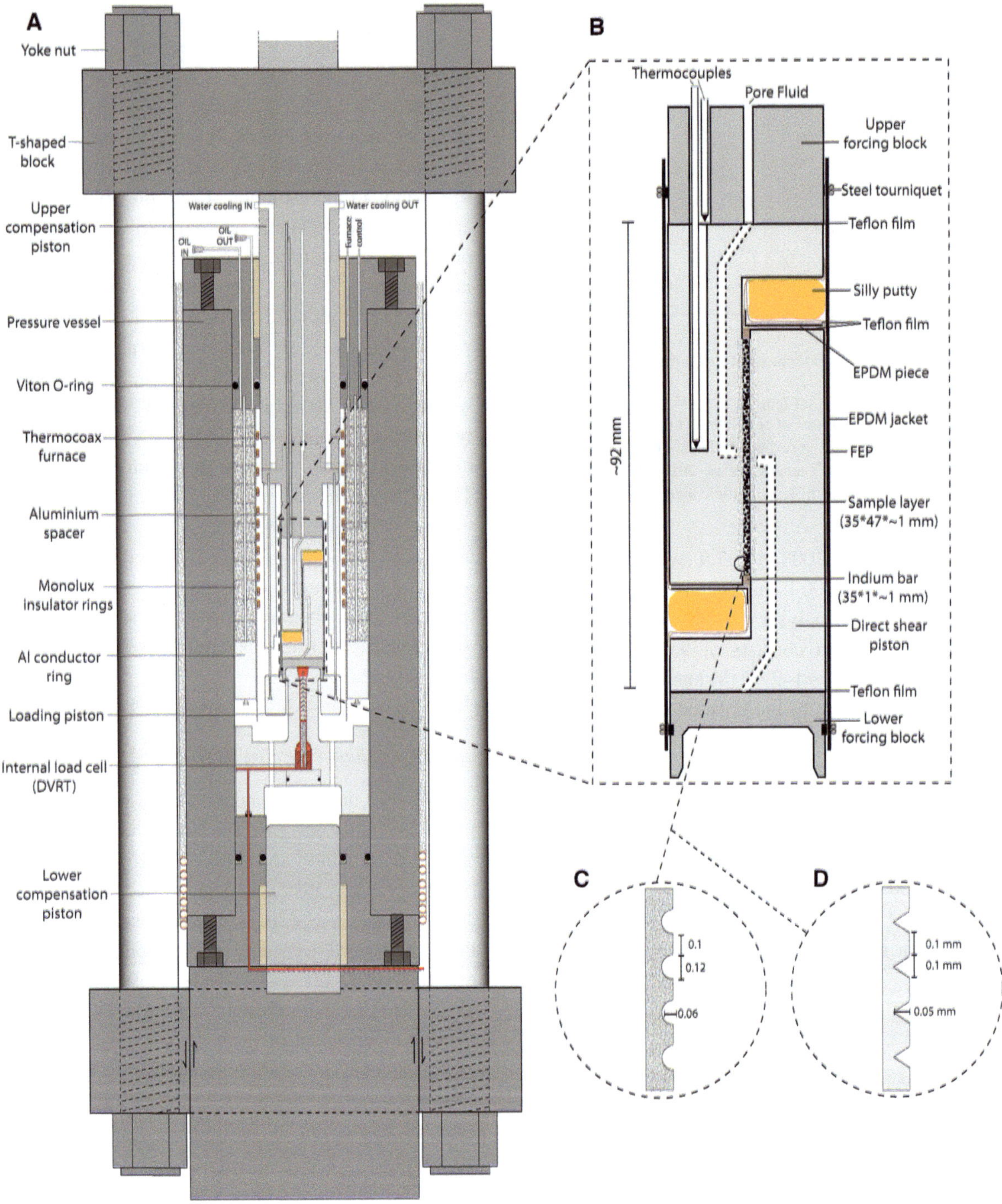

Figure 2
Schematic cross-sections of the testing apparatus (**a**) and direct shear assembly (**b**). Groove imprint on the sample-piston interface, as indicated for wet (**c**) and dry tests (**d**). Modified from Verberne *et al.* (2013)

distribution of the pore-fluid over the sample layer. The groove pattern spark-eroded in the porous frit comprised a set of 0.1 mm regularly spaced, half-circular grooves 0.12 mm in diameter, i.e. cutting 0.06 mm into the frit (Fig. 2c). For dry experiments, the piston face used supported a set of regularly spaced (0.1 mm), triangular grooves 0.1 mm wide and 0.05 mm deep (Fig. 2d).

In preparing each experiment, all powders were mixed with demineralized water in proportions of roughly 4:1 gouge-to-water, yielding a moldable paste. A purpose-built jig allowed for easy emplacement of a ~1.1 mm thick gouge paste layer onto the sample-piston interface, forming a plate measuring 35 mm wide by 47 mm long. For dry experiments, the sample was dried in an oven at ~115 °C for at least 4 h before starting the test. We confirmed dehydration of the sample by using a dummy sample-paste, and weighing before and after drying. To avoid contamination from atmospheric water vapour, the sample was assembled while the piston plus sample layer were still hot. All experiments used two Indium bars measuring 35 × 1 × 1 mm, placed at each end of the sample layer to avoid volume-loss of the gouge layer upon shearing (Fig. 2b). The displacement-accommodating voids characterizing the 69 direct shear assembly were filled with a pack of polytetrafluorethylene (PTFE), ethylene-propylene-diene monomer (EPDM) rubber, and classic white silicone putty (Fig. 2b). This design ensured that the silicone putty flowed outwards upon loading the sample assembly, offering zero resistance to the piston motion without contaminating the sample. The entire assembly, including the sample layer and the displacement-accommodating void-fills, was jacketed first in a thin fluorinated ethylene propylene (FEP) sleeve for support, and then in an EPDM rubber sleeve of ~1.4 mm thickness. The FEP sleeve was cut along the parts covering the displacement-accommodating voids. In order to reduce shear resistance between the pistons and the load-transmitting forcing blocks, circular PTFE sheets were placed onto the base of the direct shear pistons (Fig. 2b).

2.4. Experimental Procedure and Conditions

We ran our dry experiments using a confining pressure (P_c) of 50 MPa, while the wet experiments were conducted using $P_c = 60$ MPa and a pore fluid pressure of 10 MPa. Because the confining pressure equals the normal stress (σ_n) in the testing assembly used here (Fig. 2), this implies that for all experiments we used a nominal effective normal stress of 50 MPa. The sample temperatures used measured ~18, ~50, ~100, and ~150 °C. For every individual test, the jacketed sample assembly including gouge layer was loaded into the pressure vessel, which was then sealed with a closure nut and de-aerated before applying a confining pressure of ~20 MPa. For the wet experiments, the pore fluid system was filled with de-mineralized water and was subsequently de-aerated, followed by pressurization to the testing value of 10 MPa. The pore fluid may not have been fully saturated prior to each test; hence, we cannot exclude possible effects of differences in saturation states between or during our wet experiments. However, departures from saturation of the pore fluid over the duration of the experiments are likely to be small, due to the extremely fine grain size of our crushed samples (20–40 vol% <10 μm—Fig. 1), their high dislocation stored energy content due to crushing, and the further refinement of grain size and increase in mechanical damage that occurs during deformation. For experiments at elevated temperature, we switched on the internal furnace for heating to the desired test temperature. In the first minute after switching on the internal furnace, the confining pressure increased at a rate of ~1 to 2 MPa/s, followed by a slow approach (≪1 MPa/s) to a near-constant value. By intermittent bleeding of the vessel during heating, the confining pressure was at all times held below that used during the test. Equilibration of the testing apparatus with respect to pressure and temperature took from around 1 h at room temperature to 4 h at 150 °C. In some experiments at elevated temperatures, the testing apparatus was left to equilibrate overnight.

Upon starting an experiment, we used a ramp generator connected to a digital interface to apply the desired load-point displacement rate. The initial displacement rate (v) used in each test was 1 μm/s. After achieving sliding at a constant (average) strength, or else slip-weakening along a near-linear trend, the displacement rate was stepped in the sequence 0.1, 10, 1, 0.1, 10, and 1 μm/s, employing 0.2–1.0 mm displacement intervals in each step. In terminating each experiment, the axial loading ram was backed off at a rate of 10 μm/s. Upon complete removal of the axial load, the furnace was switched off in the case of experiments at elevated temperature, and the sample was depressurized to atmospheric conditions at a rate of <~0.5 MPa/s.

2.5. Data Processing and Analysis

All data were recorded at a sampling rate of 10 Hz using a 16-bit National Instruments DAQPad-6015 A/D converter. The axial force supported by the sample was calculated from the DVRT internal load signal (Fig. 2a), which was calibrated against an Instron (external) load cell with an accuracy of ±0.25 % on full scale (100 kN). The measured axial displacement was corrected for elastic machine distortions using calibrations obtained from compression tests on a steel dummy of known elastic properties. Calibrations were generated at or near the pressure and temperature conditions used in this study.

To calculate the shear stress (τ) supported by the sample layer, we divided the measured axial force by the load-bearing sample-area. However, in some experiments, the true load-bearing sample area decreased with increasing displacement due to small changes in the detailed deformation geometry. Correction for this effect proved unsuccessful due to subtle differences between each test. Therefore, to avoid over-correction and hence artifacts in the strength data, we assumed a constant load-bearing sample area for each experiment, i.e. $35 \times 47 = 1{,}645$ mm^2. In the experiments reported here, linear apparent weakening was negligible. From the ratio of the shear stress τ over the effective normal stress σ_n^{eff} we determined the apparent coefficient of sliding friction μ, defined

$$\mu = \frac{\tau - S_0}{\sigma_n - \alpha P_f} \approx \frac{\tau}{\sigma_n^{\text{eff}}} \quad (1)$$

To qualify as a friction coefficient, this widely used expression implicitly assumes that (1) the cohesion (S_0) of the sliding medium is negligible (e.g. Byerlee 1978), and (2) that α in Eq. (1) is equal to 1 (e.g. Morrow *et al.* 1992). To quantify the velocity dependence of strength, for each velocity step, i.e. including both upward and downward steps, we determined the ratio of the change in (steady-state) strength ($\mu_n \to \mu_{n+1}$) over the change in the logarithm of the sliding velocity ($\ln(v_n) \to \ln(v_{n+1})$), thus using the rate sensitivity parameter

$$(A - B) = \frac{\mu_{n+1} - \mu_n}{\ln(v_{n+1}) - \ln(v_n)} = \frac{\Delta\mu_{\text{step}}}{\Delta \ln(v)} \quad (2)$$

In the case of true steady-state frictional sliding before and after the velocity step, as was the case in most of our dry experiments, Eq. (2) yields the equivalent of the parameter (*a–b*) used in rate-and-state-friction modelling (Dieterich 1979; Ruina 1983; Marone 1998; Scholz 1998; see Fig. 3). However, some of our experiments exhibited unstable stick–slip, and in particular, the wet experiments sometimes exhibited (linear) slip hardening or weakening characteristic for a single velocity interval. In the case of stick–slips, we used the change in the maximum strengths reached at the corresponding sliding velocities (Fig. 3). In the case of linear slip hardening or weakening we corrected using non-progressive linear detrending of the data, i.e. characteristic for a single velocity interval (following Blanpied *et al.* 1995; Den Hartog *et al.* 2012) (Fig. 3). In this way, each value of (*A–B*) characterizes the change in strength associated with a single velocity step while assuming that linear weakening or hardening trends are independent of the rate-sensitivity of frictional strength. A positive value of (*A–B*) indicates stable velocity strengthening and a negative value indicates (potentially) unstable velocity weakening (see Scholz 1998).

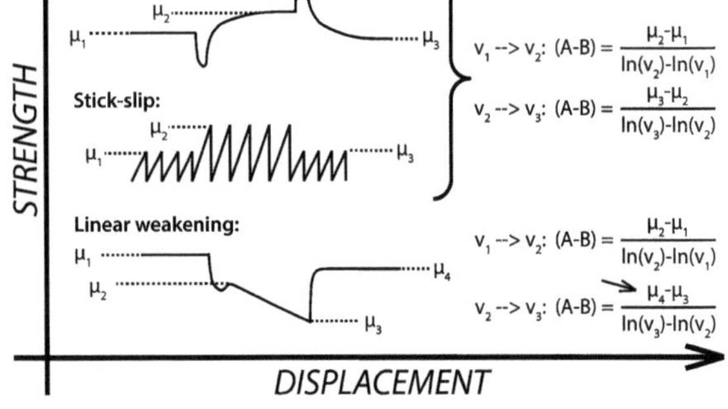

Figure 3
Method for determining the rate sensitivity (*A–B*) (Eq. 2) from velocity stepping data in our experiments

Table 1

Frictional strength data. The error to the absolute frictional strength μ is $\leq$0.5 %

Experiment	T (°C)	d_{fnl}^{a} (mm)	Gouge strength			Λ^{e} (%)	Type of motion[f]
			μ_{pk}^{b}	μ_{fnl}^{c}	$\Delta\mu^{d}$		
Simulated XJ limestone gouge							
XJ-RT-dry	17.8	1.05	0.71	0.70	0.01	–	Stable vs
XJ-50-dry	50.4	1.10	0.75	0.73	0.02	–	Stable vs
XJ-100-dry	103	0.90	0.79	0.78	0.01	–	Unstable vw (0.1); osc. vw (1.0); stable vw (10)
XJ-150-dry	150	1.00	0.79	0.78	0.01	–	Unstable vw
XJ-RT-wet	18.3	0.95	0.72	0.70	0.02	0	Stable vs
XJ-50-wet	52.0	1.00	0.71	0.66	0.05	10	Stable vs
XJ-100-wet	102	0.90	0.72	0.61	0.11	22	Unstable vw (0.1); stable vs. (1.0, 10)
XJ-150-wet	149	1.05	0.74	0.61	0.13	22	Unstable vw (0.1); osc. vw (1.0); stable versus (10)
Simulated pure $CaCO_3$ gouge							
$CaCO_3$-RT-dry	17.9	1.10	0.77	0.73	0.04	–	Stable vs
$CaCO_3$-50-dry	49.9	0.80	0.76	0.72	0.04	–	Stable vs
$CaCO_3$-100-dry	101	0.80	0.78	0.76	0.02	–	Unstable vw (0.1, 1.0); stable vs (1.0, 10)
$CaCO_3$-150-dry	150	0.95	0.80	0.78	0.02	–	Osc/unstable vw (0.1); stable vw (1.0, 10)
$CaCO_3$-RT-wet	18.6	0.80	0.69	0.59	0.10	19	Stable vs
$CaCO_3$-50-wet	50.6	0.80	0.71	0.60	0.11	17	Stable vs
$CaCO_3$-100-wet	99.6	0.75	0.71	0.58	0.13	24	Stable vs
$CaCO_3$-150-wet	149	0.70	0.74	0.57	0.17	27	Unst. vw (0.1); stable versus (1.0); stable vw (10)

vs velocity strengthening, *vw* velocity weakening, *osc* oscillations

[a] d_{fnl} = thickness of the sample layer after shear deformation

[b] μ_{pk} = peak strength (v = 1 μm/s). In the case of stick–slip, we picked the failure strength

[c] μ_{fnl} = near-steady-state final strength (x > 4 mm, v = 1 μm/s). In the case of stick–slip, we picked the failure strength

[d] $\Delta\mu = \mu_{pk} - \mu_{fnl}$

[e] $\Lambda = [\mu_{fnl}(dry) - \mu_{fnl}(wet)]/\mu_{fnl}(dry) \times 100$ %, i.e. the relative decrease of μ_{fnl} in wet experiments compared with dry tests

[f] Slip stability and velocity dependence. In parentheses the reference velocity in μm/s

2.6. Sample Recovery and Microstructure Analyses

After each experiment, the sample assembly was unloaded from the pressure vessel and dismantled. Fragments of some sheared samples were used for investigation using an incident light microscope, but for most experiments, the complete sheared gouge layer was recovered, and subsequently impregnated with Araldite 2020 epoxy resin. The completeness of sample recovery depended strongly on the splitting fabric that developed upon separation of the direct shear pistons. Especially for the wet experiments, this often prevented preservation of the complete microstructure. After hardening for several days, the samples were cut in an orientation normal to the shear plane and parallel to the shear direction. Selected samples were also cut in an orientation normal to the shear plane and normal to the shear direction. The cut sections were used to prepare ultrathin sections (~5 to 10 μm thickness).

All transmitted and incident light analyses were performed using a Leica DMR polarizing light microscope. Selected samples were analysed in more detail using a Universal (U-) stage mounted on a Leitz Wetzlar microscope, using a FEI Nova Nanolab 600 focused ion beam-scanning electron microscope (FIB-SEM), and using a Technai 20F transmission electron microscope (TEM). FIB–SEM cross-sections were normally imaged in BackScatter Electron (BSE) mode. We used FIB–SEM to prepare electron transparent foils for TEM study, i.e. measuring ~30 μm wide by ~15 μm deep by ~50 to 100 nm thick. Further details on the electron microscope methods used are described by Verberne *et al.* (2013).

3. Mechanical Results

3.1. Frictional Strength Data

A list of the experiments performed and the key frictional strength parameters obtained is given in Table 1. We plot the apparent coefficient of sliding friction (Eq. 1) against displacement (x) for all tests in Fig. 4. All experiments showed rapid, near-linear loading in the first $x \approx 0.3$–0.5 mm of displacement, followed by apparent yield and subsequent hardening towards a peak frictional strength (μ_{pk}) attained at $x \approx 1$–2 mm. After reaching the peak strength, sliding continued at a near-constant average strength in the dry experiments, while the wet experiments showed gradual slip weakening (except experiment XJ–RT-wet). Experiment XJ–RT-wet showed a similar frictional strength evolution with displacement as the dry experiments. Our observations on wet simulated XJ limestone gouge are consistent with those reported by Verberne *et al.* (2010) on the same material, sheared wet under similar testing conditions of effective normal stress and temperature. Although some tests did not reach steady-state at maximum displacements (notably XJ-50-wet and $CaCO_3$-RT-wet), in all experiments a near-steady-state final value of the frictional strength (μ_{fnl}) can be defined for $x \geq 4$ mm (Table 1; Fig. 4). Comparison of μ_{fnl} between tests will provide insight into the magnitude of displacement weakening with increasing temperature.

In response to velocity stepping, all experiments showed stable sliding at $T \approx 18$–50 °C, at all sliding velocities at all displacements (Fig. 4; Table 1). In experiments performed at $T \geq 100$ °C, we observed velocity weakening behaviour for all dry experiments plus for wet experiments using simulated XJ limestone gouge. Wet experiments using simulated pure $CaCO_3$ gouge showed stable velocity strengthening at all sliding velocities for $T \approx 18$–100 °C, but unstable velocity weakening at $T \approx 150$ °C at $v = 0.1$ µm/s. In general, experiments showing velocity weakening first exhibited stick–slips at the lowest sliding velocity ($v = 0.1$ µm/s), at the higher strains imposed (Fig. 4;

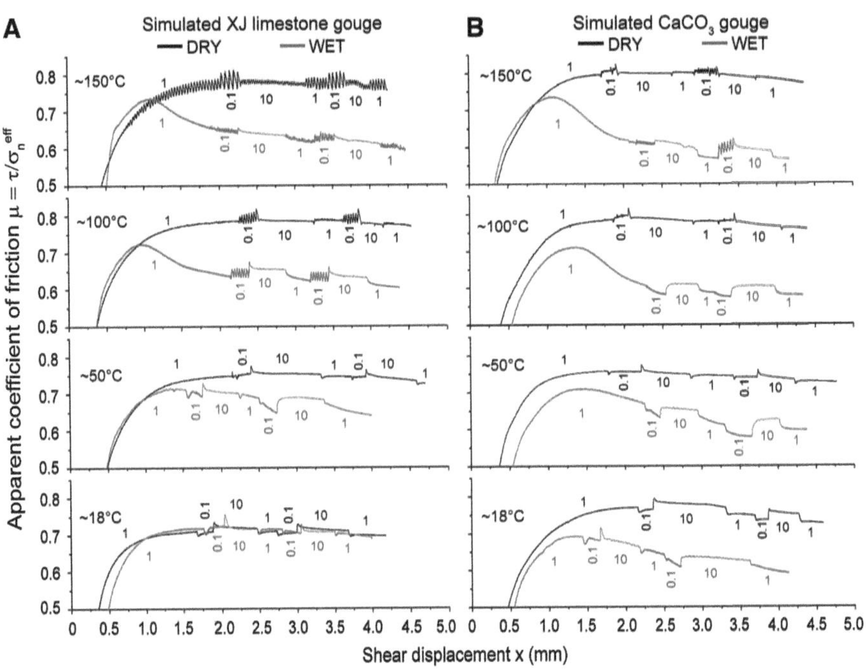

Figure 4
Plots of the apparent coefficient of sliding friction (μ = shear strength/effective normal stress = τ/σ_n^{eff}) against shear displacement (x) for dry (*black lines*) and wet (*grey lines*) experiments using simulated Xujiahe (XJ) limestone (**a**) and pure $CaCO_3$ gouge (**b**), at temperatures of ~18, ~50, ~100, and ~150 °C, as indicated. All experiments used an initial shear displacement rate of 1 µm/s, which was then stepped following the velocity stepping scheme indicated under the curves (in µm/s). A list of the experiments performed and key frictional strength parameters obtained is given in Table 1

Table 1). Experiment XJ-150-dry showed spontaneous stick-slip motion at $v = 1$ µm/s at $x \approx 0.7$ mm (Fig. 4). In this experiment, in the displacement interval $x \approx 0.7$–1.9 mm, the frictional strength drop associated with the 'slip'-phase increased with increasing displacement, before reaching a near-constant maximum or 'peak' strength of $\mu_{pk} \approx 0.79$ at $x \approx 1.9$ mm. Stick–slip was observed at all subsequent sliding velocities imposed, with the frictional strength drop consistently larger at lower sliding velocities. In experiment XJ-150-wet, gradual displacement weakening was followed by spontaneous nucleation of stick–slips at $v = 1$ µm/s at $x \approx 1.7$ mm.

The peak and near-steady-state values of the apparent coefficient of friction determined at $v = 1$ µm/s, and the finite difference $\Delta\mu = \mu_{pk} - \mu_{fnl}$, are plotted against temperature in Fig. 5. For both samples tested, dry and wet, the enveloping surfaces defined by the values of μ_{pk} showed a weakly increasing trend with increasing temperature (Fig. 5a). However, for the near-steady-state strengths, we found a (weakly) increasing trend with increasing temperature for the dry experiments, but a decreasing trend with increasing temperature for the wet experiments, especially for tests using simulated XJ limestone gouge (Fig. 5b). Thus, for all dry experiments, $\Delta\mu < 0.05$ (Table 1), while for the wet experiments, $\Delta\mu$ increased with increasing temperature (Fig. 5c). The decrease in the near-steady-state strength of the wet gouges relative to that of the dry gouges, i.e. $\Lambda = [\mu_{fnl}(\text{dry}) - \mu_{fnl}(\text{wet})]/\mu_{fnl}(\text{dry}) \times 100$ %, at corresponding testing temperatures, measured $\Lambda = 0$ % at 18 °C, 10 % at 50 °C, and 22 % at 100 and 150 °C for the simulated XJ limestone gouge, and $\Lambda = 19$ % at 18 °C, 17 % at 50 °C, 24 % at 100 °C, and 27 % at 150 °C for the simulated pure $CaCO_3$ gouge (Table 1).

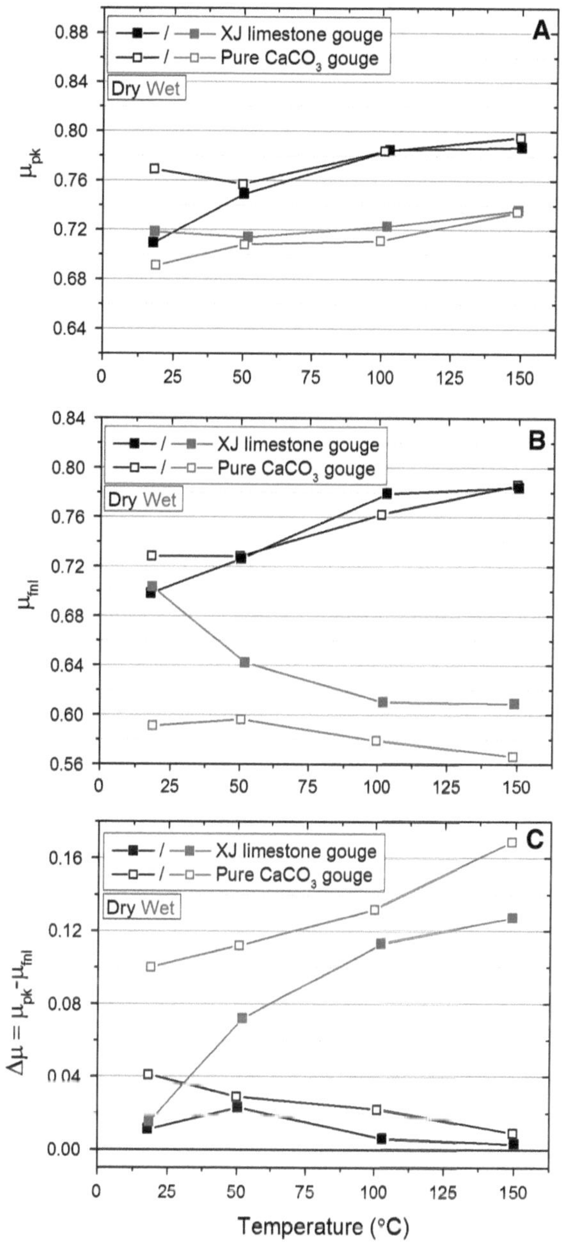

Figure 5
Plots of frictional strength picked at $v = 1$ µm/s, for each test, against temperature. *Black lines* show results from dry experiments, *grey lines* those from wet experiments. The relative error to the frictional strength μ is ≤0.5 %. **a** Peak frictional strength (μ_{pk}). **b** Near-steady-state final ($x > 4$ mm) strength (μ_{fnl}). **c** The finite difference $\Delta\mu = \mu_{pk} - \mu_{fnl}$

3.2. Velocity Dependence Data

A list of the rate sensitivity parameter values (*A–B*) determined for all experiments, at each velocity step applied, is given in Table 2. All values of (*A–B*) are plotted against temperature in Fig. 6. The scatter seen between the data points at each temperature, i.e. for each experiment, indicates the spread in the values of (*A–B*) with increasing displacement and/or type of velocity step (up-step or down-step) in the experiment (see also Table 2). For experiments showing velocity weakening, minima in the values of (*A–B*) are typically associated with strength data exhibiting stick-slip

Table 2

Rate sensitivity parameter (A–B) determined for each velocity step (1–6), with the reference sliding velocities as indicated in μm/s. The error to each value of (A–B) is ~±0.0017

Experiment	T (°C)	$(A–B)$ (see Eq. 2)					
Simulated XJ limestone gouge		Step 1 1 → 0.1	Step 2 0.1 → 10	Step 3 10 → 1	Step 4 1 → 0.1	Step 5 0.1 → 10	Step 6 10 → 1
XJ-RT-dry	17.8	0.0013	0.0022	0.0039	0.0017	0.0022	0.0052
XJ-50-dry	50.4	−0.0022	0.0000	0.0022	0.0004	−0.0002	0.0035
XJ-100-dry	103	−0.0091	−0.0048	−0.0017	−0.0074	−0.0054	−0.0022
XJ-150-dry	150	−0.0104	−0.0061	−0.0052	−0.0065	−0.0046	−0.0061
XJ-RT-wet	18.3	−0.0013	−0.0007	0.0004	0.0017	0.0011	0.0022
XJ-50-wet	52.0	0.0022	0.0037	0.0013	0.0061	0.0091	0.0048
XJ-100-wet	102	−0.0087	0.0002	0.0061	−0.0100	−0.0013	0.0056
XJ-150-wet	149	−0.0009	−0.0009	−0.0013	−0.0061	−0.0017	0.0000
Simulated pure $CaCO_3$ gouge							
$CaCO_3$-RT-dry	17.9	0.0026	0.0037	0.0083	0.0052	0.0033	0.0096
$CaCO_3$-50-dry	49.9	0.0009	−0.0009	0.0013	0.0017	−0.0002	0.0022
$CaCO_3$-100-dry	101	−0.0052	−0.0017	0.0004	−0.0009	−0.0026	0.0009
$CaCO_3$-150-dry	150	−0.0065	−0.0026	−0.0013	−0.0039	−0.0039	−0.0017
$CaCO_3$-RT-wet	18.6	0.0013	0.0009	0.0035	0.0056	0.0059	0.0056
$CaCO_3$-50-wet	50.6	0.0074	0.0056	0.0069	0.0087	0.0093	0.0117
$CaCO_3$-100-wet	99.6	0.0039	0.0063	0.0083	0.0026	0.0063	0.0083
$CaCO_3$-150-wet	149	−0.0013	0.0015	0.0078	−0.0161	−0.0037	0.0078

(Tables 1, 2). For both materials tested and for all velocity steps, the dry experiments showed a near-linear decrease of $(A–B)$ with increasing temperature (Fig. 6a, b). For dry experiments using simulated XJ limestone gouge, linear interpolation of the mean values of $(A–B)$, i.e. averaged over the individual velocity steps, showed that a transition from positive to negative $(A–B)$ occurs at $T \approx 70$ °C. Similarly, for dry experiments using simulated pure $CaCO_3$ gouge this transition occurs at $T \approx 80$ °C. Compared with the dry experiments, the wet experiments showed much more scatter in the values of $(A–B)$ at each temperature. Nonetheless, the results showed that the values of $(A–B)$ first increased with increasing temperature up to $T \approx 50$ °C, and then decreased toward $T \approx 150$ °C. Negative values of $(A–B)$ were first observed at $v = 0.1$ μm/s, at $T \approx 100$ °C for wet simulated XJ limestone gouge, and at $T \approx 150$ °C for wet simulated pure $CaCO_3$ gouge (Fig. 6a, b).

4. *Microstructures*

4.1. *Inspection of Samples Upon Retrieval*

When separating the 69 piston blocks to expose the sheared gouge within, all samples split along a shear-induced fabric developed near the sample-piston interfaces, as well as obliquely to this fabric. Split surfaces that developed parallel to the shear plane displayed shiny, elongate patches, measuring ~0.1 to 0.5 mm wide and up to 3 mm long, aligned with their long axis parallel to the shear direction (Fig. 7a). Incident light microscopy revealed these shiny patches to be striated at the sub-μm scale, but topographically flat at the μm–mm scales (Fig. 7b). While the shiny surfaces were mainly exposed on planes that split parallel to the shear plane, they were also observed on obliquely split planes. In the former case, they are located directly beneath the imprint of the piston groove pattern (Fig. 7a, c). The total number and dimensions of individual shiny surfaces exposed varied from experiment to experiment. However, this variability strongly depended on the splitting fabric produced upon sample recovery, and was therefore not investigated further in this study.

4.2. *Optical Analysis of Thin Sections*

From thin sections prepared parallel to the shear direction and normal to the shear plane, transmitted light optical analysis revealed that all sheared gouge microstructures developed horizontal and inclined shear bands, comprising mainly boundary and R_1 shears and, much less common, P-shears, Y-shears

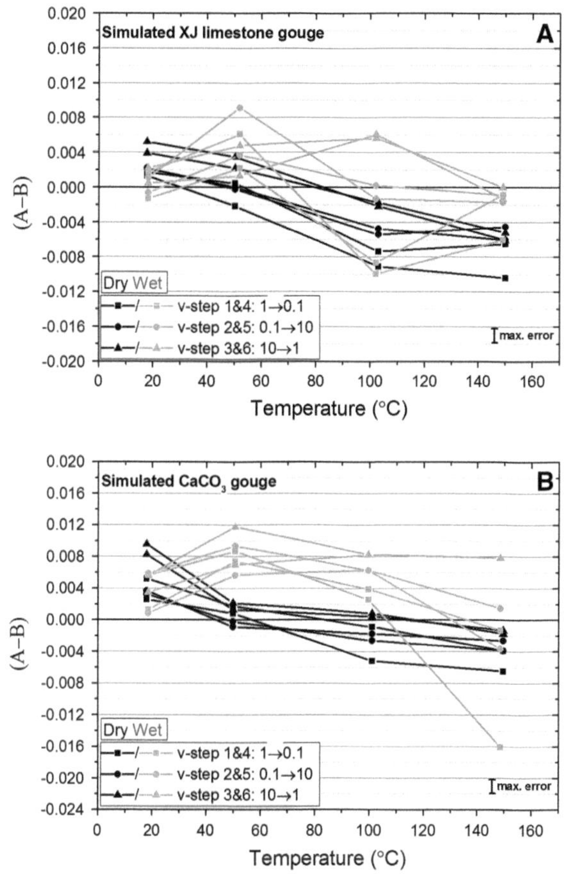

Figure 6
Plots of the rate sensitivity parameter *(A–B)* (Eq. 2) determined for each velocity step, against temperature, from experiments using simulated XJ limestone (**a**) and pure $CaCO_3$ gouge (**b**). The spread between the data points (*grey* or *black*) at each temperature indicates the spread in the values of *(A–B)* with increasing displacement and/or type of velocity step (up-step or down-step) in the experiment. *Standard error* propagation shows that the error for each value of *(A–B)* is $\sim\pm 0.0017$, as indicated in each plot. A list of all *(A–B)* values is given in Table 2

and R_2-shears (terminology following Logan *et al.* 1979; see also Figs. S1–2 and Verberne *et al.* 2013). In all samples, shear bands are seen as sharply defined zones of strongly reduced grain size cutting the gouge matrix (Fig. 8a, b). The maximum grain size in the matrix is similar to the median grain size of the starting material ($\sim$15 to 20 μm—cf., Fig. 1). Deformation twins are common in the larger matrix grains. Close to a shear band the grain size abruptly decreases, and the larger grains are often seen truncated, and sometimes bent, in a sense consistent with the imposed shear direction (Fig. 8c, d). Fractured grains seen close to or within a shear band are often clustered, reflecting the shape of the parent grain. The mean shear band width is $\sim$30 to 40 μm, but varies from $\sim$10 to $\sim$150 μm. Boundary shears developed in all samples, and are continuous and straight, located immediately adjacent to the grooved surface of at least one of the sample-piston interfaces (Figs. 8a, b; S1–2). The wider parts of a boundary shear have an internal structure characterized by an anastamosing pattern of shears parallel and at an angle to the shear plane (Fig. 8a). Some of these segments are seen to emanate from the piston groove apices at an angle $\sim$10° to 20° to the shear plane, to connect with the anastamosing structure underneath. Inclined shear bands (R_1-shears, R_2-shears, P-shears) consist of bands of comminuted grains cross-cutting the gouge matrix. They are often seen to interfere with each other, and with horizontal shears, to form branching patterns (Figs. S1–2). The angle between R_1-shears and the shear plane is usually largest ($\sim$18° to 24°) in the centre of the gouge, gradually decreasing to zero nearing the sample-piston interface, where they merge with the boundary shear. The only systematic difference between dry and wet sheared gouge microstructures was found in the number of (#) of R_1-shears normalized to the length of the recovered gouge sample (*l*). For the dry experiments using simulated $CaCO_3$ gouge, the ratio $\#R_1/l$ measured $\sim 0.6\ mm^{-1}$ for all temperatures tested, while for the wet experiments $\#R_1/l \approx 0.8\ mm^{-1}$ at $T \approx 18$ and 100 °C, but $0.1\ mm^{-1}$ at 150 °C (Figs. S1–2). However, the value of $\#R_1/l$ may depend strongly on sample recovery, and on the location in the sheared gouge of the cross-section chosen for thin section preparation.

For all sheared gouge microstructures, for both inclined and boundary shears, we observed that grains in the shear band core cannot be distinguished with an optical light microscope; hence, they are < 1 μm in size. When using crossed nicols, these aggregates of grains show near-uniform optical interference, i.e. blue or orange when using the gypsum plate (Figs. 8a, b; S1–2). Rotation of the microscope stage over 45° causes the interference color to change near uniformly, i.e. from blue to orange or vice versa, over the entire width of the

Figure 7
a Patchy shiny surfaces marking the shear plane in sample $CaCO_3$-50-dry. **b**, **c** Incident light micrographs of shiny patches found in sample $CaCO_3$-RT-dry

shear band core. This suggests a distinct crystallographic preferred orientation (CPO) of the sub-micron grains present here. To investigate this in more detail, we used a U-stage to determine the average orientation of the optical indicatrix of well-exposed segments of boundary shears found in sample $CaCO_3$-150-dry. From roughly square-shaped shear band segments with sides measuring ~50 to 100 μm in length (Fig. 9a, b), we determined the angles at which > 90–100 % of the grains present here went into optical extinction. In this way we measured the average optic axis of shear band segments, and hence the average calcite *c* axes orientations (calcite is uniaxial optically negative). We performed this analysis on sample $CaCO_3$-150-dry, on thin sections prepared in a direction normal to the shear plane, and normal (Fig. 9a) and parallel (Fig. 9b) to the shear direction, i.e. from mutually perpendicular sections of boundary shears developed in the same sample. All optic axis data were plotted in a lower hemisphere equal-area (Schmidt) stereographic projection, and then rotated to a common orientation using the shear plane and shear direction as reference (Fig. 9c). The results revealed a strong clustering of all optic axes hence *c* axes, at 120°–130° to the shear direction, i.e. at 30°–40° to the shear plane normal.

4.3. Electron Microscope Analysis

To determine the internal sub-micron structure of the shear bands, we used FIB–SEM to mill cross-sections of boundary shear bands formed in experiments $CaCO_3$-RT-dry (Fig. 10a) and $CaCO_3$-150-dry

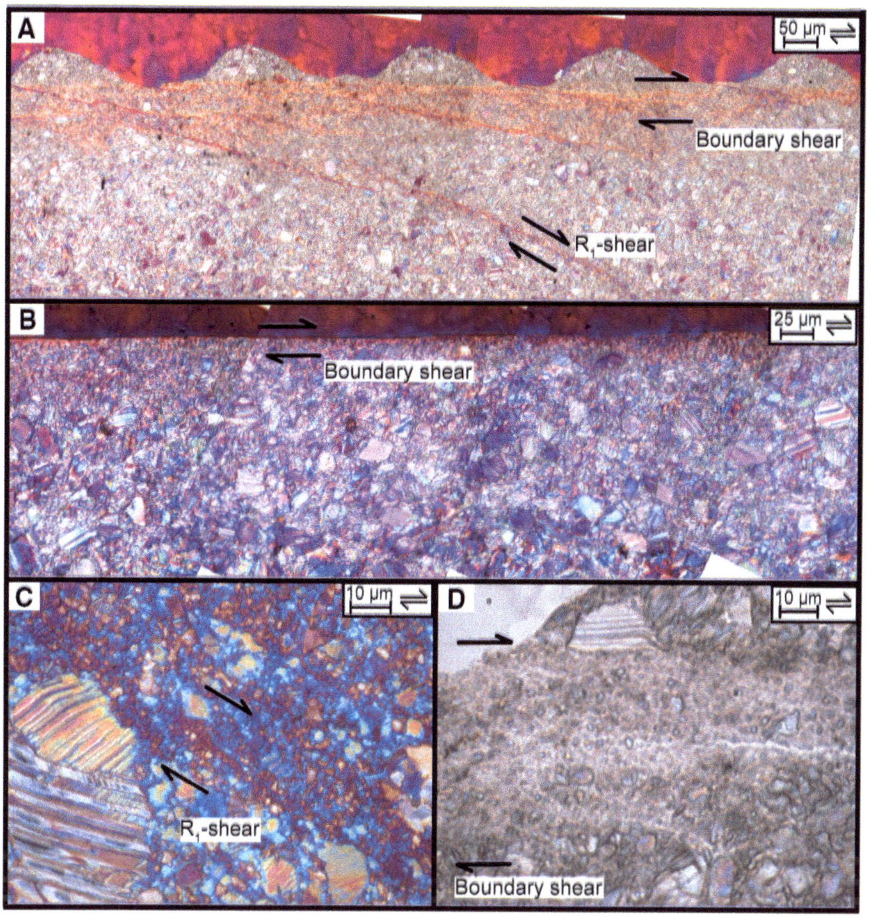

Figure 8
Transmitted light micrographs of thin sections prepared from sheared gouge samples taken using a gypsum plate inserted in a polarizing light microscope. A list of all experiments is given in Table 1. **a** Sample $CaCO_3$-RT-dry, taken from VERBERNE *et al.* (2013). **b** Sample $CaCO_3$-150-wet. **c** Sample XJ-150-dry. **d** Sample $CaCO_3$-100-dry. Complete photomosaics of micrographs of sheared simulated $CaCO_3$ gouge, from dry and wet experiments performed at $T \approx 18$, 100, and 150 °C, are given in supplement S1 and S2

(Fig. 10b). In this way, we obtained boundary shear cross-sections in an orientation normal to the shear plane and the shear direction, for both velocity strengthening and velocity weakening samples. Regardless of testing conditions, the shear band cross-sections showed a localized structure with grains decreasing in size from ~10 to 20 μm at the outer margins to < 1 μm in the shear band core (Fig. 10a, b). Coarser grains (0.3–20 μm in size) at the margins, and those present in the shear band core, having survived grain size reduction, are angular, whereas the finer particles (< 300 nm) in the shear band core are rounded and clustered in dense aggregates. From sample $CaCO_3$-150-dry, we also prepared cross-sections of boundary shear bands parallel to the shear direction and normal to the shear plane (Fig. 10c). This revealed an alignment of porosity and a shape preferred orientation (SPO) of elongate particles ~0.3 to 2 μm in length, oriented at 140°–150° to the shear direction (Fig. 10c). Based on all boundary shear cross-sections investigated, the width of the shear band core, i.e. the zone containing dense, sub-micron-sized particle aggregates, is estimated to range from 1 to 10 μm.

All samples involved dilation of the shear bands resulting from sample depressurization and handling, e.g. seen as the resin-filled crack splitting the boundary shear band core (Fig. 10). To determine the orientations of the observed CPO and SPO in the shear bands with respect to the shear plane during

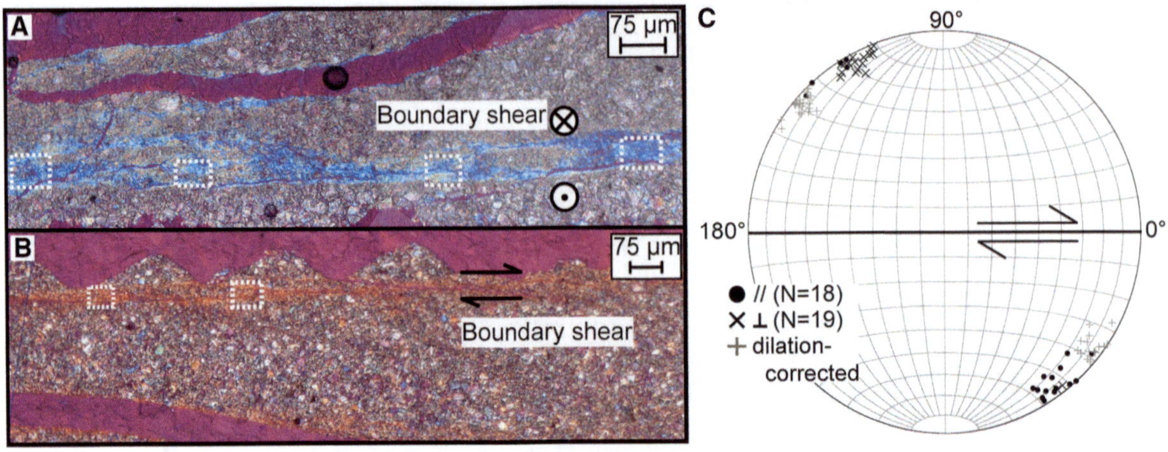

Figure 9
Optic axis data from shear band segments collected with U-stage. **a** Sample $CaCO_3$-150-dry, normal to the shear plane and normal to the shear direction. **b** Sample $CaCO_3$-150-dry, normal to the shear plane and parallel to the shear direction. *Dashed boxes* in **a** and **b** show examples of shear band segments used for measuring the average optic axis. **c** Average optic axes from **a** to **b** plotted in a lower hemisphere Schmidt net projection, with the shear plane and shear direction oriented E–W. Also plotted are the optic axis data corrected for post-test sample dilation

deformation, this must be corrected for. Using the linear intercept method and Fig. 10c, the strain perpendicular to the shear zone resulting from post-shear dilation was estimated to be ~23 %. Taking zero pre-dilation porosity, and using the above value for strain, a (maximum) estimate of the orientation of the shear band c-axes and SPO yields 130°–135° (Fig. 9c) and 150°–155°, respectively.

To further investigate the shear band core nano-structure and crystallography, we used FIB–SEM to prepare electron-transparent foils from boundary shear band cross-sections for study using TEM. The sections were selected on the basis of observations made using the light microscope, i.e. TEM foils were prepared from samples showing a strong preferred optical birefringence (see VERBERNE *et al.* 2013). Bright-field and high resolution TEM imaging showed that dense particle aggregates seen in the shear band cores of samples $CaCO_3$-RT-dry and $CaCO_3$-150-dry (Fig. 10) comprise nanocrystalline calcite grains down to ~5–20 nm in size (Fig. 11a–c). Selected-area electron diffraction patterns taken from the nanocrystallite aggregates showed ring-shaped diffraction patterns, with distinct arcs developed on the (104) reciprocal plane (hexagonal structural cell; Fig. 11d-inset), suggesting a polycrystalline aggregate with a CPO. Using any of the higher-intensity arcs to form a dark-field image, i.e. satisfying the individual Bragg condition, resulted in a preferred illumination of grains in the shear band core (Fig. 11d, e). This suggests a preferred alignment of the (104) calcite (*r*-)planes parallel to the shear plane.

5. Discussion

The frictional strength and velocity dependence data obtained in this study showed important differences between experiments at different temperatures, but also between dry and wet experiments, and between experiments using simulated XJ limestone gouge and those using simulated pure $CaCO_3$ gouge (Figs. 4, 5, 6). Although the different testing materials showed contrasts in frictional behaviour, wet and dry, all sheared gouge microstructures showed essentially the same shear-induced features, i.e. localization into boundary and inclined shear bands, characterized by cataclasis and grain size reduction, plus a CPO (Figs. 8, 9; S1–2). In both velocity strengthening and weakening samples, FIB–SEM cross-sections of boundary shear bands showed a localized internal structure with grains decreasing from 10 to 20 μm in size in the outer margins to dense nanoparticle aggregates in the shear band core (Fig. 10). TEM showed that these aggregates consist of nanocrystallites down to ~5 to 20 nm in size (Fig. 11). In the following, we focus on the

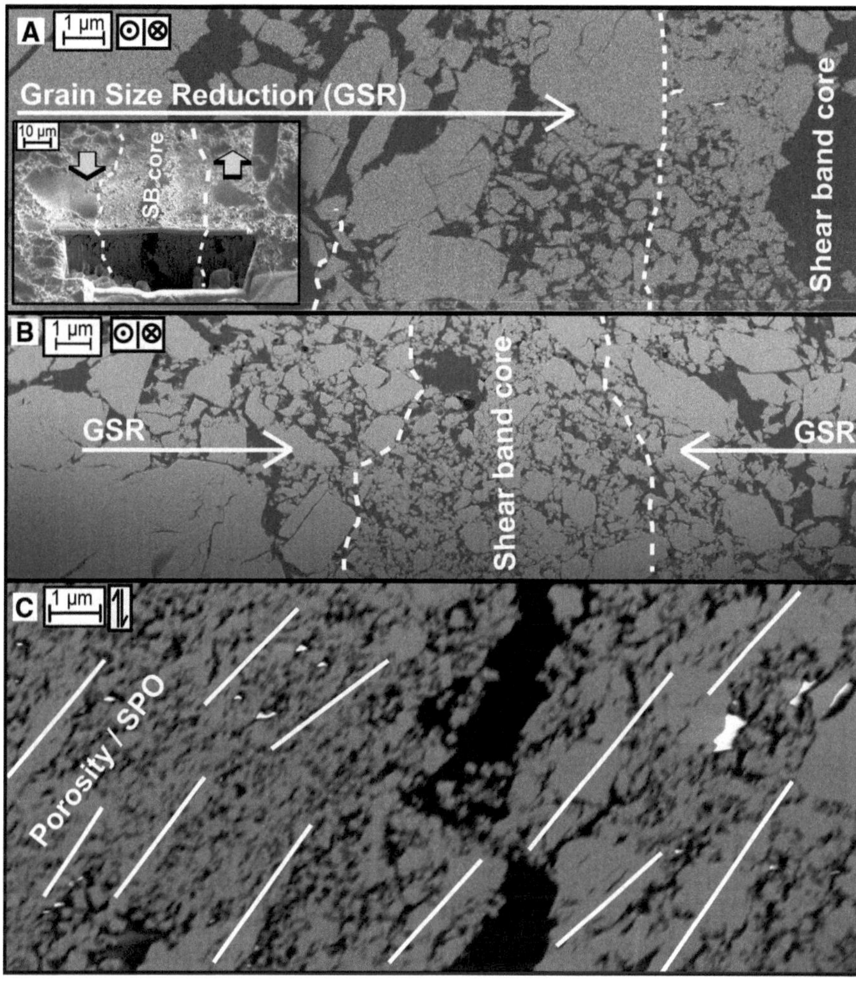

Figure 10
BSE micrographs of cross-sections of boundary shear bands prepared using FIB–SEM. *Black* regions are resin-filled porosity, mainly due to post-test dilation. **a** Sample $CaCO_3$-RT-dry, normal to the shear direction. Inset shows the FIB-cut on a smaller scale. **b** Sample $CaCO_3$-150-dry, normal to the shear direction. *Dashed lines* in **a** and **b** highlight the shear band core. **c** Sample $CaCO_3$-150-dry, parallel to the shear direction. The alignment of porosity and SPO as indicated. **a** and **c** are modified from Verberne *et al.* (2013)

mechanisms controlling the frictional strength and velocity dependence in simulated calcite fault gouges, and discuss the possible origin of the contrasts in the frictional behaviours observed, especially with regard to the role of water. We also discuss the implications of our results in the context of natural faulting in tectonically active carbonate terrains.

5.1. Microphysical Mechanisms Operating in Dry and Wet Samples

From the extreme grain size reduction seen in all sheared gouge microstructures, and the fractured grains and angular shapes of larger grains surviving in and at the margins of shear bands (Figs. 8, 9, 10), we infer that cataclasis must have played a role in all our experiments. However, the finer grains within the shear band cores are rounded rather than angular, and cataclastic flow alone is not expected to produce the c-axis preferred orientation observed here (Figs. 9, 10, 11). Rather, our observations of the preferred alignment of calcite r-planes in the shear band core parallel to the shear plane (Fig. 11) suggests a role of r-glide crystal plasticity in the shear bands. Together with e-twinning, glide on r is the dominant crystal plastic (dislocation) deformation mechanism in

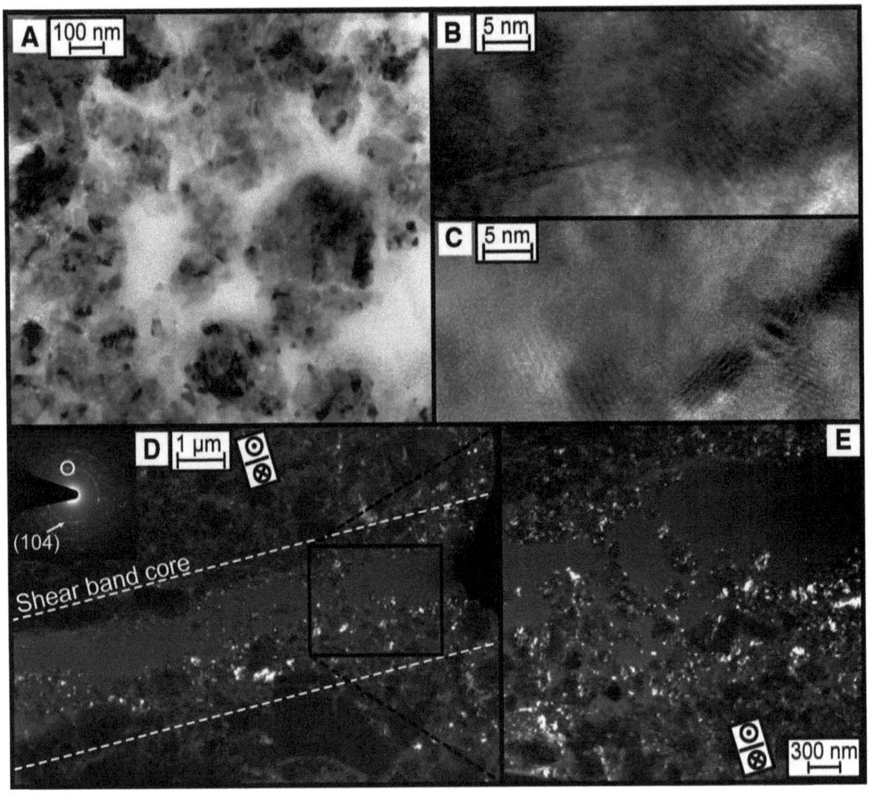

Figure 11
TEM micrographs of foils prepared from boundary shear bands cross-sections in an orientation normal to the shear plane and shear direction. **a** Bright-field image of sub-micron particle aggregates found in the boundary shear band core of sample $CaCO_3$-150-dry. HRTEM images from rounded sub-micron particles in the shear band cores of samples $CaCO_3$-RT-dry (**b**) and $CaCO_3$-150-dry (**c**). **d** *Dark*-field image of the shear band core of sample $CaCO_3$-150-dry, generated using a tilt angle corresponding with the (104) (r-) plane of calcite (hexagonal structural cell). *Inset* shows a ring-shaped selected-area diffraction pattern taken from a nanoparticle aggregate, with distinct arcs representing (104) planes. **e** Magnification of part of **d**. The apparent preferred illumination of nanograins in this image is enhanced due to better resolution at higher magnification. **a**, **c–e** taken from Verberne *et al.* (2013)

calcite at low temperatures (De Bresser and Spiers 1997). We investigated the role of r-glide and e-twinning in our experiments by postulating that the e- or r-planes lie parallel to the shear plane, and by using a stereographic projection to compare the predicted angles to the c-axis with our measured bulk c-axis preferred orientation (Fig. 12). From the known crystal structure of calcite (e.g. De Bresser and Spiers 1993, 1997), the conic section (small circle) that describes the angles between the hypothetical calcite *c* axes and the e- (108) or r- (104) planes predicts angles of 116° and 135° to the shear direction, respectively (Fig. 12). The dilation-corrected c-axes orientation in the shear bands measured ~130° to 135° to the shear direction, i.e. consistent with that of a projected c-axis orientation when the r-plane lies parallel to the shear plane (Fig. 12). Together with our dark-field micrographs from TEM (Fig. 11d, e), this convincingly shows that the r-glide system $(104)[\bar{2}01]$ is aligned (sub-)parallel to the shear plane and direction, suggesting that r-glide crystal plasticity has played a role in the shear band system. Because we found a CPO in all our samples—that is, because all samples showed essentially the same sheared-induced microstructures and nanostructures—we infer that r-glide plasticity played a role in all our experiments, wet and dry, including those sheared at room temperature.

Although r-glide may account for the CPO in the shear band core, and accommodated at least part of the strain here, it does not explain the alignment of nanopores and the SPO defined by angular grains

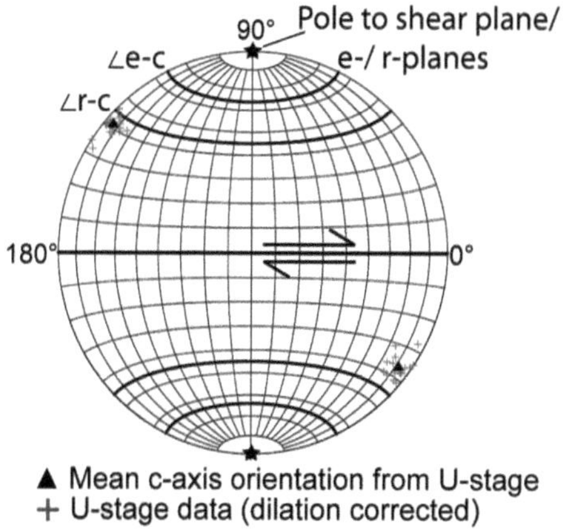

Figure 12
Lower hemisphere Schmidt net projection representing the shear plane and shear direction E–W. Assuming the e-plane or r-plane lies parallel to the shear plane, the conic sections (*small circles*) plot the possible angles (∠) of (the pole of) the shear plane to the calcite *c* axis. Also plotted are the optic axis data (*c* axes) from our samples (Fig. 9), corrected for post-test dilation assuming zero shear band porosity

(>0.3–2 μm) as seen in FIB sections cut normal to the shear plane and parallel to the shear direction (Fig. 10c). The dilation-corrected angle of ~150 to 155° observed between the SPO/nanopore alignment and the shear direction is roughly consistent with an R_1-/T-shear band configuration within the boundary shear (see LOGAN *et al.* 1979). However, we cannot discriminate whether the nanopore alignment reflects slip on R_1 -shears that dilated after sample extraction, or else elongated dilatational cavities formed at grain boundaries lying roughly parallel to the incremental shortening direction, implying some type of grain boundary sliding of the nanoparticles within the shear bands (cf. MA 2004; SHAN *et al.* 2004).

Returning to the origin of the CPO, which would seem inconsistent with the operation of significant grain boundary sliding, we note that the strain rates characterizing the shear bands in our experiments (10^{-2}–10^{1} s^{-1}) are similar to those needed for plasticity-induced formation of cellular equiaxed nanograins in experiments on metals (10^{0}–10^{3} s^{-1}) at room temperature (UMEMOTO 2003; LI *et al.* 2008; YANG *et al.* 2010). This supports crystal plasticity as the most likely explanation for the formation of both the nanograins and their CPO. However, we cannot exclude the possibility that after nanograin formation, some kind of orientation-dependent, diffusive sintering or grain growth process is responsible for the CPO seen in the nanocrystalline shear bands. Such a process could lead to a CPO, even during grain boundary sliding dominated flow (POWER and TULLIS 1989; ZHANG *et al.* 2010a). In the case of our dry experiments, reasonable measures were taken to avoid contamination by water prior to shear, but in view of the much lower (21 %) surface energy of hydrous versus anhydrous nanocrystalline calcite (FORBES *et al.* 2011), nanograins in the shear bands may have adsorbed atmospheric water during our experiments. The presence of such a thin water film (≤1 nm) can strongly facilitate surface-diffusion processes (RUTTER 1983; ZHANG *et al.* 2010b; KOELEMEIJER *et al.* 2012). Clearly, the microphysical processes controlling strain accommodation in the shear band system is as yet incompletely understood, and more research is needed to elucidate the underlying mechanisms of nanograin formation and deformation, and their role in controlling frictional strength and slip stability.

5.2. *Mechanisms Controlling the Velocity Dependence of Strength*

Both materials investigated showed a transition from velocity strengthening to velocity weakening slip with increasing temperature, regardless of whether or not (pressurized) pore fluid water was present. Because crystal plasticity is strongly thermally activated, we infer that this process is responsible for the observed decrease of (*A*–*B*) with increasing temperature (Fig. 6). Taking an (empirical) exponential creep law used to describe dislocation glide-controlled flow of calcite at 20–500 °C at laboratory strain rates, the plastic strain rate ($\dot{\varepsilon}$) depends on absolute temperature (*T*) at fixed stress (σ) via the relation (RUTTER 1974; SCHMID *et al.* 1980; DE BRESSER *et al.* 2002)

$$\dot{\varepsilon}(T) \propto \exp\left(-\frac{Q}{RT}\right).\exp\left(\frac{\sigma}{\sigma_0}\right) \qquad (3)$$

In this relation, Q is the activation energy for flow, R is the gas constant, and σ_0 is the flow stress at 0 K. To evaluate the acceleration in glide-controlled creep between our experiments at room temperature and 150 °C, we calculated the ratio of the strain rate

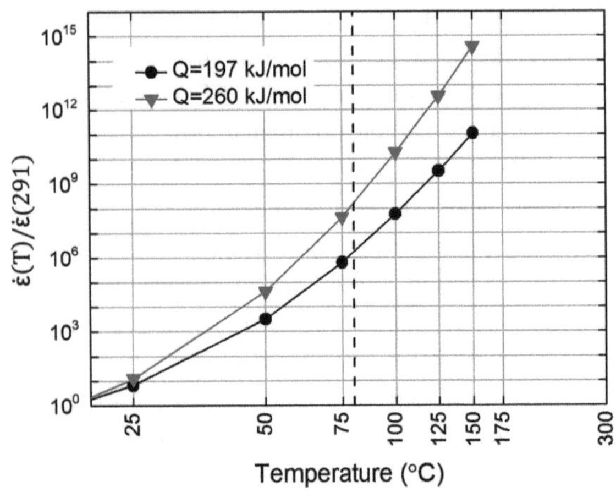

Figure 13
Plot of $\dot{\varepsilon}(T)/\dot{\varepsilon}(291)$ against temperature, using activation energies (Q) characteristic for exponential-type crystal plastic flow in Solnhofen limestone (197 kJ/mol) and Carrara Marble (260 kJ/mol) (RUTTER 1974; DE BRESSER *et al.* 2002)

at a given temperature T relative to that at 18 °C (291 K), i.e.

$$\frac{\dot{\varepsilon}(T)}{\dot{\varepsilon}(291)} = \exp\left[\frac{Q}{R}\left(\frac{1}{291} - \frac{1}{T}\right)\right] \quad (4)$$

Equation (4) is plotted against temperature in Fig. 13, using activation energies reported for exponential-type crystal plastic flow in Solnhofen Limestone (197 kJ/mol) and Carrara Marble (260 kJ/mol) at 400–500 °C (RUTTER 1974; DE BRESSER *et al.* 2002). These graphs demonstrate that at $T \approx 80$ °C, i.e. nearing the transition to velocity weakening slip seen in our dry experiments, dislocation plasticity in calcite is 10^6–10^8 times faster than at 18 °C, while at 150 °C it is 10^{11}–10^{15} times faster.

However, plasticity alone cannot explain velocity weakening slip or a transition thereto, since pure crystal plastic flow is intrinsically velocity strengthening ($\partial\sigma/\partial\dot{\varepsilon} > 0$ in Eq. 3). Our interpretation is that both cataclastic/granular flow and dislocation glide-controlled crystal plasticity have been operative at all temperatures in our experiments, but that their relative contribution to strain-accommodation in the shear band system varies with temperature, and, presumably, in the presence of water. Because of the acceleration of crystal plastic creep with increasing temperature (Fig. 13), time-dependent decrease in porosity and increase in grain contact area becomes increasingly important towards higher temperatures. At lower sliding velocities, pores then have more time to compact and grain contacts have more time to spread and possibly "bond" by plastic creep, ultimately imparting higher strength at lower sliding velocities, i.e. producing velocity weakening. In this framework, we propose that it is competition between time-dependent (e.g. crystal plasticity) versus relatively time-insensitive (e.g. granular flow) strain accommodation mechanisms that controls the sign of the velocity dependence. When velocities are high enough to maintain some porosity, and temperatures are sufficiently high, then creep processes allowing compaction and contact area increase are possible, and velocity weakening behaviour will be observed. Such a conceptual model is similar to the classical asperity model for friction of BOWDEN and TABOR (1950) and to that based on competition between compaction creep and dilatational granular flow of gouge as proposed by NIEMEIJER and SPIERS (2007). When temperatures are lower, creep processes will ultimately be too slow to allow significant porosity reduction and contact area increase, and critical-state granular/cataclastic flow will dominate, with friction being controlled by processes operating within grain contacts (cf. NIEMEIJER and SPIERS 2007), i.e. much finer scale asperity processes. These in turn are likely to involve micro/nanoplasticity (BHUSHAN 2005), as opposed to grain-contact-scale plastic creep, and will therefore be velocity strengthening, producing macroscopic velocity strengthening.

5.3. *The Role of Water and Impurities*

All wet experiments using simulated pure $CaCO_3$ gouge showed a well-defined peak strength after apparent yield, i.e. slip hardening and subsequent slip weakening (Fig. 4). In contrast, the dry experiments using this material showed a near-constant frictional strength in the post-yield region (Fig. 4). For experiments using simulated XJ limestone gouge, we observed the same difference in frictional behaviour between the dry and wet tests, except at room temperature. The magnitude of slip weakening, i.e. the value of $\Delta\mu = \mu_{pk} - \mu_{fnl}$, increased with increasing temperature for both materials tested (Fig. 5). In addition, the differences between the simulated XJ limestone and pure $CaCO_3$ gouge were only substantial in the presence of (pressurized) pore water, i.e. revealed in the frictional strength evolution with displacement seen at room temperature, but also in the sign of (*A–B*) with increasing temperature (Figs. 4, 5, 6). In view of the similarity between the shear-induced microstructures of the simulated XJ limestone and pure $CaCO_3$ gouges, the contrast observed in the frictional behaviour with displacement in the wet experiments are likely to be due to the compositional differences between the sample materials. In the following, we discuss which processes are likely to be active in the wet experiments and not in the dry, and qualitatively assess their effect on the frictional behaviours observed.

The frictional strength evolution with displacement seen in most of our wet experiments has the same broad peak as seen in shear experiments on over-consolidated soils (e.g. Yong *et al.* 2012). Gouge compaction during equilibration of the testing apparatus prior to each test may have produced over-consolidated samples such that the gouge must dilate, and hence do work against the normal stress to reach the critical state porosity that is associated with steady-state frictional sliding (Morrow and Byerlee 1989). Indeed, compaction of granular calcite is accelerated in the presence of water, mainly due to fluid-enhanced deformation mechanisms such as subcritical cracking (e.g. Rice 1978; Atkinson 1984) and/or pressure solution (e.g. Rutter 1983; Spiers *et al.* 2004). Moreover, the total pre-test compaction times in our tests have been longer for experiments performed at higher temperatures; hence, this may explain the increase of μ_{pk} seen with increasing temperature (Fig. 5), in particular for the wet experiments. However, if the presence of water only promotes consolidation of the gouge prior to shear, than the near-steady-state strengths (at $x > 4$ mm) of the dry experiments are expected to be the same as those of the wet experiments (cf. Morrow and Byerlee 1989; Yong *et al.* 2012). Clearly this is not the case for both materials investigated here (Figs. 4, 5; Table 1). Therefore, pre-test over-consolidation of the gouge may have produced the broad shape of the strength-displacement curves (Fig. 4) in the wet experiments; but other, water-assisted dynamic chemical and/or mechanical weakening effects must have played a role as well.

In the presence of water, the calcite fracture surface energy is 2–3 times lower compared with that of a dry calcite fracture surface (Røyne *et al.* 2011). This means that the energy needed for grain failure is much higher in dry experiments compared with that needed in wet experiments (Rice 1978; Atkinson 1984). The microstructural evidence for cataclasis and grain size reduction seen in the gouge microstructures indeed suggests that grain fracturing played an important role in all our experiments, wet and dry. Thus, reduced fracture surface energy may explain the lower strengths of wet samples compared with dry samples. However, Logan *et al.* (1992) showed that for (calcite) gouge shear deformation, the post-yield region of the curve describing frictional strength with displacement corresponds to initial shear band formation in the sample microstructure (see also Gu and Wong 1994; Marone 1998). At higher displacements the comminution rate decreases substantially (Marone and Scholz 1989). Therefore, a decrease in the calcite fracture surface energy may explain the lower peak strengths of the wet gouges compared with the dry, but not the major differences seen between the near-steady-state strengths (μ_{fnl}; Fig. 5).

On the other hand, intergranular pressure solution likely became more important in the shear band system at higher displacements, i.e. once it had fully developed, especially in view of the very fine grains present here. Assuming diffusion-controlled pressure solution for calcite (Zhang *et al.* 2010b), the rate of pressure solution creep ($\dot{\varepsilon}_{ps}$) is proportional to the inverse of the grain size (d) cubed, i.e. $\dot{\varepsilon}_{ps} \propto 1/d^3$. A high intergranular pressure solution creep rate will lead to porosity reduction hence increasing grain contact area (Bos *et al.* 2000; Bos and Spiers 2002a; Niemeijer and Spiers 2006), as well as cohesion through cementation (e.g. Zhang *et al.* 2010b; Gratier *et al.* 2013). Recent experimental work showed that pressure solution creep rates in calcite decrease with increasing temperature, probably because the solubility of calcite decreases with increasing temperature (Liteanu *et al.* 2012). This is consistent with our data, which show that the near-steady state strength of wet gouges decreases with increasing temperature (Fig. 5), i.e. suggesting that porosity reduction and growth of asperity contact-areas due to compaction by pressure solution will be less effective at higher temperatures, resulting in lower frictional strength. Note that this effect would be enhanced if the pore fluid was more undersaturated with calcite at room temperature than at 150 °C, as the driving force for compaction by pressure solution would be increased. However, if pressure solution is the only water-assisted mechanism active, the increased gouge cohesion in wet experiments should lead to a higher steady-state frictional strength (μ_{fnl}) than in dry tests, which is the opposite of that seen here (Fig. 5; Table 1). Thus, albeit likely that pressure solution played a substantial role in the shear bands in the presence of water, some other water-assisted process must have played a role as well, i.e. in addition to subcritical cracking.

A third mechanism that may have been operative is intergranular lubrication due to water-adsorption to (nano-)grains. In view of the very large surface area of grains in the shear bands, this may have been very effective in decreasing intergranular friction and/or adhesion. Water-adsorption is mainly controlled by the grain surface electrokinetic potential, which in the case of calcite is complexly dependent on the presence of impurities (Cicerone *et al.* 1992) and on pressure and temperature (Rodríguez and Araujo 2006), and is generally non-neutral (Stipp 1999). In this study, we observed that the compositional difference between the simulated XJ limestone and pure $CaCO_3$ gouge is only relevant in the wet experiments (Figs. 4, 5, 6; Tables 1, 2), suggesting that the type or presence of impurities only influences the frictional behaviour of our gouges in the presence of (pressurized) pore water. At the same time, it is well established that the presence of clay or quartz (XJ limestone gouge) compared with Mg impurities ($CaCO_3$ gouge) may have profound effects on the electrokinetic potential of the calcite-water interface (Cicerone *et al.* 1992; Dunning *et al.* 1994). Therefore, we suggest that water–adsorption also facilitated an intergranular (grain-boundary) lubrication mechanism in our wet experiments, which effectively lowered the steady-state frictional strength (μ_{fnl}) of wet gouges compared with dry gouges. Moreover, in view of the lower near-steady-state strengths (μ_{fnl}) of the wet gouges compared with dry (Table 1; Fig. 5), weakening due to intergranular lubrication must substantially outweigh that of strengthening due to compaction (hence porosity decrease and grain contact increase) by pressure solution.

Regarding the effect of (pressurized) pore water on the velocity dependence of strength, the values of (*A*–*B*) from the wet experiments showed more complex trends with increasing temperature than those of the dry experiments (Fig. 6). For both materials tested wet, the values of (*A*–*B*) first increased with increasing temperature (i.e. from 18 to 50 °C), followed by a decrease toward higher temperatures. Velocity weakening first occurred at the lowest sliding velocities imposed, at $T = 100$ °C for the simulated XJ limestone gouge but not until $T = 150$ °C for the simulated $CaCO_3$ gouge. In the preceding discussion, we inferred that both pressure solution and intergranular lubrication by grain-surface water-adsorption must have played a role in controlling the steady-state frictional strength of our wet gouges. These micromechanisms likely also affected the velocity dependence of strength. For wet samples, in the case of an upward step in sliding velocity in the velocity weakening regime, dilatation will result in net weakening of the gouge at steady state. However, compared with dry gouges, the net dilation and hence net strength decrease at steady state will be less, because net compaction rates are faster due to pressure solution. Also, in the wet experiments, the effect of changing porosity, and hence contact area, on strength will be less compared with dry samples, due to intergranular lubrication by water. Therefore, in the velocity weakening regime of our wet gouges, the value of (*A*–*B*) should be less negative compared with dry samples. At the same time, for both dry and wet samples, velocity strengthening can only be explained if grain boundary friction is velocity strengthening, as discussed above (as in the case of phyllosillicates—Ikari *et al.* 2011; Den Hartog and Spiers 2013), and if this effect outweighs any velocity weakening tendency due to competition between dilation and compaction. In the limiting case of high velocity and low temperature, compaction creep effects of any kind will be negligible, so that critical state granular flow will occur, in which case the strength and the value of (*A*–*B*) are entirely controlled by grain boundary friction (i.e. nanoscale processes acting within grain contacts). In the wet gouges, in the velocity strengthening regime, intergranular lubrication by water would imply that grain boundary friction will be less velocity strengthening than in dry samples, hence yielding lower values of $(A–B)$ compared with dry gouges. We cannot rule out other explanations for the trends in $(A–B)$ seen in our wet versus dry experiments, but this interpretation is consistent with all of our results, i.e. with the trends in both frictional strength and $(A–B)$ seen with increasing temperature, for both dry and wet experiments. The detailed microphysical processes controlling frictional sliding in wet calcite gouges, especially in the case of transients in the frictional strength, remains a subject for further research.

5.4. Geological Implications

Our results have important implications for understanding faulting and seismicity in tectonically-active carbonate terrains, such as those commonly found in rifts and the external zones of orogenic belts. Examples of recent destructive earthquakes in such terrains include the 1995 Aigion event (M_w 6.2) in the Corinth Rift zone of Greece, and the 2009 L'Aquila (M_w 6.3) and 2012 Emilia-Romagna earthquakes (M_w 6.0–6.1) in the Apennines of Italy. Assuming a typical geotherm of ~20 to 25 °C/km, faults cutting calcite-rich rocks can be expected to show the onset of velocity weakening and hence possible seismogenesis at depths of ~4–8 km. This, coupled with the high frictional strength measured ($\mu \approx 0.7–0.8$), suggests that tectonically active limestone terrains may be particularly prone to shallow-focus seismicity characterized by large stress drops, compared to e.g. clay-rich upper crustal faults, which show much lower strength ($\mu \approx 0.3$) and velocity strengthening behaviour up to 250 °C (Verberne *et al.* 2010; Den Hartog *et al.* 2012). Indeed, earthquakes that have occurred in the Apennines or the Corinth Rift zone are commonly associated with shallow focus (≤10 km) hypocentral depths of the main shock and aftershock sequences (Bernard *et al.* 2006; Mirabella *et al.* 2008; Valoroso *et al.* 2013).

In addition, we note that the shiny surfaces seen in our experiments are similar to the smooth, strongly reflecting slip surfaces recently reported from high-velocity friction experiments on carbonate rocks and gouges (Tisato *et al.* 2012; Smith *et al.* 2013; Fondriest *et al.* 2013), and from natural samples taken from the slip zone of faults cutting limestones (so called 'fault mirrors'—Siman-Tov *et al.* 2013). Our microstructural and nanostructural analyses are consistent with these authors' conclusions that these surfaces comprise nanocoatings marking a principal slip zone. However, we emphasize that our results demonstrate that these nanocoated surfaces are not necessarily diagnostic for co-seismic slip, and that natural calcite gouge microstructures showing such features must be interpreted with caution.

6. Conclusions

We have obtained data on frictional strength and velocity dependence parameters for simulated fault gouges prepared from natural Xujiahe (XJ) limestone (96 % $CaCO_3$) and pure $CaCO_3$ single crystals. Dry and wet, direct-shear, friction experiments were conducted on both materials at an effective normal stress of 50 MPa, at temperatures in the range 18–150 °C, at sliding velocities of 0.1–10 μm/s. Final shear displacements of 4–5 mm were imposed in each experiment. The mechanical observations were coupled with microstructural analysis, at the mm- to the nm-scales, using optical and electron microscopy methods. Our main conclusions are summarized below:

1. The gouges studied exhibited peak frictional strengths of $\mu_{pk} \approx 0.7–0.8$. Dry samples, and wet XJ limestone samples sheared at room temperature, subsequently showed minor slip weakening, reaching steady-state frictional strengths (μ_{fnl}), measured at the end of each experiment at $v = 1$ μm/s, that differed from the peak strength by <0.05. In contrast, most wet experiments showed a marked post-peak displacement-weakening, resulting in a steady-state frictional strength of around 0.6. This strength drop was generally more pronounced in the pure calcite samples (17–27 % compared to dry samples) than in those derived from XJ limestone (0–22 %). At the same time, the different

composition of the samples had no significant effect on frictional strength in the dry state.

2. The velocity dependence of strength of our samples, quantified using the rate sensitivity parameter (*A–B*), decreased nearly linearly with increasing temperature for the dry gouges, with a transition to negative values (i.e. velocity weakening slip) occurring at around 70–80 °C. For the wet samples, (*A–B*) values showed a similar overall trend but with more scatter, and with velocity weakening first occurring at ~100 °C for the simulated XJ limestone and at ~150 °C for the simulated pure $CaCO_3$ gouge. This implies that the presence of (pressurized) pore water is not crucial for promoting or inhibiting velocity weakening in calcite(-rich) gouges with increasing temperature. In other words, the presence of water has relatively little effect on velocity dependence, but influences the friction coefficient significantly.
3. All recovered gouge samples revealed a microstructure characterized by boundary and inclined shear bands (cf. Logan *et al.* 1979). These shears showed extreme internal grain size reduction to ~5–20 nm and the development of a crystallographic preferred orientation (CPO). The latter involved alignment of the $(104)[\bar{2}01]$ or r-slip system of the calcite nanocrystallites (sub-) parallel to the shear direction. On this basis, we suggest that crystal plasticity involving r-glide played a key role in shear band development, alongside cataclasis (i.e. cataclastic/granular flow). We found no microstructural indicators specific to samples showing (stable) velocity strengthening versus (potentially unstable) velocity weakening slip, or specific to dry versus wet samples.
4. In the wet experiments, the slip weakening observed relative to dry samples suggests that water-assisted chemical and mechanical processes must have played a role, in addition to cataclastic/granular flow and crystal plasticity. These processes likely involved pressure solution, fluid-assisted sub-critical cracking, and intergranular lubrication due to water adsorption on grain surfaces.
5. All recovered sheared gouge samples showed patchy, mirror-like surfaces exposed in the boundary shear planes, with (sub-) micron-scale striations aligned parallel to the shear direction. These mirror-like surfaces correspond to the nanocrystalline shear bands seen in cross-section. Contrary to recent claims, these observations demonstrate that nanocrystalline, mirror-like slip surfaces in calcite(-rich) faults are not exclusively diagnostic for seismic slip.

Acknowledgments

This work has benefited from discussions with Jianye Chen and Colin Peach. Matt Ikari and an anonymous reviewer are thanked for constructive comments on the manuscript. We further thank Peter van Krieken, Gert Kastelein, Hans Meeldijk, Vu Hoang Trung, Changrong He, Otto Stiekema, and Zhao Yue. B.A.V. was supported by the Netherlands Research Centre for Integrated Solid Earth Sciences (ISES) under Grant 2011-75. A.R.N. was supported by VENI grant 863.09.013, awarded by the Netherlands Organization for Scientific Research (NWO).

References

Atkinson, B.K. (1984), *Subcritical crack growth in geological materials*, J. Geophys. Res. *89*, 193–222.

Bernard, P., and 19 others (2006), *Seismicity, deformation and seismic hazard in the western rift of Corinth: New insights from the Corinth Rift Laboratory* (CRL), Tectonophysics *426*, 7–30, doi:10.1016/j.tecto.2006.02.012.

Bhushan, B. (2005), *Nanotribology and nanomechanics*, Wear *259*, 1507–1531.

Blanpied, M. L., Lockner, D. A., and Byerlee, J. D. (1995), *Frictional slip of granite at hydrothermal conditions*, J. Geophys. Res. *100*, B7, 13045–13064.

Bos, B., Peach, C. J., and Spiers, C. J. (2000), *Slip behavior of simulated gouge-bearing faults under conditions favoring pressure solution*, J. Geophys. Res. *105*, B7, 16699–16717.

Bos, B., and Spiers, C. J. (2002a), *Frictional-viscous flow of phyllosillicate-bearing fault rock: Microphysical model and implications for crustal strength profiles*, J. Geophys. Res. *107*, B2, 2028, doi:10.1029/2001JB000301.

Bos, B., and Spiers, C. J. (2002b), *Fluid-assisted healing processes in gouge-bearing faults: Insights from experiments on a rock analogue system*, Pure Appl. Geophys. *159*, 2537–2566, doi:10.1007/s00024-002-8747-2.

Bowden, F. P. and Tabor, D.,The Friction and Lubrication of Solids, Part I (Clarendon Press, Oxford 1950).

BRACE, W. F., and BYERLEE, J. D. (1966), *Stick-slip as a mechanism for earthquakes*, Science *153*, 990–992.

BURKHARD, M. (1990), *Ductile deformation mechanisms in micritic limestones naturally deformed at low temperatures (150–350 °C)*, Geol. Soc. London Spec. Pub. *54*, 241–257.

BURKHARD, M. (1993), *Calcite twins, their geometry, appearance and significance as stress-strain markers and indicators of tectonic regime; a review*, J. Struct. Geol. *15*, 351–368.

BYERLEE, J. D. (1978), *Friction of rocks*, Pure Appl. Geophys. *116*, 615–626.

CARPENTER, B. M., MARONE, C., and SAFFER, D. M. (2011), *Weakness of the San Andreas Fault revealed by samples from the active fault zone*, Nat. Geosci. *4*, 251–254.

CHESTER, F. M., and CHESTER, J. S. (1998), *Ultracataclasite structure and friction processes of the Punchbowl fault, San Andreas system, California*, Tectonophysics *295*, 199–221.

CICERONE, D. S., REGAZZONI, A. E., and BLESA, M. A. (1992), *Elektrokinetic properties of the calcite/water interface in the presence of Magnesium and organic matter*. J. Colloid Interf. Sci. *154*, 423–433.

COLLETTINI, C., NIEMEIJER, A. R., VITI, C., SMITH, S. A. F., and MARONE, C. (2011), *Fault structure, frictional properties and mixed-mode fault slip behavior*, Earth Planet. Sc. Lett. *311*, 316–327, doi:10.1016/j.epsl.2011.09.020.

DE BRESSER, J. H. P., and SPIERS, C. J. (1993), *Slip systems in calcite single crystals deformed at 300–800 °C*, J. Geophys. Res. *98*, 6397–6409.

DE BRESSER, J. H. P., and SPIERS, C. J. (1997), *Strength characteristics of the r, f, and c slip systems in calcite*, Tectonophysics *272*, 1–23.

DE BRESSER, J. H. P., EVANS, B., and RENNER, J., On estimating the strength of calcite rocks under natural conditions, In Deformation mechanisms, Rheology and Tectonics: Current status and future perspectives (eds. DE MEER, S., DRURY, M. R., DE BRESSER, J. H. P., AND PENNOCK, G. M.) (Geological Society, London 2002) Spec. Pub. 200, 309–329.

DEN HARTOG, S. A. M., and SPIERS, C. J. (2013), *Influence of subduction zone conditions and gouge composition on frictional slip stability of megathrust faults*, Tectonophysics *600*, 75–90.

DEN HARTOG, S. A. M., PEACH, C. J., DE WINTER, D. A. M., SPIERS, C. J., and SHIMAMOTO, T. (2012), *Frictional properties of megathrust fault gouges at low sliding velocities: New data on effects of normal stress and temperature*, J. Struct. Geol. *38*, SI, 156–171.

DIETERICH, J. H. (1979), *Modeling of rock friction I: Experimental results and constitutive equations*, J. Geophys. Res. *84*, 2161–2168.

DUNNING, J., DOUGLAS, B., MILLER, M., and MCDONALD, S. (1994), *The role of the chemical environment in frictional deformation: stress corrosion cracking and comminution*, Pure Appl. Geophys. *143*, 151–178.

FONDRIEST, M., SMITH, S. A. F., CANDELA, T., NIELSEN, S. B., MAIR, K., and DI TORO, G. (2013), *Mirror-like faults and power dissipation during earthquakes*, Geology *41*, 1175–1178.

FORBES, T. Z., RADHA, A. V., and NAVROTSKY, A. (2011), *The energetics of nanophase calcite*, Geochem. Cosmochem. Ac. *75*, 7893–7905.

GRATIER, J.-P., DYSTHE, D. K., and RENARD, F. (2013), *The role of pressure solution creep in the ductility of earth's upper crust*, Adv. Geophys. *54*, 47–179.

GU, Y., and WONG, T.-F. (1994), *Development of shear localization in simulated quartz gouge: effect of cumulative slip and gouge particle size*, Pure Appl. Geophys. *143*, 387–423.

HE, C., WANG, Z., and YAO, W. (2007), *Frictional sliding of gabbro gouge under hydrothermal conditions*, Tectonophysics *445*, 353–362.

HE, C., LUO, L., HAO, Q.-H., and ZHOU, Y. (2013), *Velocity-weakening behavior of plagioclase and pyroxene gouges and stabilizing effect of small amounts of quartz under hydrothermal conditions*, J. Geophys. Res. *118*, 3408–3430.

IKARI, M. J., MARONE, C., and SAFFER, D. M. (2011), *On the relation between fault strength and frictional stability*, Geology *39*, 83–86.

IKARI, M. J., NIEMEIJER, A. R., SPIERS, C. J., KOPF, A., and SAFFER, D. (2013), *Experimental evidence linking slip stability with seafloor lithology and topography at the Costa Rica convergent margin*, Geology *41*, 891–894.

KOELEMEIJER, P. J., PEACH, C. J., and SPIERS, C. J. (2012), *Surface diffusivity of cleaved NaCl crystals as a function of humidity: impedance spectroscopy measurements and implications for crack healing in rock salt*, J. Geophys. Res *117*, doi:10.1029/2011JB008627.

LI, Y. S., TAO, N. R., and LU, K. (2008), *Microstructural evolution and nanostructure in copper during dynamic plastic deformation at cryogenic temperatures*, Acta Mater *56*, 230–241.

LITEANU, E., NIEMEIJER, A. R., SPIERS, C. J., PEACH, C. J., and DE BRESSER, J. H. P. (2012), *The effect of CO_2 on creep of wet calcite aggregates*, J. Geophys. Res. *117*, B03211, doi:10.1029/2011JB008789.

LOGAN, J. M., FRIEDMAN, M., HIGGS, N., DENGO, C. A., and SHIMAMOTO, T., Experimental studies of simulated gouge and their application to studies of natural fault zones, In Proceedings of Conference VIII - Analysis of Actual Fault Zones in Bedrock (eds. Speed., R., and Sharp., R.) (United States Geological Survey, Menlo Park 1979) pp. 305–343.

LOGAN, J. M., DENGO, C. A., HIGGS, N. G., and WANG, Z. Z. Fabrics of experimental fault zones: Their development and relationship to mechanical behavior. In Fault mechanics and transport properties of rocks (eds. Evans, B., and Wong, T. F.) (Academic Press, London 1992) pp. 33–67.

MA, E. (2004), *Watching the nanograins roll*, Science *305*, 623–624.

MARONE, C. (1998), *Laboratory-derived friction laws and their application to seismic faulting*, Annu. Rev. Earth Pl. Sc. *26*, 643–696.

MARONE, C., and SCHOLZ, C. H. (1989), *Particle-size distribution and microstructures within simulated fault gouge*, J. Struct. Geol. *7*, 799–814.

MIRABELLA, F., BARCHI, M., LUPATTELLI, A., STUCCHI, E., and CIACCIO, M. G. (2008), *Insights on the seismogenic layer thickness from the upper crust structure of the Umbria-Marche Apennines (central Italy)*, Tectonics, *27*, TC1010, doi:10.1029/2007TC002134.

MORROW, C. A., and BYERLEE, J. D. (1989) *Experimental studies of compaction and dilatancy during frictional sliding on faults containing gouge*, J. Struct. Geol. *7*, 815–825.

MORROW, C. A., RADNEY, B., and BYERLEE, J., Frictional strength and the effective pressure law of montmorillonite and illite clays, In Fault mechanics and transport properties of rocks (eds. Evans, B., and Wong, T. F.) (Academic Press, London 1992) pp. 33–67.

Nakatani, M. (2001), *Conceptual and physical clarification of rate and state friction: Frictional sliding as a thermally activated rheology*, J. Geophys. Res. *106*, B7, 13347–13380.

Nakatani, M., and Scholz, C. H. (2004), *Frictional healing of quartz gouge under hydrothermal conditions: 1. Experimental evidence for solution transfer healing mechanism*, J. Geophys. Res. *109*, B07201, doi:10.1029/2001JB001522.

Niemeijer, A. R., and Spiers, C. J. (2006), *Velocity dependence of strength and healing behaviour in simulated phyllosillicate-bearing fault gouge*, Tectonophysics *427*, 231–253.

Niemeijer, A. R., and Spiers, C. J. (2007), *A microphysical model for strong velocity weakening in phyllosillicate-bearing fault gouges*, J. Geophys. Res. *112*, B10405, doi:10.1029/2007JB005008.

Power, W. L., and Tullis, T. E. (1989), *The relationship between slickenside surfaces in fine-grained quartz and the seismic cycle*, J. Struct. Geol. *11*, 879–893.

Rice, J. R. (1978), *Thermodynamics of the quasi-static growth of Griffith cracks*, J. Mech. Phys. Solids *26*, 61–78.

Rodríguez, K., and Araujo, M. (2006), *Temperature and pressure effects on zeta potentials of reservoir minerals*, J. Colloid Interf. Sci. *300*, 788–794.

Røyne, A., Bisschop, J., and Dysthe, D. K. (2011), *Experimental investigation of surface energy and subcritical crack growth in calcite*, J. Geophys. Res. *116*, B024204, doi:10.1029/2010JB008033.

Ruina, A. (1983), *Slip instability and state variable friction laws*, J. Geophys. Res. *88*, B12, doi:10.1029/JB088iB12p10359.

Rutter, E. H. (1974), *The influence of temperature, strain rate and interstitial water in the experimental deformation of calcite rocks*, Tectonophysics *22*, 311–334.

Rutter, E. H. (1983), *Pressure solution in nature, theory and experiment*, J. Geol. Soc. London *140*, 725–740.

Samuelson, J. E., and Spiers, C. J. (2012), *Fault friction and slip stability not affected by CO_2: evidence from short-term laboratory experiments on North Sea reservoir sandstones and caprocks*, Int. J. Greenh. Gas Con. *11*, S78–S90, doi:10.1016/j.ijggc.2012.09.018.

Schmid, S. M., Paterson, M. S., and Boland, J. N. (1980), *High temperature flow and dynamic recrystallization in Carrara Marble*, Tectonophysics *65*, 245–280.

Scholz, C. H. (1998), *Earthquakes and friction laws*, Nature *391*, 37–42.

Shan, Z., Stach, E. A., Wiezorek, J. M. K., Knapp, J. A., Follstaedt, D. M., and Mao, S. X. (2004), *Grain boundary-mediated plasticity in nanocrystalline nickel*, Science *305*, 654–657.

Sibson, R. H. (1986), *Earthquakes and rock deformation in crustal fault zones*, Annu. Rev. Earth Pl. Sc. *14*, 149–175.

Siman-Tov, S., Aharonov, E., Sagy, A., and Emmanuel, S. (2013), *Nanograins from carbonate "fault mirrors"*, Geology *41*, 703–706.

Smith, S. A. F., Di Toro, G., Kim, S., Ree, J.-H., Nielsen, S., Billi, A., and Spiess, R. (2013), *Coseismic recrystallization during shallow earthquake slip*, Geology *41*, 63–66.

Spiers, C. J., de Meer, S., Niemeijer, A. R., and Zhang, X. Kinetics of rock deformation by pressure solution and the role of thin aqueous films. In Physicochemistry of water in geological and biological systems - structures and properties of thin aqueous films (eds. Nakashima, S., Spiers, C. J., Mercury, L., Fenter, P. A. and Hochella, M. F., Jr.) (Universal Academy Press, Tokyo 2004) p. 129–158.

Stipp, S. L. (1999), *Toward a conceptual model of the calcite surface: hydration, hydrolysis and surface potential*, Geochim. Cosmochim. Ac. *63*, 3121–3131.

Tisato, N., Di Toro, G., De Rossi, N., Quaresimin, M., and Candela, T. (2012), *Experimental investigation of flash weakening in limestone*, J. Struct. Geol. *38*, 183–199.

Umemoto, M. (2003), *Nanocrystallization of steels by severe plastic deformation*, Mater. Trans. *44*, 1900–1911.

Valoroso, L., Chiaraluce, L., Piccinini, D., Di Stefano, R., Schaff, D., and Waldhauser, F. (2013), *Radiography of a normal fault system by 64,000 high-precision earthquake locations: the 2009 L'Aquila (central Italy) case study*, J. Geophys. Res., doi:10.1002/jgrb.50130.

Verberne, B. A., He, C., and Spiers, C. J. (2010), *Frictional properties of sedimentary rocks and natural fault gouge from the Longmenshan Fault Zone, Sichuan, China, B.* Seismol. Soc. Am. *100*, 2767–2790.

Verberne, B. A., De Bresser, J. H. P., Niemeijer, A. R., Spiers, C. J., De Winter, D. A. M., and Plümper, O. (2013), *Nanocrystalline slip zones in calcite fault gouge show intense crystallographic preferred orientation: Crystal plasticity at sub-seismic slip rates at 18–150 °C*, Geology *41*, 863–866.

Xu, X. W., Wen, X. Z., Yu, G. H., Chen, G. H., Klinger, Y., Hubbard, J., and Shaw, J. (2009), *Coseismic reverse- and oblique-slip surface faulting generated by the 2008 M_w 7.9 Wenchuan earthquake, China*, Geology *37*, 515–518.

Yang, D. K., Cizek, P., Hodgson, P. D., and Wen, C. E. (2010), *Microstructure evolution and nanograin formation during shear localization in cold-rolled titanium*, Acta Mater. *58*, 4536–4548.

Yong, R. N., Nakano, M., and Pusch, R. Environmental soil properties and behaviour (CRC press Taylor & Francis Group, Boca Raton 2012), pp. 446.

Zhang, J., Huang, F., and Lin, Z. (2010a), *Progress of nanocrystalline growth kinetics based on oriented attachment*, Nanoscale *2*, 18–34.

Zhang, L., and He, C. (2013), *Frictional properties of natural gouges from Longmenshan fault zone ruptured during the Wenchuan Mw 7.9 earthquake*, Tectonophysics *594*, 149–164.

Zhang, X., Spiers, C. J., and Peach, C. J. (2010b), *Compaction creep of wet granular calcite by pressure solution at 28 °C to 150 °C*, J. Geophys. Res. *115*, B09217, doi:10.1029/2008JB005853.

(Received May 6, 2013, revised December 1, 2013, accepted December 3, 2013, Published online December 15, 2013)

Pure Appl. Geophys. 171 (2014), 2641–2664

DOI 10.1007/s00024-013-0759-6

Pure and Applied Geophysics

Frictional Properties of a Low-Angle Normal Fault Under In Situ Conditions: Thermally-Activated Velocity Weakening

ANDRÉ R. NIEMEIJER[1] and CRISTIANO COLLETTINI[2,3]

Abstract—The Zuccale fault is a regional, low-angle, normal fault, exposed on the Isle of Elba in central Italy that accommodated a total shear displacement of 6–8 km. The fault zone structure and fault rocks formed at <8 km crustal depth. The present-day fault structure is the final product of several deformation processes superposed during the fault history. In this study, we report results from a series of rotary shear experiments performed on 1-mm thick powdered gouges made from several fault rock types obtained from the Zuccale fault. The tests were done under conditions ranging from room temperature to in situ conditions (i.e., at temperatures up to 300 °C, applied normal stresses up to 150 MPa, and fluid-saturated.) The ratio of fluid pressure to normal stress was held constant at either $\lambda = 0.4$ or $\lambda = 0.8$ to simulate an overpressurized fault. The samples were sheared at a constant sliding velocity of 10 μm/s for at least 5 mm, after which a velocity-stepping sequence from 1 to 300 μm/s was started to determine the velocity dependence of friction. This can be represented by the rate-and-state parameter (*a*–*b*), which was determined by an inversion of the data to the rate-and-state equations. Friction of the various fault rocks varies between 0.3 and 0.8, similar to values obtained in previous studies, and decreases with increasing phyllosilicate content. Friction decreases mildly with temperature, whereas normal stress and fluid pressure do not affect friction values systematically. All samples exhibited velocity strengthening, conditionally stable behavior under room temperature conditions and (*a*–*b*) increased with increasing sliding velocity. In contrast, velocity weakening, accompanied by stick–slips, was observed for the strongest samples at 300 °C and sliding velocities below 10 μm/s. An increase in fluid pressure under these conditions led to a further reduction in (*a*–*b*) for all samples, so that they exhibited unstable, stick–slip behavior at low sliding velocity. The results suggest that phyllosilicate-bearing fault rocks can deform by stable, aseismic creep at low, resolved shear stress and low shear rate conditions. An increase in fluid pressure or loading of stronger portions could lead to a runaway instability. The runaway instability might be limited in size because of (1) the fault heterogeneity, (2) the observed strengthening at higher sliding velocities, and (3) a co-seismic drop in pore-fluid pressure.

Key words: Friction, sliding stability, hydrothermal conditions, LANF.

1. Introduction

Exhumed fault zones typically display complex geometries and mineralogies, which presumably were also present at depth during the activity of the fault (e.g. SIBSON 1977; CHESTER *et al.* 1993; FAULKNER *et al.* 2003; COLLETTINI and HOLDSWORTH 2004; JEFFERIES *et al.* 2006, see also IMBER *et al.* 2008 for a review). In order to understand how faults accommodate deformation, i.e., what the evolution of slip and slip velocity of a particular fault is, we need to perform detailed field and experimental studies on exhumed faults. Despite the fact that outcrop samples are problematic due to weathering and uncertainties in the pressure, temperature, and fluid conditions pertaining to their formations, studying outcrop samples can be very useful to appreciate the complexity of natural fault zones and to attempt to unravel this complexity.

Here, we present results from an experimental study on powdered samples of the Zuccale fault, an exhumed, low-angle, normal fault (LANF), currently exposed on the Isle of Elba (e.g., COLLETTINI and HOLDSWORTH 2004; SMITH *et al.* 2008). We performed systematic experiments using a hydrothermal ring shear apparatus under conditions pertaining to the upper crust (~4–10 km) to characterize sliding friction values for a sample suite taken from the fault core and to determine the stability of sliding through velocity-step tests.

Low-angle normal faults are peculiar structures because they accommodate deformation at an angle

[1] Faculty of Geosciences, HPT Laboratory, Utrecht University, Utrecht, The Netherlands. E-mail: a.r.niemeijer@uu.nl

[2] Dipartimento di Scienze della Terra, Università degli Studi di Roma "La Sapienza", Rome, Italy.

[3] Istituto Nazionale di Geofisica e Vulcanologia, Rome, Italy.

to σ_1 larger than 60°, which according to Anderson's theory is not possible unless friction is low (<0.3) or fluid pressure is high (Collettini 2011). In the case of the Zuccale fault, it seems that the presence of talc and/or fluid overpressure is sufficient to lower the strength, and to make slip on a gently dipping plane possible. However, an outstanding question is whether LANFs are capable of generating large seismic events. A compilation of dip estimates of moderate to large, global shallow earthquakes ($M > 5.5$) on normal faults shows that no earthquakes with positively discriminated rupture planes occur on planes that dip <~30° (Collettini and Sibson 2001; Collettini 2011 for a review). However, large earthquakes occurring on low-angle structures have been suggested in the Woodward Basin, Papua New Guinea (Abers *et al.* 1997).

Beyond this debate, it has been established that microseismicity occurs on low-angle structures such as the Alto Tiberina fault, ATF in central Italy (Chiaraluce *et al.* 2007), and possibly beneath the Gulf of Corinth, Greece (Rigo *et al.* 1996). It has also been suggested that weak fault rocks ($\mu < 0.5$) are only capable of velocity strengthening (Ikari *et al.* 2010) and should not be able to nucleate a seismic event, which raises the question whether the microseismicity on the ATF is truly happening within the low-angle fault core. Here, we investigate the question of potentially unstable slip in weak materials by characterizing the frictional properties of the samples from the fault core of the Zuccale fault under hydrothermal conditions. An additional question to be answered is how frictional stability is affected by changes in fluid pressure during the seismic cycle.

2. Background

2.1. Regional Geology and the Zuccale Fault

The locus of active extension in central Italy has migrated eastward from the Tyrrhenian Sea to the Appenine chain over the last 18 Ma (Elter *et al.* 1975; Jolivet *et al.* 1998; Collettini and Holdsworth 2004, Fig. 1). In the active sector of the Northern Apennines, the currently active Alto-Tiberina fault (ATF) was imaged in the seismic reflection profile CROP 03 as a gently eastward-dipping plane, at an angle of ~15°–20° (Fig. 1 and Barchi *et al.* 1998), and is highlighted by micro-seismicity ($M < 3.0$) at depths of 4–16 km (Boncio *et al.* 2000; Chiaraluce *et al.* 2007). Composite focal mechanism solutions of repeating earthquakes ($M \leq 2.3$) are consistent with normal fault motion on a gently east-dipping plane (Chiaraluce *et al.* 2007). Analysis from data of continuous GPS stations shows that the ATF is currently slipping aseismically at an average rate of ~2.4 mm/a at depths below 4 km (Hreinsdottir and Bennet 2009). In addition, the region is characterized by widespread CO_2 degassing with a deep mantle origin inferred from geochemical analysis. Deep boreholes indicate CO_2 pressures at 85 % of the lithostatic load, at depths of 3–5 km (Chiodini *et al.* 2000, 2004; Trippetta *et al.* 2013).

Within the Apenninic LANF system, the Zuccale fault is one of the most westerly located structures (Fig. 1) and is exposed on the Isle of Elba. The Zuccale fault was active between 13 and 4 Ma at depths of <8 km and accommodated about 6–8 km of displacement (Collettini and Holdsworth 2004). Within the fault core, exposed at the Punta di Zuccale, a significant amount of the deformation seems to have been accommodated within a pervasively foliated, carbonate-rich and phyllosilicate-rich basal horizon up to 5 m thick that formed in the early stage of the fault activity (Smith *et al.* 2007). Here, initial cataclastic deformation increased permeability and allowed the influx of silica-rich fluids, which in turn caused the reaction of the dolostone host rock to produce talc, calcite, and CO2 through the reaction $3MgCa(CO_3)_2 + 4SiO_2 \rightarrow Mg_3Si_4O_{10}(OH)_2 + 3CaCO_3 + 3CO_2$ (Collettini *et al.* 2009a). Our samples are only derived from the lithologically heterogeneous basal horizon of the foliated fault core (Fig. 2), and consist mostly of calcite, tremolite and talc in various proportions (Table 1). One sample is from a non-foliated portion of the fault core and consists mostly of dolomite.

2.2. Rate and State Friction

In experimental rock friction studies, friction changes instantaneously when the driving velocity

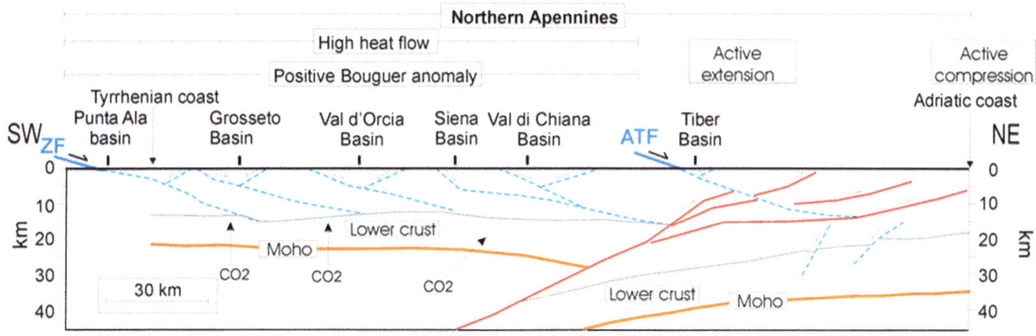

Figure 1
Cross section (Southwest to Northeast) across the Northern Appenines showing the series of low-angle, normal faults that have sequentially accommodated extension over the past 13 Ma. The Alto Tiburina fault (ATF) is the currently active fault, whereas the Zuccale fault (ZF) is exposed on the Isle of Elba and was active between 13 and 4 Ma

Figure 2
Photograph of the fault zone structure of the Zuccale fault on the Isle of Elba (Italy). The samples used in this study were from the foliated fault core

Table 1

Semi-quantitative X-ray diffraction analysis results for the compositions of our sample suite

Sample	Calcite	Tremolite	Talc	Clays	Dolomite
ZF01	42	0	56	2	0
ZF02	23	42	22	13	0
ZF03	78	0	16	5	0
ZF04	43	46	0	6	0
ZF05	29	0	7	0	64

Clays include smectite/illite and chlorite (see also Table 3 and Collettini *et al.* 2011)

is changed and evolves towards a new steady state value over a characteristic displacement. This behavior is well-described by a set of constitutive equations, known as the Rate-and-State Friction (RSF) equations (Dieterich 1978; Ruina 1983).

$$\mu = \mu_0 + a \cdot \ln(V/V_0) + b \cdot \ln(V_0 \cdot \theta/d_c) \qquad (1)$$

$$\frac{d\theta}{dt} = 1 - \frac{\theta \cdot V}{d_c} \qquad (2a)$$

or

$$\frac{d\theta}{dt} = -\frac{V\theta}{d_c}\ln\left(\frac{V\theta}{d_c}\right), \qquad (2b)$$

where μ is the coefficient of friction (=shear stress/effective normal stress), μ_0 is the coefficient of friction at a reference velocity, V_0, a is a parameter characterizing the direct effect, V is the current velocity, b is a parameter characterizing the evolution effect, and d_c is a characteristic or critical displacement over which this evolution effect occurs. θ is a state variable that evolves upon a velocity-step according to Eq. 2a ("Dieterich or aging law") or Eq. 2b ("Ruina or slip law"). In a rock friction experiment, the simulated fault or fault gouge is driven at a certain load-point velocity and when this load-point velocity is quasi-instantaneously changed, the sliding velocity in the simulated fault is not equal to the load-point velocity due to the elastic distortion of the experimental apparatus and non-sliding portions of the fault. This can be expressed by.

$$d\mu/dt = K(V - V_{lp}), \quad (3)$$

where K is the stiffness of the actively shearing fault surroundings and V_{lp} is the load-point velocity.

Here, we solve Eqs. (2) and (3) simultaneously, with Eq. (1) as a constraint, using a fifth-order Runga-Kutta method. The values for a, b, and d_c can then be obtained as the solutions of a nonlinear inverse problem using an iterative least-squares method (see e.g., Reinen and Weeks 1993; Blanpied *et al.* 1998; Ikari *et al.* 2009).

Since the RSF equations were proposed, numerous authors have sought to find a physical clarification for the individual parameters of a, b, and d_c. The direct effect, a, is typically interpreted to be the strain rate-dependent shear strength of the contacting asperities (e.g., Ruina 1983; Baumberger *et al.* 1999; Nakatani 2001; Beeler *et al.* 2007 and many others). In the case of granular fault gouges, an additional contribution to the value of a can be expected from the dilational work done against normal stress upon a change in velocity (e.g. Sammis and Steacy 1994; Marone 1998; Niemeijer *et al.* 2010a). The evolution effect, b, is interpreted to be related to the change in total true contact area and/or strengths of the contacts upon a change in sliding velocity. The distance over which the contact area or contact strength evolves is defined as the characteristic sliding distance, d_c, and the average age of individual contacts can be calculated as $t_c = d_c/v_1$. Again, for the case of granular fault gouges, a portion of b is taken up by the change in gouge volume (i.e., ϕ) associated with a change in velocity (e.g., Sammis and Steacy 1994). The deformation mechanism that determines the strength of the contacts has not been generalized, but some authors speculate that dislocation glide is the main mechanism controlling the sliding strength of highly stressed asperities (Reinen *et al.* 1994; Beeler *et al.* 2007). In these models, all contacts are critically stressed, i.e., their size is determined by the yield strength of the material and the applied normal stress. However, in natural situations, fluids are likely to be present and solution-transfer processes can be fast under these conditions. The operation of normal fault compaction (shear-perpendicular) through time-dependent, fluid-assisted mechanisms will tend to increase the size of the contacts so that the contact stresses will be less than the yield strength of the material. In addition, subcritical stress corrosion cracking is likely to operate, reducing average grain sizes and further enhancing solution-transfer processes (e.g., Liteanu *et al.* 2012; Gratier *et al.* 2013). Besides this effect of increasing contact areas, the strength of the contacts is likely to change as well. Because of the complex interplay between all these different thermally-activated, fluid-assisted, time-dependent mechanisms in controlling the contact area evolution, contact strengths, and steady state volume of fault gouge, we can expect complex changes in the values of a, b, and d_c with temperature, effective normal stress, sliding velocity, shear strain, and perhaps even fluid pressure (aside from its mechanical effect on effective normal stress). This will be further complicated because of the heterogeneous mineralogy of natural fault rocks, as well as the possibility of mineral transformations (e.g., the talc-producing reaction in the Zuccale fault, Collettini *et al.* 2009a). Work under hydrothermal conditions on simulated fault gouges of westerly granite (Blanpied *et al.* 1995), quartz (Chester 1994), calcite (Verberne *et al.* 2010, in press, this volume), gabbro (He *et al.* 2007), quartz/illite (Den Hartog *et al.* 2012a, b), and quartz/muscovite mixtures (Den Hartog *et al.* 2013) has demonstrated that the value of (a–b) changes with temperature, and in fact probably also shows a dependence on sliding velocity and effective normal stress. It is clear from the above that there is a need for a mechanism-based microphysical model that would allow us to predict values of a, b, and d_c under hydrothermal conditions. However, this is not a trivial task and a first necessary step is to experimentally investigate possible relations between experimental parameters (P, T, P_f, etc.) and the individual RSF parameters, and attempt to link these to rates of the operating deformation mechanisms. Such a study is not the purpose of the present manuscript, but we will attempt to discuss possible trends in (a–b) and link

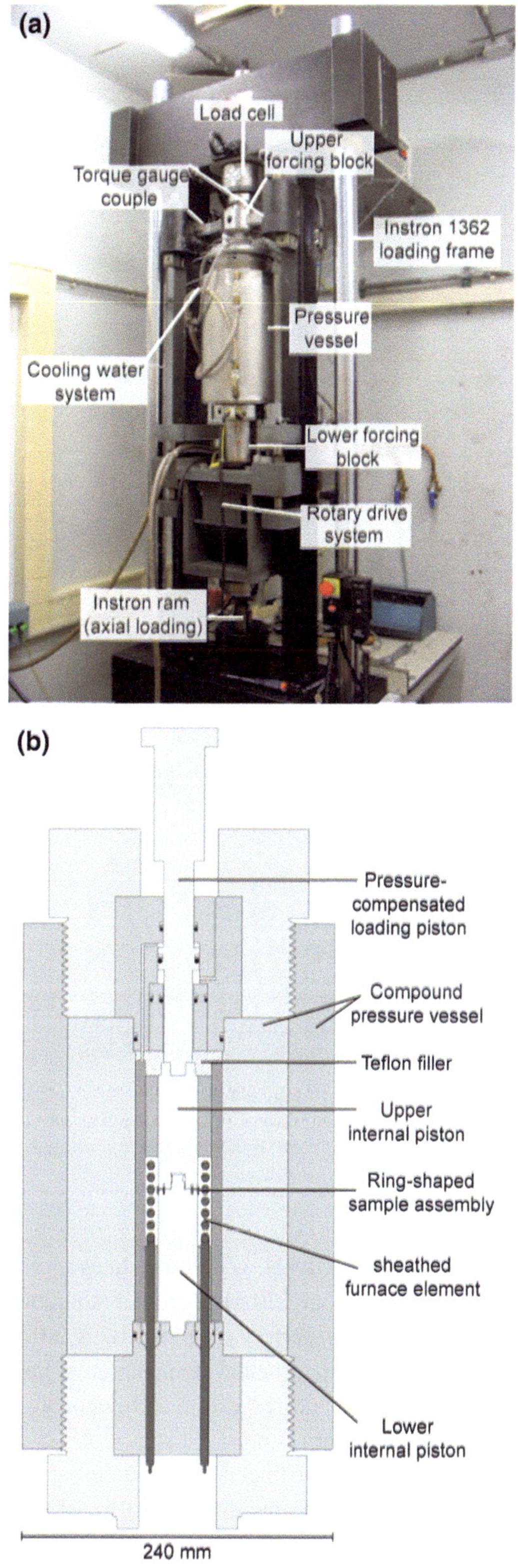

Figure 3
a Photograph of the hydrothermal pressure vessel inside the Instron loading frame. **b** Schematic drawing of the components of the pressure vessel (after DEN HARTOG *et al.* 2012b)

these to deformation mechanisms, which, in the absence of microstructures, will have to remain rather speculative, but, we hope, will direct future efforts towards understanding the controls on a, b, and d_c.

3. Experimental Methods

All experiments reported here were performed in the hydrothermal ring shear apparatus at the HPT laboratory at Utrecht University. The apparatus consists of an internally-heated pressure vessel located inside a 1362 Instron loading frame and has been described in detail elsewhere (NIEMEIJER *et al.* 2008a; VAN DIGGELEN *et al.* 2010; DEN HARTOG *et al.* 2012b, Fig. 3). In brief, the ring-shaped sample (22/28-mm inner/outer diameter) is sandwiched between two René 41 pistons (Fig. 4) and held in place by precision-cut René 41 inner and outer rings, coated with a thin layer of low-friction molykote spray. Normal load is applied to the upper piston through the dynamic O-ring seals that seal the pressure medium, which is demineralized water in all experiments reported here. The piston is pressure-compensated, so that the applied normal load, when corrected for the dynamic seal friction, equals the effective normal load on the sample. The pressure vessel is pressurized through a manually operated, high-pressure pump, capable of reaching 300 MPa fluid pressure. The entire vessel can be rotated at a prescribed velocity using a 0.17 kW electromotor in line with two gear boxes with a reduction factor of 1:100 each (i.e. 1:10,000 total). The upper pressure-compensated piston is gripped in an external forcing block, which houses an axial load cell (100 kN full scale, 0.01 kN resolution) for normal force measurement and a load-cell couple for torque measurements. Two different sets of load-cell couples with different full scales (one for up to 1.2 kN each and a second one for up to 15 kN each) are available for varying loading conditions. Most experiments reported here were performed with the set that has a full scale of 1.2 kN and a

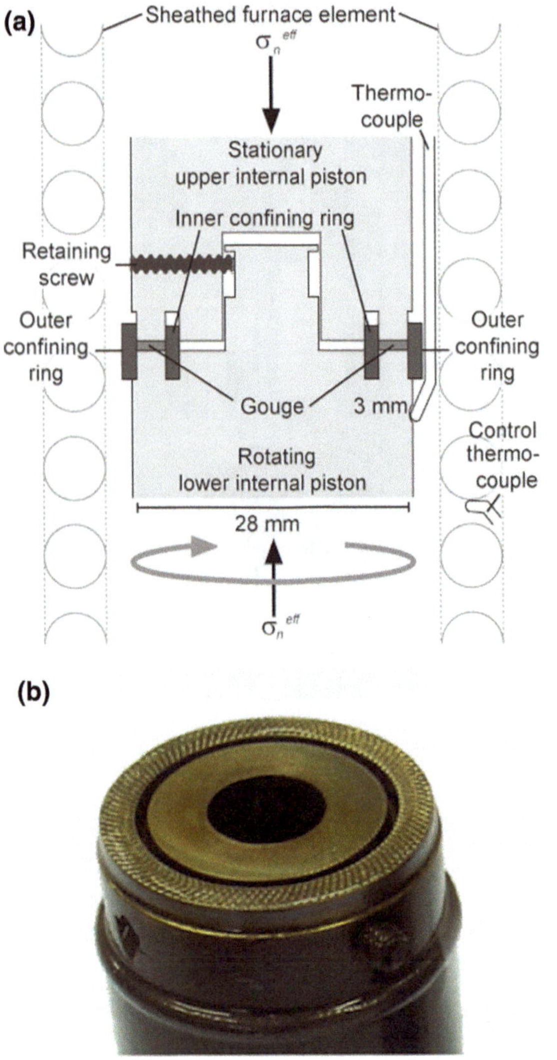

Figure 4
a Schematic drawing of the piston assembly (after Niemeijer *et al.* 2008; den Hartog *et al.* 2012b). **b** Picture of the top of one of the pistons

resolution of ~0.0001 kN, which translates to a full scale in shear stress of 120 MPa and a resolution of ~0.015 MPa (when perfectly balanced). A 3.0-kW furnace enables experimental temperatures up to 700 °C, which is monitored with a thermocouple situated inside the lower piston at a distance of ~5 mm from the sample. Layer thickness is monitored using two external linear variable differential transformers (LVDT's), one with a full scale of 50 mm, resolution 5 μm, and one with a full scale of 1 mm and a resolution of 0.1 μm.

3.1. *Sample Preparation and Experimental Procedure*

Collected fault rocks were crushed by hand using a pestle and mortar and sieved to obtain a grain-size fraction smaller than 125 μm. Each sample was analyzed using XRD and the results are shown in Table 1. For each experiment, approximately 0.5 g of material were loaded onto the bottom piston, fitted with the coated confining rings. The top piston was then placed gently on top of the sample material and rotated by hand for two revolutions to distribute the sample material evenly. The piston was then attached to the upper pressure-compensated piston, which was subsequently lowered into the pre-filled pressure vessel, locking the lower piston into the key-hole at the bottom of the pressure vessel. The vessel was closed, pore fluid tubing and cooling water hoses were attached, and the entire vessel was placed into the Instron loading frame. The desired normal load was applied, followed by the fluid pressure. The furnace was switched on and heating started. Typically, the desired test conditions of temperature and fluid pressure were reached within 30 min and the sample was left to equilibrate for 1–2 h. In all experiments, initial shear loading was at 10 μm/s for at least 5 mm, but typically 10 mm of displacement, after which a velocity-stepping sequence was initiated (see Table 2 for all experimental conditions). Tests were terminated in reverse order, i.e., shear load was removed, the furnace switched off, the fluid pressure was released and normal stress was removed. The piston set was dried at 50 °C overnight and carefully taken apart. The sheared samples in almost all cases broke along inclined surfaces, leaving only small parts (typically pieces of ~5–10 mm length over the entire width of the ring) of the sample intact. The largest pieces were impregnated under a vacuum with Araldite 20/20 epoxy resin, and standard thin sections for light and electron microscopy were cut, the analysis of which will be reported in a separate contribution.

4. *Results*

4.1. *Steady-State Friction*

In Fig. 5, we show the coefficient of friction as a function of displacement for experiments at room

Table 2

List of experimental conditions for all experiments reported here

Sample	Exp. no.	T (°C)	$\sigma_n^{eff,applied}$ (MPa)	P_f (MPa)	x (mm)	h_{final} (μm)	γ	SS?
ZF01	u064	Room	90–120–150	60–80–100	61.1	518	96.6	No
ZF02	u071	Room	90–120–150	60–80–100	58.8	636	82.4	No
ZF03	u083	Room	90–120–150	60–80–100	61.8	458	108.5	No
ZF04	u084	Room	90–120–150	60–80–100	61.9	492	99.2	No
ZF05	u082	Room	90–120–150	60–80–100	62.0	643	80.0	No
ZF01	u068	150	90–120–150	60–80–100	62.8	179	199.6	No
ZF02	u069	150	90–120–150	60–80–100	62.7	530	91.0	No
ZF03	u087	150	90–120–150	60–80–100	61.7	498	100.9	No
ZF04	u029	150	60–75–90–105–120	40–50–60–70–80	90.7	519	146.5	Yes
ZF05	u080	150	90–120–150	60–80–100	66.9	445	100.6	Yes
ZF01	u063	300	90–120–150	60–80–100	61.3	566	80.9	No
ZF02	u073	300	90–120–150	60–80–100	67.4	690	86.6	Yes
ZF03	u026	300	120–150	80–100	49.2	643	66.4	No
ZF03	u085	300	90–120–150	60–80–100	61.5	691	75.8	No
ZF04	u025	300	120–150	80–100	48.4	671	64.0	Yes
ZF05	u074	300	90–120–150	60–80–100	68.7	595	78.7	Yes
ZF03	u053	Room	20–40–30	80–160–120	55.7	827	64.7	No
ZF04	u052	Room	20–40–60	80–160–200	54.5	718	64.3	No
ZF01	u060	300	20–40–60	80–120–160	68.5	591	100.3	No
ZF02	u078	300	20–40	80–160	33.4	636	39.7	No
ZF02	u079	300	20–40–60	80–160–240	61.7	472	91.0	No
ZF03	u033	300	20–40–60	80–160–240	55.1	270	174.1	Yes
ZF04	u028	300	20–40–60	80–160–240	56.2	716	70.1	Yes
ZF05	u077	300	20–40–60	80–160–240	61.5	725	60.7	Yes
ZF01	u061	300	20–40–60	13.3–26.6–40	61.3	870	62.7	No
ZF04	u086	300	60	240–120–60–30–15	73.5	511	129.2	Yes
ZF04	u088	300	60	15–30–60–120–240	72.1	611	81.5	Yes
ZF01	u075	300	90–120–150	Dry	61.8	773	74.1	No

x is the total shear displacement, h_{final} is the final layer thickness after shearing but before removal of temperature, fluid pressure, and normal stress, γ is total shear strain, SS signifies whether sliding was stable ("no") or unstable ("yes") during at least one part of the experiment

temperature on all samples tested. The coefficient of friction plotted here was calculated as the shear stress divided by the effective normal stress, thereby ignoring cohesion, and corrected for seal friction using calibration tests performed earlier (e.g. Den Hartog *et al.* 2012b). The uncertainty in seal friction leads to some uncertainty in the absolute level of friction, but the relative coefficients of friction of individual experiments performed under the same conditions of effective normal stress and fluid pressure can be compared with each other with confidence. In addition, temperature is not expected to have an effect on seal friction, since the seals are kept at relatively constant temperature by the circulation of the cooling water. Figure 5 shows for all samples a similar initial quasi-elastic load-up with displacement, resulting in a peak strength followed by a slight drop in friction and some (minor) strain hardening. The variation in steady state sliding friction for all tested samples from the foliated fault core is quite large, with values as low as ~0.3 (ZF01) to as high as ~0.8 (ZF04), with friction roughly increasing with decreasing total phyllosilicate content. The sample from outside of the foliated fault core (ZF05) shows slightly lower friction than the strongest samples from the fault core (ZF04), although the phyllosilicate content of these samples is similarly low.

In Fig. 6, we plot the steady-state friction values obtained at the end of the run-in at 10 μm/s, at 120 MPa effective normal stress and 80 MPa fluid pressure as a function of temperature for all samples tested. The two samples with <10 % phyllosilicates (ZF04 and ZF05) show a monotonic, almost linear

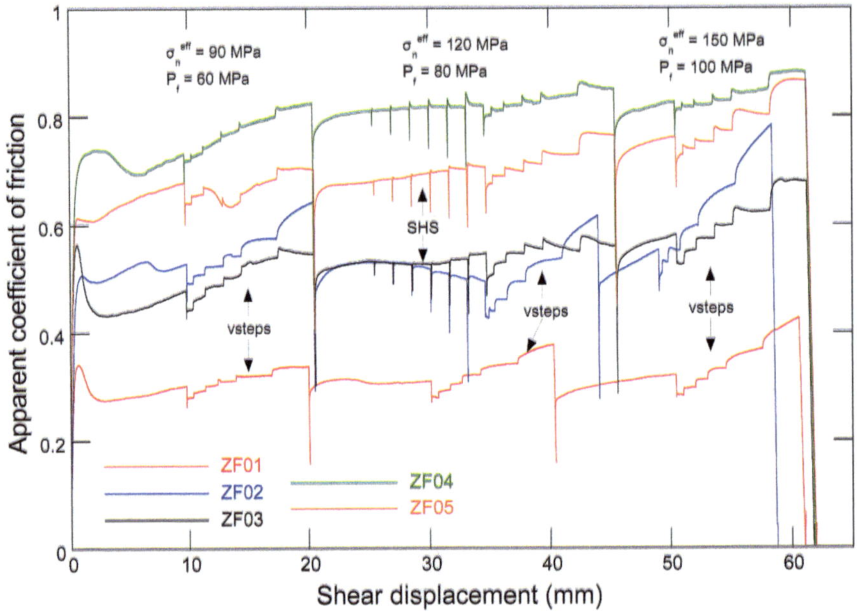

Figure 5

Plot of the coefficient of friction (≈shear stress/effective normal stress), corrected for seal friction, as a function of shear displacement for the five different fault rock units (see Table 2). The experiments were done at room temperature, with applied effective normal stresses of 90, 120, and 150 MPa and fluid pressures of 60, 80, and 100 MPa. Samples were sheared initially for 10 mm at 10 μm/s, followed by a velocity-step sequence of 0.3–1–3–10–30–100 μm/s with displacement increments of 0.6–1–1.5–1.5–3–3 mm, respectively (vsteps). Normal stress and fluid pressure were subsequently increased and samples were sheared at 10 μm/s for 5 mm, followed by a slide-hold-slide series (SHS), with hold periods of 3–10–30–100–300–1,000 s and 1-mm displacement in between (with the exception of sample ZF01), and a velocity-stepping sequence. After the final normal stress and fluid pressure increment, this procedure was repeated without the SHS

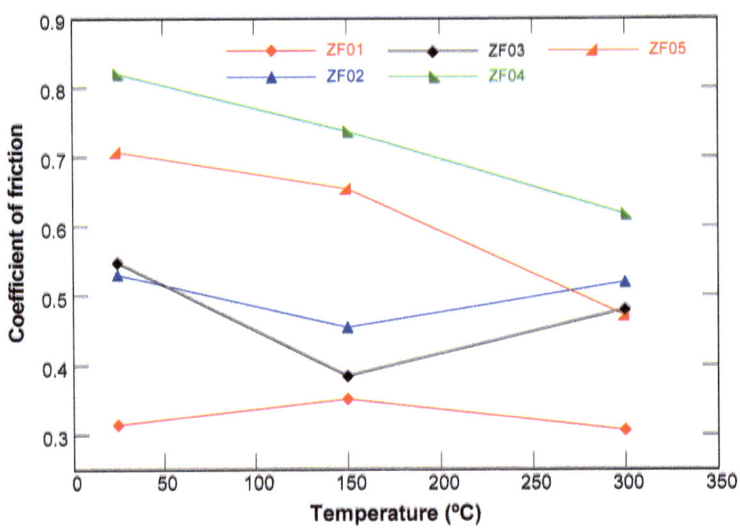

Figure 6

Plot of the coefficient of friction (corrected for seal friction) as a function of temperature for the entire sample suite. The values of friction were determined by averaging the values during the last 0.5 mm of run-in sliding at 10 μm/s, 120 MPa effective normal stress and 80 MPa fluid pressure decrease of friction with temperature. The two samples with intermediate phyllosilicate-content (ZF02 and ZF03) show a minimum in friction at 150 °C, whereas the sample with the highest phyllosilicate content (ZF01) shows a maximum in friction at the same temperature.

4.2. Velocity-Dependence of Friction

In Fig. 7, we show the velocity-dependence of friction, expressed as the rate parameter (*a–b*) as a function of upstep sliding velocity for all samples tested at an applied effective normal stress of 120 and 80 MPa fluid pressure. As mentioned earlier, a positive value of (*a–b*) is indicative of stable sliding (creep), whereas a negative value of (*a–b*) indicates the possibility of unstable slip. Figure 7 shows that (*a–b*) increases mildly with sliding velocity for all samples tested. The amount of increase in (*a–b*) depends on mineralogy as well as temperature, and is smallest (virtually absent) for sample ZF01 at room temperature. In addition, (*a–b*) shows a general decrease with increasing temperature at low sliding velocities for all samples, leading to negative values at the lowest sliding velocities (0.3 μm/s) and highest temperature (300 °C) tested, with the exception of sample ZF03. In the case of samples ZF04 and ZF05, the combination of negative (*a–b*) and machine stiffness resulted in the occurrence of unstable stick-slip events, with stress drops of up to 0.1 in friction units at a sliding velocity of 0.3 μm/s.

In Fig. 8, we show the values of (*a–b*) as a function of sliding velocity for a series of experiments with different fluid pressure conditions. Figure 8a shows that, for sample ZF01, in the absence of water but at elevated temperature (300 °C), the value of (*a–b*) does not change with sliding velocity as opposed to the experiment with water present that shows a clear increase in (*a–b*) with sliding velocity. Figure 8b compares the values of (*a–b*) for room temperature experiments performed under an elevated fluid pressure ($\lambda = 0.8$) with experiments done at normal fluid pressures ($\lambda = 0.4$) for samples ZF03 and ZF04. Interestingly, the effect of elevated fluid pressure and low effective normal stress is reversed for the two samples: for sample ZF03, (*a–b*) is smaller at increased fluid pressure at all sliding velocities, whereas (*a–b*) is generally larger at increased fluid pressure for

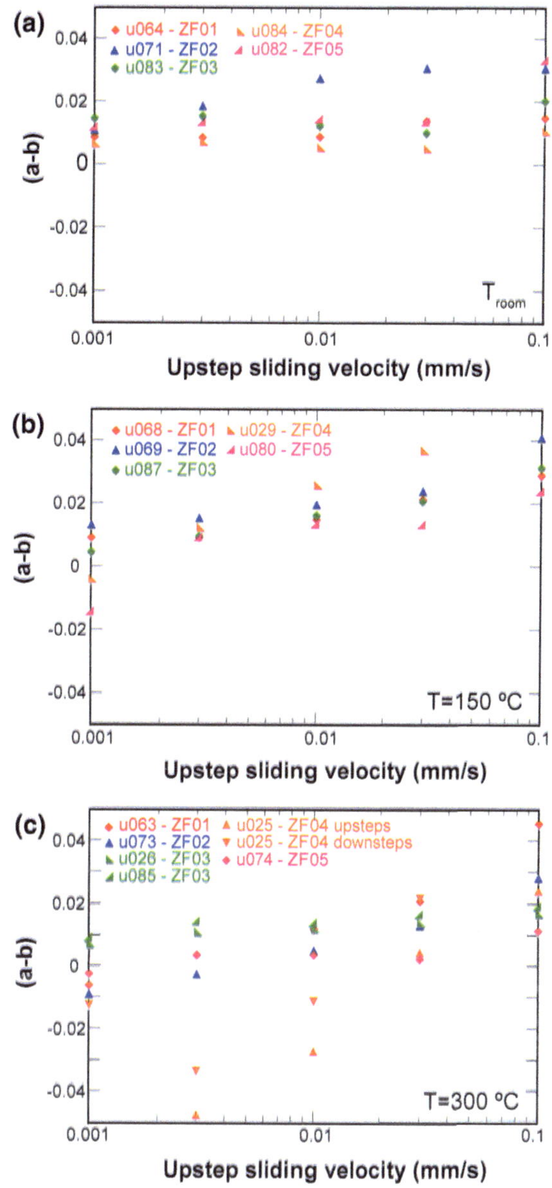

Figure 7
Semi-log plots of the rate parameter (*a–b*) as a function of upstep sliding velocity for the three temperatures investigated, 120 MPa effective normal stress and 80 MPa fluid pressure. All plots have the same scales to facilitate comparison. **a** Room temperature. **b** 150 °C. **c** 300 °C. Values for duplicate experiments are shown for sample ZF03 and values for both up- and downsteps are shown for sample ZF04

sample ZF04, with the exception of the step from 3 to 10 μm/s. Figure 8c compares the (*a–b*) values for two experiments performed on sample ZF01 at 300 °C and the same effective normal stresses (20–40–60 MPa, see Table 2), but with different fluid pressures. The values of

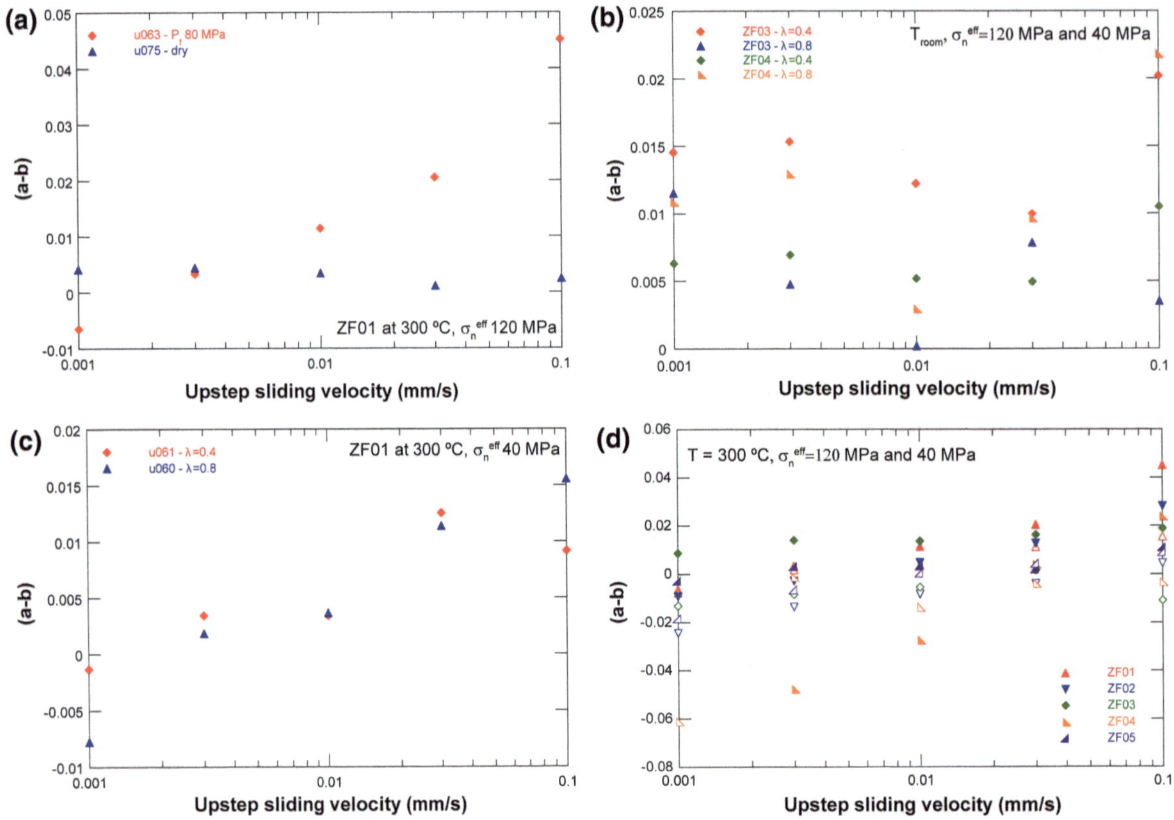

Figure 8

Semi-log plots of the rate parameter (*a–b*) as a function of upstep sliding velocity. **a** Comparison between wet and dry experiments using sample ZF01 at an effective normal stress of 120 MPa and a fluid pressure of 80 MPa (wet) and a temperature of 300 °C. **b** Comparison between four room temperature experiments using samples ZF02 and ZF04 for a hydrostatic fluid pressure case (effective normal stress of 120 and 80 MPa fluid pressure, λ = 0.4) and an overpressurized case (effective normal stress of 40 MPa and 160 MPa fluid pressure, λ = 0.8). **c** Comparison between two experiments using sample ZF01 at a temperature of 300 °C at the same effective normal stress of 40 MPa, but different fluid pressures (26.6 MPa, λ = 0.4 and 160 MPa, λ = 0.8). **d** Comparison between experiments at a temperature of 300 °C under hydrostatic conditions (*closed symbols*, effective normal stress of 120 MPa and a fluid pressure of 80 MPa, λ = 0.4) and overpressurized conditions (*open symbols*, effective normal stress of 40 MPa and a fluid pressure of 160 MPa, λ = 0.8) for the entire sample suite

(*a–b*) are smaller for high pore-fluid pressures at low sliding velocities, but larger at high sliding velocities with the transition occurring at the 3–10 μm/s velocity-step. In Fig. 8d, we show (*a–b*) as a function of upstep sliding velocity for all experiments done at 300 °C and with different pore fluid pressure ratios. For almost all samples and velocities, the value for (*a–b*) is somewhat lower for the over pressurized case, i.e. at low effective normal stress and high pore-fluid pressures. Notable exceptions are sample ZF04 at upstep velocities below 30 μm/s and sample ZF05 only at the 10–30 μm/s velocity-step. Also note that for sample ZF03, the value of (*a–b*) switches from positive to negative for all velocities when fluid pressure is high and normal stress low, and sliding was unstable in this experiment at the lowest sliding velocity. Figure 9 shows the value of (*a–b*) as a function of fluid pressure for different velocity-steps for two experiments performed on sample ZF04 at 60 MPa effective normal stress and 300 °C (Table 2). The two experiments were done under the same conditions, except that in one experiment the fluid pressure was stepped up and in the other it was stepped down. The difference in (*a–b*) for the two experiments is largest at the lowest sliding velocity and fluid pressures. Despite the difference in (*a–b*) for the two experiments, a general trend of increasing (*a–b*) with increasing fluid pressure can be seen in Fig. 9a at sliding velocity-steps of 0.3–1 and 1–3 μm/s. This trend is reversed, i.e., (*a–b*) decreases with increasing fluid pressure for velocity steps at higher sliding velocities (Fig. 9b).

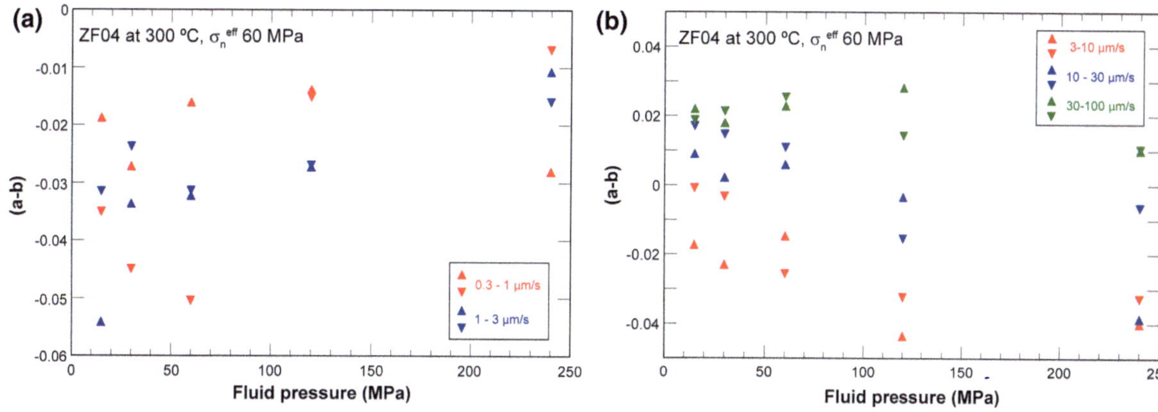

Figure 9

Plots of the rate parameter (*a*–*b*) as a function of fluid pressure for two experiments using sample ZF04. *Upward triangles* indicate an increasing fluid pressure with displacement, *downward triangles* indicate a decreasing fluid pressure with displacement. **a** Velocity-steps of 0.3–1 and 1–3 μm/s. **b** Velocity-steps of 3–10, 10–30 and 30–100 μm/s

Table 3

*Reported compositions for Zuccale fault rocks from previous friction studies (*Collettini *et al. 2009a, 2011;* Smith *and* Faulkner *2010)*

Paper	Name	Calcite	Tremolite	Talc	Smectite	Chlorite	Our sample
Collettini (2011)	L1	39	26	14	20	1	ZF02
Collettini (2011)	L2	43	36	6	14	1	ZF02
Smith	L3.1	60	39				ZF04
Smith	L3.2	20	10	70			ZF01
Smith	L3.3	55	17	16		12	ZF02

Note that samples L1 and L2 from Collettini *et al.* (2011) are denoted L2 and L3 in Collettini *et al.* (2009a). Further, note that the composition of sample L3.2 was determined from image analysis, all other compositions were obtained using semi-quantitative XRD

5. Discussion

5.1. Comparison with Previous Studies

There have been a number of experimental studies on the frictional properties of Zuccale fault rocks (Collettini *et al.* 2009b, 2011; Smith and Faulkner 2010; Tesei *et al.* 2012). In these studies, samples from the entire fault zone were studied, whereas here we only report on samples from the foliated fault core plus one sample of non-foliated fault rock, which was not studied in the previous studies. We list the relevant samples names and compositions in Table 3 to facilitate comparison with our data.

Collettini *et al.* (2009b, 2011) reported on dry, room-temperature, double-direct shear experiments on powdered samples, as well as on intact fault rock samples sheared parallel to their foliation. They reported values of friction for their powdered sample L1 (L3 in the 2009 paper) of 0.60–0.63 and 0.50–0.55 for their powdered sample L2 at normal stresses of 10–150 MPa. The intact fault rocks sheared parallel to their foliation showed a decrease of about 0.2 friction units. Our powdered sample ZF02 showed values of 0.50–0.55 at room temperature and normal stresses of 90–150 MPa, but under wet conditions, which is comparable to their sample L3 (which is most similar in composition to our sample ZF02, see Tables 1, 3). Similar values for steady-state friction of samples with similar mineralogy were obtained by Tesei *et al.* (2012) under room-dry conditions and at slightly lower normal stress (20 and 50 MPa).

Smith and Faulkner (2010) performed experiments at room temperature and effective normal stresses of between 25 and 75 MPa on fluid-saturated, simulated gouges of powdered samples from the entire suite of fault rocks present in the Zuccale fault zone. They report a value of 0.25–0.31 for their sample L3.2, comparable to our sample ZF01 (see Tables 1, 3), which showed friction values of ~0.3 at room temperature. Their sample L3.1 has a friction value of ~0.79 and is in composition most comparable to our sample ZF04, which showed friction values of 0.75–0.80 at room temperature. Finally, their sample L3.3 has an intermediate friction of 0.40–0.42 and is most similar to our sample ZF02, which shows somewhat higher friction values of 0.45–0.50. It must be noted that we report steady-state friction values at 10 μm/s and after at least 5 mm of displacement, whereas Smith and Faulkner (2010) report values after 500 μm of slip, presumably at a velocity of 0.3 μm/s. Given that all materials display velocity-strengthening behavior, we expect to find slightly higher values than those reported by Smith and Faulkner. Despite this, the values of friction that we report here are very similar to those obtained by Smith and Faulkner (2010) for their zone L3 samples.

We obtained only positive values for the velocity-dependence of friction expressed as $(a\text{–}b)$, ranging between ~0.004 and ~0.033 for all samples at room temperature. These values are similar to the values shown for zone L3 in Fig. 9 of Smith and Faulkner (2010), with the exception of the one small negative value (~−0.001) obtained for sample L3.1 at 75 MPa effective normal stress. Our $(a\text{–}b)$ values for sample ZF02 range between 0.01 and 0.045, which are much larger than the range of values obtained by Collettini *et al.* (2011) ($-0.005 < (a\text{–}b) < 0.009$). In addition, we observed a nonlinear increase in $(a\text{–}b)$ with increasing sliding velocity as opposed to the decrease in $(a\text{–}b)$ with sliding velocity observed for powders by Collettini *et al.* (2011). Interestingly, their intact sheared samples demonstrate a similar increase in $(a\text{–}b)$ with increasing sliding velocity, although the magnitude is about 20 % lower. It is possible that the presence of pressurized fluid combined with the high shear-strain obtained in our experiments ($\gamma > 10$ before velocity-stepping started) reorganized the initially homogeneously distributed phyllosilicates into a through-going weak foliation. However, if this was the case, we would expect to see significant strain weakening during the initial sliding at constant velocity, which was not observed. To truly evaluate the possibility of foliation development and how this might affect the velocity-dependence of friction, a systematic experimental and microstructural study would be required, which is beyond the scope of the current work.

5.2. *Temperature Dependence of Steady-State Friction*

We observe a steady decrease in frictional strength with temperature for the two samples with the least amount of phyllosilicates (ZF04 and ZF05, Fig. 6). For these samples, friction drops from 0.81 to 0.61 and 0.70 to 0.47, respectively. The amount of decrease in strength is surprisingly similar, considering that the mineralogy of the samples is quite different (Table 1). To the best of our knowledge, there are no friction studies on similar mineralogies under hydrothermal conditions.

However, recent experiments on pure calcite as well as natural limestone gouges up to a temperature of 150 °C at an effective normal stress of 50 MPa have shown some weakening with increasing temperature (Verberne *et al.* 2010, 2013, this volume). Steady-state friction in those experiments decreased from 0.59 to 0.56 (for pure calcite gouge) and 0.70 to 0.61 (for natural limestone gouge). The decrease in friction for their natural limestone gouge with about 5 wt% of impurities (quartz and clays) is similar to what we observe for our two samples with respectively 43 wt% calcite and 29 wt% carbonates. They observe that deformation is localized into boundary and Riedel shear bands where grain size is significantly reduced, and infer that the presence of water allows for some lubrication effect, lowering contact friction with respect to dry-sheared gouges. However, it is not clear what role temperature plays in this mechanism. Scuderi *et al.* (2013) report on double-direct shear experiments on simulated dolomite fault gouges at a temperature of 75 °C, effective normal stresses of 10, 20, and 35 MPa, with a pore-fluid pressure of 2 MPa. At the lowest normal stress, friction of those samples was similar to the dry, room-temperature values (~0.65), but some weakening, which was absent in the room-temperature experiments, was observed at the higher normal stresses. The lowest friction observed was 0.5 at 35 MPa and

75 °C, which is much lower than what we observed at 150 °C. However, the other dolomite sample of Scuderi *et al.* (2013) showed a friction of 0.62 under the same conditions, which is more comparable to our value at 150 °C and 120 MPa effective normal stress. Since the conditions of our experiments, the experimental configuration, as well as the sample compositions are quite different from previous studies, a direct comparison of the amount of weakening and the absolute level of friction is not possible. However, we can conclude that the thermal weakening that we observe for our two samples with low phyllosilicate content has also been observed in almost pure carbonate samples, and can thus potentially be ascribed to the presence of calcite and dolomite in our samples. The fact that the sample with only 43 wt% calcite shows similar weakening might imply that: (1) its sliding strength is controlled by the calcite, or (2) the amphibole (tremolite) deforms by a similar, thermally-weakened mechanism as the calcite. The frictional properties of amphiboles as well as the deformation mechanisms that might operate in tremolite are pretty much unknown, particularly under hydrothermal conditions.

We can conclude that there is significant thermal weakening in the samples with low phyllosilicate content that is mostly absent in the samples with higher phyllosilicate content. The range of frictional strengths of the different fault rocks is thus decreased from 0.31 to 0.82 at room temperature, similar to that observed by Smith and Faulkner (2010), to a narrower range of 0.31–0.62 at 300 °C.

5.2.1 Velocity-Dependence of Friction (a–b): Reproducibility

Numerous friction studies have reported values for (*a*–*b*) for a range of materials over the last several decades. Interestingly, little attention has been given to the reproducibility of the values of (*a*–*b*), nor to possible changes in (*a*–*b*) as a function of experimentally-controlled parameters such as effective normal stress, sliding velocity, or fluid pressure, with the exception of the effect of temperature (e.g. Blanpied *et al.* 1995; He *et al.* 2007; Verberne *et al.* 2010, 2013, this volume, den Hartog *et al.* 2012a, b; Den Hartog *et al.* 2013). One exception here is a series of experiments by several authors on bare granite surfaces at room temperature and humidity conditions that showed clear changes in (*a*–*b*) with applied normal stress and sliding velocity (Blanpied *et al.* 1987; Marone *et al.* 1990; Kilgore *et al.* 1993).

Here, we show one example of the reproducibility of the values of (*a*–*b*) in Fig. 7c, for sample ZF03 at 120 MPa effective normal stress, 80 MPa fluid pressure, and 300 °C. Considering the strain weakening and hardening that is observed in some of the velocity steps, it shows a surprisingly good reproducibility with a maximum absolute difference in (*a*–*b*) of <0.003, and both experiments show the same trend with sliding velocity. This lends some confidence to the observed trends in (*a*–*b*) as a function of mineralogy, sliding velocity, effective normal stress, fluid pressure, and temperature, which we will discuss in the next sections.

5.2.2 Velocity-Dependence of Friction, (a–b): Effect of Mineralogy and Temperature

A first glance, Fig. 7 shows that for all mineralogies, (*a*–*b*) at low sliding velocities goes down with temperature, yielding negative, potentially unstable values for (*a*–*b*) at 300 °C for all but one sample (ZF03). Focusing on the lowest sliding velocity step (0.3–1 μm/s), we see a decrease in (*a*–*b*) from 0.008 to −0.006 for sample ZF01, 0.019 to −0.009 for ZF02, 0.0136 to 0.007 for ZF03, 0.006 to −0.013 for ZF04, and 0.011 to −0.003 for ZF05. The rate parameter (*a*–*b*) for this velocity step does not show a continuous decrease with temperature for all samples, but is in fact lower at 150 °C than at 300 °C for samples ZF03, ZF04, and ZF05 and higher at 150 °C than at room temperature for samples ZF01 and ZF02. This observation indicates that it is likely that several deformation mechanisms are operating simultaneously, and that it is the competition between the different mechanisms at any given temperature and sliding velocity that determines the friction, and thus the velocity dependence of friction.

In terms of mineralogy, (*a*–*b*) decreases with a decrease in total phyllosilicate content. This observation is similar to what is generally observed in room-temperature experiments on natural fault gouges as well as artificial mixtures (e.g., Saffer and Marone 2003; Ikari *et al.* 2009; Numelin *et al.* 2007; Moore and Lockner 2007; Tembe *et al.* 2010; Carpenter *et al.* 2009, 2011; Boulton *et al.* 2012). The explanation for this behavior is that the platy nature of phyllosilicate minerals allows for an alignment of these particles, maximizing the contact area and minimizing the potential for contact area change with changes in sliding velocity, and thus yielding a near-zero value for the evolution effect, *b*. However, it is not clear what the threshold content of phyllosilicates is for this to be an efficient deformation accommodation mechanism (e.g., Niemeijer *et al.* 2010b). Our data suggest that the "contact saturation effect" (Saffer and Marone 2003; Ikari *et al.* 2009) is complicated by thermally-activated mechanisms that probably operate in the secondary phases. For example, sample ZF03, which has an intermediate phyllosilicate content (~21 wt%, see Table 2) shows larger (*a–b*) values than the two samples with higher phyllosilicate contents (ZF01 and ZF02), at least at the slowest velocity steps.

5.2.3 Velocity-Dependence of Friction, (a–b): Effect of Sliding Velocity

All samples show an increase in (*a–b*) with increasing sliding velocity at all temperatures and under hydrostatic stress conditions. This effect has been noted previously for sliding of bare granite surfaces and granite gouges at room-temperature conditions and various normal stresses (Blanpied *et al.* 1987; Marone *et al.* 1990; Kilgore *et al.* 1993). In the present experiments, we observed a change in the dependence of (*a–b*) on sliding velocity with increasing temperature. For samples with <25 wt% phyllosilicates, the dependence of (*a–b*) with sliding velocity at room temperature is erratic, as shown by the low R^2-values for a log-linear fit through the data (Table 4). In contrast, the (*a–b*) data under hydrothermal conditions can be well-described by a log-linear fit. In addition, the samples with more than 25 wt% phyllosilicates (ZF01 and ZF02) show a clear increase in the log-linear dependence of (*a–b*) on sliding velocity with increasing temperature (see Table 4). A similar statement can be made for sample

Table 4

Log-linear fits of (a–b) at a function of sliding velocity at 120 applied (effective) normal stress and 80 MPa fluid pressure

Sample	Slope at 25 °C	R^2 at 25 °C	Slope at 150 °C	R^2 at 150 °C	Slope at 300 °C	R^2 at 300 °C
ZF01	0.00355	0.792	0.01045	0.932	0.02416	0.940
ZF02	0.01037	0.880	0.01290	0.851	0.01803	0.969
ZF03	0.00122	0.006	0.01299	0.979	0.00451[a]	0.963[a]
ZF04	0.00130	0.209	0.03118	0.984	0.04873	0.988
ZF05	0.00875	0.589	0.01609	0.803	0.00530	0.715

Note the poor fits at room temperature for samples with a low amount of phyllosilicates

[a] The repeat experiment at these conditions yielded the exact same slope, but a slightly lower goodness of fit ($R^2 = 0.893$)

ZF04, which at 300 °C shows the strongest log-linear dependence of (*a–b*) on sliding velocity observed (a log-linear slope of 0.048), although the fit at room temperature is poor. The remaining two samples show the largest slopes at the intermediate temperature of 150 °C. An increase of (*a–b*) with increasing sliding velocity has previously been attributed either to frictional heating (for bare surfaces, Blanpied *et al.* 1987; Kilgore *et al.* 1993) or to an increased contribution of dilational work with increasing sliding velocity (for gouges, Marone *et al.* 1990). However, these explanations were for room-temperature, dry experiments where the rates of thermally-activated, fluid-assisted processes are negligible. In our wet and hot experiments, several different mechanisms can be expected to operate simultaneously, but at different rates under different conditions of temperature, stress, and sliding velocity. We will attempt to explain the observed temperature-dependent velocity-dependence of (*a–b*) in a subsequent paragraph, in which we discuss the possible deformation mechanisms operating in our simulated fault gouges.

5.2.4 Velocity-Dependence of Friction, (a–b): Effect of Fluid Pressure and Effective Normal Stress

Our results indicate that there are subtle differences in the frictional behavior of simulated fault gouges sheared at high effective normal stress (i.e., representative of hydrostatic pressure conditions in the Earth's crust) and those sheared at low effective normal stress (i.e., representative of overpressurized conditions). Our experimental stress and fluid pressure conditions are representative of depths ~4–10 km, regardless of the fluid pressure condition. Our results indicate that the development and maintainment of fluid overpressures will not only change the shear stress supported by the fault through the lowering of the effective normal stress, but it can in fact also change the velocity dependence of friction, (*a–b*), and thus the potential for the occurrence of instabilities (i.e. earthquakes). However, our results also indicate that the change in (*a–b*) with increasing fluid pressure and lower effective normal stress depends on fault gouge mineralogy, temperature, and sliding velocity in a very complex manner.

At room temperature, an increase in fluid pressure leads to an increase in (*a–b*) in sample ZF04, but to a decrease in (*a–b*) for sample ZF03, which contains only a slightly higher amount of phyllosilicates. At the same time, the amount of change in (*a–b*), irrespective of whether it is an increase or decrease, changes considerably with the magnitude of the upstep sliding velocity. However, all (*a–b*)-values remain positive for these room-temperature conditions and these two samples.

In contrast, under hydrothermal conditions of 300 °C, we observe a more or less consistent decrease in (*a–b*) for overpressured conditions for all samples (Fig. 8d). The observed decrease in (*a–b*) results in a switch in sign, indicating potentially unstable behavior when fluid pressure is high (or equivalently, when effective normal stress is low). In order to test whether the high pore pressure or the low effective normal stress is responsible for the observed changes, we sheared sample ZF01 at the same low effective normal stress, but different pore pressures (Fig. 8c). This shows that the change in (*a–b*) is mostly related to the change in effective normal stress, at least for this sample. Lower effective normal stress will significantly reduce the amount of cataclasis and grain-size reduction, which in turn will lower the rate of any grain-size sensitive processes that might operate (with the notable exception of stress corrosion cracking, but this process is even more strongly enhanced by increasing stress, e.g., Atkinson 1982). The potential non-mechanical effect of a higher fluid pressure is expected to only affect the solubility of minerals and/or the reaction of minerals, and would thus significantly alter the velocity dependence of friction, which was not observed for sample ZF01. In contrast, our results at constant effective normal stress with steps in fluid pressure for sample ZF04 do show slight changes in (*a–b*) with changing fluid pressure (Fig. 9). Here, we also observe that changes in (*a–b*) with pore pressure are more pronounced at low sliding velocity. This fits with the assumption that fluid pressure will only affect chemical reactions (including dissolution), since these are time-dependent processes, such that changes in the rates of these processes will have different effects at different sliding velocities (i.e., possibly making them more effective at low sliding velocities but still rendering them too slow for the higher sliding velocities). The consistent decrease in (*a–b*) seen at almost all sliding velocities for all samples at 300 °C for the overpressured case (Fig. 8d) is most likely due to the low effective normal stress rather than the high pore pressure, possibly with a more pronounced decrease of (*a–b*) at the lower sliding velocities due to the effect of pore-fluid pressure on chemical reaction rates.

5.2.5 Deformation Mechanisms and Potential Effects on (a–b): Phyllosilicate-Rich Sample

Our sample with the largest amount of phyllosilicates (~58 wt%) shows almost exclusively positive values of (*a–b*), with a notable exception for the lowest sliding velocity step at a temperature of 300 °C, and all investigated normal stress and fluid pressure conditions (Figs. 7c, 8c, d). In addition, we observe a consistent log-linear increase in (*a–b*) with sliding velocity, the slope of which changes with increasing temperature (see Table 4). The observation of solely velocity-strengthening behavior at room-temperature conditions is consistent with previous studies on phyllosilicate-rich gouges (e.g., Saffer and Marone 2003; Numelin *et al.* 2007; Moore and Lockner 2008; Moore 2011; Ikari *et al.* 2009; Ikari and Saffer 2011; Carpenter *et al.* 2011; Boulton *et al.* 2012). In all of these studies, as in ours here, the values for *b* are close to zero or even negative. A zero *b* value has been attributed to "contact saturation"

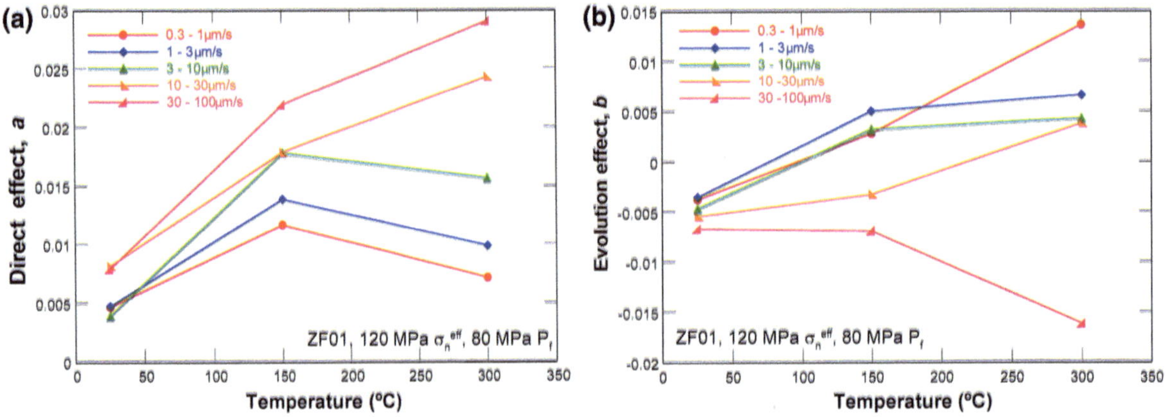

Figure 10
Plots of the individual rate parameters *a* and *b*, as a function of temperature for different velocity-steps for sample ZF01

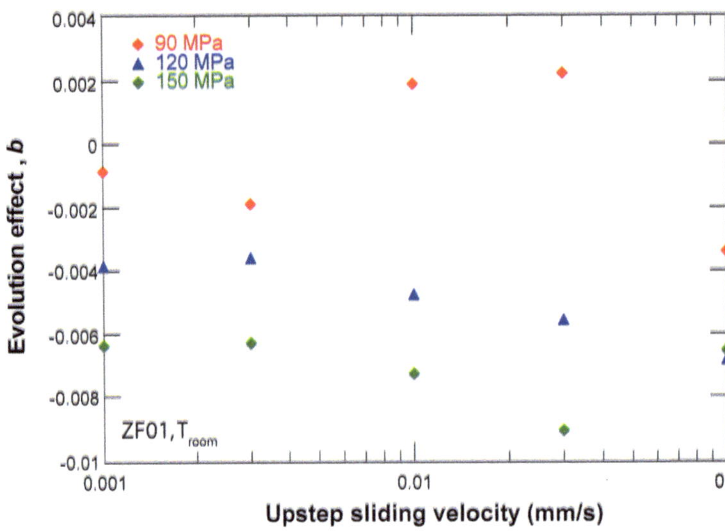

Figure 11
Semi-log plot of the evolution effect parameter, *b*, as a function of upstep sliding velocity for different effective normal stresses at room temperature on sample ZF01 (experiment u064, see Table 2)

(Saffer and Marone 2003; Ikari *et al.* 2009), which means that the contact area is maximized due to alignment of the platy phyllosilicates so that there is no evolution of contacts upon changes in sliding velocity. However, we observe mostly negative values for *b* in our room temperature experiments (Fig. 8) that become more negative with increasing sliding velocity. A negative *b*-value cannot easily be explained in terms of contact area evolution. Alternative explanations invoke kinking of the phyllosilicate grains (Ikari *et al.* 2009), a mechanical effect of an undrained pore-fluid pressure drop as a result of dilation (Ikari *et al.* 2009) or frictional heating, thus causing "thermal strengthening" (Kilgore *et al.* 1993). Frictional heating seems unlikely at these low sliding velocities, although the more negative *b*-values with higher velocity are consistent with this idea.

However, the thermal strengthening that is necessary to achieve this, is not consistent with our observation of lower friction at elevated temperature. In contrast, at higher temperature we observe an increase in *b*-values (Fig. 10b), indicating weakening during the evolution of friction at higher temperatures. A similar reasoning can be applied to the possible effect of local pore-fluid pressure decrease. Calculations by Ikari *et al.* (2009) show that the diffusion time is much too short for this to be an effective mechanism at the low sliding velocities employed here. Our sample size is only ~10 % of that of Ikari *et al.* (2009), so that diffusion times will be even shorter in our experiments. Kinking of talc is rarely observed and only at very high temperature (Escartín *et al.* 2008; Viti and Collettini 2009). Instead, talc typically accommodates strain by delamination and interlayer slip, due to the weak Van Der Waals bonds that link adjacent layers (Viti and Collettini 2009). In summary, the proposed explanations for a negative *b*-value do not appear to be valid for our data set. As an alternative explanation, we might consider the thickness and nature of the adsorbed water films. Since the sliding of wet sheet silicates seems to be controlled by the presence of thin water films on the (001) planes (Moore and Lockner 2004), the properties of the water–solid interface and/or those of the water film itself will control friction and its velocity dependence. If we assume that dilation always occurs upon a velocity step (e.g., Ikari *et al.* 2009) and that fluid flow is fast, the thickness of the fluid film will transiently increase, leading to a lower strain rate in the fluid film. The viscosity of a thin fluid film might change when the film becomes very thin (Israelachvili 1986). Based on the model of Renard and Ortoleva (1997), the thickness of a fluid film in between two quartz grains would be on the order of 5–10 Å at 50–100 MPa, which could decrease viscosity (Israelachvili 1986). Regardless of a possible viscosity change, a reduction of the thickness of the fluid film after the instantaneous thickening would increase the strain rate in the fluid film, and thus lead to strengthening. The decrease in *b* with increasing normal stress at room temperature (Fig. 11) fits with this explanation, since the change in strain rate and thus the change in stress will be relatively larger for thinner fluid films (at higher normal stresses).

As temperature increases, we observe different trends in *a* and *b* for different velocities. For low sliding velocities, the value of *a* is largest at 150 °C, whereas for high sliding velocities, we see a monotonic increase in *a* with temperature (Fig. 10a). For the *b*-value, we observe a more or less continuous increase with temperature, with the exception of the fastest velocity-step (30–100 μm/s), which shows a decrease in *b* with increasing temperature. As mentioned in Sect. 2.2, in granular gouges, dilation contributes to the value of *a*. We speculate that the velocity-dependent changes in *a* with temperature are most likely related to changes in pressure solution compaction rates of the calcite grains. The strain rate-dependent shear strength of the sliding contacts (which in this case are mostly likely the phyllosilicates) are expected to decrease with increasing temperature, which is contrary to the observed increase in *a*. Therefore, we might argue that the increase in *a* is due to an increased contribution of dilation due to enhanced compaction rates of the calcite grains at elevated temperature. Despite the fact that almost all sliding is most likely accommodated by the phyllosilicates, the calcite grains are probably involved in the dilation upon a change in sliding velocity to some extent. The contribution of dilational work to *a* counters the expected decrease of *a* as a result of thermal lowering of the contact shear strengths, which might explain why *a* does not increase further at 300 °C. This explanation fits with the observed increase of the *b*-value with temperature, at least for upstep velocities lower than 100 μm/s, and also fits with the observed decrease in *b* with sliding velocity (note that all *b*-values except that at 100 μm/s are positive at 300 °C).

5.2.6 *Deformation Mechanisms and Potential Effects on (a–b): Intermediate Phyllosilicate Content*

Here, we will discuss the possible deformation mechanisms that might operate in our samples ZF02 and ZF03 with an intermediate phyllosilicate content. Both samples have considerable amounts of phyllosilicates, as is also demonstrated by their friction values at room temperature, which are lower than typical Byerlee values. However, the velocity dependence of friction of these samples is quite different at elevated temperature conditions (Fig. 7). Whereas sample ZF02 shows a similar decrease in (*a*–*b*) with increasing temperature as sample ZF01, the (*a*–*b*)-values for sample ZF03 hardly change with temperature. In addition, the velocity dependence of (*a–b*) for sample ZF02 is log-linear with an increasing slope with increasing temperature, whereas the velocity dependence of (*a–b*) for ZF03 is absent at T_{room}, similar to that of ZF02 at 150 °C and lower at 300 °C. The main difference between the two samples is the absence of the amphibole tremolite in sample ZF03. The absence of tremolite in sample ZF03 might allow this material to deform via frictional-viscous flow, at least under hydrothermal conditions.

Frictional-viscous flow involves the sliding of competent, strong, but dissolvable material over a frictionally-weak, anastomosing phyllosilicate foliation with accommodation of geometrical incompatibilities via dissolution, diffusion, and precipitation (Bos and Spiers 2001, 2002; Niemeijer and Spiers 2005, 2006, 2007). However, if these coupled processes operate so effectively in sample ZF03, why would it not operate in sample ZF02, which has higher contents of phyllosilicates? A possible explanation is the presence of tremolite (in ZF02), which might change the pore-fluid chemistry in such a way that dissolution rates of calcite are slowed down. In addition, the presence of a third phase will inevitably interfere with the development of an anastomosing network of talc grains with intervening calcite clasts, as we envision for sample ZF03. If this is the case, then deformation in sample ZF02 is most likely very similar to that in ZF01, i.e., sliding occurs mostly over and between phyllosilicate grains, with the calcite and tremolite grains acting to promote dilation, and thus producing (*a–b*)-values that increase with increasing sliding velocity and decrease with temperature.

An additional argument for the operation of frictional-viscous flow in sample ZF03 might be that the values of (*a–b*) decrease with decreasing effective normal stress. A lower normal stress will reduce pressure solution rates, so that shear accommodation via this process is no longer possible and the gouge has to dilate to allow for slip. The steady-state porosity will then depend on the competition between compaction and dilation, and will thus increase with sliding velocity. Assuming a direct link between porosity, contact area and contact angle, and friction, this will lead to velocity-weakening at lower effective normal stresses.

5.2.7 *Deformation Mechanisms and Potential Effects on (a–b): Low Phyllosilicate Content*

Our samples ZF04 and ZF05 contain <10 wt% phyllosilicates, which renders it unlikely that the phyllosilicates affect the frictional properties of the simulated gouges substantially. Both samples show room-temperature values of friction typical of Byerlee's rule (0.7–0.8, Byerlee 1978), but with significant weakening with increasing temperature, down to values of 0.5–0.6. This is similar to what has been observed for wet limestone (Verberne *et al.* 2010), as discussed in Sect. 5.2. In addition, (*a–b*) decreases dramatically with increasing temperature, leading to unstable stick–slips in both samples at the lower sliding velocities. Lower effective normal stress and higher pore-fluid pressure have only a small effect on the (*a–b*)-values, whereas with increasing sliding velocity we observe a log-linear increase in (*a–b*) under hydrothermal conditions. All these observations indicate that deformation most likely involves a competition between cataclastic granular flow-type processes and thermally-activated, time-dependent processes. The fact that effective normal stress only has a small effect on (*a–b*) probably shows that grain-size reduction is negligible after the initial 10-mm run-in. This could be the result of an early onset of localization into Y- and B-shear bands, as was observed in calcite and limestone gouges (Verberne *et al.* 2010). The operation of dislocation and pressure solution creep processes in calcite particularly, but probably also in dolomite, are effective at temperatures of 150 and 300 °C (e.g., Zhang *et al.* 2010; Liteanu *et al.* 2012), and will thus affect friction and the velocity dependence of friction. The thermal weakening observed could reflect the increase in the rates of these processes and their contribution to the accommodation of the imposed shear deformation. In addition, these processes will lead to densification of the gouge, promoting localization of deformation into Y- and B-shears (e.g., Giger *et al.* 2008; Niemeijer *et al.* 2010a). The formation of a through-going boundary shear might lead to the occurrence of velocity-weakening (Scruggs and Tullis 1998; Niemeijer *et al.* 2010a), and could thus potentially explain the observed decrease in (*a–b*) with temperature. The observed increase in

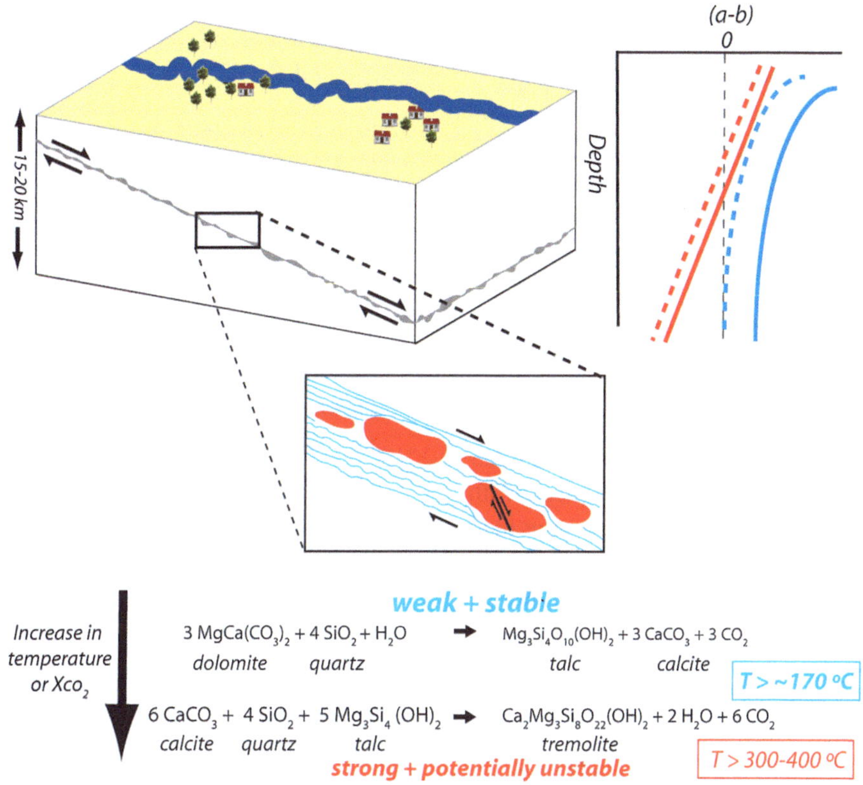

Figure 12
Schematic diagram illustrating the proposed structure, frictional properties, and mineralogical reactions of the Zuccale fault core

(*a*–*b*) with increasing sliding velocity could then be linked to an increasing width of the actively deforming zone. In order to evaluate this hypothesis, systematic experiments with microstructural observations at constant sliding velocity are needed.

5.3. *Implications for Low-Angle, Normal Faulting in Italy*

In Fig. 12, we show a schematic fault zone model, modified after Collettini *et al.* (2011). During the initial phases of deformation on the Zuccale fault, brittle deformation of the dolomitic host rock allowed the influx of hot, silica-rich fluids, resulting in the production of talc (Viti and Collettini 2009; Collettini *et al.* 2009a). The absence of a coexistence of tremolite and dolomite in the same fault rocks suggests that the CO_2 fraction in the fluid remained below ∼0.77 (e.g., Holness *et al.* 1997), suggesting a continuous fluid flux, since CO_2 is produced in the reaction. Fluctuations in temperature of as little as 20–50 °C, depending on the amount of CO_2, could cause the preferential production of tremolite over talc. However, in order to satisfy the constraint posed by the high angle of the Zuccale Fault to σ_1, displacement would only be possible on weak portions of the fault that contain significant amounts of talc (Smith and Faulkner 2010). Our experiments under basically in situ conditions confirm that friction is low enough for slip to occur, and that the slip will be conditionally stable (i.e., aseismic creep). The variability in the amounts of tremolite present in the exhumed fault core combined with our data, suggest that significant parts of the fault core would be too strong to slip, unless significant pore-fluid overpressures were developed. We suggest that this occurred in the Zuccale fault core as a result of the deformation on the talc-rich portions, trapping fluids. With increasing fluid pressure, slip on the portions of the fault poor in talc, but rich in tremolite becomes possible, which is potentially unstable (sample ZF02 and ZF04 at 300 °C, see Fig. 7c). If a seismic event occurs, the resulting rupture will be small because of two potential reasons: (1) fluid pressure might drop due to an increase in permeability, and (2) the rupture would inevitably have to propagate into the surrounding, seismically-creeping and velocity-strengthening talc-rich portions. The possibility of a rupture to propagate into weak, velocity-strengthening material has been suggested on the basis of models and experiments (Niemeijer and Spiers 2007; Faulkner *et al.* 2011; Noda and Lapusta 2013), but in order for this to occur, the rupture has to be energetic, which seems to be negated by the observed increase in (*a–b*) with increasing sliding velocity (see Table 4). However, we also observed a further decrease in the values of (*a–b*) for experiments with low effective normal stress and high pore-fluid pressure (Fig. 8), which suggests that larger ruptures might occur if pore pressure can remain high during the dynamic event.

This conceptual model for slip along the Zuccale fault is consistent with the observed behavior of the currently active Alto Tiberina fault, which is characterized by abundant microseismicity ($M < \sim 2.5$, Chiaraluce *et al.* 2007; Piccinini *et al.* 2009) and aseismic creep (Hreinsdottir and Bennet 2009). Larger earthquakes in the area ($M > 4.5$) occur on high-angle, normal faults in the hanging wall of the ATF, either antithetic or synthetic with the dip of the ATF (e.g. Collettini *et al.* 2006). It is not clear how slip on the ATF affects the nucleation of these large earthquakes, but our data suggests that the non-foliated portion of the Zuccale fault is potentially unstable under hydrothermal conditions (sample ZF05, Fig. 7b, c). If these strong fault rocks are hosted in the faults located in the hanging wall of the ATF, this could explain the occurrence of seismicity on high-angle structures that have their base in the low-angle ATF (e.g., Chiaraluce *et al.* 2007; Piccinini *et al.* 2009).

6. Conclusions

We performed a series of experiments under hydrothermal conditions on five powdered samples of fault rocks from the foliated fault core of the Zuccale low-angle, normal fault, to investigate the effect of hydrothermal conditions on frictional stability and strength. Based on our mechanical data, we can conclude that

1. Frictional strength of the foliated fault core is highly variable ($\mu = 0.3$–0.8) and precludes slip on a low-angle structure for all but one sample. The variation in friction decreases slightly with increasing temperature up to 300 °C due to weakening of the strongest samples.
2. The velocity-dependence of friction, expressed as the rate parameter (*a–b*), is positive for all samples under room-temperature conditions. With increasing temperature, (*a–b*) decreases, resulting in negative values for the strongest samples, which are low in phyllosilicate content, and causing unstable stick–slip events in our apparatus at 150 and 300 °C.
3. Under hydrothermal conditions (*a–b*) increases with sliding velocity for all samples, resulting in inherently stable, velocity-strengthening behavior at higher sliding velocity (10–100 μm/s, depending on the sample). This behavior could potentially limit rupture nucleation and/or propagation.
4. Low effective normal stress and/or high fluid pressure, i.e., overpressurized conditions, generally decrease (*a–b*)-values further under hydrothermal conditions, but the effect is small compared to the effect of temperature.
5. Negative values for the *b* parameter in the sample with the highest phyllosilicate content might be explained by transient changes in the thickness of the water layer supported by the sheet silicates.
6. We speculate that the velocity-dependence of (*a–b*) results from the competition between frictional (i.e., displacement-dependent) processes and the operation of time-dependent, thermally-activated, and fluid-assisted mechanisms operating in the strong, soluble mineral phases (calcite, dolomite, and perhaps tremolite, though little is known about deformation mechanisms in tremolite).
7. Displacement along the Zuccale fault most likely occurred in portions of the fault core with high talc-content, probably causing permeability reduction and the potential development of fluid overpressures, allowing potentially unstable slip in stronger parts of the fault core. The size of these potentially

unstable events is most likely restricted due to: (1) the velocity-strengthening behavior of the surrounding talc-rich fault rocks, (2) the observed increase in (*a–b*) with increasing sliding velocity, and (3) a coseismic decrease in fluid pressure.

Acknowledgments

We wish to thank Eimert de Graaff, Gert Kastelein, Thony van de Gon-Netscher, and Peter van Krieken for technical support, and an anonymous reviewer for his/her comments that helped to improve this paper. A. Niemeijer is/was supported by the Dutch Foundation for Scientific Research (NWO) through a VENI grant (nr. 863.09.013) and a VIDI grant (nr. 854.12.011), and by the ERC starting grant SEISMIC (nr. 335915). C. Collettini is supported by the ERC Starting Grant GLASS (Grant 259256).

References

Abers, G., Mutter, C., & Fang, J. (1997). *Shallow dips of normal faults during rapid extension: Earthquakes in the Woodlark–D'Entrecasteaux rift system, Papua New Guinea.* Journal of Geophysical Research, *102, B7*, 15301–15317.

Atkinson, B. K. (1982). *Subcritical crack propagation in rocks: theory, experimental results and applications.* Journal of Structural Geology, *4*, 41–56.

Barchi, M., Minelli, G., & Pialli, G. (1998). *The CROP 03 Profile: A synthesis of results on deep structures of the Northern Appennines.* Memorie della Societa Geologica Italiana, *52*, 383–400.

Baumberger, T., Berthoud, P., & Caroli, C. (1999). *Physical analysis of the state- and rate-dependent friction law. II. Dynamic friction.* Physical Review B, *60*(6), 3928–3939.

Beeler, N. M., Tullis, T. E., & Kronenberg, A. M. (2007). *The instantaneous rate dependence in low temperature laboratory rock friction and rock deformation experiments.* Journal of Geophysical Research, *112*(B07310). doi:10.1029/2005JB003772.

Blanpied, M. L., Tullis, T. E., & Weeks, J. D. (1987). *Frictional behavior of granite at low and high sliding velocities.* Geophysical Research Letters, *14*(5), 554-557.

Blanpied, M. L., Lockner, D. A., & Byerlee, J. D. (1995). *Frictional slip of granite at hydrothermal conditions.* Journal of Geophysical Research, *100*(B7), 13045–13064.

Blanpied, M. L., Tullis, T. E., & Weeks, J. D. (1998). *Effects of slip, slip rate, and shear heating on the friction of granite.* Journal of Geophysical Research, *103*(B1), 489–511.

Boncio, P., Brozzetti, F., & Lavecchia, G. (2000). *Architecture and seismotectonics of a regional low-angle normal fault zone in central Italy.* Tectonics, *19*(6), 1038–1055.

Bos, B., & Spiers, C. J. (2001). *Experimental investigation into the microstructural and mechanical evolution of phyllosilicate-bearing fault rock under conditions favouring pressure solution.* Journal of Structural Geology, *23*(8), 1187–2002.

Bos, B., & Spiers, C. J. (2002). *Frictional-viscous flow of phyllosilicate-bearing fault rock: microphysical model and implications for crustal strength profiles.* Journal of Geophysical Research, *107*(B2), doi:10.1029/2001JB000301.

Boulton, C., Carpenter, B. M., Toy, V., & Marone, C. (2012). *Physical properties of surface outcrop cataclastic fault rocks, Alpine Fault, New Zealand.* Geochemistry Geophysics Geosystems, *13*(1), 1-13. doi:10.1029/2011GC003872.

Byerlee, J. (1978). *Friction of rocks.* Pure and Applied Geophysics, *116*, 615–627.

Carpenter, B M, Marone, C., & Saffer, D. M. (2009). *Frictional behavior of materials in the 3D SAFOD volume.* Geophysical Research Letters, *36*(L05302), doi:10.1029/2008GL036660.

Carpenter, B. M., Marone, C., & Saffer, D. M. (2011). *Weakness of the San Andreas Fault revealed by samples from the active fault zone.* Nature Geoscience, *4*(4), 251–254. Nature Publishing Group. doi:10.1038/ngeo1089.

Chester, F. M., Evans, J. P., & Biegel, R. L. (1993). *Internal structure and weakening mechanisms of the San Andreas Fault.* Journal of Geophysical Research, *98*(B1), 771–786.

Chester, F. M. (1994). *Effects of temperature on friction: Constitutive equations and experiments with quartz gouge.* Journal of Geophysical Research, 99(B4), 7247–7261.

Chiaraluce, L., Chiarabba, C., Collettini, C., Piccinini, D., & Cocco, M. (2007). *Architecture and mechanics of an active low-angle normal fault: Alto Tiberina Fault, northern Apennines, Italy.* Journal of Geophysical Research, *112*(B10). doi:10.1029/2007JB005015.

Chiodini, G., Frondini, F., Cardellini, C., Parello, F. and Peruzzi, L. (2000). *Rate of diffuse carbon dioxide Earth degassing estimated from carbon balance of regional aquifers: The case of central Apennine, Italy.* Journal of Geophysical Research, *105*, 8423–8434.

Chiodini, G., Cardellini, C., Amato, A., Boschi, E., Caliro, S., Frondini, F., & Ventura, G. (2004). *Carbon dioxide Earth degassing and seismogenesis in central and southern Italy.* Geophysical Research Letters, *31*(7), L07615. doi:10.1029/2004GL019480

Collettini, C & Sibson, R. H. (2001). *Normal faults normal friction?* Geology, *29*, 927–930.

Collettini, C., & Holdsworth, R. E. (2004). *Fault zone weakening and character of slip along low-angle normal faults: insights from the Zuccale Fault, Isle of Elba, Italy.* Journal of Geological Society of London, *161*(6), 1039–1051. doi:10.1144/0016-764903-179.

Collettini, C., De Paola, N, Holdsworth, R.E. & Barchi, M.R (2006). *The development and behaviour of low-angle normal faults during Cenozoic asymmetric extension in the Northern Apennines, Italy.* Journal of Structural Geology, 28, 333–352. doi:10.1016/j.jsg.2005.10.003.

Collettini, C., Viti, C., Smith, S. A. F., & Holdsworth, R. E. (2009a). *The development of interconnected talc networks and weakening of continental low-angle normal faults.* Geology, *37*, 567–570. doi:10.1130/G25645A.1.

Collettini, C., Niemeijer, A., Viti, C., & Marone, C. (2009b). *Fault zone fabric and fault weakness.* Nature, *462*(7275),

907–910. Macmillan Publishers Limited. doi:10.1038/nature08585.

Collettini, C. (2011). *The mechanical paradox of low-angle normal faults: Current understanding and open questions.* Tectonophysics, *510*(3-4), 253–268. doi:10.1016/j.tecto.2011.07.015.

Collettini, C., Niemeijer, A., Viti, C., Smith, S. a. F., & Marone, C. (2011). *Fault structure, frictional properties and mixed-mode fault slip behavior.* Earth and Planetary Science Letters, *311*(3-4), 316–327. Elsevier B.V. doi:10.1016/j.epsl.2011.09.020.

Den Hartog, S. A. M., Peach, C. J., De Winter, D. A. M., Spiers, C. J., & Shimamoto, T. (2012a). *Frictional properties of megathrust fault gouges at low sliding velocities: New data on effects of normal stress and temperature.* Journal of Structural Geology, *38*, 156–171. doi:10.1016/j.jsg.2011.12.001.

Den Hartog, S. A. M., Niemeijer, A. R., & Spiers, C. J. (2012b). *New constraints on megathrust slip stability under subduction zone P–T conditions.* Earth and Planetary Science Letters, *353–354*, 240–252. doi:10.1016/j.epsl.2012.08.022.

Den Hartog, S. A. M., Niemeijer, A. R., & Spiers, C. J. (2013). *Friction on subduction megathrust faults: Beyond the illite–muscovite transition.* Earth and Planetary Science Letters, *373*, 8–19. doi:10.1016/j.epsl.2013.04.036.

Dieterich, J. H. (1978). *Time-dependent friction and the mechanics of stick-slip.* Pure and Applied Geophysics, *116*, 790–806.

Elter, P., Giglia, G., Tongiorgi, M., & Trevisan, L. (1975). *Tensional and compressional areas in recent (Tortonian to Present) evolution of the Northern Apennines.* Bolletino di Geofisica Teorica ed Applicata, *17*.

Escartín, J., Andreani, M., Hirth, G., & Evans, B. (2008). *Relationships between the microstructural evolution and the rheology of talc at elevated pressures and temperatures.* Earth and Planetary Science Letters, *268*(3–4), 463–475. doi:10.1016/j.epsl.2008.02.004.

Faulkner, D. R., Lewis, A. C., & Rutter, E. H. (2003). *On the internal structure and mechanics of large strike-slip fault zones: field observations of the Carboneras fault in southeastern Spain.* Tectonophysics, *367*(3-4), 235–251.

Faulkner, D. R., Mitchell, T. M., Behnsen, J., Hirose, T., & Shimamoto, T. (2011). *Stuck in the mud? Earthquake nucleation and propagation through accretionary forearcs.* Geophys. Res. Lett., *38*(18), L18303. doi:10.1029/2011gl048552.

Giger, S. B., Cox, S. F., & Tenthorey, E. (2008). *Slip localization and fault weakening as a consequence of fault gouge strengthening – Insights from laboratory experiments.* Earth and Planetary Science Letters, *276*(1-2), 73–84.

Gratier, J-P., D. Dysthe, F. Renard (2013). *The role of pressure solution creep in the ductility of the Earth's upper crust.* Advances in Geophysics, *54*. doi: 10.1016/B978-0-12-380940-7.00002-0.

He, C., Yao, W., Wang, Z., & Zhou, Y. (2007). *Strength and stability of frictional sliding of gabbro gouge at elevated temperatures.* Tectonophysics, *427*(1-4), 217–229. doi:10.1016/j.tecto.2006.05.023.

Holness, M. (1997). *Fluid flow paths and mechanisms of fluid infiltration in carbonates during contact metamorphism: the Beinn an Dubhaich aureole, Skye.* Journal of Metamorphic Geology, 59–70.

Hreinsdottir, S., & Bennett, R. a. (2009). *Active aseismic creep on the Alto Tiberina low-angle normal fault, Italy.* Geology, *37*(8), 683–686. doi:10.1130/G30194A.1.

Ikari, M. J., Saffer, D. M., & Marone, C. (2009). *Frictional and hydrologic properties of clay-rich fault gouge.* Journal of Geophysical Research, *114*(B5), B05409. doi:10.1029/2008JB006089.

Ikari, M. J., Marone, C., & Saffer, D. M. (2010). *On the relation between fault strength and frictional stability.* Geology, *39*(1), 83–86. doi:10.1130/g31416.1.

Ikari, M. J., & Saffer, D. M. (2011). *Comparison of frictional strength and velocity dependence between fault zones in the Nankai accretionary complex.* Geochemistry, Geophysics, Geosystems, *12*(4). doi:10.1029/2010GC003442.

Israelachvili, J. N. (1986). *Measurement of the viscosity of liquids in very thin films.* Journal of Colloid and Interface Science, *110*(1), 263–271. doi:10.1016/0021-9797(86)90376-0.

Imber, J., Holdsworth, R. E., Smith, S. a. F., Jefferies, S. P., & Collettini, C. (2008). *Frictional-viscous flow, seismicity and the geology of weak faults: a review and future directions.* Geological Society, London, Special Publications, *299*(1), 151–173. doi:10.1144/SP299.10.

Jefferies, S. P., Holdsworth, R. E., Wibberley, C. A. J., Shimamoto, T., Spiers, C. J., Niemeijer, A. R., & Lloyd, G. E. (2006). *The nature and importance of phyllonite development in crustal-scale fault cores: an example from the Median Tectonic Line, Japan.* Journal of Structural Geology, 1–16.

Jolivet, L., Faccenna, C., Goffé, B., Mattei, M., Rossetti, F., Brunet, C., Storti, F., et al. (1998). *Midcrustal shear zones in postorogenic extension: Example from the northern Tyrrhenian Sea.* Journal of Geophysical Research, *103*(97).

Kilgore, B. D., Blanpied, M. L., & Dieterich, J. H. (1993). *Velocity dependent friction of granite over a wide range of conditions.* Geophysical Research Letters, *20*(10), 903–906.

Liteanu, E., Niemeijer, a., Spiers, C. J., Peach, C. J., & De Bresser, J. H. P. (2012). *The effect of CO_2 on creep of wet calcite aggregates.* Journal of Geophysical Research, *117*(B3), B03211. doi:10.1029/2011JB008789.

Marone, C., Raleigh, B., & Scholz, C. H. (1990). *Frictional Behavior and Constitutive Modeling of Simulated Fault Gouge.* Journal of Geophysical Research, *95*(B5), 7007–7025.

Marone, C. (1998). *Laboratory-derived friction laws and their application to seismic faulting.* Annual Reviews Earth and Planetary Science Letters, *26*, 643–696.

Moore, D. E., & Lockner, D. A. (2004). *Crystallographic controls on the frictional behavior of dry and water-saturated sheet structure minerals.* Journal of Geophysical Research, *109*(B03401). doi:10.1029/2003JB002582.

Moore, D. E., & Lockner, D. A. (2007). *Comparative deformation behavior of minerals in serpentinized ultramafic rock: application to the slab-mantle interface in subduction zones.* International Geology Review, *49*(5), 401–415.

Moore, D. E., & Lockner, D. a. (2008). *Talc friction in the temperature range 25°–400 °: relevance for fault-zone weakening.* Tectonophysics, *449*(1-4), 120–132. doi:10.1016/j.tecto.2007.11.039.

Moore, D. E., & Lockner, D. a. (2011). *Frictional strengths of talc-serpentine and talc-quartz mixtures.* Journal of Geophysical Research, *116*(B1), 1–17. doi:10.1029/2010JB007881.

Nakatani, M. (2001). *Conceptual and physical clarification of rate and state friction: Frictional sliding as a thermally activated rheology.* Journal of Geophysical Research, *106*(B7), 13347–13380.

Niemeijer, A.R. & Spiers, C. J. (2006). *Velocity dependence of strength and healing behaviour in simulated phyllosilicate-*

bearing fault gouge. Journal of Geophysical Research, *427*(1-4), 231–253. doi:10.1016/j.tecto.2006.03.048.

Niemeijer, A. R. & Spiers, C. J. (2005). Influence of phyllosilicates on fault strength in the brittle-ductile transition: Insights from rock analogue experiments. In D. Bruhn & L. Burlini (Eds.), *High-Strain Zones: Structure and Physical Properties* (Vol. 245, pp. 303–327). Nice: Geological Society of London. doi:10.1144/GSL.SP.2005.245.01.15.

Niemeijer, A. R., & Spiers, C. J. (2007). *A microphysical model for strong velocity weakening in phyllosilicate-bearing fault gouges*. Journal of Geophysical Research, *112*(B10405). doi:10.1029/2007JB005008.

Niemeijer, A. R., Spiers, C. J., & Peach, C. J. (2008). *Frictional behaviour of simulated quartz fault gouges under hydrothermal conditions: Results from ultra-high strain rotary shear experiments*. Tectonophysics, *460*(1-4), 288–303.

Niemeijer, A., Marone, C., & Elsworth, D. (2010a). *Frictional strength and strain weakening in simulated fault gouge: Competition between geometrical weakening and chemical strengthening*. Journal of Geophysical Research, *115*(B10), B10207. doi:10.1029/2009jb000838.

Niemeijer, A., Marone, C., & Elsworth, D. (2010b). *Fabric induced weakness of tectonic faults*. Geophys. Res. Lett., *37*(3), L03304. AGU. doi:10.1029/2009gl041689.

Piccinini, D., Piana Agostinetti, N., Roselli, P., Seht, M. I., & Braun, T. (2009). *Analysis of small magnitude seismic sequences along the Northern Apennines (Italy)*. Tectonophysics, *476*(1–2), 136–144. doi:10.1016/j.tecto.2009.04.005.

Noda, H., & Lapusta, N. (2013). *Stable creeping fault segments can become destructive as a result of dynamic weakening*. Nature. doi:10.1038/nature11703.

Numelin, T., Marone, C., & Kirby, E. (2007). *Frictional properties of natural fault gouge from a low-angle normal fault, Panamint Valley, California*. Tectonics, *26*(2), TC2004. AGU. doi:10.1029/2005tc001916.

Reinen, L. A., Weeks, J. D., & Tullis, T. E. (1994). *The frictional behavior of lizardite and antigorite serpentinites: Experiments, constitutive models and implications for natural faults*. Pure and Applied Geophysics, *143*, 317–358.

Reinen, L. A., & Weeks, J. D. (1993). *Determination of rock friction constitutive parameters using an iterative least squares inversion method*. Journal of Geophysical Research, *98*(B9), 15937. doi:10.1029/93JB00780.

Renard, F., & Ortoleva, P. (1997). *Water films at grain–grain contacts: Debye-Hückel, osmotic model of stress, salinity, and mineralogy dependence*. Geochimica et Cosmochimica Acta, *61*(10), 1963–1970.

Rigo, A., Lyon-Caen, H., Armijo, R., Deschamps, A., Hatzfeld, D., Makropoulos, K., Papadimitriou, P., et al. (1996). *A microseismic study in the western part of the Gulf of Corinth (Greece): Implications for large-scale normal faulting mechanisms*. Geophysical Journal International, 663–688.

Ruina, A., (1983). *Slip instability and state variable friction laws*. Journal of Geophysical Research, *88* (B12), 10,310–359,370.

Saffer, D. M., & Marone, C. (2003). *Comparison of smectite- and illite-rich gouge frictional properties: application to the updip limit of the seismogenic zone along subduction megathrusts*. Earth and Planetary Science Letters, *215*(1-2), 219–235.

Sammis, C. G., & Steacy, S. J. (1994). *The Micromechanics of Friction in a Granular Layer*. Pure and Applied Geophysics, *142*(3/4), 778–794.

Scuderi, M. M., Niemeijer, A. R., Collettini, C., & Marone, C. (2013). *Frictional properties and slip stability of active faults within carbonate–evaporite sequences: The role of dolomite and anhydrite*. Earth and Planetary Science Letters, 1–13. doi:10.1016/j.epsl.2013.03.024.

Scruggs, V. J., & Tullis, T. E. (1998). *Correlation between velocity dependence of friction and strain localization in large displacement experiments on feldspar, muscovite and biotite gouge*. Tectonophysics, *295*, 15–40.

Sibson, R. H. (1977). *Fault rocks and fault mechanisms*. Journal of the Geological Society of London, *133*(3), 191–213. doi:10.1144/gsjgs.133.3.0191.

Smith, S. A. F., Holdsworth, R. E., Collettini, C., & Imber, J. (2007). *Using footwall structures to constrain the evolution of low-angle normal faults*. Journal of the Geological Society, *164*(6), 1187–1191. doi:10.1144/0016-76492007-009.

Smith, S. A. F., Collettini, C., & Holdsworth, R. E. (2008). *Recognizing the seismic cycle along ancient faults: CO_2-induced fluidization of breccias in the footwall of a sealing low-angle normal fault*. Journal of Structural Geology, *30*(8), 1034–1046. doi:10.1016/j.jsg.2008.04.010.

Smith, S. A. F., & Faulkner, D. R. (2010). *Laboratory measurements of the frictional properties of a natural low-angle normal fault: The Zuccale fault, Elba Island, Italy*. Journal of Geophysical Research. doi:10.1029/2008JB006274.

Tembe, S., Lockner, D. A., & Wong, T.-F. (2010). *Effect of clay content and mineralogy on frictional sliding behavior of simulated gouges: Binary and ternary mixtures of quartz, illite, and montmorillonite*. Journal of Geophysical Research, *115*(B3), B03416. doi:10.1029/2009JB006383.

Tesei, T., Collettini, C., Carpenter, B. M., Viti, C., & Marone, C. (2012). *Frictional strength and healing behavior of phyllosilicate-rich faults*. Journal of Geophysical Research, *117*(B9), B09402. doi:10.1029/2012JB009204.

Trippetta, F., Collettini, C., Barchi, M. R., Lupattelli, a., & Mirabella, F. (2013). *A multidisciplinary study of a natural example of a CO_2 geological reservoir in central Italy*. International Journal of Greenhouse Gas Control, *12*, 72–83. doi:10.1016/j.ijggc.2012.11.010.

Van Diggelen, E. W. E., De Bresser, J. H. P., Peach, C. J., & Spiers, C. J. (2010). *High shear strain behaviour of synthetic muscovite fault gouges under hydrothermal conditions*. Journal of Structural Geology, *32*(11), 1685–1700. doi:10.1016/j.jsg.2009.08.020.

Verberne, B.A.V., He, C., & Spiers, C. J. (2010). *Frictional Properties of Sedimentary Rocks and Natural Fault Gouge from the Longmen Shan Fault Zone, Sichuan, China*. Bulletin of the Seismological Society of America, *100*(5B), 2767–2790. doi:10.1785/0120090287.

Verberne, B. A., de Bresser, J. H. P., Niemeijer, A. R., Spiers, C. J., de Winter, D. A. M., Plümper, O. (2013). *Nanocrystalline slip zones in calcite fault gouge show intense crystallographic preferred orientation: Crystal plasticity at sub- seismic slip rates at 18 – 150° C*, 18–21. doi:10.1130/G34279.1.

Viti, C., & Collettini, C. (2009). *Growth and deformation mechanisms of talc along a natural fault: a micro/nano-structural investigation*, 529–542. doi:10.1007/s00410-009-0395-4.

Zhang, X., Spiers, C. J., & Peach, C. J. (2010). *Compaction creep of wet granular calcite by pressure solution at 28°C to 150°C*. Journal of Geophysical Research, *115*(B9), B09217. doi:10.1029/2008jb005853.

(Received June 14, 2013, revised October 11, 2013, accepted December 3, 2013, Published online December 15, 2013)

Pure Appl. Geophys. 171 (2014), 2665–2684
© 2013 Springer Basel
DOI 10.1007/s00024-013-0721-7

Pure and Applied Geophysics

Frequency–Magnitude Distribution of $-3.7 \leq M_W \leq 1$ Mining-Induced Earthquakes Around a Mining Front and b Value Invariance with Post-Blast Time

MAKOTO NAOI,[1] MASAO NAKATANI,[1] SHIGEKI HORIUCHI,[2] YASUO YABE,[3] JOACHIM PHILIPP,[4] THABANG KGARUME,[5] GILBERT MOREMA,[6] SIFISO KHAMBULE,[7] THABANG MASAKALE,[7] LUIZ RIBEIRO,[6] KOJI MIYAKAWA,[1] ATSUSHI WATANABE,[1] KENSHIRO OTSUKI,[8] HIROKAZU MORIYA,[9] OSAMU MURAKAMI,[10] HIRONORI KAWAKATA,[11] NANA YOSHIMITSU,[12] ANTHONY WARD,[6] RAY DURRHEIM,[5] and HIROSHI OGASAWARA[11]

Abstract—We investigated frequency-magnitude distribution (FMD) of acoustic emissions (AE) occurring near an active mining front in a South African gold mine, using a catalog developed from an AE network, which is capable of detecting AEs down to M_W −5. When records of blasts were removed, FMDs of AEs obeyed a Gutenberg–Richter law with similar b values, not depending on post-blasting time from the initial 1-min interval through more than 30 h. This result denies a suggestion in a previous study (RICHARDSON and JORDAN Bull Seismol Soc Am, 92:1766–1782, 2002) that new fractures generated by blasting disturb the size distribution of background events, which they interpreted as slip events on existing weak planes. Our AE catalog showed that the GR law with $b \sim 1.2$ was valid between M_W −3.7 and 0 for AEs around the mining front. Further, using the mine's seismic catalog, which covers a longer time period of the same area, we could extend the validity range of the GR law with the same b value up to M_W 1.

Key words: Acoustic emission, Gutenberg Richter law, Mining-induced earthquake.

[1] Earthquake Research Institute, The University of Tokyo, 1-1-1 Yayoi, Bunkyo-ku, Tokyo 113-0032, Japan. E-mail: naoi@eri.u-tokyo.ac.jp

[2] Home Seismometer Corp., 4-36, Uenohara, Shirakawa, Fukushima 961-0026, Japan.

[3] Research Center for Prediction of Earthquakes and Volcanic Eruptions, Graduate School of Science, Tohoku University, 6-6 Aza-Aoba, Aramaki, Aoba-ku, Sendai 980-8578, Japan.

[4] Gesellschaft für Materialprüfung und Geophysik mbH, Dieselstraße 6a, 61239 Ober-Mörlen, Germany.

[5] Council for Scientific and Industrial Research, P.O. Box 395, Pretoria 0001, South Africa.

[6] SeismoGen cc, P.O. Box 1177, Carletonville 2500, South Africa.

[7] Open House Management Solutions, P.O. Box 200, Potchefstroom 2520, South Africa.

[8] Department of Earth Science, Graduate School of Science, Tohoku University, 6-3 Aza-Aoba, Aramaki, Aoba-ku, Sendai 980-8578, Japan.

[9] Graduate School of Engineering, Tohoku University, 6-3, Aza-Aoba, Aramaki, Aoba-ku, Sendai 980-8578, Japan.

[10] Research Organization of Science and Engineering, Ritsumeikan University, 1-1-1 Noji Higashi, Kusatsu 525-8577, Japan.

[11] College of Science and Engineering, Ritsumeikan University, 1-1-1 Noji Higashi, Kusatsu 525-8577, Japan.

[12] Graduate School of Science and Engineering, Ritsumeikan University, 1-1-1 Noji Higashi, Kusatsu 525-8577, Japan.

1. *Introduction*

It is well known that the frequency-magnitude distribution (FMD) of earthquakes follows a power law (ISHIMOTO and IIDA 1939; GUTENBERG and RICHTER 1944). Such a relationship, commonly known as the Gutenberg–Richter (GR) law, indicates there is no characteristic scale for seismic events. In the meantime, characteristic scales are inherent features of friction laws, which are used to understand seismic source processes (e.g., IDA 1972; ANDREWS 1976; DIETERICH 1986; SCHOLZ 1988). This has led to the speculation there may be a lower limit to the magnitude range where the GR law remains valid (AKI 1987).

Some authors said they observed breakdowns of the GR law in real data below M 3 (AKI 1987; RYDELEK and SACKS 1989; MALIN *et al.* 1989; TAYLOR *et al.* 1990; UMINO and SACKS 1993), while others said they found the GR law valid down to around M 0 (ABERCROMBIE and BRUNE 1994; ABERCROMBIE 1996). There is a need to observe larger numbers of microearthquakes to establish whether the GR law breaks down for earthquakes of smaller magnitudes.

Gold mines in South Africa are home to a large number of mining-induced earthquakes, with the moment magnitude M_W seldom exceeding 3. They allow seismic stations to be installed near seismic sources, and therefore provide vantage points from where to observe microearthquakes more easily and more precisely than by surface-based observations. A number of previous studies have used that environment to search for any deviations from the GR law (e.g., Richardson and Jordan 2002; Boettcher *et al.* 2009; Kwiatek *et al.* 2010).

To investigate size distributions in gold mines in South Africa, it is necessary to distinguish "blasts" from other seismic events. Some studies have pointed out the FMD of mining-induced earthquakes have bimodal shapes with a kink near M_W 0, where the slope is steeper for smaller earthquakes than for larger events (e.g., Finnie 1999; Richardson and Jordan 2002). Richardson and Jordan (2002) proposed that such a pattern indicated the coexistence of two different types of earthquakes; that is, "fracture-dominated" events, which represent new fractures generated in the intact rock as a more direct consequence of mining blasts, and "friction-dominated" events, which represent slippage along existing weak planes. Spottiswood and Linzer (2004) and Spottiswood (2010), however, pointed out that all or almost all of the events classified as "fracture-dominated" by Richardson and Jordan (2002) were blasts themselves, from the size, location, and temporal occurrence pattern of them.

Indeed, Boettcher *et al.* (2009) showed that no deviation from the GR law appeared in FMDs of earthquakes in a mine, when analyzing only events in blast-free hours. They used records from a seismic network in the TauTona gold mine in South Africa, one of the gold mines studied by Richardson and Jordan (2002), only with a denser sensor configuration and enhanced detectability for smaller earthquakes. They demonstrated that the FMD of earthquakes during blast-free hours (00:00–10:59 and 20:00–23:59 Monday through Friday and around the clock Sunday, local time) obeyed the GR law down to M_W −1.3, the presumable detection limit of the seismic network.

The size distribution of further smaller seismic events in a mine was studied by Kwiatek *et al.* (2010). They used records of an acoustic emissions (AE) network, installed 3.3 km beneath the ground in the Mponeng gold mine in South Africa and capable of detecting events below M_W −4. They examined two types of AEs separately. One type examined was post-blasting AEs, which occurred in the neighborhood of a mining stope within 11 h of mining blasts (7 p.m.–6 a.m. on the next day. Incidentally, blasting is typically conducted around 7 p.m. in the mine). Another type examined was aftershocks that occurred within a week of an M 2 event. Kwiatek *et al.* (2010) showed post-blasting AEs obeyed the GR law between M_W −3.5 and −1.5, and the aftershocks, which occurred in a closer area to the network, obeyed the GR law between M_W −4.4 and −1.9.

If the results of Boettcher *et al.* (2009) and Kwiatek *et al.* (2010) are taken together, it may be suggested that size distributions of seismic events in mines follow the GR law between M_W −4.4, the lower limit of the magnitude range where Kwiatek *et al.* (2010) confirmed the GR law was valid, and M_W 2, an upper limit of a magnitude range where Boettcher *et al.* (2009) studied (although they did not mention the upper limit of validity of the GR law, FMDs shown in Fig. 2 in the article seem to follow the GR law up to M_W 2). However, some points need to be clarified to unambiguously conclude the validity of the GR law over the entire range.

1. Strictly speaking, the magnitude ranges where Kwiatek *et al.* (2010) and Boettcher *et al.* (2009) confirmed the GR law was valid did not overlap. It therefore remains uncertain whether the numbers of AEs counted by Kwiatek *et al.* (2010) coincide with extrapolations of the GR law that applies to the FMD of larger events. To check this, it is necessary either to compare the event rates, as Abercrombie and Brune (1994) did with data from California's Parkfield area, or to investigate the FMD using a data set that covers a broader magnitude range.
2. Kwiatek *et al.* (2010) used AEs that occurred in 11 h after blasting to study size distributions for post-blasting AEs. This data set covers a long time interval after blasting and might consist of two types of AEs having different FMDs, "fracture-dominated events" and "friction-dominated events", as suggested by Richardson and Jordan (2002). Although the existence of these two types was not established from the observations of

Richardson and Jordan (2002), as pointed out by Spottiswood and Linzer (2004) and Spottiswoode (2010), the idea that blasting generates new fractures, which may be strongly dominated by very small ones, sounds plausible. Therefore, the time variance of FMDs after blasting needs to be examined, especially for very small events. In addition, to combine the results for post-blasting AEs of Kwiatek *et al.* (2010) and the results of Boettcher *et al.* (2009) from mining-off hours, invariability of FMD with post-blasting time has to be established.

3. Even if FMD is invariant with post-blasting time, direct comparison of the results of Boettcher *et al.* (2009) and Kwiatek *et al.* (2010) may be inappropriate, because the event selection criteria were much different spatially and temporally. Boettcher *et al.* (2009) studied events occurring for a year and over a wide (several kilometers) region, including many different mining stopes along a tabular reef; whereas Kwiatek *et al.* (2010) studied two types of transient AE sequences, aftershocks (for 5 days) of an M_W 2 earthquake and post-blasting AEs (for 11 h) around a mining stope of $\sim$150 m extent.

To unambiguously conclude whether the GR law is valid over a wide magnitude range, we focus on events around a single mining front, which can overcome the problems mentioned above. We use two seismic catalogs based on different observation networks covering this area. Our study area is along a 100-m scale mining stope at 1 km depth in the Cooke 4 shaft (previously known as the Ezulwini gold mine) in South Africa. Acoustic emissions network (AEnet) can detect very small seismic events as far down the magnitude scale as in the study of Kwiatek *et al.* (2010). A seismic network operated by the mine (Mnet) has a longer time coverage. We also investigated the temporal dependency of FMDs with post-blasting time using the catalog from AEnet.

2. *Observations*

We installed six triaxial accelerometers and 24 more sensitive AE sensors within an area spanning 100 m in diameter in the Cooke 4 shaft (Fig. 1a). All sensors were grouted inside boreholes that were drilled from horizontal tunnels. This observation network was targeted at seismic events with M_W approximately from -5 to 0. A planar mining area, advancing from south to north, lay about 20 m above the horizontal tunnels from where we deployed our sensors. Our observation targeted seismic activity induced by that mining.

Three of the six accelerometers have a flat frequency response up to 25 kHz (Wilcoxon Research Inc. Model 736), whereas the other three have a flat response up to 10 kHz (Wilcoxon Model 728). The AE sensors are products of Germany's Gesellschaft für Materialprüfung und Geophysik mbH (GMuG) and can record signals with workable signal-to-noise (S/N) ratios up to about 50 kHz for a majority of events, although that may vary according to the input signal intensity. They consist of a columnar piezoelectric element, 31.75 mm in diameter and 57.2 mm in thickness, embedded in a brass housing and grouted into the boreholes for use. During our earlier AE observations in the Mponeng gold mine (Nakatani *et al.* 2008; Kwiatek *et al.* 2010), we used sensors of the type whereby a spring presses a sensor against the borehole bottom for coupling, but in the present study, we used the grouting-type sensors because of their better installation efficiency and stability.

We used GMuG's 48-channel AE monitoring system (Manthei 2005) as our data logging system. It can record 65.5 ms worth of waveform data from all channels at a rate of 500 kS/s (32,768 samples) when a predetermined trigger condition is met in any one channel. The 16-bit analog-to-digital conversion channels of the system are strictly synchronized to each other, and have built-in anti-aliasing filters.

A trigger generates dead time in the hundreds of milliseconds (the exact length varies with recording conditions), and the waveforms of events that occur during that interval are not recorded. The waveforms are analog high-pass filtered at 50 Hz for the accelerometers and at 1 kHz for the AE sensors before they are recorded.

The Cooke 4 shaft operates a network of eight geophones (velocimeters) with a characteristic frequency of 4.5 Hz, spaced at intervals of 200–300 m (Mnet; Fig. 1b), to observe earthquakes of M_W approximately from -1 to 3, larger than those

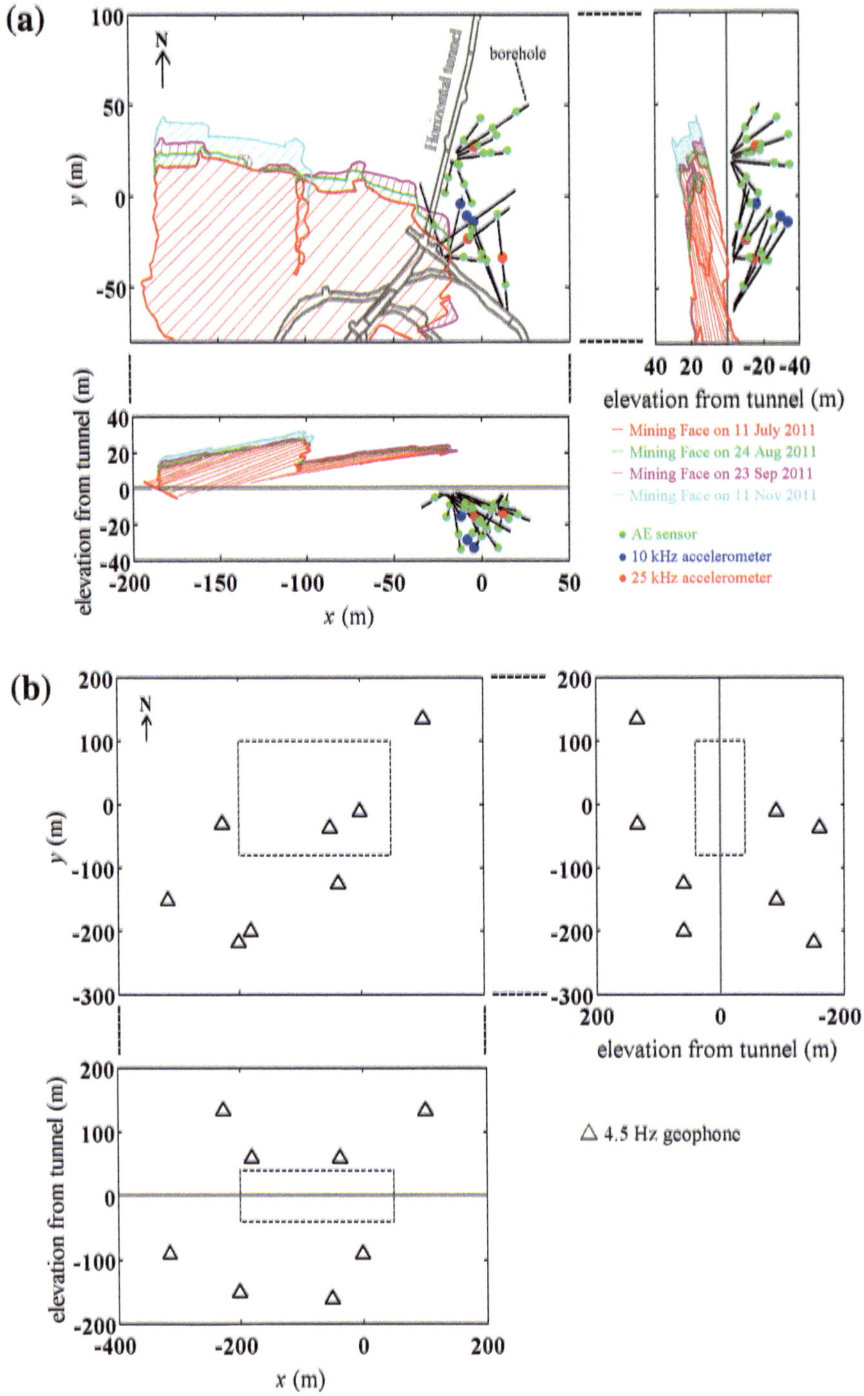

Figure 1

a Configuration of the AE monitoring system (AEnet) installed 1 km beneath the ground in the Cooke 4 shaft, South Africa. A plan view and side views from the south and the east are shown. The origin of coordinates in the plan view roughly coincides with AEnet's geometrical center of gravity. The elevations are in reference to the depth of the horizontal tunnels (*gray line*) from where AEnet was deployed. The mining face, advancing northward, was located about 20 m above our network. The hatched areas show mined-out cavities. **b** Configuration of the seismic network operated by the mine (Mnet). The *dashed contour box* corresponds to the area shown in **a**. Both **a** and **b** share the same coordinate system targeted by AEnet. They record seismic waveforms on trigger at sampling rates that vary between 200 and 2,000 Hz depending on the event size, whereby the Institute of Mine Seismology (IMS), a South African company, compiles an earthquake catalog (Mnet catalog) on a routine basis. In the present study, we used the hypocenter locations, seismic moment M_0, and M_W given in the Mnet catalog along with the hypocenter locations, M_0 and M_W in a catalog compiled from AEnet records (AEnet catalog). The IMS determines M_0 by fitting a single, theoretical displacement spectrum to displacement spectra from multiple seismic stations (Mendecki 1997). Mnet and AEnet are operated totally independently, with no clock synchronized and no waveform shared among them. We follow the clock of AEnet whenever we mention time in the present paper, unless otherwise noted.

We often refer to a seismic event detected by AEnet as "AE", and one detected by Mnet as "earthquake". We use these words just for the sake of convenience, and do not imply any difference.

3. AE Cataloguing Procedures

3.1. AE Array Data

In the present study, we used data obtained between 06:32:22 on 8 August 2011 and 09:54:58 on 10 October 2011, local time. There were no variations in the number of live sensors due to maintenance and other reasons during that period, but 30-min or longer measurement interruptions occurred on seven occasions (total 329 h). Waveform recordings were triggered on about 2.27 million occasions.

There were time intervals when an unusually high trigger rate lasted for several hours in daytime on weekdays, typically between 9:00–14:00. Similar waveforms were continuously stored during such periods. These were presumably due to working noise in the mine, such as vibration from drilling. Under virtually continuous noises exceeding the trigger level, more than 80 % of observations become dead time. We therefore identified and eliminated such noisy intervals by visually inspecting the waveforms recorded following all individual triggers during the study period. We labeled a total of 133 hours as the noisy intervals. We determined the hypocenters, M_0, and M_W for 1.52 million of the remaining triggers.

3.2. Hypocenter Determination

We read P and S wave arrival times and determined hypocenters from the waveforms following the 1.52 million triggers. We used automatic travel-time detection software of Japan's Home Seismometer Corp. (Horiuchi *et al.* 2011), which uses an array of standards to identify wave arrival times. The software is first fed with manual reading records as a learning aid to determine optimal values of parameters pertaining to the individual standards. Those optimal values are subsequently used to read other data automatically. In the present study, we used the manual reading results of Naoi *et al.* (2011) for some 25,000 events in the Mponeng gold mine as a learning aid for determining the parameters.

Of all events for which hypocenters were determined using the software, we used in the following analysis only 365,237 events that had at least ten P wave travel time readings and root-mean-square (rms) arrival-time residuals of 0.2 ms or less. We visually inspected the travel time readings, theoretical travel times and observed waveforms for several hundred among those events, and confirmed that obvious mistakes in readings led to erroneous hypocenter determinations in only <1 % of them.

We assumed a homogeneous medium with V_P = 5,700 m/s and V_S = 3,600 m/s to determine the hypocenters. Those seismic wave velocities were estimated by Kawakata *et al.* (2011) from sound transmission tests on the site.

3.3. Magnitude Estimation

We next evaluated the M_0 and the M_W for the 365,237 events by fitting theoretical spectra to ground displacement spectra derived from accelerograms and AE sensor records (see "Appendix B"). Because the frequency responses of the AE sensors were unavailable, we used an AE sensor-to-accelerometer frequency response ratio estimated in situ to estimate ground displacement spectra from AE sensor records (see "Appendix A"). The M_W was obtained for

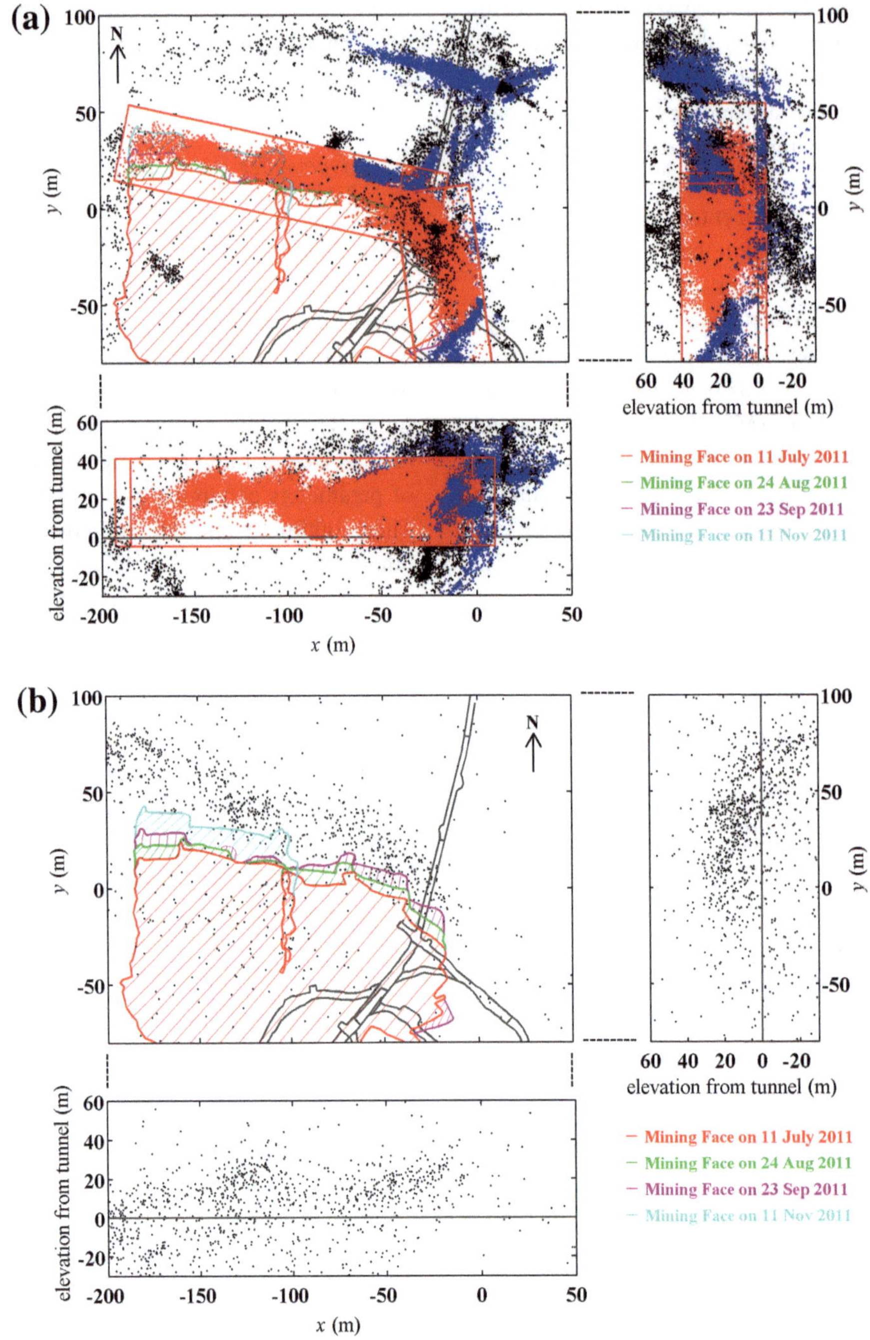

Figure 2

a Hypocenter distribution of the 358,093 events in the AEnet catalog with at least ten *P* wave travel time readings identified by the automatic travel-time detection software and with rms travel-time residuals of 0.2 ms or less. A plan view and side views from the south and the east are shown. *Blue dots* AE events that form planar clusters of two-dimensional expanse. *Red dots* the 324,873 events defined as constituting a "stope cluster" (or all events in the red boxes, which approximately cover the volumetric hypocenter distributions along the mining fronts, except for the ones shown by *blue dots*). *Black dots* the other events. **b** Hypocenter distribution of the 1,025 earthquakes detected by Mnet between 8 August and 10 October 2011

359,372, or more than 98 %, of all 365,237 events analyzed. The smallest M_W estimate was −5.3.

In the present study, we will compare the size distribution based on M_W estimates from the AEnet records (M_W^{AEnet}) and the M_W given in the Mnet catalog (M_W^{Mnet}). To check the consistency between them, we used the events that are listed in both the Mnet catalog and our AEnet catalog after correcting time difference between both system clocks ("Appendix C"). The comparison between M_W^{Mnet} and M_W^{AEnet} demonstrates that the M_W^{AEnet} tends to be underestimated with respect to the M_W^{Mnet} for events with $M_W^{\mathrm{Mnet}} \geq -1.3$ (Fig. 13). This difference was presumably caused by underestimation for large events in M_W^{AEnet} due to 50-Hz high-pass filtering for accelerograms records. The underestimation of M_W^{AEnet} was corrected using the empirical relationship (a kinked gray line in Fig. 13). We use the corrected M_W^{AEnet} whenever we mention M_W of AEs in the present paper, unless otherwise noted.

4. *Hypocenter Distribution*

Figure 2a shows the spatial distribution of the 358,093 of all 365,237 events in the AEnet catalog whose hypocenters fell within a 250 × 180 × 60 m^3 box region surrounding AEnet. Figure 2b is a plot of the 1,025 hypocenters in the Mnet catalog that were dated between 8 August and 10 October 2011, approximately the same as our study period for the AEnet data, and were located within the same box region. Both plots show that the hypocenters are aggregated near where the mining front lay at the time.

The hypocenters in the AEnet catalog appear to be located closer to the mining front than those in the Mnet catalog. But the mine's seismic network had a poorer precision of hypocenter location, and many of the events in the Mnet catalog presumably took place in the immediate vicinity of the mining front, where the AEnet catalog hypocenters are aggregated.

About 90 % of all AEs in Fig. 2a lie along the mining front. The bulk of the remaining 10 % or so form planar clusters of two-dimensional expanse, whereas very few AEs ever occur away from both the mining front and the planar clusters. Some of the planar clusters extend along existing fault planes known from geological surveys.

In the subsequent analysis, we use only the events along the mining front for the AEnet catalog, because this study aims to investigate FMDs of seismic events around the mining front, where strong influence of mining blasts are likely. We further exclude the planar clusters, which likely represent a different population and are not recognized from the Mnet catalog because they consist of only very small events below the detection limit of the Mnet. The planar clusters will be the subject of a separate, forthcoming research article. Specifically, we first extracted 340,698 events in two, partially overlapping boxes that approximately cover the regions of volumetric hypocenter distributions along the mining front (red boxes in Fig. 2a), and then removed 15,825 of them (blue dots in the red boxes) that we decided likely belong to the planar clusters. The remaining 324,873 AEs (red dots; about 95 % of the events in the red box) were defined as constituting a "stope cluster." As far as the AEnet catalog was concerned, we analyzed only the events in the stope cluster.

5. *How AE FMDs are Related with Post-Blast Time*

Figure 3 presents stope-cluster AE activity between 14:00 and 19:00 on 25 August 2011 that was likely related to blasting. Nine successive events with similar magnitudes (red circles in Fig. 3a; M_W from −1.0 to −0.2) were detected within the space of 30 s past 15:14, which falls in the typical blasting hours at the Cooke 4 shaft. The hypocenters of the nine events migrated systematically from east to west along the mining front of the time (Fig. 3a). We have noted frequent, similar sequences of M_W −1 class events followed by an immediate rise in the AE rate around 15:00 on weekdays and on some Saturdays, but never on Sundays. That pattern is consistent with the operation calendar of the mine.

The black squares in Fig. 3d represent the FMD of all stope-cluster events that were located within the blue frame of Fig. 3a. The graph is kinked at M_W −1 and M_W −1.5. Yet, such deviations from the GR law all but disappeared when the nine aforementioned events were excluded from the statistics (blue

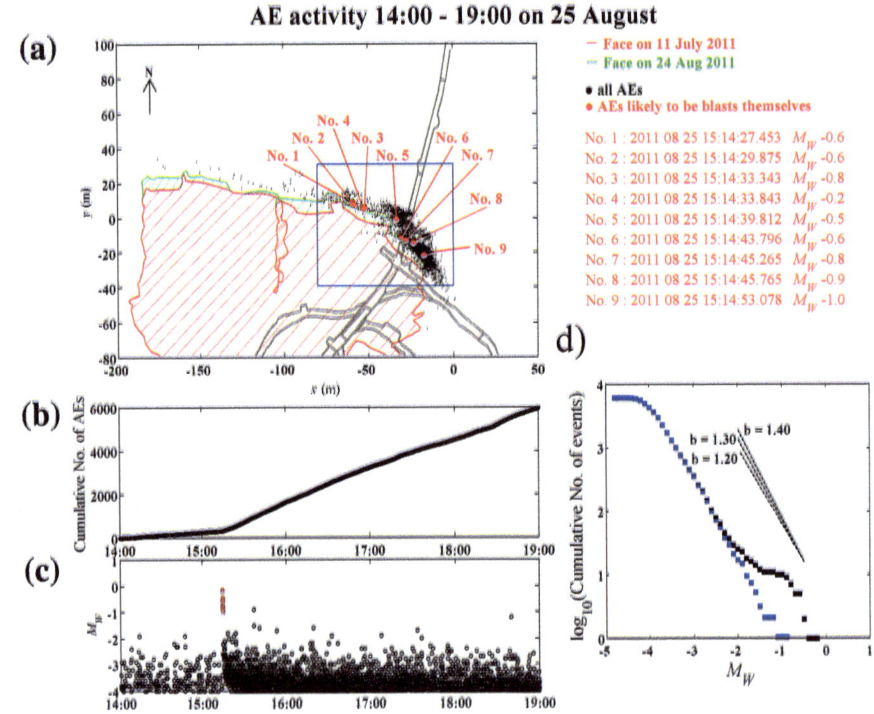

Figure 3

Typical post-blast AE activity. **a** Hypocenters of AEs in the stope cluster between 14:00 and 19:00 on 25 August 2011. Events likely representing blasts themselves have been labeled no. 1 through no. 9. Their occurrence times and M_W values are listed on the right. **b** Cumulative count of stope-cluster AEs within the blue frame in **a**. **c** Magnitude–time diagram of the same. *Red circles* indicate the nine events likely representing blasts themselves, highlighted in **a**. **d** FMD of the same. The *black squares* represent results based on all events, whereas the *blue squares* represent results with the nine events excluded

squares in Fig. 3d). This was also the case if M_W before correction ("Appendix C") was used (not shown).

To examine temporal variations in the size distributions of post-blast AEs with higher resolution, we assembled and stacked a large quantity of data pertaining to post-blast AEs within the blue frame, where detectability was particularly high due to the closeness to AEnet. We singled out 12 weekdays (including 25 August 2011, the date of the events shown in Fig. 3), on which M_W -1 class events, taking place within the region, coincided with a visible rise in event rates around 15:00. We aligned (Fig. 4a) and stacked (Fig. 4b, c) the counts of stope-cluster AEs on those 12 days in that region in terms of t_{pb}, or the time counted from the first of the successive M_W -1 class events in each sequence.

Figure 5 shows FMDs of the AEs shown in Fig. 4 grouped for different t_{pb} brackets, denoted by different colors. In addition, the FMD of stope-cluster events that occurred in the same area (i.e., the blue frame in Fig. 3a) on 21 August and 4 September 2011, which are not part of the aforementioned stacked data, are shown in gray to illustrate the size distribution of AEs with longer post-blast times. Both days are Sundays that followed blast-free Saturdays (20 August and 3 September), with the most recent blasts occurring no later than 15:30 on Friday (19 August and 2 September). (An event, likely a blast, took place in the study area at around 15:30 on 2 September, but it has yet to be learned whether any blast took place on 19 August.) That means gray represents the size distribution of AEs that occurred between 32.5 and 56.5 hours, or possibly more, after blasts. M_W used in Fig. 5 are those after correction by the curve of Fig. 13, but the following results are essentially the same if uncorrected M_W are used.

Figure 5 shows that the distributions follow the GR law without systematic dependence on post blasting time, except in the initial 1-min bracket (red

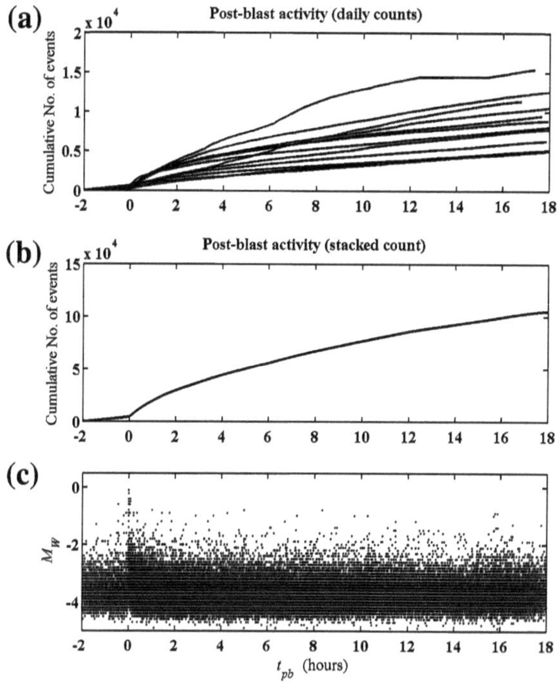

Figure 4
Stope-cluster AE activity in the blue frame of Fig. 3a on 12 days on which M_W −1 class events, likely representing mining blasts, took place inside the blue frame, and coincided with a visible rise in event rates. **a** Cumulative count of AE events on each day. The origin of the horizontal axis is set to coincide with the first of the blasts in each sequence. **b** Cumulative counts stacked over the 12 days. **c** Magnitude–time diagram stacked over the 12 days

filled triangles). We notice the initial 1-min bracket contains no events with $-1.9 < M_W < -1.4$, so the events in that category may be divided into a group of 169 events with $M_W \leq -1.9$ and another group of 39 events with $M_W \geq -1.4$. When the M_W −1.4 and greater events, which likely represent the mining blasts themselves, are excluded, the distribution in that time bracket follows the GR law with a similar b value as in the other time brackets (open red triangles in Fig. 5). In other words, all AEs induced by mining blasts have a size distribution that follows the GR law, with their b value invariant with post-blast time, from the immediate aftermath of the blasts through more than a day following them. Table 1 shows b values, their standard errors, and the minimum validity magnitude M_c of the Gutenberg–Richter law for each bracket, estimated for the FMDs based on the corrected M_W. Detailed methods of FMD analysis are shown in "Appendix D".

6. *Comparison Between FMDs of Events in the Mnet Catalog and the AEnet Catalog*

In the current section, we compare the size distributions of events in the Mnet catalog and our AEnet catalog to characterize their size distributions over a broader magnitude range.

To analyze the size distributions separately for blast-active hours and blast-free hours, we grouped AEs and earthquakes into two categories according to their occurrence times. Category 1 comprises the events that took place between 14:00 and 19:00 on weekdays, Monday through Friday, the typical blasting hours at the Cooke 4 shaft. Category 2 comprises all events that did not occur between 14:00 and 19:00, Monday through Saturday. This is roughly the same event-selection method that was used by Boettcher *et al.* (2009). We used the simple categorization because we found in the previous section that the size distribution of blast-induced AE events does not depend on post-blast time. We excluded the hours between 14:00 and 19:00 on Saturdays from both categories, because some Saturdays were free of blasts, but others were not.

From the AEnet catalog, we extracted only the events of the stope cluster that took place within the blue frame of Fig. 3a, which is the high detectability region near AEnet, to investigate the size distributions as far down the M_W scale as possible. As for the Mnet catalog, we used all events that were located across the entire region of Fig. 2 and occurred between 1 April 2011 and 1 December 2011. We used a longer time window and a larger volume window to extract events from the Mnet catalog than from the AEnet catalog to ensure a sufficient number of events are sampled, because the Mnet catalog targets events of larger magnitudes that are less frequent.

Figure 6 presents FMDs during the blast-free hours. The size distribution for AEnet catalog follows the GR law with a b value of 1.196 ± 0.003 from M_W −3.7 to M_W ∼0 (filled squares). A minor kink at $M_W \sim -1$ can be recognized in the FMD based on the uncorrected M_W (filled triangles), but it disappeared in the FMD based on the corrected M_W. The correction method of M_W ("Appendix C") seems to work appropriately.

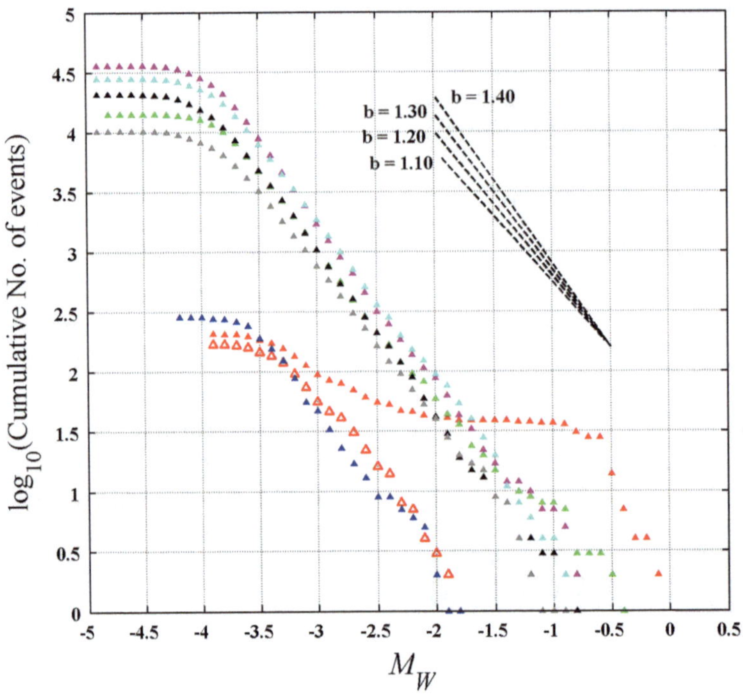

Figure 5
FMDs of post-blast AE activity, shown in stacks in Fig. 4b, c, and of activity on quiet days. The data include only stope-cluster events within the *blue frame* of Fig. 3a. *Filled red triangles* size distribution of AEs in the stacked data set for $0 \leq t_{pb} \leq 1$ min. *Open red triangles* size distribution of AEs for $0 \leq t_{pb} \leq 1$ min except for $M_W \geq -1.4$ events. *Blue triangles* 1 min $\leq t_{pb} \leq$ 2 min. *Green triangles* 2 min $\leq t_{pb} \leq$ 1 h. *Magenta triangles* 1 h $\leq t_{pb} \leq$ 6 h. *Black triangles* 6 h $\leq t_{pb} \leq$ 10 h. *Cyan triangles* 10 h $\leq t_{pb} \leq$ 18 h. *Gray triangles* FMD of events on Sundays that followed a blast-free Saturday (21 August 2011 and 4 September 2011), which corresponds to 32.5 h $\leq t_{pb} \leq$ 56.5 h or possibly more

Table 1

b values for post blasting AEs

Time from blasting	*b* value	*Mc*
$0 \leq t_{pb} \leq 1$ min except for $M_W \geq -1.4$ events	1.12 ± 0.11	−3.1
1 min $\leq t_{pb} \leq$ 2 min	1.31 ± 0.10	−3.4
2 min $\leq t_{pb} \leq$ 1 h	1.27 ± 0.02	−3.6
1 h $\leq t_{pb} \leq$ 6 h	1.34 ± 0.01	−3.7
6 h $\leq t_{pb} \leq$ 10 h	1.28 ± 0.01	−3.7
10 h $\leq t_{pb} \leq$ 18 h	1.22 ± 0.01	−3.7
32.5 h $\leq t_{pb} \leq$ 56.5 h or possibly more	1.21 ± 0.02	−3.6

The size distribution for Mnet catalog in Fig. 6 also follows the GR law with $b = 1.21 \pm 0.03$ from M_W −1.1 to M_W ~1 (open squares). Ranges where FMDs for AEnet and Mnet catalog follow the GR law with similar *b* values overlap between M_W −1.1 and M_W ~0, suggesting that events during the blast-free hours likely constitute a single population that follows the GR law with *b* ~1.2 between M_W −3.7 and M_W 1. These *b* values are also consistent with the *b* values listed in Table 1 for post-blasting AEs in different t_{pb} brackets.

As for the blast-active hours, the size distribution from the AEnet catalog (filled squares in Fig. 7a) follows the GR law with *b* ~1.2 between M_W −3.7 and M_W −2. More $M_W \geq -2$ events occurred than expected from the GR law. As we have noted in the previous section, such a pattern in the AEnet catalog could be interpreted by assuming that a population of events representing the mining blasts themselves, which are peaked near the typical blast size of M_W −1, is superposed on a distribution that follows the GR law with *b* ~1.2. The FMD plotted in terms of non-cumulative counts (occurrence frequencies) bears out that point more clearly (filled squares in Fig. 7b).

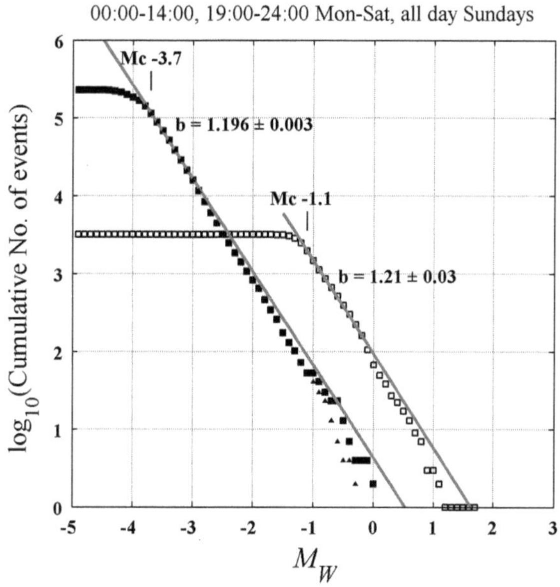

Figure 6
FMDs of events that were located near our AEnet and occurred during blast-free hours (00:00–14:00 and 19:00–24:00 Monday through Saturday and around the clock Sunday). *Filled squares* FMD of stope-cluster AEs located inside the blue frame of Fig. 3a. *Filled triangles* same as *filled squares*, except that M_W has not been corrected for likely underestimation ("Appendix C"). The *b* value for AEnet catalog was estimated for the FMD based on the corrected M_W. *Open squares* FMD of seismic events in the Mnet catalog that occurred between 1 April 2011 and 1 December 2011 and were located across the entire region of Fig. 2

A kink is found near M_W 0 in the corresponding size distribution for the Mnet catalog (open squares in Fig. 7a). The slope again indicates $b \sim 1.2$ for $M_W \geq 0$, although the uncertainty may be larger. The same interpretation for the AEnet catalog is again applicable to this case; that is, during the blast-active hours, a pattern peaked near M_W -1 is superposed on a distribution that follows the GR law with the same *b* value of about 1.2 as during the blast-free hours (open squares in Fig. 6). Incidentally, the "hump" near M_W -1 in FMD for Mnet catalog in Fig. 7b (open squares) is more consistent with the one from the corrected M_W for AEnet catalog (filled squares) than the one from the uncorrected M_W (filled triangles). This also indicates that the M_W in the AEnet catalog were corrected appropriately.

7. Discussion and Conclusion

In the present study, we installed an AE monitoring network 1 km beneath the ground in the Cooke 4 shaft, South Africa, that is capable of detecting events down to M_W -5 or less. We used two months' worth of waveform data to locate their hypocenters and estimate their M_W to study the size distribution of AEs close to a mining front in the mine.

We stacked post-blast AE activity data with reference to the blast origin times to investigate how mining blasts affect the size distribution of AEs. Detailed studies of the size distribution of blast-induced AEs showed that, when the blasts themselves are excluded from the data, the size distribution follows the GR law with similar *b* values without any systematic temporal variation, even during the minute when the blasts themselves are taking place and many new fractures may be generated in the intact rock.

We also investigated size distributions around the same mining front using two seismic catalogs, AEnet catalog and Mnet catalog. FMDs from the both catalogs were shown to follow the GR law, with a *b* value of around 1.2 across the entire magnitude range they cover ($-3.7 \leq M_W \leq \sim 0$ for the AEnet catalog and $-1.1 \leq M_W \leq \sim 1$ for the Mnet catalog, respectively), if only the mining blasts themselves were excluded from the data. The validity ranges of the GR law from the both catalogs overlap, which suggests that all events belong to a single population that obeys the GR law between M_W -3.7 and M_W ~ 1.

If most seismic events in a mine occur in the vicinity of mining fronts, we may further conclude that size distributions of seismic events in mines follow the GR law between M_W -3.7 and M_W 2, by combining results of the present study and Boettcher *et al.* (2009). However, other types of seismic events, such as events far from mining stopes or aftershocks of large events as Kwiatek *et al.* (2010) studied, may occupy a significant fraction of all seismic events in a mine. In that case, size distributions around a mining front investigated in the present study cannot be connected directly with the result of Boettcher *et al.* (2009).

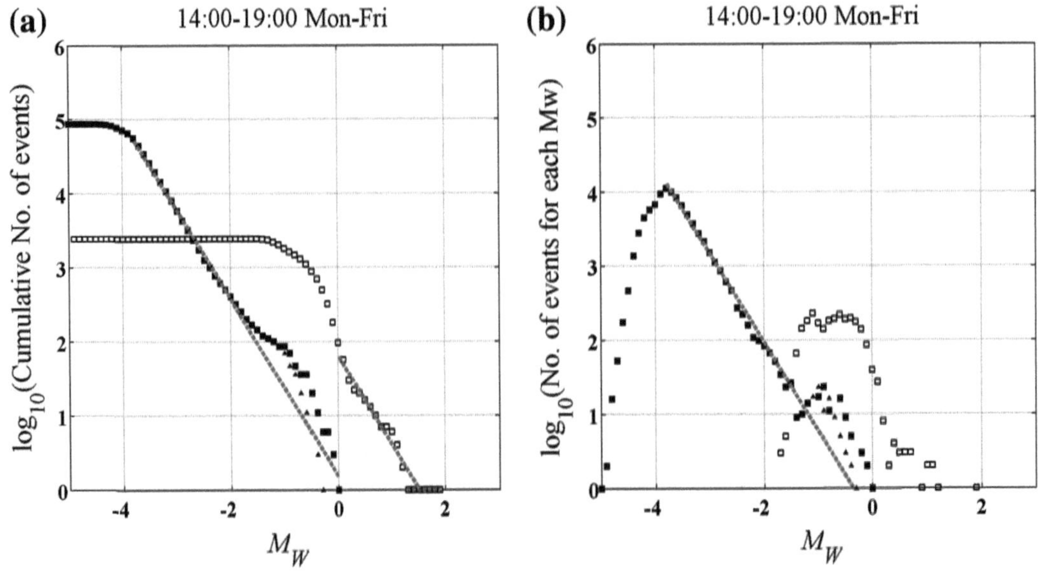

Figure 7

FMDs of events that were located near AEnet and occurred during blast-active hours (between 14:00 and 19:00 on weekdays). *Filled squares* FMDs of stope-cluster AEs located inside the blue frame of Fig. 3a. *Filled triangles* the same, except that M_W has not been corrected for likely underestimations ("Appendix C"). *Open squares* FMDs of seismic events in the Mnet catalog that occurred between 1 April 2011 and 1 December 2011 and were located across the entire region of Fig. 2. **a** Diagrams based on cumulative counts. **b** Diagrams based on occurrence frequencies (non-cumulative counts). The *dashed lines* indicate the slope of $b = 1.2$, shown for reference

Acknowledgments

We thank Gold One International Ltd., operator of the Cooke 4 shaft, for approving our publication, and also personnel at the mine and at the Institute of Mine Seismology for facilitating our project. This project was funded by Grants-in-Aid from the Ministry of Education, Culture, Sports, Science and Technology (MEXT) of Japan (21224012, 21246134). It was also funded by JST/JICA, SATREPS, and the MEXT's Observation and Research Program for Prediction of Earthquakes and Volcanic Eruptions.

Appendix A: Calibration of AE Sensor Frequency Responses

Here, we describe how we evaluated the frequency response of the AE sensors, largely following the method of Kwiatek *et al* (2011).

Let $S_i(f)$ be the amplitude spectrum of the *i*th-component ground acceleration incident on an accelerometer during a seismic event, $A_{acci}(f)$ be the *i*th-component accelerometer output spectrum, and $I_{acc}(f)$ be the accelerometer's frequency response:

$$A_{acci}(f) = I_{acc}(f)S_i(f). \tag{A1}$$

If we define the total spectrum $A_{acc}(f)$ of the three-component accelerometer output by

$$A_{acc}(f) = \sqrt{\sum_{i=1}^{3} A_{acci}^2(f)}, \tag{A2}$$

we have

$$A_{acc}(f) = I_{acc}(f)S_{acc}(f), \tag{A3}$$

where $S_{acc}(f) = \sqrt{\sum_i S_i(f)^2}$ is the total acceleration spectrum.

Meanwhile, the AE sensor output spectrum A_{AE} is given by

$$A_{AE}(f;\theta_1,\theta_2) = I_{AE}(f;\theta_1,\theta_2)S_{acc}(f), \tag{A4}$$

where $I_{AE}(f;\theta_1,\theta_2)$ is the frequency response of the AE sensor, θ_1 is the incidence angle of waves on the AE sensor with 0° pointing right to the sensor front and 180° pointing right to the sensor's back, and θ_2 is the oscillation direction of the incident waves. The AE sensors are directional, and their frequency responses are thought to depend on θ_1 and θ_2. We assume their frequency responses are axially symmetric because of the sensors' symmetry.

Let us also denote, for a trio of an AE sensor, an accelerometer and a seismic source, the source-AE sensor distance by R_{AE}, the source-accelerometer distance by R_{acc}, and the accelerometer-AE sensor distance by R. When $R \ll R_{acc}$ and $R \ll R_{AE}$, the ground acceleration spectrum of waves incident on the accelerometer, $S_{acc}(f)$, is thought to agree with its AE sensor counterpart. The ratio of the frequency responses of the AE sensor and the accelerometer, $T(f;\ \theta_1,\ \theta_2)$, can therefore be written as follows:

$$T(f;\theta_1,\theta_2) = \frac{I_{AE}(f;\theta_1,\theta_2)}{I_{acc}(f)} = \frac{A_{AE}(f;\theta_1,\theta_2)}{A_{acc}(f)}, \quad \text{(A5)}$$

so that the total AE sensor output spectrum, A_{AE}, can be converted to its accelerometer equivalent, A_{acc}, by

$$A_{acc}(f) = A_{AE}(f;\theta_1,\theta_2)/T(f;\theta_1,\theta_2). \quad \text{(A6)}$$

We used Eq. (A5) to calculate $T(f;\ \theta_1,\ \theta_2)$. For P waves, we do not need to consider the θ_2 dependency because the angle is determined uniquely from θ_1. For S waves, although the θ_2 dependency should be considered independently of θ_1, we cannot correct for the θ_2 dependency for most AE sensors that do not have nearby accelerometers without knowing the focal mechanism of individual events, even if we know the θ_2 dependency of the sensor. We therefore ignore θ_2 dependency of sensor characteristics and assume $I_{AE}(f;\ \theta_1,\theta_2) = I_{AE}(f;\ \theta_1)$ and $T(f;\ \theta_1,\ \theta_2) = T(f;\ \theta_1)$.

We used records of two AE sensors (channels 25 and 42), only 2.0 m and 3.7 m apart from the nearest accelerometer, respectively, to determine $T(f;\ \theta_1)$, whereby we took the average of the spectral ratios for both sensor pairs and applied the results to all AE sensors. We used only seismic events with hypocenters located within 200 m of the center of gravity of the AE network but at least 30 m apart from both the accelerometer and the AE sensor, and with incidence angles differing 10° or less for both types of sensors. For channel 25, we used P waves from 38,196 events and S waves from 33,413 events. For channel 42, the number of events used was 30,981 for P waves and 29,299 for S waves.

We calculated the spectra by using 2.048 ms (1,024 samples) portions of the accelerograms and the AE waveforms that include the direct P and S phases. After taking moving averages of the spectra with a window width of 2 kHz, we used equation (A5) to calculate $T(f;\ \theta_1)$, and stacked the results separately in different bins of θ_1, the incidence angle with respect to the AE sensor. The θ_1 bins overlap partially, with a bin width of 20° and an angle increment of 5°.

Figure 8 shows the stacks of $T(f;\ \theta_1)$ in the different θ_1 bins. We stacked the spectral ratios only in the frequency bands where the S/N ratios, with respect to the pre-P noise spectra, were 10 dB or more. Further, we excluded the frequency bands where <10 $T(f;\ \theta_1)$ samples were available for stacking, labeling such bands as "unfit for analysis" and suppressing them as in Fig. 8.

In Fig. 8, we find the spectral ratios to behave slightly differently for the P and S waves and to vary systematically with the wave incidence angle θ_1, but both those variations are smaller than, or at most comparable to, the differences between the two AE sensors. We therefore decided to ignore all dependency on θ_1 and all differences between the P and S waves, and to use the average of all stacked results shown in Fig. 8a–d (solid black curves) to convert the AE sensor spectra into equivalent acceleration spectra. The frequency bands labeled "unfit for analysis" were not used in the averaging.

Appendix B: Method to Estimate Moment Magnitude of AEs

We estimated M_o by fitting theoretical spectra to ground displacement spectra derived from accelerograms and AE sensor records. The AE sensor records were used after correction by $T(f)$, an AE sensor-to-accelerometer frequency response ratio estimated in situ, estimated in "Appendix A".

Let $S_i(f)$ denote the ground acceleration spectrum of waves incident on an AE sensor or an accelerometer, and let $S_{acc}(f)$ be the total ground acceleration spectrum integrated from the three components:

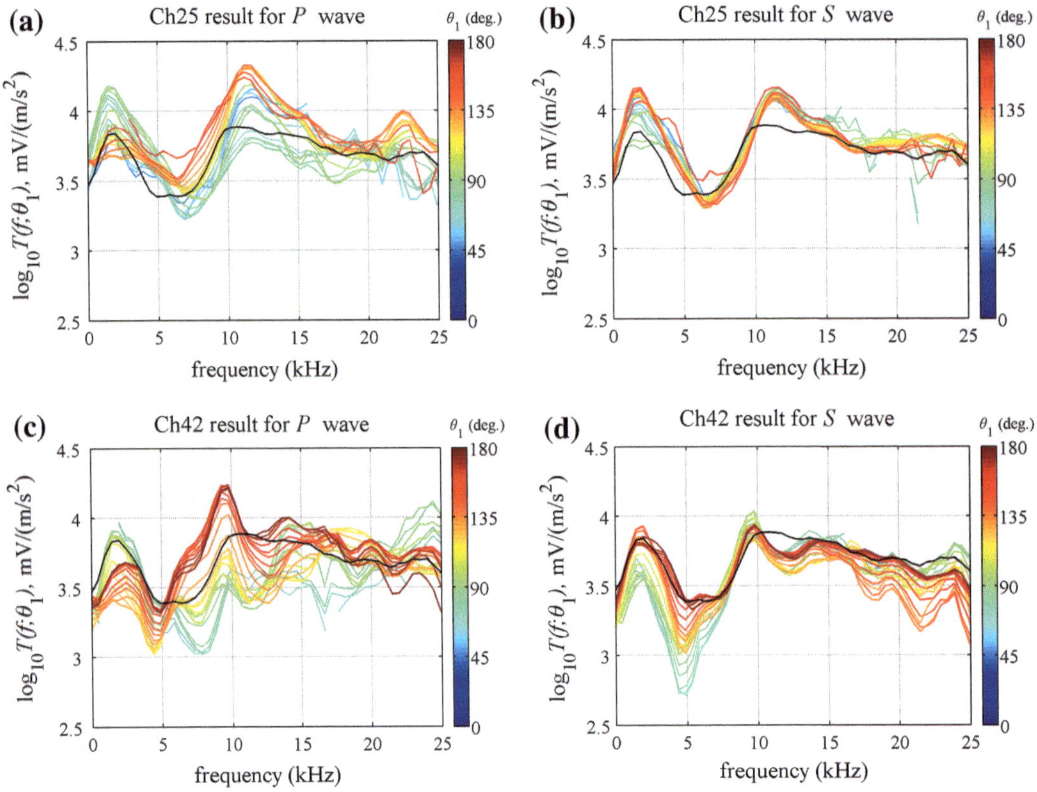

Figure 8
Frequency responses of AE sensors in terms of AE sensor-to-accelerometer spectral ratios $T(f;\,\theta_1)$, estimated for AE sensors installed adjacent to accelerometers. The $T(f;\,\theta_1)$ was stacked separately in different bins of θ_1, the incidence angle with respect to the AE sensor. The θ_1 bins overlap partially, with a bin width of 20° and an angle increment of 5°. Indicated in *color* is the central angle of each bin. Frequency bands where <10 $T(f;\,\theta_1)$ samples were available for stacking, because of poor S/N ratios, are not shown. Stacked spectral ratios for an accelerometer-AE sensor pair near the AE network center (inter-sensor distance 2.0 m): **a** results for *P* waves. **b** Results for *S* waves. Stacked spectral ratios for an accelerometer-AE sensor pair on the north side of the AE network (inter-sensor distance 3.7 m): **c** results for *P* waves. **d** Results for *S* waves. *Black curves* average of all stacked spectral ratios shown in **a–d**. We used this average spectral ratio to convert AE sensor spectra into equivalent acceleration spectra

$S_{acc}(f) = \sqrt{\sum_i S_i(f)^2}$. The total ground displacement spectrum $D(f)$ can be obtained from the total acceleration spectrum $S_{acc}(f)$ as follows:

$$D(f) = \frac{1}{(2\pi f)^2} S_{acc}(f). \tag{B1}$$

We evaluated M_o by fitting the theoretical total ground displacement spectrum $D^{th}(f)$ to the $D(f)$ thus obtained. Subsequent analysis is similar in method to Ide *et al.* (2003). From Boatwright (1978),

$$D^{th}(f) = \frac{R^c}{4\pi\rho v^3 r} \frac{M_o}{\sqrt{1 + (f/f_c)^4}} \exp\left(-\frac{\pi f r}{Qv}\right), \tag{B2}$$

where R^c is the radiation pattern coefficient, ρ is the medium density, v is the elastic wave velocity, r is the source-receiver distance, f is the frequency, f_c is the corner frequency, and Q is the quality factor. Taking the logarithms on both sides, we have

$$\log D^{th}(f) = \log M_o + \log \frac{R^c}{4\pi\rho v^3 r} - \frac{1}{2}\log\left[1 + (f/f_c)^4\right] - \frac{\pi f r}{Qv}. \tag{B3}$$

Because $\log D^{th}(f)$ is linearly related to $\log M_o$, M_o may be obtained by linear least squares fitting with the other parameters fixed. We defined the residual, Res, by

$$\text{Res} = \left[\log D(f) - \log D^{th}(f)\right]^2 \tag{B4}$$

and used a least squares method to determine M_o and a grid search method to determine f_c that produced a

$D^{th}(f)$ that minimizes the residual. We followed the definition of HANKS and KANAMORI (1979) to convert M_o to M_W:

$$M_W = \frac{2}{3}(\log_{10} M_o - 9.1). \quad \text{(B5)}$$

We analyzed the spectra of the individual sensor records near their P and S wave travel time readings (or theoretical travel times in their absence). We applied a cosine taper to the initial and final 5 % parts of the time-domain waveforms before putting them to spectral analysis.

The $D(f)$ was obtained by resampling the FFTs of time-domain waveforms at logarithmically equal intervals and taking their moving averages with a window width of 0.05 on the logarithmic frequency scale. Fitting was conducted only on frequency bands where the S/N ratios, with respect to the noise spectra obtained from pre-P signals, were 10 dB or more. We labeled the M_W estimates as unavailable when no continuous frequency band with S/N $\geq$ 10 dB was broader than a threshold width (0.4 on the logarithmic frequency scale for AE sensors and 0.5 for accelerometers).

We also labeled M_o estimates as unavailable when $\log f_c - \log f_{\min} < 0.2$, where f_c is the corner frequency estimate and $f_{\min}$ is the minimum frequency used in the spectral fitting. That is because M_o tends to be underestimated considerably when spectral fitting is done only in high frequency bands above the corner frequency.

We used a data length of 4,096 samples ($\sim$8.2 ms) for the spectral analysis, but repeated the analysis with a doubled data length of 8,192 samples ($\sim$16.4 ms) when the shorter data length produced an M_W estimate of -3 or greater. If the doubled data length produced an M_W estimate of -2 or greater, we repeated the analysis further with a tripled data length of 12,288 samples ($\sim$24.6 ms). We also repeated the analysis with similarly extended data lengths when the M_o estimate was labeled unavailable. We labeled it as "indefinite" for the sensor in question when it remained unavailable for a data length of 12,288 samples. When the S–P time was short, we reduced the data lengths appropriately to prevent the P and S phases from being mixed up in the analysis.

We used the parameters V_P = 5,700 m/s, V_S = 3,600 m/s, ρ = 2,700 kg/m^3, Q_P = 150, Q_S = 200, $R_P^c = 2/\sqrt{15}$ and $R_S^c = \sqrt{2}/\sqrt{5}$ in Eqs. (B2) and (B3). The V_P and V_S are the same values used to determine the hypocenters. The Q_P and Q_S were estimated by trial and error to allow the observed acceleration spectra to be well explained by Eq. (B3). Sound transmission tests have also produced the same Q_P value (KAWAKATA *et al.* 2011). The R_P^c and R_S^c are rms values averaged over all azimuths (AKI and RICHARDS 1980).

Figure 9 illustrates how theoretical spectra derived from Eq. (B3) were fit to the observed records. The figure shows that the theoretical spectra well explain the observed spectra for both the accelerograms (Fig. 9a, c) and the AE sensor records (Fig. 9b, d).

In this study, we used records of all six accelerometers and 11 of the 24 AE sensors in the analysis. We used only part of the available AE sensor records to save on computation time.

Of the 365,237 AE events, which were well located by the automatic picking and location software (Sect. 3.2), 17,050 had their hypocenters located within 150 m of AEnet's center and their M_o and M_W determined for both the P and S wave parts of all six accelerograms. In those analysis results, the M_W estimates from P waves (M_{WP}) exceeded their S wave counterparts (M_{WS}) by 0.24 on average. A similar tendency was confirmed for every single monitoring station, be it an accelerometer or an AE sensor. Possible causes include biases in radiation patterns and imprecision of the seismic wave velocities used.

In some cases, only one of the M_{WP} and M_{WS} was available. We therefore subtracted 0.12 from every single M_{WP} and added 0.12 to every single M_{WS}, including estimates from AE sensor records, to equilibrate the M_W estimates from both P and S wave records on average. We took the mean of all available M_{WP} and M_{WS} and rounded the outcome to the nearest tenth to determine the final M_W estimate. The standard deviation of M_W among stations and phases for event i, σ_{Mi}, is shown in a histogram of Fig. 10. For 249,017 events which have more than ten available M_W from different stations and phases, the mean of σ_{Mi} was 0.136. In calculating σ_{Mi}, the corrections of 0.12 on the M_W^P and M_W^S have been taken into account.

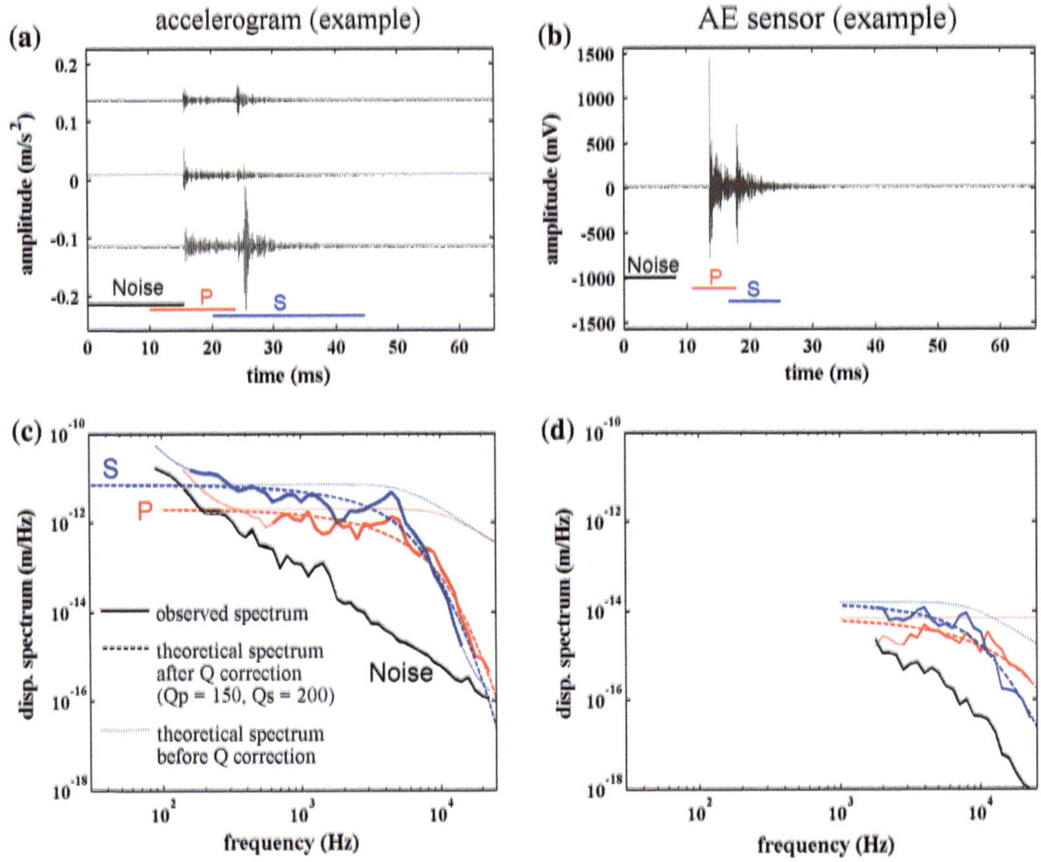

Figure 9

Examples showing how fitting was conducted on *P* and *S* wave spectra to evaluate the seismic moment M_0. **a** Waveform records of a three-component 25-kHz accelerometer for an event that occurred at 05:20:14.187 on 6 October 2011. The hypocenter was located 87.2 m from the accelerometer. The *red*, *blue*, and *black bars* beneath the waveforms show the time windows used to calculate the *P*, *S*, and noise wave spectra, respectively. **b** Waveform records of an AE sensor for an event that occurred at 06:02:16.812 on 9 August 2011. The hypocenter distance was 39.6 m. **c** Total ground displacement spectra *D*(*f*) calculated from the waveforms shown in **a**. *Red*, *blue*, and *black solid curves* *D*(*f*) of the *P*, *S*, and noise waves, respectively. The highlighted parts show the frequency bands with S/N ratios ≥ 10 dB. *Dashed curves*: theoretical spectra $D^{th}(f)$, which include the effect of *Q*, and provide the best fit for the highlighted parts of the *D*(*f*) curves. *Dotted curves* theoretical spectra without the effect of *Q*. The fitting led to a moment magnitude estimate of M_{WP} −1.9 based on the *P* waves and M_{WS} −2.0 based on the *S* waves, both before correction for the systematic differences between M_{WP} and M_{WS}. **d** Total ground displacement spectra *D*(*f*) calculated from the waveforms shown in **b** by using *T*(*f*) ("Appendix A"). See legend to **c**. M_{WP} −3.7, M_{WS} −4.0

It should be noted that no low-frequency content below 1 kHz was available for estimating M_o using AE sensor records, because they were all high-pass filtered at 1 kHz before analog-to-digital conversion. Accelerograms were available for use down to lower-frequency bands than the AE sensor records, because the accelerograms were high-pass filtered only at 50 Hz. The M_o risks being underestimated when only limited low-frequency content is available, although we took care, as we stated above, to label the M_o estimate as unavailable when only high-frequency content above the corner frequency was used in the fitting.

Figure 11 shows the relationship between the mean M_W values determined using AE sensor records, M_W^{AE}, and those determined from accelerograms, M_W^{acc}, for 287,025 events, which were located within 150 m of the center of gravity of AEnet and for which records from at least one accelerometer and at least one AE sensor were used to evaluate M_o. The corrections of 0.12 on the M_W^P and M_W^S have been taken into account. It shows that M_W^{AE} tends to fall

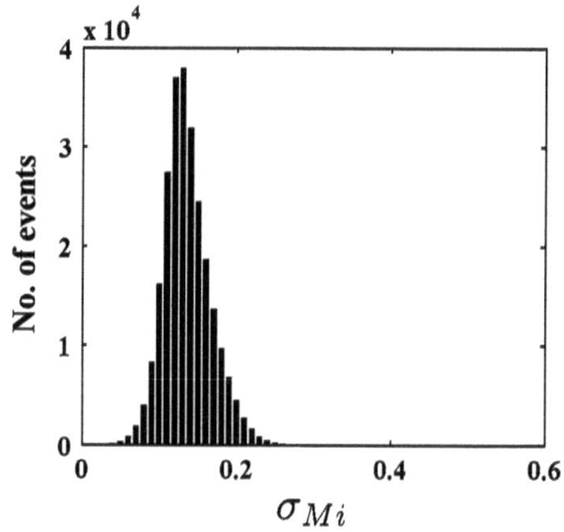

Figure 10
The histogram of σ_{Mi}, the standard deviation of M_W among stations and phases for event i, for 249,017 events which have more than ten available M_W from different stations and phases

short of M_W^{acc} and is likely underestimated for events above M_W −2. To avoid that bias, we used M_W^{acc} as the final M_W value, without using analysis results from the AE sensor records, when M_W^{acc} was larger than −2.5.

Appendix C: Consistency Between M_W Estimates from the AE Network and the Mine's Network

To compare size distributions from two different catalogs, the AEnet catalog and the Mnet catalog, we need to check the consistency between the M_W estimates from the AEnet records (M_W^{AEnet}) and the M_W given in the Mnet catalog (M_W^{Mnet}), using the events that are listed in both catalogs.

As we stated in Sect. 2, the clocks in both monitoring networks are not synchronized. Therefore, we first determine the time difference curve between both system clocks as described below, and used that result to correct them before we singled out what we thought to be pairs of identical event records in both catalogs, for which consistency of M_W estimates were checked.

To estimate the time difference curve, we first searched for candidates of such pairs. For every single event in the Mnet catalog, we picked up events in the AEnet catalog that have a proximal origin time, location and M_W estimate. We used 300 s $\leq dt \leq$ 600 s, $D \leq$ 100 m and $|dM_W| \leq 1$ for our selection criteria, where dt is the difference between the Mnet and AEnet origin times, D is the distance between the Mnet and AEnet hypocenters, and $dM_W = M_W^{\mathrm{AEnet}} - M_W^{\mathrm{Mnet}}$. We set the selection standard for dM_W somewhat loose because our final objective is to verify the consistency of the M_W estimates. The range of D was determined so as not to exclude pairs of identical event records for which mutually distant hypocenters were given by both catalogs, considering that location errors can sometimes be as large as 100 m in the Mnet catalog. As for dt, we started the actual analysis using broader ranges, such as −1,000 s $\leq dt \leq$ 1,000 s, and narrowed it by reference to the result. We show here only the final result using 300 s $\leq dt \leq$ 600 s, which turned out to be the true range of the clock difference in the analysis term, but essentially the same results were obtained when we used broader dt range.

We used only 363,672 events in the AEnet catalog whose hypocenters were located within 200 m of the

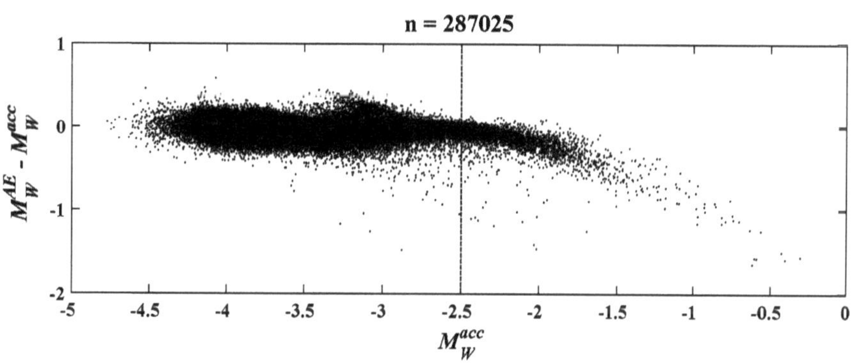

Figure 11
Relationship between M_W^{AE} (the mean of M_W values determined using AE sensor records) and M_W^{acc} (the mean of M_W values determined using accelerograms) for the 287,025 events that were located within 150 m of the center of gravity of AEnet and for which records from at least one accelerometer and at least one AE sensor were used to evaluate M_o

AEnet center. Figure 12 shows a time-series plot of *dt* for events in the AEnet catalog that satisfied the above condition, each being the only one event that could be paired with an event in the Mnet catalog. Figure 12 likely contains numerous plots of *dt* between what in fact are different events, but it does show a single, monotonously rising trend of densely populated data points. That trend likely represents the time difference between both system clocks.

We manually eliminated data that are obviously at odds with the time differences of preceding and following events (gray triangles in Fig. 12) and determined a time difference curve (solid curve in Fig. 12) by taking the moving averages of the remaining data points (open circles) with a window width of 2 days.

The origin times in the Mnet catalog were used only after correcting for the time differences according to the curve. We subsequently used stricter selection criteria to identify which pairs of events are believed to represent the same events with strong certainty and compared the M_W estimates in those pairs.

Let dt' be the residual difference between the catalogued origin times following the corrections. We singled out events in the Mnet catalog, with hypocenters located within 100 m of the AEnet center, each being associable with only one event in the AEnet catalog that satisfied $-5\ \mathrm{s} \leq dt' \leq 5\ \mathrm{s}$, $D \leq 50$ m and $|dM_W| \leq 1$. Figure 13 shows the relationship between the M_W^{Mnet} and the M_W^{AEnet} for the 117 events thus singled out. It demonstrates that the M_W^{AEnet} tends to be underestimated with respect to the M_W^{Mnet} for events with $M_W^{\mathrm{Mnet}} \geq -1.3$.

As we stated in "Appendix B", we used the accelerograms alone in estimating the M_W^{AEnet} for events with $M_W^{\mathrm{acc}} \geq -2.5$ to avoid the influence of the 1 kHz high-pass filter on the AE sensor waveforms, but the accelerograms, for their own part, have been high-pass filtered at 50 Hz and their lower-frequency content has been eliminated. That presumably accounts for why the M_W estimates from AEnet are underestimated for events upward of M_W^{Mnet} -1.3. To correct those underestimations, we used the relationship between the M_W estimates in the Mnet catalog and the dM_W, indicated by the kinked gray line in Fig. 13.

Appendix D: b Value Analysis

Here, we describe how we evaluated *b* values, their standard errors, and the minimum validity magnitude M_c of the Gutenberg–Richter law.

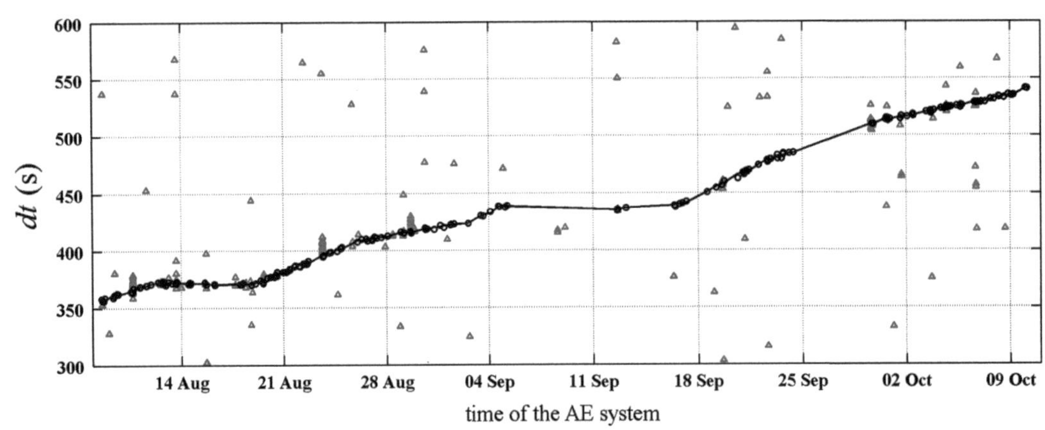

Figure 12

A plot of time differences between the AEnet and the Mnet records, with the *horizontal axis* standing for the event occurrence times according to the clock of AEnet. Each data point represents an event pair satisfying $300\ \mathrm{s} \leq dt \leq 600\ \mathrm{s}$, $D \leq 100$ m and $|dM_W| \leq 1$, where *dt* is the difference in origin times between the AEnet catalog and the Mnet catalog, *D* is the distance between the two catalogued hypocenters, and dM_W is the difference between the two catalogued M_W values. Cases where two or more events in the AEnet catalog could be paired with an event in the Mnet catalog were not included in the plot. *Open circles* data points defining a single, monotonously rising trend. *Gray triangles* data points obviously at odds with the trend. *Solid curve* time correction curve determined from the *open circles*

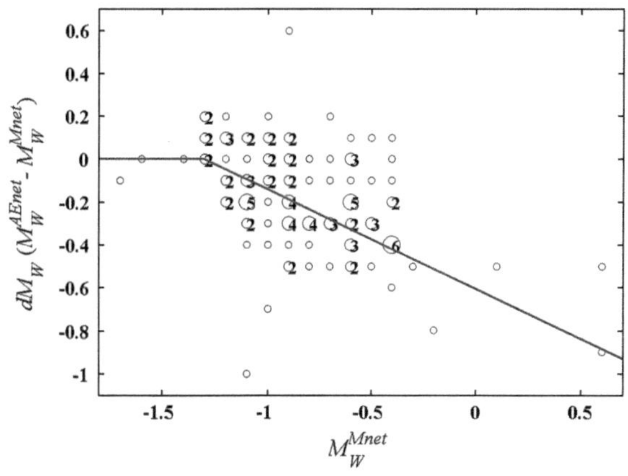

Figure 13
Differences between the M_W^{AEnet} (M_W determined by AEnet) and the M_W^{Mnet} (M_W determined by Mnet) for 117 pairs that are most certainly associated with the identical events. The sizes of the *circles* and the numeral figures superimposed on them indicate the number of events falling on each (M_W^{Mnet}, dM_W) grid point. The *gray line* approximating the trend was used to correct the underestimation of M_W^{AEnet}

We calculated the b values with the maximum likelihood method (Aki 1965; UTSU 1965) by using the formula:

$$b = \frac{1}{\overline{M_W} - (M_c - \Delta M/2)} \log_{10} e, \qquad \text{(D1)}$$

where $\overline{M_W}$is the mean magnitude of all M_c and greater events, and ΔM is the sampling increment for M_W in the catalog. Following Shi and Bolt (1982), we used the following formula to evaluate e_b, or the standard error of a b value estimate:

$$e_b = 2.30b^2 \sqrt{\sum_{i=1}^{n} \left(M_i - \overline{M_W}\right)^2 / n(n-1)}, \qquad \text{(D2)}$$

where M_i represents the magnitude of the ith event and n is the total number of events.

We used the maximum curvature method of Wiemer and Wyss (2000) to estimate M_c, the minimum validity magnitude of the Gutenberg-Richter law in a given data set. For a given threshold magnitude $M_{\min}$, that method uses a "goodness of fit", R, defined by

$$R(a, b, M_i) = 100 - \frac{\sum_{M_i = M_{\min}}^{M_{\max}} |B_i - S_i|}{\sum_i B_i} 100, \qquad \text{(D3)}$$

to evaluate the discrepancy between the size distribution predicted by the Gutenberg–Richter law and the observed size distribution, where a and b are parameters of the Gutenberg–Richter law, M_i is the M_W of the ith magnitude bin, and B_i and S_i are the observed and predicted cumulative numbers of events, respectively, in the ith magnitude bin. We evaluated $R(a, b, M_i)$ for different choices of M_{min}, and designated the smallest M_{min} that gave R in excess of a certain threshold value (typically 90) as our estimate of M_c, the magnitude where the frequency-magnitude distribution begins to deviate from the power law.

Woessner and Wiemer (2005) said this method tends to slightly underestimate M_c, and recommended adding 0.2 to the M_c estimates. Following their recommendation, we added 0.2 to the minimum M_{min} with $R > 90$ to obtain our final M_c estimate.

References

Abercrombie, R. E. (1996), *The magnitude-frequency distribution of earthquakes recorded with deep seismometers at Cajon Pass, southern California*, Tectonophysics, *261*, 1–7, doi:10.1016/0040-1951(96)00052-2.

Abercrombie, R. E., and J. N. Brune (1994), *Evidence for a constant b-value above magnitude 0 in the southern San Andreas, San Jacinto and San Miguel Fault Zones, and at the Long Valley Caldera, California*, Geophys. Res. Lett., *21*, 1647–1650, doi:10.1029/94GL01138.

Aki, K. (1965). *Maximum likelihood estimate of b in the formula* log $N = a - bM$ *and its confidence limits*, Bull. Earthq. Res. Inst. Univ. Tokyo, *43*, 237–239.

Aki, K (1987), *Magnitude-frequency relation for small earthquakes: A clue to the origin of f_{max} of large earthquakes*, J. Geophys. Res., *92*, 1349–1355, doi:10.1029/JB092iB02p01349.

Aki, K., and P. G. Richards, *Quantitative Seismology*, (W. H. Freeman and company, New York 1980).

Andrews, D. J. (1976), *Rupture propagation with finite stress in antiplane strain*, J. Geophys. Res., *81*, 3575–3582, doi:10.1029/JB081i020p03575.

Boatwright, J. (1978), *Detailed spectral analysis of two small New York state earthquakes*, Bull. Seismol. Soc. Am., *68*, 1117–1131.

Boettcher, M. S., A. McGarr, and M. Johnston (2009), *Extension of Gutenberg-Richter distribution to M_W −1.3, no lower limit in sight*, Geophys. Res. Lett., *36*, L10307, doi:10.1029/2009GL038080.

Dieterich, J. H., *A model for the nucleation of earthquake slip, In Earthquake Source Mechanics*, Geophys. Monogr. Ser., vol. *37*, (ed. Das, S., J. Boatwright, and C. H. Scholz) (American Geophysical Union, 1986) pp. 37–47.

FINNIE, G. J. (1999), *Using neural networks to discriminate between genuine and spurious seismic events in mines*, Pure Appl. Geophys., *154*, 41–56, doi:10.1007/s000240050220.

GUTENBERG, B., and C. F. RICHTER (1944), *Frequency of earthquakes in California*, Bull. Seismol. Soc. Am., *34*, 185–188.

HANKS, T. C., and H. KANAMORI (1979), *A moment magnitude scale*, J. Geophys. Res., *84*, 2348–2350, doi:10.1029/JB084iB05p02348.

HORIUCHI, S., Y. HORIUCHI, Y. IIO, Y. SAWADA, S. SEKINE, H. NAKAMURA, T. OKADA, M. NAKATANI, and M. NAOI (2011), *Automatic P and S wave arrival time picking compared to manually picking*, IUGG General Assembly, 3430, Melbourne, Australia.

IDA, Y. (1972), *Cohesive force across the tip of a longitudinal-shear crack and Griffith's specific surface energy*, J. Geophys. Res., *77*, 3796–3805, doi:10.1029/JB077i020p03796.

IDE, S., G. C. BEROZA, S. G. PREJEAN, and W. L. ELLSWORTH (2003), *Apparent break in earthquake scaling due to path and site effects on deep borehole recordings*, J. Geophys. Res., *108*, 2271, doi:10.1029/2001JB001617.

ISHIMOTO, M., and K. IIDA (1939), *Observations sur les séismes enregistrés par le microsismographe construit dernièrement (in Japanese with French abstract)*, Bull. Earthquake Res. Inst. Univ. Tokyo, *17*, 443–478.

KAWAKATA, H., N. YOSHIMITSU, M. NAKATANI, J. PHILIPP, I. DOI, M. NAOI, T. WARD, V. VISSER, G. MOREMA, S. KHAMBULE, T. MASAKALE, A. MILEV, R. J. DURRHEIM, L. RIBEIRO, M. WARD, and H. OGASAWARA (2011), *Monitoring transmitted waves across a fault with a high potential for mining induced earthquakes—The Ezulwini gold mine in South Africa*, AGU Fall Meeting, S31C-2257, San Francisco.

KWIATEK, G., K. PLENKERS, G. DRESEN, and JAGUARS Research Group (2011), *Source parameters of picoseismicity recorded at Mponeng Deep Gold Mine, South Africa: Implications for scaling relations*, Bull. Seismol. Soc. Am., *101*, 2592–2608, doi:10.1785/0120110094.

KWIATEK, G., K. PLENKERS, M. NAKATANI, Y. YABE, G. DRESEN, and JAGUARS-GROUP (2010), *Frequency-magnitude characteristics down to magnitude −4.4 for induced seismicity recorded at Mponeng gold mine, South Africa*, Bull. Seismol. Soc. Am., *100*, 1165–1173, doi:10.1785/0120090277.

MALIN, P. E., S. N. BLAKESLEE, M. G. ALVAREZ, and A. J. MARTIN (1989), *Microearthquake imaging of the Parkfield Asperity*, Science, *244*, 557–559, doi:10.1126/science.244.4904.557.

MANTHEI, G. (2005), *Characterization of acoustic emission sources in a rock salt specimen under triaxial compression*, Bull. Seismol. Soc. Am., *95*, 1674–1700, doi:10.1785/0120040076.

MENDECKI, A. J., *Seismic monitoring systems, In Seismic Monitoring in Mines* (ed. Mendecki, A. J.) (Chapman and Hall, 1997).

NAKATANI, M., Y. YABE, J. PHILIPP, G. MOREMA, S. STANCHITS, G. DRESEN and JAGUARS GROUP (2008), *Acoustic emission measurements in a deep gold mine in South Africa—Project overview and some typical waveforms*, Seismol. Res. Lett. *79*, 311.

NAOI, M., M. NAKATANI, Y. YABE, G. KWIATEK, T. IGARASHI, and K. PLENKERS (2011), *Twenty thousand aftershocks of a very small (M 2) earthquake and their relation to the mainshock rupture and geological structures*, Bull. Seismol. Soc. Am., *101*, 2399–2407, doi:10.1785/0120100346.

RICHARDSON, E., and T. H. JORDAN (2002), *Seismicity in deep gold mines of South Africa: Implications for tectonic earthquakes*, Bull. Seismol. Soc. Am., *92*, 1766–1782, doi:10.1785/0120000226.

RYDELEK, P. A., and I. S. SACKS (1989), *Testing the completeness of earthquake catalogues and the hypothesis of self-similarity*, Nature, *337*, 251–253, doi:10.1038/337251a0.

SCHOLZ, C. H. (1988), *The critical slip distance for seismic faulting*, Nature, *336*, 761–763, doi:10.1038/336761a0.

SHI, Y., and B. A. BOLT (1982). *The standard error of the magnitude-frequency b value*, Bull. Seismol. Soc. Am., *72*, 1677–1687.

SPOTTISWOODE, S. M. (2010), *Mine seismicity: prediction or forecasting?*, J.S.A. Inst. Min. Metall., *110*, 11–20.

SPOTTISWOODE, S. M., and L. M. LINZER (2004), *Improved seismic locations and location techniques, SIM 020304 Final Project Report (Appendix 1)*, Mine Health and Safety Council, Johannesburg, South Africa.

TAYLOR, D. W. A., J. A. SNOKE, I. S. SACKS, and T. TAKANAMI (1990), *Nonlinear frequency-magnitude relationships for the Hokkaido corner, Japan*, Bull. Seismol. Soc. Am., *80*, 340–353.

UMINO, N., and I. S. SACKS (1993), *Magnitude-frequency relations for northeastern Japan*, Bull. Seismol. Soc. Am., *83*, 1492–1506.

UTSU, T. (1965), *A method for determining the value of b in a formula* $\log n = a - bM$ *showing the magnitude-frequency relation for earthquakes (in Japanese with English abstract)*, Geophys. Bull. Hokkaido Univ., *13*, 99–103.

WIEMER, S., and M. WYSS (2000). *Minimum magnitude of completeness in earthquake catalogs: Examples from Alaska, the western United States, and Japan*, Bull. Seismol. Soc. Am., *90*, 859–869, doi:10.1785/0119990114.

WOESSNER, J., and S. WIEMER (2005). *Assessing the quality of earthquake catalogues: Estimating the magnitude of completeness and its uncertainty*, Bull. Seismol. Soc. Am., *95*, 684–698, doi:10.1785/0120040007.

(Received May 30, 2013, revised September 17, 2013, accepted September 26, 2013, Published online October 23, 2013)

Pure Appl. Geophys. 171 (2014), 2685–2707

DOI 10.1007/s00024-013-0749-8

Pure and Applied Geophysics

Scaling Transition in Earthquake Sources: A Possible Link Between Seismic and Laboratory Measurements

Luca Malagnini,[1] Kevin Mayeda,[2] Stefan Nielsen,[1] Seung-Hoon Yoo,[2] Irene Munafo',[1] Christopher Rawles,[3] and Enzo Boschi[1]

Abstract—We estimate the corner frequencies of 20 crustal seismic events from mainshock–aftershock sequences in different tectonic environments (mainshocks $5.7 < M_W < 7.6$) using the well-established seismic coda ratio technique (Mayeda *et al.* in Geophys Res Lett 34:L11303, 2007; Mayeda and Malagnini in Geophys Res Lett, 2010), which provides optimal stability and does not require path or site corrections. For each sequence, we assumed the Brune source model and estimated all the events' corner frequencies and associated apparent stresses following the MDAC spectral formulation of Walter and Taylor (A revised magnitude and distance amplitude correction (MDAC2) procedure for regional seismic discriminants, 2001), which allows for the possibility of non-self-similar source scaling. Within each sequence, we observe a systematic deviation from the self-similar $M_0 \propto f_c^{-3}$ line, all data being rather compatible with $M_0 \propto f_c^{-(3+\varepsilon)}$, where $\varepsilon > 0$ (Kanamori and Rivera in Bull Seismol Soc Am 94:314–319, 2004). The deviation from a strict self-similar behavior within each earthquake sequence of our collection is indicated by a systematic increase in the estimated average static stress drop and apparent stress with increasing seismic moment (moment magnitude). Our favored physical interpretation for the increased apparent stress with earthquake size is a progressive frictional weakening for increasing seismic slip, in agreement with recent results obtained in laboratory experiments performed on state-of-the-art apparatuses at slip rates of the order of 1 m/s or larger. At smaller magnitudes ($M_W < 5.5$), the overall data set is characterized by a variability in apparent stress of almost three orders of magnitude, mostly from the scatter observed in strike-slip sequences. Larger events ($M_W > 5.5$) show much less variability: about one order of magnitude. It appears that the apparent stress (and static stress drop) does not grow indefinitely at larger magnitudes: for example, in the case of the Chi–Chi sequence (the best sampled sequence between M_W 5 and 6.5), some roughly constant stress parameters characterize earthquakes larger than $M_W \sim 5.5$. A representative fault slip for M_W 5.5 is a few tens of centimeters (e.g., Ide and Takeo in J Geophys Res 102:27379–27391, 1997), which corresponds to the slip amount at which effective lubrication is observed, according to recent laboratory friction experiments performed at seismic slip velocities ($V \sim 1$ m/s) and normal stresses representative of crustal depths (Di Toro *et al.* in Nature in press, 2011, and references therein). If the observed deviation from self-similar scaling is explained in terms of an asymptotic increase in apparent stress (Malagnini *et al.* in Pure Appl Geophys, 2014, this volume), which is directly related to dynamic stress drop on the fault, one interpretation is that for a seismic slip of a few tens of centimeters ($M_W \sim 5.5$) or larger, a fully lubricated frictional state may be asymptotically approached.

1. Introduction

Self-similarity has been widely accepted as a general property of earthquakes (Aki 1967; Ide and Beroza 2001; Ide *et al.* 2003; Prieto *et al.* 2004; Shearer *et al.* 2006), and implies that the dynamics of an earthquake rupture is always governed by the same physics, whether the event is small or large. In this view, though random fluctuations are expected, no systematic variations should be observed in the scaling law between large and small earthquakes.

However, in a number of recent papers (e.g., Malagnini *et al.* 2010 or 2014, and references therein), the idea of strict self-similarity of earthquakes was challenged by the results obtained using a well-established technique based on the stable properties of the seismic coda (Mayeda *et al.* 2007; Mayeda and Malagnini 2010; Yoo and Mayeda 2013). The approach was used for the calculation of apparent stresses ($\sigma_a = GE_r/M_0$) for all the events of our sequences, where G is rigidity, M_0 is seismic moment, and E_r is radiated seismic energy. Apparent stress is related to the dynamics of earthquakes, such as the radiated seismic energy.

The quantity σ_a represents the amount of seismic energy that is radiated (on average) by one square meter of fault surface, for one meter of co-seismic slip: in all the cited studies, larger earthquakes

[1] Istituto Nazionale di Geofisica e Vulcanologia, Rome, Italy. E-mail: luca.malagnini@gmail.com
[2] Weston Geophysical Corporation, Lexington, MA, USA.
[3] Berkeley Seismological Laboratory, University of California Berkeley, Berkeley, CA, USA.

seemed to radiate more normalized energy than smaller ones. One interpretation is that dynamic friction on a specific fault keeps decreasing for increasing seismic slip, until a minimum dynamic friction level is reached after a finite slip amount, or toward an asymptotic steady-state value.

Apparent stress can be directly linked to static stress drop: for example, for a radially growing, circular crack model in which radiation is from a zero-width crack edge, in the hypothesis of no overshoot or undershoot (Malagnini *et al.* 2010), gave a direct proportionality between the two parameters. Because of the physical meaning that was given to apparent stress (seismic energy radiated by a square meter of fault, in a meter of seismic slip), we prefer to use apparent stress over seismic stress drop. However, the reader must keep in mind that the two parameters are completely related.

In this study we analyze events from 20 crustal sequences that occurred worldwide (see Fig. 1). Whereas self-similarity predicts constant apparent stress and stress drop at all magnitudes, Mayeda, Malagnini, and co-workers have always found larger average apparent stresses for larger earthquakes (e.g., $M_W > 5.5$). A comparison of the results obtained for some of the sequences that we have in common with other authors will be discussed in detail in Sect. 2.2.

One way to explain the common characteristics of the earthquake source scaling observed in our 20 sequences is through slip-weakening friction experiments such as recent results from state-of-the-art lab apparatuses. We will propose a link between the scaling of parameters estimated seismologically (e.g., static stress drop, apparent stress), and the gradual weakening that is observed in laboratory experiments at seismic slip rates (1–10 m/s), and at realistic, effective normal stresses for crustal rocks at seismogenic depths (at ~ 10 km depth, for an optimally-oriented fault plane, 30° from the maximum compressional stress, σ_1, and a coefficient of friction $\mu_S = 0.75$: roughly between 30 and 400 MPa, depending on tectonic setting and pore-fluid pressure, see later).

From the point of view of seismology, earthquake source scaling remains a controversial issue and has large implications on source physics, ground motion

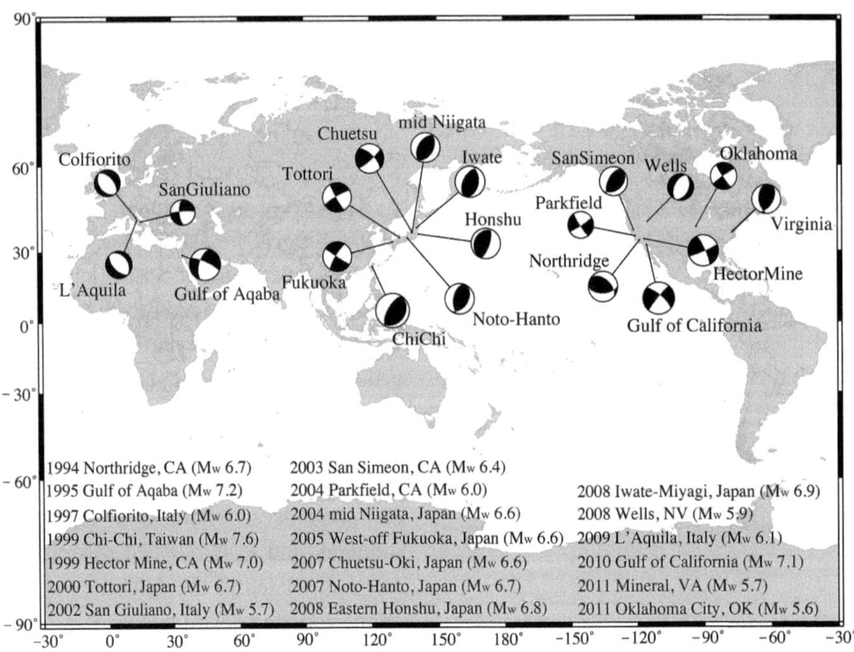

Figure 1

Map of the locations of all 20 sequences analyzed in this study. For each sequence, the focal mechanism of the main event is shown. For the Colfiorito and San Giuliano sequences, both characterized by multiple main shocks with the same fault mechanics, only the focal mechanisms of the largest earthquakes are shown prediction and estimating hazard. Several researchers described their observations in terms of self-similar earthquake scaling (constant apparent stress or stress drop, e.g., IDE and BEROZA 2001), whereas others (e.g., MAYEDA and WALTER 1996) claimed that they observed increasing apparent stresses for increasing seismic moments.

MALAGNINI *et al.* (2010) investigated the scaling of stress drop with earthquake size in six mainshock–aftershock sequences, and described their results in terms of a simple, slip-weakening frictional fault model. By using an approach that is very similar to the one used in the present paper, MALAGNINI *et al.* (2010) calculated a set of coefficients of dynamic friction for earthquakes belonging to different seismic sequences, and showed that the average coefficient of dynamic friction of a seismogenic fault may dramatically decrease when the seismic slip grows beyond a critical threshold, roughly in the range between 0.1 and 1.0 m. The definition of a plausible functional form that may be used to describe dynamic fault weakening is the subject of ongoing research. Following DI TORO *et al.* (2011), who used an exponential weakening function for their lab friction experiments, MALAGNINI *et al.* (2014, this volume) hypothesized the same exponential weakening for the description of dynamic fault's strength during seismic slip: they analyzed two sequences (L'Aquila, 2009, and Northridge, 1994), and used two best-fit, exponential weakening functions for the two faults' shear strengths. In both cases, the dynamic strengths were reduced to $1/e$ of their original values after approximately 0.5 m of slip.

2. Seismic Observations

2.1. Coda-Derived Moment-Rate Spectral Ratios

MAYEDA *et al.* (2007) developed a variant of the empirical Green's function approach that compares narrowband time-domain, coda envelope amplitudes for closely located event pairs measured at the same station, and for events that are closely located such as mainshock–aftershock sequences. For closely located events, amplitude ratios made at the same station therefore remove the common site response as well as the common path effect. In their method, dimensionless, narrowband spectral amplitude measurements (in $\log_{10}$) were made, and then source spectral ratios were formed. When compared across a network of local-to-near-regional stations at various azimuths, these source ratios were found to be very stable, roughly a factor of 3–4 times less variable than direct S-wave ratios (see Fig. 2 from MAYEDA and MALAGNINI 2010). Specifically, the method was tested on the Wells, Nevada M_W 5.9 seismic sequence, which was well recorded by the Earthscope Transportable Array (USArray), where results showed that the coda spectral ratios were virtually insensitive to both source radiation pattern and source directivity, thus making it significantly more robust than classic techniques that use direct S-waves (MAYEDA and MALAGNINI 2010; YOO and MAYEDA 2013). These stable-source spectral ratios provide an easy test to check whether self-similarity holds under the assumption of the omega-square source model (BRUNE 1970).

Coda envelopes, regardless of the region of study, size of the event, or frequency band, have been empirically fit to a very simple functional form that closely follows the single-scattering formulation of AKI (1969), where the coda amplitude A_c is dominated by an exponentially decaying function of time:

$$A_c(f,t,r) = A_O(f) \cdot H(t-t_s) \cdot (t-t_s)^{-\gamma(r)} \exp[-b(r)\cdot(t-t_s)], \quad (1)$$

where $A_O(f) = W_O(f) \cdot S(f) \cdot T(f), f$ is the center frequency, r is the epicentral distance in kilometers, t is the time in seconds from the origin time, t_s is the S-wave travel time in seconds, W_o represents the S-wave source, S is the site effect, T represents the S-to-coda transfer function effect, H is the Heaviside step function, and $\gamma(r)$ and $b(r)$ are the distance-dependent coda shape factors that controls the coda envelope shape for that particular region. Because we take amplitude ratios for roughly co-located earthquakes recorded at the same station site, for each narrowband all the constants will cancel except for the relative source terms, W_o. For each of our 20 sequences, we empirically found the envelope shape parameters that best fit our data as a function of frequency for each station. The dimensionless coda amplitude (in $\log_{10}$) is simply the D_c-difference between our empirically-derived synthetic and the $\log_{10}$ envelope for each band, at each station. We simply store these relative values and then use them to build our frequency-dependent source spectral ratios.

Thus, following the method outlined in MAYEDA *et al.* (2007), the moment-rate spectral ratio computed between two closely located generic events #1 and #2 can be written as follows:

$$\frac{\dot{M}_1(\omega)}{\dot{M}_2(\omega)} = \frac{M_{0_1}\left[1+(\omega/\omega_{c_2})^2\right]^{p/2}}{M_{0_2}\left[1+(\omega/\omega_{c_1})^2\right]^{p/2}}, \qquad (2)$$

where M_{0_i} is the seismic moment of the ith event; p is the high-frequency roll-off of the moment-rate spectra (the omega-square model dictates $p = 2$, and the simple consideration that the energy integral must be finite forces: $p > 1.5$). Figure 2 shows example coda-derived source spectral ratios between the M_W 6.7 Noto-Hanto, Japan earthquake on March 25th, 2007 and selected aftershocks using averaged ratios obtained from four regional, three-component broadband stations. Two horizontal asymptotes characterize the right-hand-side of Eq. (2): the low-frequency one is simply $\frac{M_{0_1}}{M_{0_2}}$, whereas, in the hypotheses of self-similar scaling (constant stress drop or apparent stress) and omega-square spectral model ($p = 2$), the high-frequency asymptote is: $\left(\frac{M_{0_1}}{M_{0_2}}\right)^{1/3}$ (MAYEDA *et al.* 2007).

In Fig. 2, theoretical asymptotes are plotted for each ratio. Interestingly, the data systematically suggest other asymptotes, instead of the expected ones. The high-frequency self-similar asymptotes are never reached by the empirical source ratios,

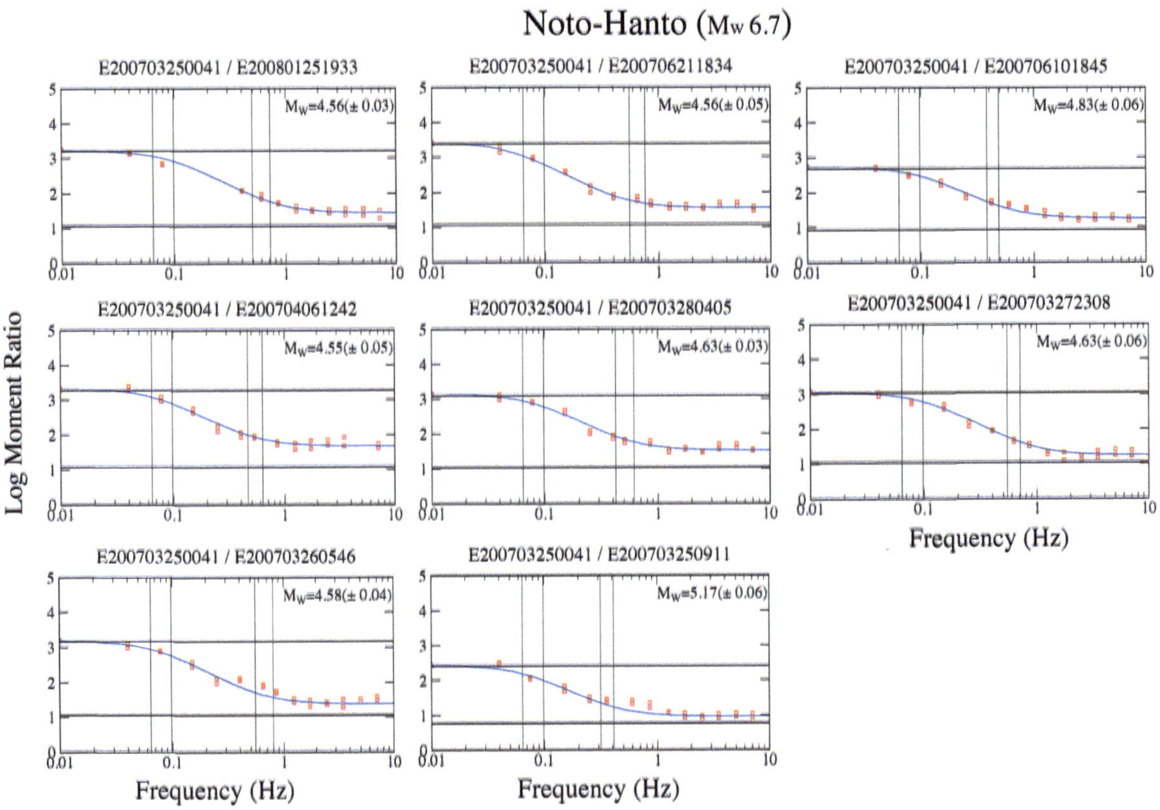

Figure 2

Comparison between the empirical spectral ratios computed on the events of the 2007 Noto-Hanto seismic sequence (*squares* indicate average ±1 SD) and the best-fitting scaling model (*solid curves*). *Solid horizontal lines* represent the asymptotes expected for self-similar scaling. *Vertical lines* indicate estimates of mainshock (same in all plots) and aftershock corner frequencies, with the related uncertainties (average ±1 SD). In each frame are indicated the main/aftershock identification numbers, and the aftershock's moment magnitude adjusted during the fitting procedure representing a clear violation of self-similarity if $p = 2$, or the need for a different spectral roll-off at high frequency ($p \neq 2$, see a later section).

To account for the possibility of non-self-similar scaling, we used the implementation of the MDAC source model by WALTER and TAYLOR (2001), which uses an omega-square source spectral model (i.e., $p = 2$) but permits non-self-similar scaling. From their study:

$$\omega_{c_i} = 2\pi f_{c_i} = \left(\frac{k_i \sigma_{a_i}}{M_{0_i}}\right)^{1/3} \tag{3}$$

is the ith event's corner frequency (ω_{c_i} in radians/s, f_{c_i} in Hz), and k_i are, for the ith earthquake, a function of the seismic velocities (α, β), corner frequencies $(\omega_{cP_i}, \omega_{cS_i})$, and radiation patterns $\left(R_{\theta\varphi P_i}, R_{\theta\varphi S_i}\right)$ of the appropriate wave types (P, S):

$$k_i = \frac{16\pi}{\beta^2\left(\frac{R^2_{\theta\varphi P_i}\omega^3_{cP_i}}{\alpha^5\omega^3_{cS_i}} + \frac{R^2_{\theta\varphi S_i}}{\beta^5}\right)}, \tag{4}$$

$$\sigma_{a_i} = G\frac{E_{r_i}}{M_{0_i}} \tag{5}$$

is the ith event's apparent stress, i.e., the ratio between the radiated seismic energy (E_{r_i}) and the seismic moment, multiplied by rock rigidity (G). For simplicity, following WALTER and TAYLOR (2001), we use an additional parameter to link the P- and S-wave corner frequencies:

$$\omega_{cP_i} = \zeta\omega_{cS_i}, \tag{6}$$

where ζ is the scale factor. For the BRUNE (1970) omega-square model, ζ is the ratio of the wave velocities, but other studies have shown that the corner frequencies are more equal. In general: $1 \leq \zeta \leq \frac{\alpha}{\beta}$. However, as suggested by WALTER and TAYLOR (2001), we used $\zeta = 1$, that is:

$$\omega_{cP_i} = \omega_{cS_i}. \tag{7}$$

Note that a different choice made for Eq. (7) would not change anything in our source spectral ratios, or in their interpretation.

The various parameters used in Eq. (4) have the following numerical values: $\alpha = 6.0$ km/s, $\beta = 3.5$ km/s; P-wave radiation pattern: $R_{\theta\varphi P} = 0.44$; S-wave radiation pattern: $R_{\theta\varphi S} = 0.60$. For completeness, in the calculations we also entered the rock densities at the source and at the receivers (respectively: $\rho_S = 2{,}700$ kg/m^3, and $\rho_r = 2{,}500$ kg/m^3).

The spectral ratio fits in Fig. 2 are all obtained using the theoretical function given in Eq. (2), using Eqs. (3)–(7), and assuming the following non-self-similar scaling for the apparent stress:

$$\sigma_{a_i} = \sigma'_a\left(\frac{M_{0_i}}{M'_0}\right)^{\psi}, \tag{8}$$

where

$$\psi = \frac{\varepsilon}{3+\varepsilon}, \tag{9}$$

σ'_a is the apparent stress at the arbitrary reference seismic moment $M'_0 = 4.0e + 16$ Nm (which corresponds to $M_W = 5.0$), and the parameter ε (KANAMORI and RIVERA 2004) reflects the deviation from self-similarity in the corner frequency-seismic moment scaling relation:

$$M_0 \propto f_c^{-(3+\varepsilon)}. \tag{10}$$

It must be clear to the reader that, even though the coda-derived estimates of seismic moments of earthquakes are extremely precise, the entire coda method needs to be accurately calibrated with ground truth information on as many events as possible (for details, see MAYEDA and WALTER 1996). This issue was discussed in detail for the M_W 7.0 Hector Mine, CA sequence (MAYEDA *et al.* 2007), and also for subsequent sequences investigated using the coda ratio method (e.g., Chi–Chi, Taiwan, MAYEDA and MALAGNINI 2009; Wells, Nevada, MAYEDA and MALAGNINI 2010; YOO and MAYEDA 2013; L'Aquila, Italy, MALAGNINI *et al.* 2011).

For what is shown in the present study, in order to avoid significant biases in corner frequency, stress drop, and inferred scaling during the spectral ratio fitting procedure, we use independent, well-constrained seismic moment estimates derived from regional moment-tensors for all 20 sequences whose mainshock events are shown in Fig. 1, so that all of our ratios are tied to their respective long-period amplitudes. This point is crucial because imprecise moment estimates can have a large effect in scaled energy and corner frequency estimates, thus biasing results and interpretations (e.g., YOO and MAYEDA 2013).

Table 1

For each seismic sequence, we give: (a) the source for the moment tensor solutions that are used for all earthquakes; (b) an estimate of the scaling parameter ψ [see Eqs. (8), (9)] (average and related standard deviation); and (c) the fault mechanism

Sequence	Source of MT solutions	⟨Best Ψ⟩	σ (Best Ψ)	Mechanism
Gulf of Aqaba (Red Sea)	Hofstetter *et al.* (2003)	0.13	0.02	S-S
Gulf of California (Mexico)	USGS	0.316	0.002	S-S
Fukuoka (Japan)	F-net regional moment tensor catalog	0.2	0.02	S-S
Hector Mine (CA, USA)	G. Ichinose, pers. comm. (2006)	0.23	0.01	S-S
Parkfield (CA, USA)	Berkeley moment tensor catalog	0.122	0.002	S-S
Tottori (Japan)	F-net regional moment tensor catalog	0.27	0.01	S-S
Oklahoma City (OK, USA)	Herrmann, pers. comm. (2012)	0.28	0.07	S–S
San Giuliano (Italy)	MedNet RCMT catalog	0.23	0.02	S-S
Chi–Chi (Taiwan)	Harvard CMT catalog, and Kao and Angelier (2001)	0.15	0.04	R
Chuetsu (Japan)	F-net regional moment tensor catalog	0.04	N/A	R
Honshu (Japan)	F-net regional moment tensor catalog	0.237	0.003	R
Iwate (Japan)	F-net regional moment tensor catalog	0.23	0.001	R
Mid Niigata (Japan)	F-net regional moment tensor catalog	0.22	0.01	R
Northridge (CA, USA)	Dreger, pers. comm. (2008)	0.23	0.01	R
Noto-Hanto (Japan)	F-net regional moment tensor catalog	0.17	0.01	R
San Simeon (CA, USA)	Berkeley moment tensor catalog	0.28	0.01	R
Virginia (USA)	Herrmann, pers. comm. (2012)	0.17	0.04	R
L'Aquila (Italy)	Herrmann *et al.* (2011)	0.23	0.02	N
Colfiorito (Umbria-Marche, Italy),	Ekstrom *et al.* (1998)	0.28	0.03	N
Wells (NV, USA)	Herrmann, pers. comm. (2012)	0.24	0.07	N

S-S strike-slip, *R* reverse, *N* normal

For each mainshock–aftershock pair, we performed a grid-search over the parameters σ_a' and ψ of Eqs. (8) and (9). Moreover, we allow the aftershock's M_W to vary by a small amount from its initial estimate (typically, ±0.1 magnitude units). After adjusting the moment magnitudes of all events, we used a bootstrap method to estimate, under the $p = 2$ assumption in Eq. (2), the corner frequency and the associated error for each event of the sequence, including the mainshock.

Details of the grid-search scheme are as follows: we choose the first spectral ratio from the set of all the available ones. We use Eqs. (2)–(9) to find a best-fit theoretical ratio from which we extract the scaling parameters σ_a' and ψ of Eqs. (8) and (9), and use Eq. (3) to obtain an estimate of the mainshock's corner frequency. We also allow small adjustments of the aftershock's seismic moment to obtain a better fit, and loop through all the spectral ratios. The first part of the procedure, just described, allows us to slightly adjust the estimate of each aftershock's seismic moment (generally, <±0.1 M_W units), and to obtain N different estimates of the mainshock's corner frequency (one for each one of the N aftershocks), to be used in order to get an average value and a standard deviation.

At the end of this first part of the processing, we have a set of independent estimates of the mainshock's corner frequency from which we obtain the average and standard deviation. After adjusting the aftershocks' seismic moments, we loop through the individual ratios again: from each mainshock/aftershock ratio, we fix the corner frequency of the mainshock, and calculate the aftershock's corner frequency, then we use the apparent scaling existing between the specific event pair, and predict the corner frequencies of all the remaining (N-1) aftershocks [again, Eqs. (2)–(9) are used for the task]. At the end of the second loop, we have a set of N estimates of corner frequency for each aftershock, which allows the calculation of average values and standard deviations.

In the individual plots of Fig. 2, error bars indicate ±1 standard deviation. In all cases, the scaling parameter ψ in Eq. (9) is non-zero, consistent with non-self-similar scaling. The parameters calculated for all events are listed in Tables 1, 2, and 3.

2.2. *Combined Seismic Results and Comparisons with Other Studies*

The entire data set analyzed in the present study is derived from local and regional broadband seismograms belonging to 20 mainshock–aftershock crustal sequences. Again, regional moment-tensor solutions were used to define the low-frequency amplitudes (for most of the events we could rely on moment-tensor solutions from regional, long-period waveform modeling), and when necessary, to calibrate the coda-derived moment magnitude (e.g., MAYEDA and WALTER 1996, 2003). For each one of the investigated sequences, we tried to select aftershocks characterized by a focal mechanism very similar to that of their respective main event.

As demonstrated by MAYEDA and MALAGNINI (2010), coda-derived source ratios are not sensitive to focal mechanism and directivity, and the possible inclusion of events with differing mechanisms (or without a focal solution available) has no deleterious effect on our results. In what follows, regardless of the stress re-distribution that followed the mainshock rupture, and of the possibility that some of the aftershocks in our data set occur on faults different from that of the main event, we show that mainshock and aftershocks within a seismic sequence (or foreshock, in the case of Colfiorito, Italy) always share great similarities in their dynamic behavior and characteristics of source scaling.

All of our spectral ratios have an "artificial" low-frequency data point computed as the ratio of the seismic moments, which is set at an arbitrary low frequency (e.g., $f_{low} = 0.01$ Hz). Again, such critical information comes from independent, regional moment-tensor studies through waveform inversion and/or coda calibration for stable M_W. For cases where we do not have adequate long-period bandwidth, we find this to be a necessary step—otherwise the grid-search is unconstrained, leading to large fluctuations and biases in corner frequency estimates, especially for the mainshock corner frequency estimate.

For the Chi–Chi sequence (Taiwan) (see Fig. 1 in MAYEDA and MALAGNINI 2009) the aftershocks' locations form a cloud of events that do not necessarily coincide with the main fault plane. Similar characteristics (aftershocks not occurring right on the main event's fault plane) may affect the sequences triggered on reverse faults. Regardless of the imperfections of the data sets, clear and self-consistent patterns of source scaling can be recognized in each individual sequence from the simple analysis of the apparent stress or corner frequency scaling (see parameters in Tables 1, 2, and 3).

Raw source spectral ratios such as those in Fig. 2 were used in our best-fit analyses only if their bandwidths allowed the recognition of the aftershocks' corner frequencies. That is, if the spectral asymptotes at high frequency were adequately pronounced. The presence of a clear, high-frequency spectral plateau (past the corner frequency of the specific aftershock), guarantees that the best-fits obtained in the regressions yield reliable estimates of apparent stress and corner frequency for the small events. In the case of insufficient bandwidth, source spectral ratios may still be used in order to obtain information on the mainshock's corner frequency, and on the aftershock's seismic moment (see, for the Hector Mine seismic sequence, the study by MAYEDA *et al.* 2007).

The entire set of (M_0, f_c) data points computed here is plotted in Fig. 3a (different sequences are indicated by different colors and symbols), whereas apparent stresses of strike-slip, normal, and thrust sequences, respectively, are plotted as a function of earthquake size in Fig. 3b–d. Clear trends of increasing apparent stress with increasing size characterize all sequences, and smaller overall scatter characterizes dip-slip sequences (normal and thrust), with respect to strike-slip ones.

Some of our sequences have already been studied by other research groups. For example, using results from the analysis of eight seismic sequences, including the ones of Parkfield, CA, USA (2004), mid Niigata (Chuetsu), Japan (2004), Iwate-Miyagi, Japan (2008), Chuetsu-Oki (2007), Hector Mine, CA, USA (1999), and Wells, NV, USA (2008), BALTAY *et al.* (2010, 2011) concluded that there was no scale dependency for the apparent stress over the investigated magnitude range. In Fig. 4, we compare our results for the Iwate-Miyagi, 2008 sequence with the results of BALTAY *et al.* (2011), as well as revised

Table 2

Parameters of all the strike-slip-fault earthquakes used in this study

Event ID	Date	Time	Lat.	Lon.	Depth (km)	M_W	Strike (°)	Dip (°)	Rake (°)	f_c (Hz)	Δf_c (Hz)	$\text{Log}_{10}\ \sigma_a$ (Pa)	$\Delta\ (\log_{10}\ \sigma_a)$
Gulf of Aqaba													
E953260410	**1995/11/22**	**04:10:00**	**28.76**	**34.66**	**10.0**	**7.15**	**196**	**59**	**−15**	**0.0601**	**0.0069**	**5.7830**	**0.1539**
E953261240	1995/11/22	12:40:00	28.45	34.75	9.0	5.20	20	89	12	0.4447	0.0657	5.4598	0.1980
E953262210	1995/11/22	22:10:00	28.55	34.72	9.0	5.34	288	88	168	0.3093	0.0379	5.2004	0.1637
E953401750	1995/12/06	17:50:00	28.70	34.77	2.0	4.64	284	90	143	0.5025	0.0726	4.7794	0.1956
E953450130	1995/12/11	01:30:00	29.20	34.86	24.0	5.04	339	79	164	0.5980	0.0838	5.6067	0.1889
E960520450	1996/02/21	04:50:00	28.86	34.71	10.0	5.14	N/A	N/A	N/A	0.6236	0.0744	5.8145	0.1593
E960570710	1996/02/26	07:10:00	28.73	34.82	9.0	4.97	188	90	1	0.9472	0.1375	6.0999	0.1986
Gulf of California													
2010094223300	**2010/04/04**	**22:40:42**	**32.25**	**−115.29**	**5.0**	**7.10**	**219**	**84**	**−17**	**0.0580**	**0.0071**	**5.6597**	**0.1661**
2010099052500	2010/04/09	05:32:15	32.20	−115.31	5.0	4.03	132	62	−124	0.8599	0.1568	4.5564	0.2631
2010101163800	2010/04/11	16.42:08	32.25	−115.32	5.0	4.63	194	89	−14	0.3819	0.0615	4.4043	0.2207
2010104223000	2010/04/14	22:34:55	32.19	−115.21	5.0	4.17	15	57	−44	0.5992	0.1013	4.2991	0.2374
2010107160500	2010/04/17	16:09:53	32.27	−115.34	11.0	4.23	291	57	−133	0.4859	0.0633	4.1230	0.1786
2010257104500	2010/09/14	10:52:18	32.04	−115.20	8.0	5.03	214	80	−33	0.3617	0.0483	4.9380	0.1813
Oklahoma													
20111106035310	**2011/11/06**	**03:53:10**	**35.54**	**−96.75**	**8.0**	**5.59**	**145**	**85**	**−5**	**0.2068**	**0.0219**	**5.0442**	**0.1384**
20111105071245	2011/11/05	07:12:45	35.57	−96.70	3.0	4.70	300	80	−10	0.4958	0.0454	4.8502	0.1197
20111106063110	2011/11/06	06:31:10	35.56	−96.87	8.0	3.69	140	45	−5	0.7878	0.0744	3.9383	0.1215
20111106093957	2011/11/06	09:39:57	35.50	−96.86	3.0	3.65	320	90	−10	1.0577	0.1011	4.2618	0.1251
20111106105236	2011/11/06	10:52:36	35.57	−96.78	3.0	3.55	140	80	−10	1.0690	0.1024	4.1256	0.1252
20111106150705	2011/11/06	15:07:05	35.54	−96.91	3.0	3.74	320	85	−15	0.8991	0.0862	4.1851	0.1252
20111106175234	2011/11/06	17:52:34	35.55	−96.82	3.0	3.43	145	75	5	1.2615	0.1216	4.1613	0.1253
Fukuoka													
	2005/03/20	**01:53:47**	**33.74**	**130.18**	**11.0**	**6.53**	**122**	**87**	**−11**	**0.1296**	**0.0174**	**5.8511**	**0.1692**
	2005/03/20	05:32:00	33.79	130.09	8.0	4.39	289	79	18	0.6134	0.1051	4.6600	0.2173
	2005/03/20	10:52:00	33.80	130.09	5.0	4.49	293	87	5	0.7759	0.1131	5.1207	0.1828
	2005/03/20	11:38:00	33.71	130.19	8.0	4.27	202	85	−173	1.1553	0.2209	5.3016	0.2356
	2005/03/21	14:59:00	33.78	130.10	8.0	4.52	49	80	−168	0.7790	0.1196	5.1699	0.1910
	2005/04/19	21:11:31	33.68	130.29	8.0	5.41	312	90	14	0.3952	0.0706	5.6163	0.2222
	2005/04/19	21:44:00	33.68	130.29	8.0	4.39	235	61	−130	0.9006	0.1427	5.1628	0.1988
	2005/04/20	00:09:00	33.68	130.28	8.0	4.74	134	90	−16	0.8618	0.1587	5.6260	0.2278
	2005/05/01	16:23:00	33.67	130.32	8.0	4.67	138	84	−8	0.7802	0.1264	5.3953	0.2031
Hector Mine													
E1319532	**1999/10/16**	**09:46:44**	**34.59**	**−116.27**	**10.0**	**7.00**	**336**	**80**	**174**	**0.0701**	**0.0219**	**5.7175**	**0.3764**
E1319565	1999/10/16	12:57:20	34.44	−116.25	5.0	5.32	287	79	25	0.3520	0.1138	5.2970	0.3870
E1319624	1999/10/16	17:38:48	34.43	−116.25	14.0	4.55	163	75	142	0.6586	0.2384	4.9445	0.4375
E1319917	1999/10/19	12:20:44	34.71	−116.34	5.0	4.24	178	85	−146	0.7288	0.2761	4.6089	0.4406
E1320111	1999/10/22	16:08:48	34.86	−116.41	5.0	4.93	267	87	−21	0.3888	0.1535	4.8203	0.4559
E1320305	1999/10/25	18:26:00	34.62	−116.24	5.0	4.38	144	55	85	0.7003	0.2289	4.7813	0.3952

Table 2 *continued*

Event ID	Date	Time	Lat.	Lon.	Depth (km)	M_W	Strike (°)	Dip (°)	Rake (°)	f_c (Hz)	Δf_c (Hz)	$\text{Log}_{10}\,\sigma_a$ (Pa)	$\Delta(\log_{10}\sigma_a)$
E1323158	1999/12/23	14:30:54	34.59	−116.26	10.0	3.74	N/A	N/A	N/A	1.2478	0.5034	4.5485	0.4792
Parkfield													
	2004/09/28	**17:15:00**	**35.81**	**−120.37**	**8.0**	**5.91**	**57**	**85**	**−7**	**0.2624**	**0.0709**	**5.8114**	**0.3725**
	2004/09/29	17:10:00	36.00	−120.54	12.0	5.01	231	90	0	0.6339	0.1765	5.6078	0.3853
	2004/11/29	01:54:00	35.94	−120.49	10.4	4.32	323	87	−176	1.2917	0.4205	5.4849	0.4519
	2005/05/16	07:24:00	35.93	−120.48	10	4.48	321	88	−170	1.2048	0.3855	5.6353	0.4496
	2007/05/22	11:34:00	35.86	−120.41	10	4.15	143	85	−172	1.4410	0.4757	5.3668	0.4827
San Giuliano													
7001	**2002/10/31**	**10:33:00**	**41.64**	**14.99**	**17.4**	**5.71**	**174**	**67**	**−8**	**0.5202**	**0.1202**	**6.4109**	**0.2948**
7002	**2002/11/01**	**15:08:00**	**41.71**	**14.94**	**16.9**	**5.69**	**170**	**75**	**−4**	**0.4788**	**0.0583**	**6.2948**	**0.1524**
7007	2002/11/01	17:21:31	41.71	14.88	N/A	4.44	N/A	N/A	N/A	2.0381	0.3037	6.2865	0.1901
7009	2002/11/02	02:37:01	41.71	14.85	N/A	4.07	N/A	N/A	N/A	1.8819	0.3465	5.6210	0.2289
7011	2002/11/02	17:20:36	41.72	14.80	N/A	3.59	N/A	N/A	N/A	2.5197	0.4776	5.2717	0.2294
7014	2002/11/04	00:35:46	41.70	14.87	N/A	4.55	N/A	N/A	N/A	1.5777	0.2243	6.1358	0.1743
7015	2002/11/04	03:26:30	41.74	14.81	N/A	4.11	N/A	N/A	N/A	2.3642	0.4587	5.9672	0.2456
7018	2002/11/05	23:10:41	41.71	14.93	N/A	3.91	N/A	N/A	N/A	2.4123	0.4971	5.6853e	0.2637
7019	2002/11/09	13:40:30	41.68	14.83	N/A	3.67	N/A	N/A	N/A	2.3389	0.4877	5.2817	0.2591
7020	2002/11/10	12:23:37	41.66	14.83	N/A	3.80	N/A	N/A	N/A	2.3266	0.4693	5.5058	0.2462
7021	2002/11/11	18:32:15	41.66	14.87	N/A	3.66	N/A	N/A	N/A	2.7349	0.6030	5.4719	0.2833
7022	2002/11/12	09:27:49	41.71	14.89	20.9	4.59	72	80	171	1.5665	0.2508	6.1545	0.2085
7023	2002/11/12	13:46:08	41.66	14.79	N/A	3.70	N/A	N/A	N/A	2.8003	0.6301	5.5472	0.2808
7024	2002/12/02	20:52:32	41.66	14.85	N/A	4.26	N/A	N/A	N/A	1.3127	0.1874	5.6041	0.1789
Tottori													
	2000/10/06	**04:30:25**	**35.30**	**133.40**	**11.0**	**6.56**	**150**	**85**	**−9**	**0.1186**	**0.0387**	**5.7372**	**0.4095**
	2000/10/06	19:59:00	32.28	133.36	5.0	4.34	50	51	125	1.0361	0.3940	5.2128	0.4775
	2000/10/06	21:38:00	35.40	133.30	5.0	4.30	156	86	−38	0.7085	0.2873	4.8034	0.4541
	2000/10/08	11:51:15	35.40	133.30	5.0	4.90	165	87	−9	0.4072	0.1351	4.8521	0.4208
	2000/10/10	12:57:00	35.37	133.30	8.0	4.51	163	90	−14	0.6270	0.2321	4.8184	0.4586
	2000/11/03	07:33:00	35.40	133.30	5.0	4.43	255	87	−174	0.7298	0.2743	4.8945	0.4629

The table contains dates and times of nucleation; latitude, longitude, and depth of hypocenter; M_W, strike, dip, rake, corner frequency with associated error, and apparent stress with associated error

The mainshock(s) of each sequence is (are) indicated in bold estimates from Oth *et al.* (2010) (published by Oth 2013) who used selected, high-quality F-net M_W's to better constrain their source spectral inversions. Oth (2013) concluded that the stress drop scale-dependence in the analyzed Japanese earthquakes was highly variable and locally non-self similar.

We see that Baltay *et al.*'s (2011) values of static stress drop are systematically higher by nearly an order of magnitude than those of the present study and of the revised estimates from Oth *et al.* (2010). Similarly, Yoo and Mayeda (2013) found roughly an order of magnitude lower static stress drops for the Wells, NV, USA, sequence using their source spectral ratio method when compared against results of Baltay *et al.* (2010), who derived their estimates from path- and site-corrected source spectra. Yoo and Mayeda (2013) suggest the discrepancy is due to large stress drops (~ 13 MPa) that were erroneously assumed a priori for the smallest Green's function events used in Baltay *et al.* (2010), and to a lesser extent, moment magnitude biases resulting from mixing NEIC M_W, m_b, and M_L. In addition, Yoo and Mayeda (2013) demonstrated that under the omega-square assumption, observed relative ground motions (PGA, PGV and PSA) between the mainshock and aftershocks could only be explained if there was a break in self-similar source scaling.

For other common sequences in Japan, we compared corner frequency estimates from revised estimates from the Oth *et al.* (2010) study (Oth 2013) and found good absolute agreement; although stress drops from two sequences, notably Fukuoka, 2005, and Noto-Hanto, 2007, were roughly a factor of two larger than ours in the revised (Oth *et al.* 2010) estimates, the trends were similar, and are the topic of an ongoing study. Finally, we point out that our inferred static stress drops, as well as those derived from Oth *et al.* (2010) for a number of the mainshocks, are in good agreement with recent finite-fault inversion estimates of average stress drop from Asano and Iwata (2011).

2.3. *To Be or Not to Be (Self-Similar), That is the Question*

Figure 3b–d describes earthquake sequences that (under the assumption of the omega-square Brune spectral model: i.e., $p = 2$ in Eq. (2)) do not follow self-similarity. Instead, within each seismic sequence, the increasing apparent stresses may be described by an increasing function of earthquake size. In all cases, with the exception of the Chi–Chi, Taiwan, earthquake sequence, apparent stresses steadily increase as a function of seismic moment. For the Chi–Chi seismic sequence, beyond roughly M_W 5.5, the monotonic increase of apparent stress with seismic moment is followed by a saturation to a value in the range of 2–3 MPa that characterizes all the large events of the sequence, including the mainshock.

We have seen in Fig. 3b–d that, within the same fault zone, earthquakes with different apparent stresses are generated; moreover, apparent stress seems to be a monotonic function of seismic moment. In order to have a larger apparent stress, more energy needs to be available for radiation and, consequently, less energy needs to be dissipated (by friction) on the fault plane as heat. In the hypothesis that the Brune omega-square spectral model holds, it appears that large earthquakes radiate more energy per square meter of fault surface, per meter of slip. If this is true, then the seismic efficiency of earthquakes (i.e., the ratio between radiated energy and total strain energy available) should gradually increase as the seismic slip gets larger.

Table 1 summarizes the results obtained on each seismic sequence in terms of the scaling parameter ψ, with an estimate of its standard deviation. The tabulated values are all positive, with values around 0.25 (i.e., $\varepsilon = 1$ in Eq. (9), very similar to the results obtained by Mayeda and Walter 1996). The indicated scaling parameters represent the average behavior of the specific sequence, and do not contain information about the possible saturation of the apparent stress at larger moment magnitudes, which we believe likely happens as seen in the Chi–Chi sequence. The results of our optimizations (corner frequencies, apparent stresses, and adjusted seismic moments) are listed in Tables 2 (strike-slip faults), 3 (normal faults), and 4 (reverse faults).

Interestingly, all the results described up this point could have an alternative explanation: all sequences may be internally self-similar (constant stress drop), provided that the high-frequency spectral fall-off deviates significantly from the classic omega-square

Table 3

Parameters of all the normal-fault earthquakes used in this study

Event ID	Date	Time	Lat.	Lon.	Depth (km)	M_W	Strike (°)	Dip (°)	Rake (°)	f_c (Hz)	Δf_c (Hz)	$\text{Log}_{10}\ \sigma_a$ (Pa)	$\Delta\ (\log_{10}\ \sigma_a)$
Colfiorito													
8190	1997/09/07	23:08:07	43.03	12.79	4.3	4.06	N/A	N/A	N/A	2.1004	0.3633	5.7434	0.2049
8001	**1997/09/26**	**00:33:18**	**42.99**	**12.78**	**10.0**	**5.69**	**152**	**46**	**−83**	**0.5236**	**0.1171**	**6.3991**	**0.2611**
8002	**1997/09/26**	**09:40:31**	**43.01**	**12.85**	**10.0**	**5.91**	**143**	**36**	**−86**	**0.4547**	**0.0834**	**6.5509**	**0.2387**
8670	1997/09/27	02:17:27	43.06	12.89	3.3	3.93	N/A	N/A	N/A	1.7534	0.2187	5.3180	0.1512
8680	1997/09/27	03:03:21	43.05	12.80	5.8	3.85	N/A	N/A	N/A	2.5937	0.3851	5.6983	0.1803
8690	1997/09/27	08:08:07	43.10	12.81	10.0	4.36	148	55	−89	1.8188	0.4178	6.0217	0.2791
8750	1997/09/27	17:31:03	43.02	12.84	6.1	4.17	N/A	N/A	N/A	1.8047	0.3524	5.6979	0.2312
9025	**1997/10/03**	**08:55:22**	**43.03**	**12.84**	**10.0**	**5.26**	**141**	**43**	**−74**	**0.5655**	**0.0985**	**5.8642**	**0.2284**
9124	**1997/10/06**	**23:24:52**	**43.02**	**12.83**	**10.0**	**5.48**	**145**	**40**	**−80**	**0.5144**	**0.0588**	**6.0791**	**0.1555**
9143	1997/10/07	01:24:34	43.02	12.94	4.3	4.27	126	26	−102	1.6999	0.4496	5.8286	0.3206
9484	**1997/10/12**	**15:23:15**	**42.96**	**13.03**	**1.2**	**5.62**	**122**	**38**	**−100**	**0.7496**	**0.0757**	**6.7831**	**0.1281**
9704	1997/10/19	16:00:17	42.98	12.90	5.7	4.29	128	44	−103	2.0701	0.3479	5.9043	0.2136
9530	1997/11/27	07:02:35	42.95	12.91	4.0	4.02	N/A	N/A	N/A	2.1147	0.2623	5.7834	0.1536
9540	1997/11/29	19:15:16	43.04	12.95	5.5	3.96	N/A	N/A	N/A	2.0394	0.2699	5.5729	0.1685
9730	1998/01/02	07:31:33	42.95	12.83	9.0	4.43	N/A	N/A	N/A	1.8686	0.6018	5.7523	0.4019
1179	2000/08/08	21:31:15	42.99	12.82	5.0	3.93	N/A	N/A	N/A	1.7475	0.2433	5.3199	0.1792
L'Aquila													
	2009/04/06	**01:32:39**	**42.33**	**13.35**	**5.0**	**6.13**	**135**	**55**	**−95**	**0.3099**	**0.0866**	**6.3444**	**0.3707**
	2009/04/06	02:37:04	42.37	13.34	10.0	5.01	134	64	−114	0.6646	0.2032	5.7246	0.3996
	2009/04/06	03.56:45	42.34	13.39	7.0	4.46	145	60	−95	1.1127	0.3325	5.5280	0.3957
	2009/04/06	07:17:10	42.35	13.37	6.0	4.13	137	56	−100	1.6224	0.5705	5.5461	0.4674
	2009/04/08	22:56:50	42.51	13.36	7.0	4.06	118	48	−109	2.5591	0.8339	6.0438	0.4476
	2009/04/09	03:14:52	42.34	13.44	12.0	4.40	150	90	55	1.7589	0.5513	5.9207	0.4293
	2009/04/09	04:32:44	42.44	13.42	7.0	4.31	5	55	−45	1.5487	0.4530	5.7653	0.3923
	2009/04/13	21:14:24	42.50	13.36	6.0	4.95	143	61	−96	0.9693	0.3003	5.9699	0.4277
Wells													
	2008/02/21	**14:16:05**	**41.15**	**−114.87**	**11.0**	**5.88**	**25**	**40**	**−90**	**0.3213**	**0.0317**	**6.0661**	**0.1449**
	2008/02/21	14:34:00	41.12	−114.89	11.0	4.38	40	60	−65	1.2127	0.1398	5.5437	0.1733
	2008/02/22	15:34:00	41.13	−114.91	10.0	3.78	20	40	−80	2.0177	0.2582	5.3042	0.1960
	2008/02/22	16:20:00	41.21	−114.86	11.0	3.96	225	55	−55	1.7921	0.2049	5.4227	0.1714
	2008/02/22	23:57:00	41.15	−114.93	9.0	4.62	225	35	−40	0.8614	0.0916	5.4597	0.1579
	2008/04/01	01:50:00	41.14	−114.92	11.0	3.99	230	55	−60	1.6051	0.2019	5.3217	0.1918
	2008/04/22	23:27:00	41.10	−114.92	12.0	4.34	225	40	−85	1.1708	0.1359	5.4377	0.1747
	2008/02/27	07:59:00	41.19	114.83	11.0	4.21	90	85	10	1.6773	0.1976	5.7107	0.1776
	2008/02/22	15:10:00	41.16	−114.93	10.0	4.07	250	40	−50	1.4212	0.1684	5.2847	0.1789
	2008/04/01	13:16:17	41.23	−114.84	12.0	4.19	85	50	−15	1.5906	0.1865	5.6117	0.1764
	2008/04/22	20:40:09	41.22	−114.81	11.0	3.90	45	30	−55	2.3780	0.3160	5.6972	0.2044

The mainshock(s) of each sequence is (are) indicated in bold

model (i.e., $p \neq 2$ in Eq. (2), while $\psi = \varepsilon = 0$ in Eqs. (8) and (9)).

Using the 2007 Noto-Hanto source ratios in Fig. 2, we found equally good fits for the individual spectral ratios when $p = 1.7$ and $\varepsilon = 0$ (self-similar scaling). The fits for the two alternative cases (constant stress drop and $p \neq 2$, or or non-constant stress drop and $p = 2$) are virtually indistinguishable. For completeness, we applied a grid-search over p where $1.5 < p < 2.5$ and found that indeed a self-similar model (i.e., $\varepsilon = 0$, constant apparent stress scaling) could equally fit all the individual spectral ratios from the other sequences, but the spectral fall-off ranged between ~ 1.55 and 1.7.

Based on our results, in the two cases mentioned above:

1. constant apparent stress and $p \neq 2$;
2. $p = 2$ and magnitude-dependent stress drop;

are mutually exclusive, virtually indistinguishable, and cannot be discriminated based on our data set. We note that the consensus in the earthquake source community heavily favors models with a fall-off parameter $p = 2.0$ (i.e., BRUNE 1970; AKI 1967; BOATWRIGHT 1980; MADARIAGA 1976) and is also assumed in nearly all strong ground motion prediction studies. Quite arbitrarily, we favor the choice of case (2), that is: the omega-square spectral model ($p = 2$), with magnitude-dependent apparent stress.

Nonetheless, we are probably unable to detect subtle differences in the fall-off parameter, and thus, are unable to choose from these two different source models (omega-square vs. a more gentle spectral fall-off). A thorough investigation over the validity of the omega-square source spectral model is not the main focus of the present paper, and can be the object of a future study, but a goal of general importance is to understand, in the specific framework of the omega-square earthquake spectral model, whether or not a constant stress drop model can be used to represent the correct scaling over this magnitude range for this given set of 20 crustal earthquake sequences.

In what follows, we thus choose a working hypothesis in which the omega-square model holds, and interpret our results in terms of non-self-similar source scaling, where the static stress drop and the apparent stress significantly increase as a function of the event's magnitude. Because the omega-square model is only one of the possible options, we search a physical explanation for what we interpret as increasing stress drop with moment magnitude by looking at the results of new, state-of-the-art laboratory friction results.

2.4. *Laboratory Experiments on Friction*

A problem of primary importance in seismology is the ability to link laboratory results on rock friction to seismic observations, mainly due to differences in scale between faults dislocating the Earth's crust or a lab specimen, and/or the work rate dissipated on seismic faults or on a lab sample. Although the scale issue seems to be a major concern for many seismologists, recent high-speed experiments were able to reproduce fracture energies of the order of 1–10 MJ/m^2 (e.g., DI TORO *et al.* 2011), of the same order of the ones estimated for large earthquakes (e.g., COCCO and TINTI 2008). Pronounced dynamic weakening has recently been observed in state-of-the-art laboratory friction experiments performed at seismic slip rates of $\sim$1 m/s or larger, under normal stress conditions overlapping those present at 10-km depths on normal and strike-slip faults, and approaching those characterizing reverse faults under sub-lithostatic pore-fluid pressure ($\lambda_f = 0.8$), at the same depth. If high-speed friction experiments can be considered representative of the dynamic behavior of a seismogenic fault during a large earthquake, we should expect a gradual increase of the earthquake stress drop for increasing earthquake magnitude, at least in a seismic sequence in which all earthquakes strictly occur over the same fault plane. Moreover, products and microstructures observed in the lab (e.g., pseudotachylytes, DI TORO *et al.* 2006), or experimentally obtained damage patterns (e.g., SMITH *et al.* 2013), are very similar to those observed on exhumed natural faults.

Most of our seismically-derived data show a gradual increase of the stress drop (or apparent stress) up to a critical magnitude, beyond which the earthquakes' stress parameter stays constant. This behavior may be the consequence of a gradual fault weakening with slip, until a steady-state value is reached at relatively large slip. Though there is no rigorous seismological proof that the observed scaling variation is related to frictional behavior, we discuss why this interpretation is compatible with the above-mentioned laboratory tests under conditions approaching those of seismogenic faults at depth.

Different mechanisms have been invoked to describe the dynamic weakening of a fault: acoustic fluidization (MELOSH 1996), frictional melt (DI TORO *et al.* 2006; HIROSE and SHIMAMOTO 2005; NIELSEN *et al.* 2008, 2010), elastohydrodynamic lubrication (BRODSKY and KANAMORI 2001), formation of gel (DI TORO *et al.* 2004), pore-fluid thermal pressurization (REMPEL and RICE 2006), flash heating (RICE 2006), decarbonation with formation of nanopowders (HAN *et al.* 2007), dehydration (HIROSE and BYSTRICKY 2007), fluid pressurization due to frictional CO_2 exsolution (FAMIN *et al.* 2008), and gouge lubrication (RECHES and LOCKNER 2010).

A number of the mentioned phenomena were reproduced in laboratory experiments (DI TORO *et al.* 2004, 2006; HIROSE and SHIMAMOTO 2005; NIELSEN *et al.* 2008, 2010; HAN *et al.* 2007; DE PAOLA *et al.*

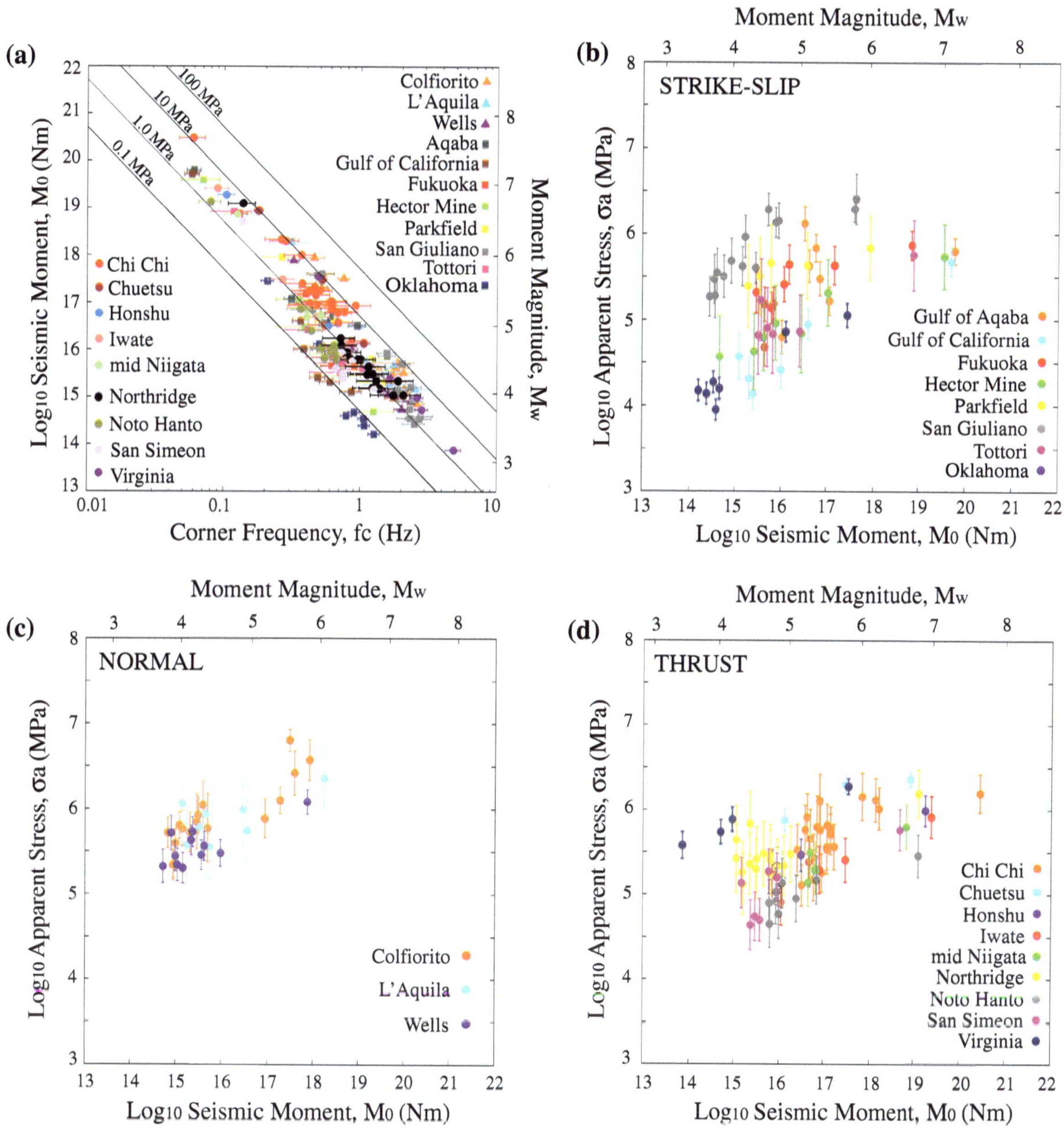

Figure 3

a Seismic moment-corner frequency data points for all the earthquakes in our data set. Different sequences are shown with *different colors*. *Triangles* indicate normal faulting, *squares* indicate strike-slip faulting, whereas *circles* represent thrust faulting; **b** apparent stresses vs. seismic moment of strike-slip earthquakes; **c** apparent stresses vs. seismic moment of normal-fault earthquakes; **d** apparent stresses vs. earthquake sizes of thrust-fault events

2011; Hirose and Bystricky 2007; Famin *et al.* 2008; Reches and Lockner 2010; Wibberley and Shimamoto 2005). Finally, since most weakening mechanisms are triggered by frictional heating, large weakening should occur whenever a large slip occurs in a short time (Rice 2006; Nielsen *et al.* 2008; Famin *et al.* 2008), that is, for large rates of frictional work.

The slip weakening distance (*D*th in Fig. 5) estimated from seismological data is between 0.05 and 1-m (e.g., Ide and Takeo 1997). Early lab

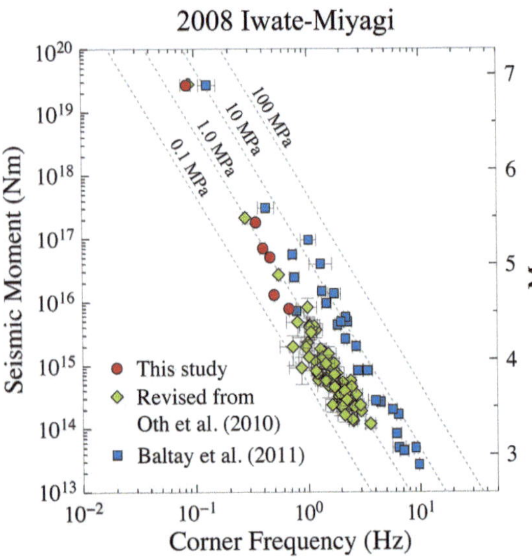

Figure 4

For the Iwate-Miyagi seismic sequence (mainshock on 06/06/2008, at 23:43:53 UTC), we provide a direct comparison between our estimates of corner frequency (*red*), those by BALTAY *et al.*'s (2011) (*blue*), and revised estimates derived originally from OTH *et al.* (2010) (*green*)

measurements at slow slip velocities have typically determined critical slip distances (d_c) through rate-and-state friction formulations that were on the order of tens of μm (e.g., MARONE 1998; IKARI *et al.* 2007; NIEMEIJER *et al.* 2010 among many others). It is not clear how this parameter should scale to natural conditions and be compared with the slip weakening distance *D*th. However, in recent years, high-velocity experiments have demonstrated significant slip weakening over distances that are comparable to, and slightly larger (1–10 m) than those estimated from seismological observations (e.g., HIROSE and SHIMAMOTO 2005; NIELSEN *et al.* 2008; DI TORO *et al.* 2011).

Recently, NIEMEIJER *et al.* (2011) performed experiments using a newly constructed, high-velocity rotatory shear apparatus (SHIVA) capable of applying higher normal stresses and slip velocities than earlier machines (up to 50 MPa normal stress and 6.5 m/s slip velocity). A series of experiments was performed at a constant slip rate of 3 m/sec on solid, ring-shaped samples of (micro) gabbro to investigate the effects of slip velocity, normal stress, acceleration, and deceleration on the evolution of friction. Under all conditions investigated, a continuous through-going melt layer was produced, associated with significant weakening. Interestingly, the combination of high normal stress and slip velocity (i.e., the power supplied to the sample) provide experimental conditions that are comparable to the inferred energy input in moderately-sized earthquakes.

Data points shown in Fig. 5 from NIEMEIJER *et al.* (2011) refer to melt-induced slip weakening of gabbro, which is measured from the distance at which complete dynamic weakening is obtained, *D*th, due to melt lubrication. They confirmed that steady-state shear stress during the melt-lubricated phase of the experiment depends on normal stress in the form of a power law equation as predicted by theoretical models (indicated in Fig. 5) by NIELSEN *et al.* (2010), who proposed that the weakening distance has the form: $D\text{th} = b\left(\sigma_n\sqrt{V}\right)^a$; $a \approx -1$. Results plotted in Fig. 5 directly verify the validity of this functional form. In the of case $V = 1$ m/s, the abscissa indicates effective normal stress.

3. *Systematic Difference in Behavior Between Aftershocks and Mainshocks*

Possible systematic differences in behavior between mainshocks and their respective aftershocks (and foreshocks) need to be mentioned at this point. According to what was stated previously, they may be due to: (a) the stress redistribution by the mainshock; (b) the reduced amount of slip in aftershocks; (c) the release or increase of pore-fluid pressure due to stress redistribution and enhanced permeability; or (d) aftershocks (foreshocks) taking place on different faults than the mainshock.

For what concerns (a), most of the studied events are $M_W < 7.0$; if the mainshock and the aftershocks have similar focal solutions, then it is not very likely that the direction of the maximum compressional stress, σ_1, has shifted significantly from an Andersonian model of faulting (vertical for normal faults, horizontal for strike-slip or reverse faults). This is evidenced by the fact that most aftershocks share the same focal mechanism as their mainshock. Note that the parameters defining all the available focal mechanisms are listed in Tables 1, 2, and 3.

Point (b) is obvious, since, as explained in more details in Fig. 6, smaller events would not reach the

Table 4

Parameters of all the reverse-fault earthquakes used in this study

Event ID	Date	Time	Lat.	Lon.	Depth (km)	M_W	Strike (°)	Dip (°)	Rake (°)	f_c (Hz)	Δf_c (Hz)	$\text{Log}_{10}\ \sigma_a$ (Pa)	$\Delta\ (\log_{10}\ \sigma_a)$
Chi Chi													
E992631747	**1999/09/20**	**17:47:35**	**24.15**	**120.80**	**21.2**	**7.60**	**37**	**24**	**96**	**0.0591**	**0.0126**	**6.4112**	**0.3021**
E992632312	1999/09/20	23:12:00	23.46	120.89	N/A	5.14	N/A	N/A	N/A	0.6757	0.1780	5.8773	0.3848
E992641812	1999/09/21	18:12:00	24.18	121.02	N/A	5.18	N/A	N/A	N/A	0.4881	0.1175	5.5238	0.3385
E992650048	1999/09/22	00:48:44	23.74	120.72	36.7	6.15	3	25	97	0.2802	0.0672	6.2545	0.3470
E992651217	1999/09/22	12:17:23	23.56	121.01	33.0	5.24	294	18	0	0.9163	0.2640	6.4150	0.4184
E992660045	1999/09/23	00:45:00	24.22	121.21	N/A	5.15	N/A	N/A	N/A	0.7585	0.1922	6.0452	0.3802
E992661244	1999/09/23	12.44:36	24.00	120.90	33.0	5.29	350	13	83	0.6154	0.1574	5.9824	0.3791
E992680842	1999/09/25	08:42:31	23.77	120.84	25.1	5.45	12	38	80	0.4983	0.1158	5.9576	0.3317
E992700355	1999/09/27	03:55:00	24.29	121.23	N/A	5.00	N/A	N/A	N/A	0.6840	0.1906	5.6796	0.3949
E992700720	1999/09/27	07:20:00	24.15	121.24	N/A	5.25	N/A	N/A	N/A	0.4717	0.1166	5.5812	0.3528
E992711537	1999/09/28	15:37:00	23.18	130.91	N/A	4.97	N/A	N/A	N/A	0.5421	0.1275	5.3468	0.3342
E992741247	1999/10/01	12:47:00	23.69	120.91	N/A	5.45	N/A	N/A	N/A	0.4149	0.1042	5.7136	0.3537
E992860139	1999/10/13	01:39:00	23.96	121.34	N/A	5.55	N/A	N/A	N/A	0.3705	0.0940	5.7156	0.3543
E992950217	1999/10/22	02:17:00	23.57	120.20	15.0	6.17	178	49	54	0.2595	0.0621	6.1865	0.3352
E992950309	1999/10/22	03:09:00	23.53	120.43	15.0	5.95	327	78	0	0.3656	0.0933	6.2955	0.3714
E993030826	1999/10/30	08:26:55	23.38	121.94	56.2	5.50	162	64	13	0.4586	0.1102	5.9212	0.3491
E993210734	1999/11/17	07:34:00	24.02	120.64	N/A	5.39	N/A	N/A	N/A	0.4658	0.1236	5.7682	0.3794
E000462132	2000/02/15	21:32:17	22.93	120.57	29.2	5.46	187	43	60	0.4706	0.1131	5.8967	0.3364
E000761310	2000/03/16	13:10:00	23.31	120.75	N/A	5.20	N/A	N/A	N/A	0.6133	0.1887	5.8261	0.4352
E001391517	2000/05/18	15:17:00	24.20	121.04	N/A	5.27	N/A	N/A	N/A	0.4213	0.0951	5.4724	0.3146
E001712155	2000/06/19	21:55:00	23.92	121.09	19.5	5.30	354	15	65	0.6071	0.1822	5.9589	0.4603
E002450924	2000/09/01	09:24:00	24.08	121.14	10.0	5.25	N/A	N/A	N/A	0.3845	0.0909	5.3182	0.3392
Chuetsu													
	2007/07/16	**01:13:29**	**37.55**	**138.60**	**8.0**	**6.58**	**49**	**80**	**215**	**0.1778**	**0.0132**	**6.3458**	**0.0971**
	2007/07/16	06:37:00	37.50	138.64	11.0	5.61	47	100	219	0.5182	0.0420	6.2841	0.1058
	2007/07/24	21:52:00	37.53	138.72	20.0	4.71	49	92	25	1.0572	0.1100	5.8618	0.1358
Honshu													
	2008/05/07	**16:45:00**	**36.22**	**141.60**	**35.0**	**6.80**	**18**	**68**	**83**	**0.1044**	**0.0147**	**5.9749**	**0.1821**
	2008/05/31	07:55:00	36.39	141.80	17.0	4.96	17	68	86	0.5804	0.0871	5.4494	0.1932
	2008/07/21	20:31:00	36.63	141.84	26.0	4.63	18	63	89	0.6189	0.0936	5.0378	0.1948
	2008/08/28	08:31:00	36.63	141.24	20.0	4.60	36	77	115	0.7975	0.1325	5.3211	0.2160
Iwate													
	2008/06/13	**23:43:53**	**39.03**	**140.90**	**12.0**	**6.89**	**209**	**51**	**104**	**0.0896**	**0.0179**	**5.9008**	**0.2464**
	2008/06/14	00:20:00	39.00	140.90	10.0	5.61	16	72	111	0.2650	0.0567	5.3906	0.2620
	2008/06/14	03:27:00	39.10	141.00	10.0	5.08	11	56	68	0.4804	0.1155	5.3652	0.2921
	2008/06/16	14:14:42	39.10	140.90	12.0	5.24	348	80	43	0.3710	0.0846	5.2707	0.2813
	2008/06/16	19:05:00	39.10	141.00	10.0	4.53	13	50	89	0.8078	0.1850	5.2183	0.2923
	2008/06/26	14:33:00	38.80	140.80	5.0	4.66	189	52	100	0.5392	0.1183	4.8900	0.2696
Mid Niigata													

Table 4 *continued*

Event ID	Date	Time	Lat.	Lon.	Depth (km)	M_W	Strike (°)	Dip (°)	Rake (°)	f_c (Hz)	Δf_c (Hz)	$\text{Log}_{10}\,\sigma_a$ (Pa)	$\Delta\,(\log_{10}\sigma_a)$
	2004/10/23	**08:56:00**	**37.28**	**138.87**	**13.0**	**6.52**	**212**	**47**	**93**	**0.1254**	**0.0249**	**5.7821**	**0.2605**
	2004/10/23	14:34:00	37.31	138.90	10.0	5.10	223	55	106	0.5129	0.1247	5.4770	0.3178
	2004/10/24	15:28:00	37.20	138.90	12.0	5.18	13	50	70	0.4032	0.0931	5.2864	0.3039
	2004/11/03	23:57:00	37.42	138.92	18.0	5.19	190	48	77	0.3893	0.0835	5.2595	0.2808
	2004/11/08	19:15:00	37.35	139.00	5.0	4.68	216	48	104	0.6513	0.1564	5.1581	0.3215
	2004/11/09	18:43:00	37.36	139.00	5.0	5.07	22	66	93	0.4021	0.0844	5.1222	0.2785
Northridge													
	1994/01/17	**04:30:55**	**34.24**	**−118.56**	**14.0**	**6.70**	**87**	**53**	**59**	**0.1380**	**0.0307**	**6.1687**	**0.2815**
	1994/01/17	13:06:28	34.25	−118.55	N/A	4.60	N/A	N/A	N/A	0.7436	0.1946	5.2021	0.3288
	1994/01/17	13:26:45	34.31	−118.45	10.0	4.70	80	58	95	0.7260	0.2069	5.3141	0.3544
	1994/01/17	14:14:30	34.33	−118.44	10.6	4.50	267	50	62	0.9602	0.2784	5.3770	0.3599
	1994/01/17	15:07:03	34.30	−118.47	11.1	4.20	271	31	58	1.8988	0.5592	5.8119	0.3770
	1994/01/17	15:54:00	34.37	−118.62	16.5	4.80	56	44	51	0.7168	0.1911	5.4531	0.3331
	1994/01/17	19:35:34	34.31	−118.45	11.7	4.00	299	19	75	2.0764	0.7162	5.6166	0.4068
	1994/01/17	22:31:53	34.34	−118.44	12.4	4.10	269	55	71	1.2309	0.3518	5.1007	0.3615
	1994/01/19	14:09:14	34.23	−118.50	14.8	4.50	74	72	61	1.0060	0.2856	5.4384	0.3613
	1994/01/21	18:39:15	34.30	−118.46	10.6	4.60	116	44	51	0.8055	0.2051	5.3087	0.3183
	1994/01/21	18:53:44	34.32	−118.48	10.7	4.30	111	29	60	1.2297	0.3832	5.3926	0.3843
	1994/01/24	05:50:24	34.36	−118.63	13.6	4.29	71	72	49	1.1278	0.3189	5.2744	0.3486
	1994/01/24	05:54:21	34.37	−118.63	16.4	4.20	84	32	35	1.3096	0.3715	5.3322	0.3593
	1994/01/27	17:19:58	34.27	−118.56	13.3	4.40	34	0	−4	1.1504	0.3722	5.4549	0.3810
	1994/02/03	16:23:35	34.30	−118.44	9.5	4.09	95	34	93	1.3818	0.3989	5.2361	0.3617
	1994/02/06	13:19:27	34.29	−118.48	9.4	4.00	118	59	90	1.7540	0.5720	5.3998	0.4033
	1994/05/25	12:56:57	34.31	−118.31	9.4	4.50	88	53	65	0.8379	0.1997	5.2143	0.3010
Noto Hanto													
	2007/03/25	**00:41:00**	**37.28**	**136.61**	**12.0**	**6.70**	**34**	**40**	**108**	**0.0795**	**0.0147**	**5.4578**	**0.2529**
	2007/03/25	09:11:00	37.41	136.72	14.0	5.20	40	53	101	0.3609	0.0756	5.1737	0.2850
	2007/03/26	05:46:00	37.20	136.50	12.0	4.59	52	71	151	0.6527	0.1301	5.0326	0.2749
	2007/03/27	23:08:00	37.20	136.80	12.0	4.68	335	62	39	0.6381	0.1352	5.1352	0.2907
	2007/03/28	04:05:00	37.30	136.70	12.0	4.63	150	61	81	0.5121	0.1063	4.7737	0.2924
	2007/04/06	12:42:00	37.10	136.40	12.0	4.50	16	62	85	0.5427	0.1097	4.6562	0.2812
	2007/06/10	18:45:00	37.30	136.50	12.0	4.89	343	65	53	0.4351	0.0883	4.9539	0.2762
	2007/06/21	18:34:00	36.90	136.70	12.0	4.49	48	53	115	0.6621	0.1324	4.9013	0.2743
	2008/01/25	19:33:00	37.30	136.80	12.0	4.59	17	58	90	0.5988	0.1188	4.9210	0.2708
San Simeon													
	2003/12/22	**19:14:00**	**37.75**	**−121.15**	**8.0**	**6.42**	**290**	**58**	**78**	**0.1373**	**0.0272**	**5.7506**	**0.2447**
	2003/12/23	03:44:00	35.69	−121.11	5.0	4.35	150	84	−170	0.6562	0.1340	4.6828	0.2522
	2003/12/23	05:28:00	35.66	−121.11	5.0	4.28	116	54	91	0.7372	0.1708	4.7235	0.2857
	2003/12/23	18:15:00	35.65	−121.05	5.0	4.60	275	47	85	0.7271	0.1719	5.1851	0.2865
	2004/01/02	10:45:00	35.68	−121.18	5.0	4.21	327	56	112	0.7412	0.1819	4.6233	0.2954
	2004/03/17	23:51:00	35.73	−121.07	5.0	4.48	302	49	91	0.8787	0.1888	5.2561	0.2653

Table 4 *continued*

Event ID	Date	Time	Lat.	Lon.	Depth (km)	M_W	Strike (°)	Dip (°)	Rake (°)	f_c (Hz)	Δf_c (Hz)	$\mathrm{Log}_{10}\,\sigma_a$ (Pa)	Δ ($\log_{10}\sigma_a$)
	2004/10/02	12:20:00	35.54	−120.81	8.0	4.08	120	48	109	1.2594	0.3059	5.1191	0.2966
Virginia													
E201108231751	**2011/08/23**	**17:51:04**	**37.94**	**−77.93**	**6.0**	**5.65**	**175**	**45**	**65**	**0.4913**	**0.0343**	**6.2655**	**0.0911**
E201108240004	2011/08/24	00:04:36	37.91	−77.95	6.0	3.77	185	35	60	2.8375	0.3105	5.7256	0.1440
E201108240445	2011/08/24	04:45:26	37.93	−77.99	7.0	3.20	175	40	80	4.8482	0.6153	5.5663	0.1610
E201108250507	2011/08/25	05:07:50	37.94	−77.90	2.0	3.94	140	60	80	2.6159	0.2874	5.8746	0.1436

The mainshock(s) of each sequence is (are) indicated in bold

weakening slip distance at which full lubrication is attained. Point (c) implies systematic effects for pore-fluid pressure-induced aftershocks that are not observed. The release of pressurized pore fluids is more important in dip-slip faulting, especially in normal structures reactivated as reverse faults in positive tectonic inversions, due to the larger gradients of pore-fluid pressure existing along steeper interfaces (SIBSON 1992a, b).

With regard to point (d), we cannot be absolutely certain that the mainshocks and their aftershocks occur on the same fault plane. However, regardless of the imperfections of the data sets, we note that clear, single, and self-consistent patterns of source scaling can be recognized in each individual sequence. Consequently, whether all earthquakes of a specific sequence occur on the same fault plane (e.g., Hector Mine and Parkfield), or on parallel faults that are spatially distributed en-echelon (e.g., the aftershocks of the Chi–Chi, Taiwan seismic sequence; SUPPE, pers. comm. 2008), the fact seems to be immaterial for what concerns the events' scaling properties that seem to depend mostly on the length of the coseismic slip.

4. Discussion

Earthquake slip occurs when the shear stress acting upon the fault's surface exceeds the fault's static frictional strength ($\tau_S = \mu_S \sigma'_n$, where $\mu_S \approx 0.7$ for most materials, see BYERLEE 1978; $\sigma'_n = \sigma_n - P_f$ is the effective normal stress, with P_f as the pore-fluid pressure). As a consequence of coseismic slip, the shear stress acting across the fault surface is partially released. During sliding, the shear stress on the fault drops to a dynamic frictional level. The characteristics of the weakening processes that is experienced by the fault during dynamic slip may be summarized by a parameter called 'breakdown work' (COCCO and TINTI 2008), which, together with the minimum level of dynamic weakening and the shape of the restrengthening curve (i.e., the energy necessary for arresting the motion) determines the amount of energy available for radiation (all energies mentioned throughout the paper are normalized to seismic moment). After sliding terminates, the fault locks itself again.

Within the given description, self-similarity implies that all earthquakes (at least the ones occurring on the same fault) share the same dynamic stress drop, the same dynamic shear stress, and, consequently, the same amount of dynamic weakening. This view comes directly from low-slip rate lab observations (OHNAKA 1992), which showed constant weakening occurring over distances in the tens of μm range, orders of magnitude smaller than the ones estimated from seismic data (e.g., IDE and TAKEO 1997).

In light of the equivalence between earthquake apparent stress and dynamic stress drop (e.g., MALAGNINI *et al.* 2010), our results show that the small aftershocks of a specific sequence (say, the ones with $M_W < \approx 5.5$) have, on average, smaller stress drops than their mainshock event, or even than the larger aftershocks of the same sequence ($M_W > \approx 5.5$). Moreover, the aftershocks of a specific seismic sequence show gradually increasing apparent stress with increasing seismic moment. If the slip-weakening frictional model is valid for crustal earthquakes, one event with larger slip would have a larger stress drop than a smaller event on the same fault (i.e., with a smaller slip). If the slip-weakening model is characterized by a steady-state frictional level beyond a certain slip (magnitude) threshold, earthquakes that occur on a specific fault that are larger than that same threshold will have the same stress drop. $M_W \approx 5.5$ may mark the transition between gradual weakening and steady-state friction, but the threshold is chosen somewhat arbitrarily.

In the present study, the average nucleation depth of the crustal events in the global data set investigated is around 10 km. Using Eq. 5a–d by (MALAGNINI *et al.* 2010), we estimated the pre-stress values corresponding to the same depth of 10 km, for optimally-oriented fault planes of the three elementary mechanisms (reverse, normal, and strike-slip) (Fig. 5). The six vertical dashed lines indicate the partially overlapping ranges of allowed effective normal stresses at 10-km depth for the three faulting mechanisms (red: reverse; blue: strike-slip; gray: normal) at the chosen nucleation depth, on a plane oriented 30° from the maximum compressional stress, σ_1, for a static coefficient of friction $\mu_S = 0.75$.

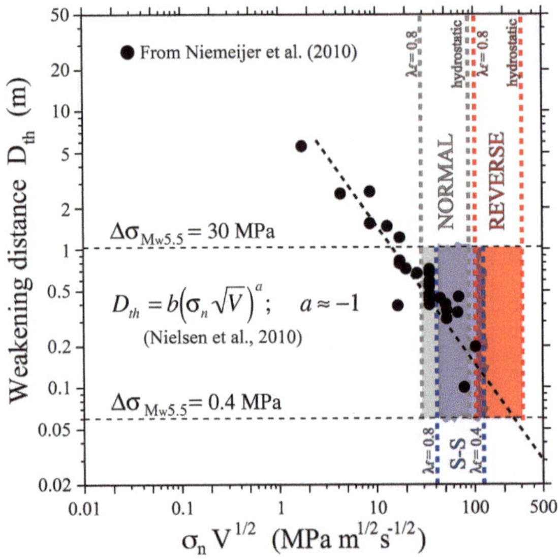

Figure 5

For a slip velocity $V = 1$ m/s, *black dots* represent frictional melt-related weakening distances as a function of normal stress (NIEMEIJER *et al.* 2011). In order to verify the formula indicated within the frame, the independent variable was chosen to be: $\sigma_n\sqrt{V}$ (NIELSEN *et al.* 2010). Data are from gabbro samples at a fixed slip rate of 1 m/s. *Vertical thick dashed lines* indicate, for unit slip rate, the allowed normal stresses at 10-km depths for optimally oriented reverse (*red*), strike-slip (*blue*), and normal faults (*gray*). Calculations of normal stresses for different fault mechanisms are performed using Eq. 5a–d of MALAGNINI *et al.* (2010) (fault plane oriented 30° from the maximum compressional stress, σ_1, and a static coefficient of friction $\mu_S = 0.75$). Upper and lower limits for the normal stress of each faulting mechanism are for: (1) hydrostatic ($\lambda = 0.4$), and, (2) overpressure conditions for pore fluids ($\lambda = 0.8$). In terms of the gradient transition depth (SUPPE and YUE 2008), and using the density of gabbro ($\rho_{gabbro} = 2.990 \times 10^3\ kg/m^3$), $\lambda = 0.8$ corresponds to a transition depth $Z_{GT} = 3{,}000$ m (the depth at which overpressured compartments seem to cluster worldwide, see HUNT 1990). *Thin horizontal dashed lines* indicate the slip corresponding to Brune stress drops of 1 (*lower line*) and 30 MPa (*upper line*), for a circular crack of dimensions corresponding to an earthquake of $M_W \sim 5.5$

The lab results indicated in Fig. 5, as well as in other recent studies, and which performed at high slip rates for ambient conditions that may be representative of the Earth's crust at 10-km depth (i.e., NIELSEN *et al.* 2010), indicate minimum weakening distances of several centimeters, of the same order of magnitude of those estimated from seismological data. Impact experiments of flash weakening (YUAN and PRAKASH 2007) also yielded results in agreement with seismological calculations (weakening distances of several cm). The high-velocity friction experiments typically show restrengthening in correspondence of slip-velocity reduction/arrest, suggesting that faults may not remain statically weak in the aftermath of a mainshock.

Possible weakening distances at $M_W \approx 5.5$ may be computed by approximating the earthquake rupture with a circular crack, and by using the following elementary expressions, linking seismic moment with rupture area and average slip:

$$M_0 = GA\Delta u \quad (11)$$

(HANKS and KANAMORI 1979) and stress drop with crack radius and slip, and

$$\Delta\sigma = \frac{7\pi G\Delta u}{16L} \quad (12)$$

(ESHELBY 1956), where $G = 30$ GPa is rock rigidity, $A = \pi L^2$ is the circular crack area of radius L, and Δu is the average slip.

In the hypothesis that apparent stress saturates for larger earthquakes, and assuming that a substantial weakening characterizes faulting at $M_W \approx 5.5$, we observe that the minimum and maximum Brune stress drop in our data set around M_W 5.5 are: $\Delta\sigma_{min} = 0.4$ MPa, and $\Delta\sigma_{max} = 30$ MPa, respectively. Thus, the range of weakening distance suggested by our data set at around 10-km depth would be:

$$\begin{aligned} \Delta u_{\Delta\sigma=0.4\,\mathrm{MPa}} &= 0.06\,\mathrm{m} \\ \Delta u_{\Delta\sigma=30\,\mathrm{MPa}} &= 1.03\,\mathrm{m}. \end{aligned} \quad (13)$$

Estimates of Eq. (13) are obtained by using the definition of seismic moment and stress drop [Eqs. 11, 12], and of moment magnitude ($M_0 = 10^{1.5(M_w+6.03)}$ Nm). Note that estimates of Eq. (13) coincide with those obtained by IDE and TAKEO (1997).

After a minimum frictional level is reached at steady-state, no further weakening occurs with additional slip, and no systematic increase in the radiated energy is observed above a saturation magnitude for which $u > D_c$. A consequence of this conceptual model is that constant stress drop and apparent stress should characterize large earthquakes occurring on the same fault ($M_W > \approx 5.5$), regardless of whether they are mainshocks or aftershocks. In terms of dynamic stress drop, our simple model is described in detail in the illustration of Fig. 6.

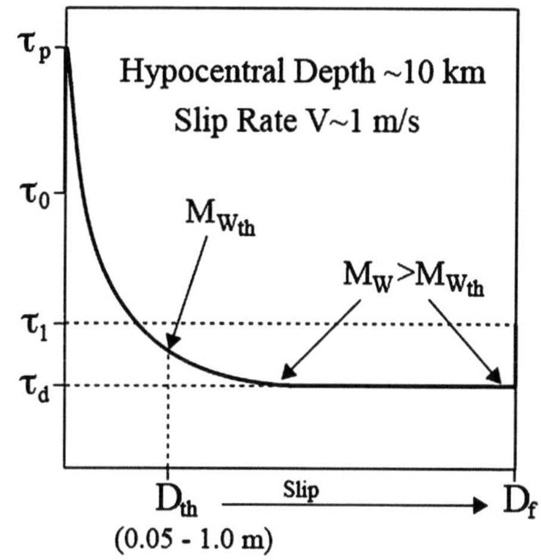

Figure 6
Cartoon explaining our interpretation of the data distributions shown in Fig. 4. The stress levels indicated as τ_0, τ_1, τ_p, and τ_d are the initial, final, peak (yield), and steady-state dynamic shear stresses, respectively. The 10-km nucleation depth was chosen in order to be consistent with the results shown in Fig. 4, in terms of the allowed range of normal stresses and critical weakening distance (Dth), with the average depth of our data set. M_{wth} indicates the threshold moment magnitude, beyond which the fault reaches full lubrication. Our combined data set indicates that M_{wth} is between M_W 5.5 and 5.8. The presented physical model predicts increasing partial lubrication for the small earthquakes, up to the threshold seismic slip (or moment magnitude), until full lubrication is attained

Important fluctuations in the absolute level of apparent stress appear to take place at $M_W > 5.5$ (see Fig. 3b–d), but it is the events below the $M_W \approx 5.5$ threshold that show the largest scatter. A possible explanation in the difference in scatter above and below the $M_W \approx 5.5$ threshold is the following: on the one hand, small events sample the frictional properties of different small patches of the fault, and their radiation characteristics may be very heterogeneous; large earthquakes, on the other hand, average over large fault areas. As a consequence, small events may show a larger variability of their characteristic physical quantities, including those on the stress parameters τ_0, τ_p, τ_1, τ_d, on the weakening distance, Dth, and on the final slip, D_f.

From the inspection of Fig. 3 we note that the apparent stress of strike-slip earthquakes is characterized by a larger inter-sequence variability, when compared to the apparent stress of dip-slip sequences, although the individual sequences generally (both strike-slip and dip-slip) show clear trends of increasing apparent stress with increasing magnitude. In other words, it appears that the apparent stress of all events belonging to a specific strike-slip sequence are dc-shifted by a factor that depends on the specific fault (or tectonic province), and thus reflect differences in the weakening curve of the individual fault planes.

Fault maturity may also be key in explaining the scattering difference existing between strike-slip and dip-slip sequences, because, as noted by Choy *et al.* (2006), very immature strike-slip faults are the high end-members in the apparent stress space (specifically, Choy *et al.* 2006, classified the least mature faults to be oceanic transform faults). On the contrary, very mature faults (subduction thrusts, but also large crustal features like the San Andreas fault) may represent the low end-member in the apparent stress space.

Low apparent stresses characterize the set of earthquakes that occurred in Baja California, for which the low normalized radiated energy was probably due to low coupling of the fault surfaces. On the opposite side of the apparent stress space are the San Giuliano earthquakes, which, as hypothesized by Malagnini and Mayeda (2008), ruptured either near-intact or well-healed crustal rocks (considering the anomalous depth of these crustal earthquakes), and so the apparent stresses of the sequence are expected to be large. Other anomalous events that have been studied previously are the 2011 M_W 5.9 Virginia, a high-stress drop event with a P-wave duration of ~2 s, and the 2011 M_W 5.6 Oklahoma, likely an induced earthquake (e.g., Keranen *et al.* 2013) that exhibited anomalously long P-wave durations of nearly 5 s (Mayeda *et al.* 2011).

The reasons why other strike-slip seismic sequences are characterized by minimal apparent stresses (i.e., Fukuoka, Tottori, Hector Mine) are to be investigated in dedicated studies. However, if fault maturity is key, given the fact that no sequences from subduction thrust faults are in our data set (i.e., sequences that occurred on faults characterized by extreme maturity), only intermediate values of apparent stress are to be expected for the dip-slip earthquakes sampled in this study, in agreement with the smaller overall scatter of the values plotted in Fig. 3.

5. Conclusions

We investigated 20 large-magnitude, crustal seismic sequences from different faulting regimes and tectonic environments using the same coda spectral ratio method, all of which were constrained at long-periods by independent regional moment estimates. Assuming the omega-square source model (Brune 1970), each individual seismic sequence exhibits a gradual increase in the earthquakes' apparent stress (or static stress drop) with increasing size (seismic moment). Within each sequence, we observe a systematic deviation from the self-similar $M_0 \propto f_c^{-3}$ line, with all data sets preferring a steeper dependence $M_0 \propto f_c^{-(3+\varepsilon)}$, with $\varepsilon > 0$ (Kanamori and Rivera 2004).

Additionally, the non-self-similar interpretation of our seismic observations is compatible with the results of recent state-of-the-art laboratory friction studies on fault rocks under normal crustal stress conditions, realistic seismic-slip velocities (~1 m/s), and amounts of fault slip corresponding to (or larger than) those of an $M_W \sim 5.5$ earthquake. Moreover, these results can explain ground motion parameters over a wide range of magnitudes (e.g., Malagnini *et al.* 2007, 2008; Malagnini and Mayeda 2008; Yoo and Mayeda 2013). Although trends of increasing apparent stress with earthquake size may be seen in each individual seismic sequence, at the smaller magnitudes ($M_W \leq 5.5$), the overall data set of apparent stresses is distributed over almost three orders of magnitude, mostly due to the scatter that affect strike-slip sequences. Larger events ($M_W > \approx 5.5$) show less variability in the stress-drop parameter, but more than one order of magnitude, and thus might represent the degree of tectonic activity, whether it be fault maturation, heat flow, or state of stress.

Finally, we show that the $\omega^{-1.7}$ law represents an alternative to the ω-square fall-off of seismic spectra. Such an alternative may represent an explanation for the observed "non-self-similar" trends shown in this study, in case we want to keep the range of possible explanations within the frames of the "ω^{-p}" class of models. This must be said despite the consensus of the ω-square law in the earthquake source community, or its mentioned success with ground motion data. For instance, ALLMANN and SHEARER (2009) observed the $\omega^{-1.7}$ law over a great number of global data.

Acknowledgments

K. Mayeda and S.-H. Yoo were supported by the Air Force Research Laboratory under Contract No. FA8718-09-C-0014.

REFERENCES

AKI, K. (1967). *Scaling law of seismic spectrum*, J. Geophys. Res., *72*, 1217–1231.

AKI, K. (1969). *Analysis of the seismic coda of local earthquakes as scattered waves*, J. Geophys. Res., *74*(2), 615–631, doi:10.1029/JB074i002p00615.

ALLMANN, B. P., and P. M. SHEARER (2009). *Global variations of stress drop for moderate to large earthquakes*, J. Geophys. Res., *114*, B01310, doi:10.1029/2008JB005821.

ASANO, K., and T. IWATA (2011). *Characterization of Stress Drops on Asperities Estimated from the Heterogeneous Kinematic Slip Model for Strong Motion Prediction for Inland Crustal Earthquakes in Japan*, Pure Appl. Geophys. *168* (2011), 105–116, doi:10.1007/s00024-010-0116-y.

BALTAY, A., PRIETO, G., and BEROZA, G.C. (2010). *Radiated seismic energy from coda measurements and no scaling in apparent stress with seismic moment*. J. Geophys. Res. *115*, B08314, doi:10.1029/2009JB006736.

BALTAY, A., IDE, S., PRIETO, G., and BEROZA, G. (2011). *Variability in earthquake stress drop and apparent stress*. Geophys. Res. Lett. *38*, L06303, doi:10.1029/2011GL046698.

BOATWRIGHT, J. (1980). *A spectral theory for circular seismic sources; simple estimates of source dimension, dynamic stress drop, and radiated seismic energy*, Bull. Seism. Soc. Am., *70*, 1 27.

BRODSKY, E. E. and H. KANAMORI (2001). *The elastohydrodynamic lubrication of faults*, J. Geophys. Res., *106*, 16357–16374.

BRUNE, J. (1970). *Tectonic stress and the spectra from seismic shear waves from earthquakes*, J. Geophys. Res., *75*, 4997–5009.

BYERLEE, J.D. (1978). *Friction of rocks*. Pure Appl. Geophys. *116*, 615–626.

CHOY, G.L., A. MCGARR, S.H. KIRBY, and J. BOATWRIGHT (2006). An Overview of the Global Variability in Radiated Energy and Apparent Stress, Earthquakes: Radiated Energy and the Physics of Faulting (Geophysical monograph, ISSN 0065-8448; 170), by Rachel Abercrombie, Art McGarr, Giulio Di Toro, Hiroo Kanamori (Editors), pp 43–58.

COCCO M., and TINTI E., (2008). *Mechanical work partitioning during dynamic earthquake propagation*, Earth and Planetary Science Letters, *273*, 123–131.

DE PAOLA, N., T. HIROSE, T. MITCHELL, G. DI TORO, T. TOGO, and T. SHIMAMOTO (2011). *Fault lubrication and earthquake propagation in thermally unstable rocks*. Geology, vol. *39*, pp. 35–38.

DI TORO, G., D.L. GOLDSBY, and T.E. TULLIS (2004). *Friction falls towards zero in quartz rock as slip approaches seismic rates*. Nature *427*, 436–439.

DI TORO, G., R. HAN, N. DE PAOLA, S. NIELSEN, K. MIZOGUCHI, F. FERRI, M. COCCO, and T. SHIMAMOTO (2011). *Fault lubrication during earthquakes*, Nature, in press.

DI TORO, G., HIROSE, T. NIELSEN, S., PENNACCHIONI, G. and T. SHIMAMOTO (2006). *Natural and experimental evidence of melt lubrication of faults during earthquakes*, Science, *311*, 647–649, doi:10.1126/science.1121012.

EKSTROM, E., A. MORELLI, E. BOSCHI, and A. M. DZIEWONSKI (1998), *Moment tensor analysis of the central Italy earthquake sequence of September–October 1997*, Geophys. Res. Lett., *25*(11), 1971–1974.

ESHELBY, J. D. (1959), *The elastic field outside an ellipsoidal inclusion*, Proc. R. Soc. London, Ser. A, *252*, 561–569.

FAMIN, V., S. NAKASHIMA, A.M. BOULLIER, K. FUJIMOTO and T. HIRONO (2008). *Earthquakes produce carbon dioxide in crustal faults*, Earth and Planetary Science Letters *265* (3–4), pp. 487–497, doi:10.1016/j.epsl.2007.10.041.

HAN, R., T. SHIMAMOTO, T. HIROSE, J.-H. REE, and J. ANDO (2007). *Ultralow friction of carbonate faults caused by thermal decomposition*. Science *316*, 878–881.

HANKS, T.C., and H. KANAMORI (1979). *A moment magnitude scale*, J. Geophys. Res., *84*, B5, 2348–2350.

HERRMANN, R.B., L. MALAGNINI, and I. MUNAFO' (2011). *Regional moment tensors of the 2009 L'Aquila earthquake sequence*, Bull. Seism. Soc. Am., Vol. *101*, No. 3, doi:10.1785/0120100184.

HIROSE, T., and M. BYSTRICKY (2007). *Extreme weakening of faults during dehydration by coseismic shear heating*, Geophys. Res. Lett., *34*, L14311, doi:10.1029/2007GL030049.

HIROSE, T. and SHIMAMOTO, T. (2005). *Growth of molten zone as a mechanism of slip weakening of simulated faults in gabbro during frictional melting*, J. Geophys. Res. *110*, doi:10.1029/2004JB003207.

HOFSTETTER, A., H.K. THIO, and G. SHAMIR (2003). *Source mechanism of the 22/11/1995 Gulf of Aqaba earthquake and its aftershock sequence*, Journal of Seismology *7*: 99–114.

HUNT, J.M. (1990). *Generation and migration of petroleum from abnormally pressured fluid compartments*, Am. Ass. Petroleum Geol. Bull., 74, 1–12.

IDE, S., and M. TAKEO (1997). *Determination of constitutive relations of fault slip based on seismic wave analysis*. J. Geophys. Res. *102*, 27379–27391.

IDE, S., and G.C. BEROZA (2001). *Does apparent stress vary with earthquake size?* Geophys. Res. Lett., *28*, 3349–3352.

IDE, S., G. C. BEROZA, S. G. PREJEAN, and W. L. ELLWORTH (2003). *Apparent break in earthquake scaling due to path and site effects on deep borehole recordings*, J. Geophys. Res., *108*(B5), 2271, doi:10.1029/2001JB001617.

IKARI, M. J., D. M. SAFFER, and C. MARONE (2007). *Effect of hydration state on the frictional properties of montmorillonite-based fault gouge*, J. Geophys. Res., *112*, B06423, doi:10.1029/2006JB004748.

KANAMORI, H., and RIVERA, L. (2004). *Static and dynamic scaling relations for earthquakes and their implications for rupture speed and stress drop*. Bull. Seismol. Soc. Am. *94*, 314–319.

KAO, H., and J. ANGELIER (2001). *Stress tensor inversion for the Chi–Chi earthquake sequence and its implications on regional collision*, Bull. Seismol. Soc. Am., *91*, 1028–1040.

Keranen, K.M, H.M. Savage, G.A. Abers, and E.S. Cochran (2013). *Potentially induced earthquakes in Oklahoma, USA: Links between wastewater injection and the 2011 M_w 5.7 earthquake sequence*, Geology, G34045.1, doi:10.1130/G34045.1.

Madariaga, R. (1976). *Dynamics of an Expanding Circular Fault*, Bull. Seism. Soc. Am., *66*, 639–666.

Malagnini, L., and K. Mayeda (2008). *High-stress strike-slip faults in the Apennines: An example from the 2002 San Giuliano earthquakes (southern Italy)*, Geophys. Res. Lett., *35*, L12302, doi:10.1029/2008GL034024.

Malagnini, L., L. Scognamiglio, A. Mercuri, A. Akinci, and K. Mayeda (2008). *Strong evidence for non-similar earthquake source scaling in Central Italy*, Geophys. Res. Lett., *35*, LXXXXXX, doi:10.1029/2008GL034310.

Malagnini, L., K. Mayeda, R. Uhrhammer, A. Akinci, and R.B. Herrmann (2007). *A regional ground motion excitation/attenuation model for the San Francisco region*, Bull. Seism. Soc. Am., *97*(3), 843–862, doi:10.1785/0120060101.

Malagnini, L., S. Nielsen, K. Mayeda, and E. Boschi (2010). *Energy radiation from intermediate to large magnitude earthquakes: implications for dynamic fault weakening*, J. Geophys. Res., doi:10.1029/2009JB006786.

Malagnini, L., A. Akinci, K. Mayeda, I. Munafo', R.B. Herrmann, and A. Mercuri (2011). *Characterization of Earthquake-Induced Ground Motion from the L'Aquila Seismic Sequence of 2009, Italy*, Geophys. J. Int., *184*, 325–337, doi:10.1111/j.1365-246X.2010.04837.x.

Malagnini, L., I. Munafo', M. Cocco, S. Nielsen, K. Mayeda, and E. Boschi (2014). *Gradual fault weakening with seismic slip: inferences from the seismic sequences of L'Aquila, 2009 and Northridge, 1994*, Pure and Appl. Geophys. (this volume).

Marone, 1998, *Laboratoray derived friction laws and their application to seismic faulting*, Annual Reviews Earth and Planetary Science Letters, *26*, pp. 643–696.

Mayeda, K. M., and W. R. Walter (1996), *Moment, energy, stress drop, and source spectra of western U.S. earthquakes from regional coda envelopes*, J. Geophys. Res., *101*, 11,195–11,208.

Mayeda, K., A. Hofstetter, J. L. O'Boyle, and W. R. Walter (2003), *Stable and transportable regional magnitudes based on coda-derived moment-rate spectra*, Bull. Seism. Soc. Am., *93*, 224–239.

Mayeda, K., L. Malagnini, and W.R. Walter (2007). *A new spectral ratio method using narrowband coda envelopes: evidence for departure in self-similarity in the Hector Mine sequence*, Geophys. Res. Lett., *34*, L11303, doi:10.1029/2007GL030041.

Mayeda, K., and L. Malagnini (2009), *Apparent stress and corner frequency variations in the 1999 Taiwan (Chi–Chi) sequence: Evidence for a step-wise increase at Mw ~ 5.5*, Geophys. Res. Lett., *36*, L10308, doi:10.1029/2009GL037421.

Mayeda, K., and L. Malagnini (2010). *Source Radiation Invariant Property of Local and Near-Regional Shear-Wave Coda: Application to Source Scaling for the Mw 5.9 Wells, Nevada Sequence*, Geophys. Res. Lett., doi:10.1029/2010GL042698.

Mayeda, K., L. Malagnini, and S.-H. Yoo (2011). Apparent Stress of the M5.8 Mineral, Virginia Sequence and Comparison with Other Crustal Sequences using the Coda Ratio Methodology, Abstract S11B-2238, presented at 2011 Fall Meeting, AGU, San Francisco, Calif., 5–9 Dec.

Melosh, H.J. (1996). *Dynamical weakening of faults by acoustic fluidization*, Nature *379*, 601–606, doi:10.1038/379601a0.

Nielsen, S., G. Di Toro, T. Hirose, and T. Shimamoto (2008). *Frictional melt and seismic slip*. J. Geophys. Res. *113* doi:10.1029/2007JB0051222008.

Nielsen, S., P. Mosca, G. Giberti, G. Di Toro, T. Hirose, and T. Shimamoto (2010). *On the transient behavior of frictional melt during seismic slip*, J. Geophys. Res. *115*, B10301, doi:10.1029/2009JB007020.

Niemeijer, A., C. Marone, and D. Elsworth (2010). *Frictional strength and strain weakening in simulated fault gouge: Competition between geometrical weakening and chemical strengthening*, J. Geophys. Res., *115*, B10207, doi:10.1029/2009JB000838.

Niemeijer, A., G. Di Toro, S. Nielsen, and F. Di Felice (2011). *Frictional melting of gabbro under extreme experimental conditions of normal stress, acceleration, and sliding velocity, J. Geophys. Res., 116*, B07404, doi:10.1029/2010JB008181.

Ohnaka, M. (1992). *Earthquake source nucleation: a physical model for short-term precursors*. Tectonophysics *211*, 149–178.

Oth, A., D. Bindi, S. Parolai, D. Di Giacomo (2010). *Earthquake scaling characteristics and the scale-(in)dependence of seismic energy-to-moment ratio: Insights from KiKnet data in Japan*, Geophys. Res. Lett., *37*, L19304, doi:10.1029/2010GL044572.

Oth, A. (2013). *On the characteristics of earthquake stress release variations in Japan*. Earth Planet. Sci. Lett., *377–378*, 132–141, doi:10.1016/j.epsl.2013.06.037.

Prieto, G. A., P. M. Shearer, F. L. Vernon, and D. Kilb (2004). *Earthquake source scaling and self-similarity estimation from stacking P and S spectra*, J. Geophys. Res., *109*, B08310, doi:10.1029/2004JB003084.

Reches, Z., and D.A. Lockner (2010). *Fault weakening and earthquake instability by powder lubrication*, Nature *467*, 452–455, doi:10.1038/nature09348.

Rempel, A. W., and J. R. Rice (2006). *Thermal pressurization and onset of melting in fault zones*, J. Geophys. Res., *111*, B09314, doi:10.1029/2006JB004314.

Rice, J. R. (2006). *Heating and weakening of faults during earthquake slip*, J. Geophys. Res., 111, doi:10.1029/2005JB004006.

Shearer, P. M., G. A. Prieto, and E. Hauksson (2006). *Comprehensive analysis of earthquake source spectra in southern California*, J. Geophys. Res., *111*, B06303, doi:10.1029/2005JB003979.

Sibson, R. H. (1992a). *Implications of fault valve behaviour for rupture nucleation and recurrence*, Tectonophysics, *211*, 283–293, doi:10.1016/0040-1951(92)90065-E.

Sibson, R. H. (1992b). *Fault valve behavior and the hydrostatic lithostatic fluid pressure interface*, Earth Sci. Rev., *32*, 141–144, doi:10.1016/0012-8252(92)90019-P.

Smith, S.A.F., G. Di Toro, S. Kim, J.-H. Ree, S. Nielsen, A. Billi and R. Spiess (2013). *Coseismic recrystallization during shallow earthquake slip*, Geology, *41*, 63–66, doi:10.1130/G33588.1.

Suppe, J., and L. F. Yue (2008). Pore-fluid pressures and crustal strength, in Geophysical Research Abstracts, vol. 10, EGU2008-A-01970, 2008 SRef-ID: 1607–7962/gra/EGU2008-A-01970, EGU General Assembly 2008.

Walter, W. R. and S. R. Taylor (2001). A revised Magnitude and Distance Amplitude Correction (MDAC2) procedure for regional seismic discriminants, Lawrence Livermore National Laboratory UCRL-ID-146882: http://www.llnl.gov/tid/lof/documents/pdf/240563.pdf.

WIBBERLEY, C.A.J., and T. SHIMAMOTO (2005). *Earthquake slip weakening and asperities explained by thermal pressurization*, Nature *436*, 689–692, doi:10.1038/nature03901.

YOO, S.-H. and K. MAYEDA (2013). *Validation of non-self-similar source scaling using ground motions from the 2008 Wells, Nevada, earthquake sequence*, Bull. Seism. Soc. Am., Vol. *103*, No. 4, doi:10.1785/0120120327.

YUAN, F., and V. PRAKASH (2007). *Slip weakening in rocks and analog materials at co-seismic slip rates*, J. Mech. Phys. Solids *56* (2008) 542–560, doi:10.1016/j.jmps.2007.05.007.

(Received September 3, 2013, revised November 13, 2013, accepted November 22, 2013, Published online December 13, 2013)

Pure Appl. Geophys. 171 (2014), 2709–2730
© 2013 Springer Basel
DOI 10.1007/s00024-013-0752-0

Pure and Applied Geophysics

Gradual Fault Weakening with Seismic Slip: Inferences from the Seismic Sequences of L'Aquila, 2009, and Northridge, 1994

LUCA MALAGNINI,[1] IRENE MUNAFO',[1] MASSIMO COCCO,[1] STEFAN NIELSEN,[1] KEVIN MAYEDA,[2] and ENZO BOSCHI[3]

Abstract—We estimate seismological fracture energies from two subsets of events selected from the seismic sequences of L'Aquila (2009), and Northridge (1994): 57 and 16 selected events, respectively, including the main shocks. Following ABERCROMBIE and RICE (Geophys J Int 162: 406–424, 2005), we postulate that fracture energy (G) represents the post-failure integral of the dynamic weakening curve, which is described by the evolution of shear traction as a function of slip. Following a direct-wave approach, we compute mainshock-/aftershock-source spectral ratios, and analyze them using the approach proposed by MALAGNINI *et al.* (Pure Appl. Geophys., this issue, 2014) to infer corner frequencies and seismic moment. Our estimates of source parameters (including fracture energies) are based on best-fit grid-searches performed over empirical source spectral ratios. We quantify the source scaling of spectra from small and large earthquakes by using the MDAC formulation of WALTER and TAYLOR (A revised Magnitude and Distance Amplitude Correction (MDAC2) procedure for regional seismic discriminants, 2001). The source parameters presented in this paper must be considered as point-source estimates representing averages calculated over specific ruptured portions of the fault area. In order to constrain the scaling of fracture energy with coseismic slip, we investigate two different slip-weakening functions to model the shear traction as a function of slip: (i) a power law, as suggested by ABERCROMBIE and RICE (Geophys J Int 162: 406–424, 2005), and (ii) an exponential decay. Our results show that the exponential decay of stress on the fault allows a good fit between measured and predicted fracture energies, both for the main events and for their aftershocks, regardless of the significant differences in the energy budgets between the large (main) and small earthquakes (aftershocks). Using the power-law slip-weakening function would lead us to a very different situation: in our two investigated sequences, if the aftershock scaling is extrapolated to events with large slips, a power law (a la Abercrombie and Rice) would predict unrealistically large stress drops for large, main earthquakes. We conclude that the exponential stress evolution law has the advantage of avoiding unrealistic stress drops and unbounded fracture energies at large slip values, while still describing the abrupt shear-stress degradation observed in high-velocity laboratory experiments (e.g., DI TORO *et al.*, Fault lubrication during earthquakes, Nature 2011).

Key words: Earthquake source scaling, fault friction, fracture energy, breakdown work, radiated seismic energy.

1. Introduction

Faulting is a complex combination of fracture and frictional phenomena (e.g., OHNAKA 2003; SIBSON 1974, 1990) in which slip occurs, either seismically or through aseismic creep and other transients, on pre-existing faults (or on surfaces of weakness of other kinds) that pervade the Earth's crust. The driving mechanism for seismic slip is the regional tectonic stress exceeding the shear strength of favorably oriented fault planes. In fact, the fault strength strongly depends on the mechanical conditions before the onset of a dynamic rupture, and therefore it also depends on the effective normal stress across its plane, and on the ambient pore-fluid pressure (SIBSON 1992a, b, 1994). Under favorable conditions, shear stress builds up on the fault until it overcomes the effective shear strength, and a slip instability can initiate and lead to a dynamic rupture (an earthquake).

Reproducing the extreme conditions undergone by crustal rocks during large, natural earthquakes (i.e., realistic effective normal stress conditions at seismogenic depths, and a sliding velocity of the order of meters per second) is one of the most important challenges faced by state-of-the-art laboratory apparatuses. In fact, seismic slip velocities may exceed 10 m/s during large, natural earthquakes (e.g., YOMOGIDA and NAKATA 1994), under very large, effective normal stresses. The effective normal stresses possibly range over several orders of magnitudes depending on the local values of normal stress

[1] Istituto Nazionale di Geofisica e Vulcanologia, Rome, Italy. E-mail: luca.malagnini@gmail.com
[2] Weston Geophysical Corporation, Lexington, MA, USA.
[3] Dipartimento di Fisica, Universita' degli Studi di Bologna, Bologna, Italy.

and pore-fluid pressure. For the special case of an optimally oriented fault with a coefficient of static friction $\mu_S = 0.75$, MALAGNINI *et al.* (2014, this issue) estimated that the effective normal stresses at $\sim$10-km depth may range between 30 and 300 MPa. The mentioned minimum value of 30 MPa refers to normal faulting under severe pore-fluid overpressure ($\lambda_f = \frac{P_f}{\rho g z} = 0.8$), whereas the maximum mentioned value (300 MPa) refers to reverse faulting in hydrostatic conditions ($\lambda_f = 0.4$, i.e., minimum pore-fluid pressure). In the equations above, λ_f is the ratio between the pore-fluid pressure, P_f, and the rock overburden: ρ is rock density, z is depth, and g is gravity. Only recently, rotary shear lab apparatuses were designed to operate within such a range of normal stresses, allowing slip velocities of several m/s and final offsets of several meters (e.g., NIEMEIJER *et al.* 2010; DI TORO *et al.* 2011). Recent experiments on these machines have shown that friction drops to extremely low values ($\mu_d \approx 0.2$) and shear stress depicts an evident weakening.

Dynamic fault weakening during an earthquake is represented by the decay of shear stress with increasing coseismic slip (see RICE and COCCO 2006, and references therein). Mechanical work is dissipated on the fault during the propagation of a dynamic rupture through frictional processes (COCCO *et al.* 2006), off-fault rock damage, and other dissipative mechanisms (including the so-called "radiation loss": the energy radiated from the crack as high-frequency stress waves that are damped before getting to the far field; KOSTROV and DAS 1988).

Very large amounts of frictional work (of the order of $\sim$1 MJ/m^2) are dissipated in short (finite) time intervals within a narrow slip zone (e.g., SHIPTON *et al.* 2006; DI TORO *et al.* 2011). During these time intervals (named breakdown times), the coseismic stress evolution is characterized by a dramatic weakening that occurs over a finite slip amount (e.g., MALAGNINI *et al.* 2014; RICE 2006; NIELSEN *et al.* 2008; FAMIN *et al.* 2008; COCCO *et al.* 2009), named as D_C. Seismological estimates of the critical weakening distance are between 0.05 and 1 m (e.g., IDE and TAKEO 1997; COCCO *et al.* 2009), whereas experiments on rotary shear apparatuses like SHIVA (DI TORO *et al.* 2011; NIEMEIJER *et al.* 2010, 2011) show that sliding under realistic slip rates trigger dramatic weakening for offsets in excess of several centimeters.

Note that the parameter named D_C in some of the cited studies may represent a constitutive parameter. However, in the present paper, D_C is just an empirical constant used to describe an exponential function (see below).

Different physical mechanisms can explain dynamic fault weakening; they might occur simultaneously and compete with each other in controlling the stress evolution on the fault plane during coseismic slip on natural (real) and experimental (lab) faults. They have been extensively studied in the literature, such as acoustic fluidization (MELOSH 1996), frictional melt (DI TORO *et al.* 2006; HIROSE *et al.* 2006; HIROSE and SHIMAMOTO 2005; WIBBERLEY and SHIMAMOTO 2005; NIELSEN *et al.* 2008, 2010), elastohydrodynamic lubrication (BRODSKY and KANAMORI 2001), formation of gel (Di Toro *et al.* 2004), thermal pressurization of pore fluids (FERRI *et al.* 2010; BIZZARRI and COCCO 2006; REMPEL and RICE 2006), flash heating (RICE 2006), decarbonation with formation of nanopowders (HAN *et al.* 2007), dehydration (HIROSE and BYSTRICKY 2007), fluid pressurization due to frictional CO_2 exsolution (HAN *et al.* 2009; FAMIN *et al.* 2008), and gouge lubrication (RECHES and LOCKNER 2010). YUAN and PRAKASH (2007) used a high-velocity, impact mechanism (Kolsy bar) to study frictional slip under extreme velocity ($\sim$70 m/s) and normal stress ($\sim$200 MPa).

The main goal of the present study is to investigate the evolution of shear stress as a function of coseismic slip on the fault plane during a large, crustal earthquake using seismological observations. For this reason we adopt a phenomenological description of friction, (see COCCO and TINTI 2008); our observation do not allow us to distinguish among all the different physical mechanisms governing shear stress evolution and dynamic fault weakening. We focus here on the mechanical work absorbed on the fault plane during a crustal earthquake and, in particular, on its seismologically measurable portion, called fracture energy. In order to be coherent with previous studies (ABERCROMBIE and RICE 2005), we call the parameter estimated here the seismological fracture energy *(G')*.

We use a high-quality data set of seismograms collected during the 2009 L'Aquila sequence (foreshocks, main event, and aftershocks all included). The L'Aquila data set was selected because of: (i) the great number of high-quality recordings that are available for the earthquakes of the sequence and, (ii) the huge amount of scientific work that has already been carried out to characterize the sequence. In order to compare our approach to previous studies, we present results from the data set of earthquakes collected during the 1994 Northridge sequence. We interpret our results in order to constrain the scaling of shear stress with coseismic slip. We use seismic recordings from both mainshocks and aftershocks, and discuss differences in fracture energies between large and small/moderate magnitude earthquakes.

2. *Earthquake Energy Budget and Apparent Stress*

We assume that the total mechanical work absorbed during an earthquake on the causative fault is partitioned into: (1) dissipation within the fault zone, and (2) kinetic energy radiated through seismic waves (Scholz 1990). We include into the dissipation term all sources of dissipation that may not be directly related to the sliding friction (off-fault damage, radiation loss, etc.) as well as surface energy; we neglect the effect of gravity. Because we follow a phenomenological (seismological) description of friction on the fault plane (Cocco and Tinti 2009), our fault is represented by a virtual mathematical plane of zero thickness.

Our analysis of coseismic stress evolution is given in terms of fault-averaged values, which implies that we consider the source parameters to be scale-independent (see Cocco and Tinti 2008). In other words, we neglect the scale dependence of the constitutive dynamic parameters, and consider stress and slip heterogeneity following the assumption that the physical processes controlling dynamic fault weakening are scale-independent.

Differently from the Brune (1970) and Orowan models (1960) who used a simplified, instantaneous drop of the shear stress on the crack surface (zero breakdown time), we do not assume instantaneous stress drop, but use the slip-weakening model sketched in Fig. 1, depicting the stress evolution during sliding: the evolution of shear stress as a function of seismic slip represents a description of the breakdown process at a single point on the fault plane and involves a finite duration. Assuming stress evolution at a single point as an average representing the dynamic weakening process on the whole fault is a strong assumption, whose consequences are discussed by Noda and Lapusta (2012). Although still too simplistic for a complete description of the dynamics of faulting, the proposed frictional model provides a simple framework, suitable for the objectives of the present study, from which to quantify the average parameters associated with dynamic weakening.

The integral of shear stress over a given value of slip represents macroscopic frictional work normalized to the fault area (see Cocco *et al.* 2006; Cocco and Tinti 2008), in units of J/m^2. A common misunderstanding stems from the fact that only the area below the steady-state frictional level is attributed to frictional heat, whereas the expression fracture

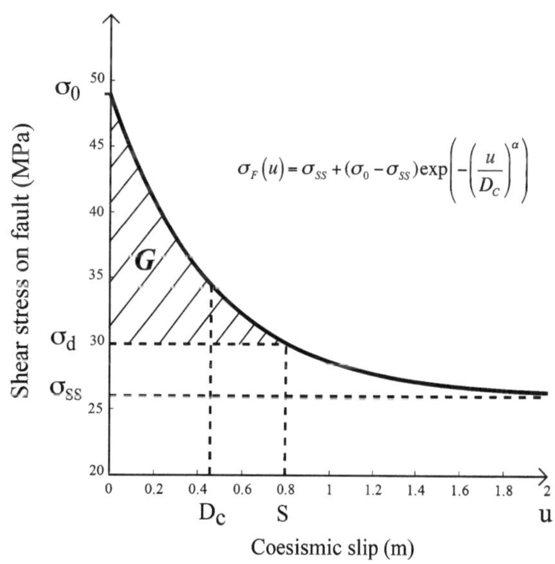

Figure 1
Weakening model used in the present study. Here σ_0 indicates the initial shear stress on the fault, and σ_{SS} is the dynamic stress experienced by the fault after undergoing full lubrication. D_C is the distance at which, in case of exponential weakening, the dynamic shear stress is 1/*e* of its initial value. The weakening *curve* in the figure was obtained with the parameters calculated for the L'Aquila seismic sequence

energy (generally indicated as G) commonly misleads the readers towards understanding that such a portion of the energy budget is totally expended for the creation of a new surface area within the fractured rocks (for a detailed discussion on the topic, see Cocco and Tinti 2008). We point out that a significant portion of the area below the weakening curve of Fig. 1 may actually represent heat (i.e., dissipation of mechanical work). More correctly, the expression breakdown work, introduced by (Tinti *et al.* 2005), describes the quantity G without interpreting what proportion of it pertains to frictional heat. The breakdown work can be considered equivalent to the seismological fracture energy (G') introduced by Abercrombie and Rice (2005). In this study, we adopt a different functional form from Abercrombie and Rice (2005) to represent the shear stress (σ_F) evolution as a function of slip (Fig. 1), and we compute fracture energy following the usual relation: $G(S) = \int_0^S [\sigma_F(u) - \sigma_d]du$, where S is the final slip for each earthquake, and σ_d is the final stress level at the last slip increment.

An important quantity that may be used for studying the dynamics of faulting is the earthquake's apparent stress: the product of rigidity (here we use: $\mu = 3 \times 10^{10}$ Pa) and radiated energy normalized to seismic moment. In other words, rigidity times the amount of energy radiated per unit fault surface, per unit slip:

$$\sigma_a = \mu\left(\frac{E_R}{M_0}\right). \tag{1}$$

Radiated energy may be quantified with a space–time integration of the observed squared ground velocity. The kinetic energy that radiated away from the fault is, in large part, damped and converted to heat, but the remainder may be detected by seismometers: in particular, at a low frequency the anelastic attenuation is less significant. In order to account for the attenuation process, assumptions must be made for the source scaling in the high-frequency portion of the spectrum. By using the approach described by (Walter and Taylor 2001), and the omega-square spectral model (Aki 1967), Mayeda and Malagnini (2009, 2010), Malagnini and Mayeda (2008), Malagnini *et al.* (2008), and Mayeda *et al.* (2007), showed that, for the Hector Mine (California), San Giuliano (Southern Italy), Chi–Chi (Taiwan), and Colfiorito (Italy) seismic sequences, the apparent stress monotonically increases with increasing earthquake size. It is easy to show (see Malagnini *et al.* 2014, in the present volume) that the same seismic sequences can be described equally well (i.e., with the same misfit) by a constant apparent stress, after dropping the omega-square hypothesis of high-frequency spectral decay, in favor of a slightly gentler fall-off of the seismic spectra. More explicitly, if p indicates the exponent of the spectral fall-off at high frequency [described explicitly in Eq. (7)], and $p = 2$ is the fall-off of the omega-square spectral model, Malagnini *et al.* (2014, this volume) found that $p = 1.6{:}1.9$ would describe their results with a constant stress-drop model, producing the same quantitative misfit given by $p = 2$, coupled with a size-dependent stress drop. Recall that, in order to keep the radiated energy finite, p must be >1.5.

In the present paper, the energy measurements relative to the events of the L'Aquila seismic sequence are obtained on the entire wavetrain of S-waves radiated by the seismic source. We analyze events from one individual seismic sequence at a time, starting with all events that have a moment tensor solution (e.g., all events of the L'Aquila seismic sequence with $M_L \geq 3.0$, see Herrmann *et al.* 2011).

3. The Working Hypotheses

In the framework of our interpretation (i.e., the slip-weakening function does reach a steady-state value), our working hypotheses are simple, yet stringent: (a) all earthquakes occur on the same fault surface; (b) all earthquakes share the same faulting mechanism; (c) all earthquakes share the same initial stress (pre-stress) and this is equal to the yield stress ($\sigma_0 = \sigma_y$); (d) seismic slip of all earthquakes share the same stress curve, so that the final slip/stress of two different earthquakes can be considered as two points on the same weakening curve; (e) spectral averages performed over the available recordings of specific events are by all means equivalent to an integration carried out over the entire focal sphere.

As for the validity of hypothesis (a) for the L'Aquila seismic sequence, the data set is clearly made of two sub-sets of earthquakes (see Fig. 2): the

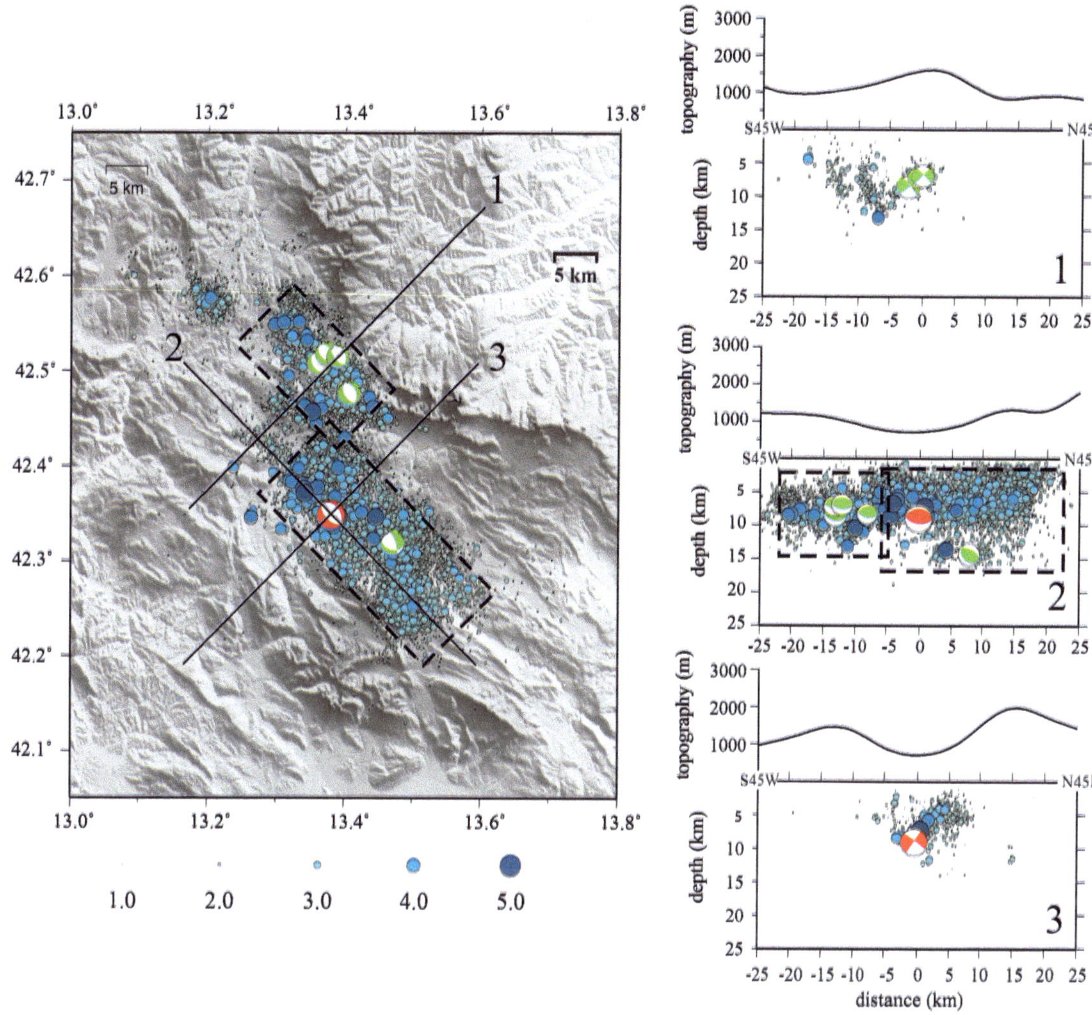

Figure 2
Map of the L'Aquila seismic sequence (modified from MALAGNINI *et al.* 2012), with cross section along and across the activated faults. *Red beach balls* indicate the focal mechanism of the main event (M_W 6.13); *green beach balls* indicate the four largest aftershocks ($M_W > 5$)

ones that occurred on the main fault (Paganica), and the ones that occurred on the secondary, adjacent fault (Monti della Laga). For both faults, we note that the L'Aquila aftershocks define sharp planes of seismic activity (e.g., CHIARABBA *et al.* 2009) having similar dip angles. VALOROSO *et al.* (2013) have shown that seismicity defines very clear fault planes on the Paganica and Monti della Laga faults. Moreover, in this study we use aftershocks larger than magnitude 3.0, which are mostly clustered on the imaged fault planes.

To support our hypothesis (b), all events of both the L'Aquila and the Northridge data sets have reliable focal mechanisms that allowed us to choose a subset of events for which the working hypothesis may be reasonable (see parameters in Tables 1 and 2). The spatial extension of the entire fault system for the L'Aquila area, elongated in the NW–SE direction, is roughly 45 km (MALAGNINI *et al.* 2012), and the fault planes (see Fig. 2) have very similar dip angles (between 40° and 50°). Also the events of the Northridge data set have homogeneous focal mechanisms (see Table 2).

We understand that hypothesis (c) might be considered too simplistic and that different processes can yield pre-stress variations on the fault plane. For example, if thermal pressurization occurs during the

Table 1

Parameters of the events of the L'Aquila seismic sequence that are computed in this study, with their standard errors: corner frequency (f_c), apparent stress (σ_a), fracture energy (G) and slip (S)

Event ID (date)	Lat.	Lon.	Depth (Km)	M_w	M_0 (Nm)	Strike (°)	Dip (°)	Rake (°)	f_c (Hz)	Δf_c (Hz)	σ_a (Pa)	$\Delta \sigma_a$ (Pa)	G (J/m^2)	ΔG (J/m^2)	S (m)	ΔS (m)
L'Aquila																
0904060132	42.334	13.334	6.0	6.13	1.95e+18	135	55	−95	0.3347	0.0769	2.90e+06	0.38e+06	1.29e+07	0.36e+07	1.78e+00	0.16e+00
0903301343	42.303	13.364	10.0	3.86	7.67e+14	148	56	−97	1.7201	0.4662	1.50e+05	0.24e+05	6.95e+03	2.33e+03	0.19e−01	0.19e−02
0903301905	42.316	13.373	11.0	3.59	3.02e+14	292	71	−137	2.2581	0.6075	1.34e+05	0.21e+05	4.20e+03	1.40e+03	0.13e−01	0.13e−02
0903302157	42.316	13.375	9.0	3.65	3.72e+14	149	55	−93	1.9658	0.4843	1.10e+05	0.16e+05	3.24e+03	0.99e+03	0.12e−01	0.11e−02
0904030444	42.327	13.358	11.0	3.41	1.62e+14	303	57	−130	2.3308	0.6164	0.79e+05	0.12e+05	1.43e+03	0.47e+03	0.72e−02	0.72e−03
0904052048	42.332	13.372	10.0	4.21	2.57e+15	126	66	−129	1.3904	0.3796	2.66e+05	0.41e+05	2.70e+04	0.89e+04	0.41e−01	0.42e−02
0904060227	42.374	13.342	11.0	4.09	1.70e+15	175	65	−75	1.7921	0.4720	3.78e+05	0.57e+05	4.21e+04	1.36e+04	0.45e−01	0.44e−02
0904060739	42.324	13.371	7.0	3.66	3.85e+14	160	50	−90	2.0400	0.5312	1.26e+05	0.19e+05	4.13e+03	1.13e+03	0.13e−01	0.13e−02
0904061036	42.343	13.402	6.0	3.66	3.85e+14	133	61	−132	1.9981	0.5730	1.16e+05	0.19e+05	3.65e+03	1.28e+03	0.13e−01	0.13e−02
0904061255	42.372	13.353	6.0	3.60	3.13e+14	345	60	−80	2.0439	0.5313	1.03e+05	0.16e+05	2.76e+03	0.89e+03	0.11e−01	0.11e−02
0904061414	42.365	13.338	9.0	3.59	3.02e+14	131	52	−117	2.1219	0.5483	1.12e+05	0.17e+05	3.11e+03	0.99e+03	0.11e−01	0.11e−02
0904062003	42.213	13.479	7.0	3.63	3.47e+14	135	60	−109	1.9555	0.5205	1.00e+05	0.15e+05	2.71e+03	0.88e+03	0.11e−01	0.11e−02
0904062247	42.349	13.293	10.0	3.88	0.82e+15	113	58	−116	1.9407	0.4821	2.35e+05	0.34e+05	1.48e+04	0.46e+04	0.25e−01	0.23e−02
0904081135	42.355	13.328	7.0	3.75	0.53e+15	15	55	−15	1.8937	0.5058	1.38e+05	0.21e+05	5.29e+03	1.72e+03	0.15e−01	0.16e−02
0904082256	42.507	13.364	7.0	4.18	2.32e+15	118	48	−109	2.0636	0.5479	0.79e+06	0.12e+06	1.58e+05	0.52e+05	0.81e−01	0.81e−02
0904090341	42.510	13.332	7.0	3.66	3.85e+14	142	51	−98	2.1875	0.6538	1.52e+05	0.26e+05	5.70e+03	2.06e+03	0.15e−01	0.17e−02
0904090429	42.504	13.345	7.0	3.48	2.07e+14	330	55	−70	3.2711	1.0071	2.69e+05	0.48e+05	1.22e+04	0.46e+04	0.18e−01	0.21e−02
0904090443	42.506	13.366	7.0	4.02	1.33e+15	133	56	−97	2.0709	0.5734	4.53e+05	0.72e+05	5.29e+04	1.80e+04	0.47e−01	0.49e−02
0904090610	42.491	13.397	5.0	3.35	1.32e+14	200	65	−55	2.4355	0.8513	0.68e+05	0.14e+05	1.08e+03	0.47e+03	0.64e−02	0.84e−02
0904090931	42.361	13.390	5.0	3.66	0.38e+15	143	61	−99	1.7128	0.4598	0.74e+05	0.12e+05	1.71e+03	0.57e+03	0.92e−02	0.93e−03
0904091518	42.308	13.495	5.0	3.78	5.82e+14	126	67	−110	1.7112	0.4646	1.12e+05	0.18e+05	3.90e+03	1.30e+03	0.14e−01	0.14e−02
0904092047	42.493	13.305	7.0	3.56	2.72e+14	320	60	−75	2.8233	0.9054	2.26e+05	0.41e+05	1.00e+04	0.39e+04	0.18e−01	0.21e−02
0904092109	42.510	13.335	6.0	3.44	0.18e+15	126	52	−117	2.8962	0.8986	1.63e+05	0.29e+05	5.01e+03	1.89e+03	0.12e−01	0.14e−02
0904101918	42.344	13.358	9.0	3.41	1.62e+14	118	58	−116	2.5929	0.7315	1.08e+05	0.18e+05	2.40e+03	0.83e+03	0.89e−02	0.95e−03
0904110539	42.386	13.402	7.0	3.82	6.68e+14	131	52	−102	1.8364	0.5280	1.57e+05	0.26e+05	7.23e+03	2.54e+03	0.18e−01	0.20e−02
0904110613	42.469	13.417	5.0	3.52	2.37e+14	145	55	−90	2.2641	0.6612	1.04e+05	0.17e+05	2.59e+03	0.92e+03	0.99e−02	0.11e−02
0904110657	42.387	13.406	8.0	3.64	0.36e+15	322	45	−85	2.0009	0.5300	1.11e+05	0.17e+05	3.26e+03	1.06e+03	0.12e−01	0.12e−02
0904110704	42.391	13.406	3.0	3.68	4.12e+14	135	75	90	1.8743	0.4925	1.05e+05	0.16e+05	3.11e+03	0.99e+03	0.12e−01	0.12e−02
0904111542	42.523	13.316	7.0	3.60	3.13e+14	136	57	−103	2.3588	0.7455	1.52e+05	0.27e+05	5.41e+03	2.05e+03	0.14e−01	0.17e−02
0904120329	42.536	13.316	7.0	3.54	2.54e+14	141	52	−102	2.2681	0.6511	1.13e+05	0.19e+05	3.01e+03	1.05e+03	0.11e−01	0.12e−02
0904131909	42.360	13.348	9.0	4.11	1.82e+15	142	61	−118	1.3525	0.3524	1.75e+05	0.26e+05	1.19e+04	0.38e+04	0.27e−01	0.27e−02
0904141928	42.536	13.307	7.0	3.79	6.03e+14	345	50	−70	1.9169	0.5791	1.59e+05	0.27e+05	7.19e+03	2.65e+03	0.18e−01	0.21e−02
0904142053	42.549	13.302	6.0	3.66	3.85e+14	150	45	−90	1.8603	0.5170	0.95e+05	0.15e+05	2.58e+03	0.87e+03	0.11e−01	0.11e−02
0904151955	42.466	13.365	8.0	3.44	1.80e+14	301	77	−128	3.0741	0.9104	1.97e+05	0.33e+05	6.84e+03	2.46e+03	0.14e−01	0.16e−02
0904160544	42.289	13.404	8.0	3.72	4.73e+14	133	56	−113	1.7741	0.5195	0.99e+05	0.17e+05	3.02e+03	1.11e+03	0.12e−01	0.13e−02
0904200713	42.407	13.349	9.0	3.51	2.30e+14	355	65	−65	2.1955	0.6353	0.92e+05	0.15e+05	2.07e+03	0.73e+03	0.90e−02	0.99e−03
0904211544	42.330	13.366	10.0	3.82	6.68e+14	143	56	−113	1.8617	0.4695	1.68e+05	0.25e+05	7.95e+03	2.48e+03	0.19e−01	0.18e−02
0904231514	42.247	13.492	7.0	4.21	2.57e+15	121	67	−110	1.3217	0.3390	2.31e+05	0.34e+05	2.12e+04	0.66e+04	0.37e−01	0.36e−02

Table 1 *continued*

Event ID (date)	Lat.	Lon.	Depth (Km)	M_w	M_C (Nm)	Strike (°)	Dip (°)	Rake (°)	f_c (Hz)	Δf_c (Hz)	σ_a (Pa)	$\Delta \sigma_a$ (Pa)	G (J/m^2)	ΔG (J/m^2)	S (m)	ΔS (m)
0904241338	42.519	13.348	5.0	3.72	4.73e+14	131	52	−117	1.5558	0.4167	0.69e+05	0.11e+05	1.61e+03	0.53e+03	0.94e−02	0.95e−03
0904261756	42.456	13.378	8.0	3.47	0.20e+15	108	56	−113	2.0686	0.5533	0.68e+05	0.10e+05	1.19e+03	0.38e+03	0.70e−02	0.70e−03
0905030514	42.365	13.390	5.0	3.68	0.41e+15	141	57	−103	1.6276	0.4442	0.68e+05	0.11e+05	1.52e+03	0.51e+03	0.89e−02	0.92e−03
0905090909	42.350	13.359	9.0	3.69	4.27e+14	136	57	−103	2.0451	0.5365	1.41e+05	0.22e+05	5.14e+03	1.66e+03	0.15e−01	0.15e−02
0905111659	42.491	13.366	7.0	3.56	2.72e+14	316	64	−106	2.2490	0.7437	1.13e+05	0.21e+05	3.18e+03	1.28e+03	0.11e−01	0.14e−02
0905140630	42.501	13.406	6.0	3.87	7.94e+14	155	55	−85	1.7765	0.5255	0.17e+05	0.29e+05	8.59e+03	0.51e+04	0.21e−01	3.46e−03
0905140632	42.481	13.410	6.0	3.72	4.73e+14	151	55	−87	2.0873	0.6941	1.57e+05	0.30e+05	6.59e+03	2.68e+03	0.17e−01	0.21e−02
0905231724	42.501	13.303	6.0	3.41	1.62e+14	315	65	−90	2.6797	0.8577	1.15e+05	0.21e+05	2.74e+03	1.07e+03	0.95e−02	0.12e−02
0905300255	42.356	13.341	9.0	3.83	6.92e+14	137	61	−118	1.7314	0.4404	1.40e+05	0.21e+05	5.92e+03	1.86e+03	0.17e−01	0.16e−02
0906161852	42.297	13.391	12.0	3.34	1.27e+14	127	63	−121	2.6013	0.6972	0.86e+05	0.14e+05	1.52e+03	0.51e+03	0.70e−02	0.71e−03
0906191947	42.302	13.383	10.0	3.41	1.62e+14	327	51	−98	3.1529	0.8774	1.95e+05	0.31e+05	6.41e+03	2.17e+03	0.13e−01	0.14e−02
0906200541	42.332	13.377	9.0	3.38	1.46e+14	146	57	−103	2.4992	0.7058	0.87e+05	0.14e+05	1.62e+03	0.56e+03	0.75e−02	0.80e−03
0906252100	42.570	13.206	5.0	4.02	1.33e+15	145	50	−80	1.4344	0.3864	1.51e+05	0.24e+05	8.49e+03	2.80e+03	0.22e−01	0.23e−02
0906252331	43.256	13.085	6.0	3.59	3.02e+14	175	55	−65	2.9888	0.9618	2.96e+05	0.55e+05	1.63e+04	0.64e+04	0.22e−01	0.27e−02
0906260714	42.567	13.189	5.0	3.74	5.07e+14	141	52	−102	1.8278	0.4779	1.20e+05	0.18e+05	4.14e+03	1.34e+03	0.14e−01	0.14e−02
0907030943	42.328	13.361	10.0	3.73	4.90e+14	139	54	−110	2.1808	0.5797	1.96e+05	0.30e+05	9.32e+03	3.05e+03	0.19e−01	0.19e−02
0907120838	42.338	13.378	9.0	4.35	4.17e+15	139	54	−110	1.4457	0.3600	4.91e+05	0.72e+05	8.74e+04	2.72e+04	0.71e−01	0.67e−02
0907122214	42.338	13.398	9.0	4.05	1.48e+15	131	57	−103	1.4842	0.3900	1.87e+05	0.28e+05	1.25e+04	0.40e+04	0.27e−01	0.26e−02
0908022239	42.365	13.391	5.0	3.68	4.12e+14	135	53		1.8516	0.5267	0.99e+05	0.16e+05	2.87e+03	1.00e+04	0.12e−01	0.12e−02

Depths and strike, dip and rake angles indicated here are from Herrmann *et al.* (2011), whereas the seismic moments of the aftershocks were slightly adjusted in this study (see main text). The seismic moment of the main event was published by Herrmann *et al* (2011) (M_W 6.13). Magnitude corrections for the aftershocks are usually of the order of 0.05 M_w units

Table 2

Parameters of earthquakes of the Northridge seismic sequence (for details, see caption of Table 1)

Event ID (date)	Lat.	Lon.	Depth (Km)	*Mw*	M_0 (Nm)	Strike (°)	Dip (°)	Rake (°)	f_c (Hz)	Δf_c (Hz)	σ_a (Pa)	$\Delta\sigma_a$ (Pa)	G (J/m^2)	ΔG (J/m^2)	S (m)	ΔS (m)
Northridge																
940117043055	34.24	−118.56	14.0	6.70	1.396e+19	87	53	59	0.1380	0.0307	1.47e+06	1.80e+05	1.12e+07	0.27e+07	2.17e+00	0.18e+00
940117130628	34.25	−118.55	12.8	4.60	9.886e+15	N/A	N/A	N/A	0.7436	0.1946	1.59e+05	0.23e+05	0.26e+05	0.71e+04	0.45e−01	0.44e−02
940117132645	34.37	−118.55	8.9	4.70	1.396e+16	279	30	51	0.7260	0.2069	2.06e+05	0.32e+05	0.46e+05	0.14e+05	0.60e−01	0.65e−02
940117141430	34.33	−118.44	10.6	4.50	6.998e+15	267	50	62	0.9602	0.2784	2.38e+05	0.37e+05	0.46e+05	0.14e+05	0.53e−01	0.58e−02
940117150703	34.30	−118.47	11.1	4.20	2.483e+15	271	31	58	1.8988	0.5592	6.48e+05	1.06e+05	1.77e+05	0.55e+05	0.73e−01	0.81e−02
940117155410	34.37	−118.62	16.5	4.80	1.972e+16	56	44	51	0.7168	0.1911	2.84e+05	0.41e+05	0.85e+05	0.24e+05	0.83e−01	0.83e−02
940117193534	34.31	−118.45	11.7	4.00	1.245e+15	299	19	75	2.0764	0.7162	4.14e+05	0.73e+05	0.70e+05	0.25e+05	0.44e−01	0.57e−02
940117223153	34.34	−118.44	12.4	4.10	1.758e+15	269	55	71	1.2309	0.3518	1.26e+05	0.20e+05	0.11e+05	0.31e+04	0.22e−01	0.24e−02
940119140914	34.23	−118.50	14.8	4.50	6.998e+15	74	72	61	1.0060	0.2856	2.74e+05	0.43e+05	0.58e+05	0.18e+05	0.58e−01	0.62e−02
940121183915	34.30	−118.46	10.6	4.60	9.886e+15	116	44	51	0.8055	0.2051	2.04e+05	0.28e+05	0.38e+05	0.10e+05	0.52e−01	0.50e−02
940121185344	34.32	−118.48	10.7	4.30	3.508e+15	111	29	60	1.2297	0.3832	2.47e+05	0.41e+05	0.40e+05	0.13e+05	0.43e−01	0.51e−02
940124055024	34.36	−118.63	13.6	4.29	3.388e+15	71	72	49	1.1278	0.3189	1.88e+05	0.29e+05	0.24e+05	0.72e+04	0.35e−01	0.38e−02
940124055421	34.37	−118.63	16.4	4.20	2.483e+15	84	32	35	1.3096	0.3715	2.15e+05	0.34e+05	0.28e+05	0.83e+05	0.35e−01	0.37e−02
940127171958	34.27	−118.56	13.3	4.40	4.954e+15	34	0	−4	1.1504	0.3722	2.85e+05	0.47e+05	0.58e+05	0.19e+05	0.54e−01	0.65e−02
940203162335	34.30	−118.44	9.5	4.09	1.698e+15	95	34	93	1.3818	0.3989	1.72e+05	0.27e+05	0.17e+05	0.51e+04	0.27e−01	0.29e−02
940206131927	34.29	−118.48	9.4	4.00	1.245e+15	118	59	90	1.7540	0.5720	2.51e+05	0.44e+05	0.30e+05	0.10e+05	0.31e−01	0.39e−02
940525125657	34.31	−118.31	9.4	4.50	6.998e+15	88	53	65	0.8379	0.1997	1.64e+05	0.21e+05	0.23e+05	0.59e+04	0.40e−01	0.36e−02

For the main event of the Northridge sequence, the estimate of seismic moment is from Wald *et al* (1996). The estimates of seismic moments from the aftershocks are slightly adjusted (after the grid-search performed to fit the empirical source spectral ratios) from the moment tensor solutions given by the Berkeley Seismological Laboratory, and are available from the Northern California Earthquake Data Center (NCEDC) mainshock, the pore pressure on the fault may increase. As a consequence, aftershocks may occur in an environment with strongly variable pore-fluid pressure. However, we emphasize that the largest events of the sequence have centroid depths around 9 km (Herrmann *et al.* 2011), whereas most earthquakes have depths between 7 and 12 km. In what follows, however, each earthquake enters the calculations using the hypocentral depth of the mainshock (10 km).

We point out that, in the hypothesis of a transition to a lithostatic gradient for the pore-fluid pressure below a certain depth, the effective stresses below the gradient transition become depth-independent (Suppe and Yue 2008). Therefore, hypothesis (c) may be valid either if all earthquakes are characterized by very similar focal depths and pore-fluid pressure, or they all are below a certain depth at which the crust is hydraulically compartmentalized, and the pore-pressure gradient within the pressurized compartments changes to lithostatic. In the latter case, the effective stresses are constant below the depth where the pore-fluid pressure gradient transitions from hydrostatic to lithostatic, see (Suppe and Yue 2008), and Malagnini *et al.* (2010).

It is also important to point out that, due to the enhanced permeability that may characterize a fresh seismic rupture (an example of the latter case is the Dobi extensional earthquake sequence in Central Afar, where Noir *et al.* (1997), found a permeability $k \approx 10^{-8}$ m^2), dynamic increments of pore-fluid pressure due to thermal pressurization are expected to be short-lived, when compared to other diffusion-driven phenomena like the ones observed in the area of the L'Aquila sequence. In fact, before the seismic rupture, Malagnini *et al.* (2012) estimated the rock permeability along the Monti della Laga fault system to be in the range of $k = 4 \times 10^{-13}$ m^2 to $k = 6 \times 10^{-13}$ m^2.

With regard to the validity of hypothesis (c) for the Northridge aftershocks analyzed here, the events that we use for our computations have similar centroid depths (see Table 2), and similar effective normal stresses are expected across the fault plane(s).

All our energy measurements relative to the L'Aquila events are performed on direct, S-wave-based source spectral ratios (see below for details). We are aware that working hypothesis (d) is a strong assumption. Assuming that all points of the fault plane (that is, the slipping portion) follow the same stress curve implies that σ_0, D_c, and σ_{SS} (the initial stress, the critical slip-weakening distance, and the steady-state stress level, respectively) do not vary on the fault plane. We emphasize that, for the purpose of this work (i.e., the computation of seismological fracture energies), this assumption might be reasonable taking into account that: (i) we do not consider extended faults, and the stress-slip curve represents the average shear stress for that source; (ii) we adopt an asymptotic (exponential) decay of shear stress versus slip, which implies that D_c is simply a normalizing constant and not a constitutive parameter; and (iii) the steady-state value of stress is reached asymptotically in our formulation. Hypothesis (e) has been successfully used by Malagnini *et al.* (2011) and it is well-justified by the excellent distribution of recording stations in our data set (see Herrmann *et al.* 2011).

Under the hypothesis of Andersonian faulting (Anderson 1905), one of the principal stresses acting on a specific fault is vertical), an estimate of the regional stress field at failure may be obtained using the expressions provided by Malagnini *et al.* (2010, 2012). The estimate requires assumed or measured values for dip angle, hypocentral depth, static coefficient of friction and pore-fluid pressure at hypocentral depth.

In the hypotheses (a–e) described above, and that of Andersonian faulting, the quantification of the stress-slip evolution on a specific fault plane may be obtained by quantifying the stress drop of each one of the available earthquakes, and subtracting it to a common prestress value. In turn, in the general case of a non-optimally oriented fault (at an angle θ with the direction of maximum compression, σ_1), the normal and the shear stresses on a specific fault plane (σ_n and τ, respectively) are calculated by solving the following system:

$$\begin{cases} \sigma_0 = \mu_s(\sigma_n - P_f) \\ \sigma_0 = \frac{\sigma_1 - \sigma_3}{2} \sin(2\theta) \\ \sigma_n = \frac{\sigma_1 + \sigma_3}{2} - \frac{\sigma_1 - \sigma_3}{2} \cos(2\theta) \end{cases} \quad (2)$$

where σ_3 is the minimum compressional stress.

Malagnini *et al.* (2010) wrote a specific solution of system (2) for each tectonic environment. For the case of a normal fault like the Paganica one, the hypothesis of Andersonian faulting prescribes that the maximum compressional stress is vertical. Specifically: $\sigma'_1 = \rho g z - P_f$, where ρ is the rock density, g is gravity, z is the depth of nucleation, and P_f is pore-fluid pressure. On the contrary, in a thrust fault it is the minimum compressional stress that must be vertical: $\sigma'_3 = \rho g z - P_f$. Finally, strike-slip faults (not investigated here) are characterized by a vertical, intermediate principal stress $\sigma'_2 = \sigma_2 - P_f$. In all cases, we can write: $\left(\frac{\sigma'_1+\sigma'_3}{2}\right) = \rho g z - P_f$.

Based on the work by Sibson (1974), Malagnini *et al.* (2010) provided the necessary equations for the task; particularly, the case of a thrust fault corresponds to their Eq. 5. By using the parameter $\lambda_f = \left(\frac{P_f}{\rho g z}\right)$, which represents the pore-fluid pressure in units of lithostatic load, their Eq. 5b may be rewritten as:

$$\begin{cases} \sigma'_n = \dfrac{g z \rho_r (1-\lambda_f)\cos(\theta)}{\cos(\theta) - \mu_S \sin(\theta)} . \\ \tau_y = \mu_S \sigma'_n \end{cases} \quad (3a)$$

For a normal fault, Eq. 5 of Malagnini *et al.* (2010) may be rewritten as:

$$\begin{cases} \sigma'_n = \dfrac{g z \rho_r (1-\lambda_f)\sin(\theta)}{\mu_S \cos(\theta) + \sin(\theta)} . \\ \tau_y = \mu_S \sigma'_n \end{cases} \quad (3b)$$

For completeness (we do not use strike-slip earthquakes in this study), for a strike-slip fault, Eq. 5 by Malagnini *et al.* (2010) may be reformulated as:

$$\begin{cases} \sigma'_n = \dfrac{g z \rho_r (1-\lambda_f)\sin(2\theta)}{\mu_S \cos(\theta) + \sin(2\theta)} . \\ \tau_y = \mu_S \sigma'_n \end{cases} \quad (3c)$$

In what follows, for the L'Aquila earthquakes we use the normal fault Eq. (3b) and the following parameters: $\theta = 40°$, $z = 10$ km. For the Northridge earthquakes we use the thrust fault Eq. (3a) and the following parameters: $\theta = 37°$, $z = 14$ km. For both sequences we used the following values for the static friction and normalized pore-fluid pressure: $\mu_S = 0.7$, and $\lambda_f = 0.65$ (moderate overpressure).

The choice of the right set of parameters is important for the estimated values of shear stress on the fault. A different set of parameters would not change the meaning of our results, but simply shift our estimates of shear stress toward larger (smaller λ_f larger μ_S, larger θ, larger z) or smaller (larger λ_f smaller μ_S, smaller θ, smaller z) absolute values.

4. *Calculation and Analysis of Source Spectral Ratios*

Mayeda *et al.* (2007) defined a technique for the analysis of coda-based source spectral ratios; their goal was to reduce the uncertainties of the retrieved source parameters in their study of the Hector Mine seismic sequence of 1999. Over the years, Mayeda and co-workers focused on source scaling: they used the cited technique in order to analyze a number of well-recorded seismic sequences that have occurred in different tectonic environments: Colfiorito, (Central Italy, Malagnini *et al.* 2008), San Giuliano (Southern Italy, Malagnini and Mayeda 2008) Chi–Chi (Taiwan, Mayeda and Malagnini 2009), Wells (NV, Mayeda and Malagnini 2010; Yoo and Mayeda 2013), and Iwate (Japan, Malagnini *et al.* 2010). They also investigated a number of other Japanese sequences (Malagnini *et al.* 2014, this volume). In the present study, we use the same approach based on source spectral ratios that is described in Mayeda *et al.* (2007).

4.1. The L'Aquila Data Set

Instead of obtaining the source spectra of the L'Aquila earthquakes from the analysis of the seismic coda (i.e., with the technique initially proposed by Mayeda and Walter 1996, and subsequently modified by Mayeda *et al.* 2003), we calculated them through the series of spectral inversions that Malagnini and co-workers have been using since the late 1990's for the quantification of ground motion (Raoof *et al.* 1999; Malagnini *et al.* 2000a, b, 2002; Malagnini and Herrmann 2000; Akinci *et al.* 2001; Bay *et al.* 2003; Bodin *et al.* 2004; Scognamiglio *et al.* 2005; Morasca *et al.* 2006; Malagnini *et al.* 2007, 2011; Meirova *et al.* 2008; Akinci *et al.* 2010; D'Amico *et al.* 2012; Urguhan *et al.* 2012).

The advantages of the direct-wave approach in the calculation of the seismic spectra are multiple. The most important one is that, because all recordings belong to an individual seismic sequence of a moderate-sized mainshock event, the spatial extent of the epicentral areas (~25 and ~15 km along-strike for the Paganica and the Monti della Laga faults, respectively) can be considered small, when compared to the regional distances traveled by the direct shear-waves (distances within 200 km were used in the regressions presented here, similarly to that described by MALAGNINI *et al.* 2011, to which the reader is referred for more details). As a consequence, the same crustal path is sampled over and over, in a very similar manner, by the seismic waves radiated by a large number of earthquakes (171, in our case) that are all located in the same volume, and the averaging action performed by the inversion procedure makes the results very stable. Due to the large amount of recordings that are available for the L'Aquila seismic sequence, the accuracy of our indirect spectral measurements is comparable to coda-based spectra, or better (Fig. 3). For coda-based spectra, the averaging action is due to the oversampling of the crust by scattered waves (AKI 1967; AKI and CHOUET 1975). Additional advantages include: (i) the entire process is fully automated; (ii) the procedure allows for the simultaneous inversion of many seismic spectra in one single step.

The technique based on the regression of the peak values of the filtered time histories (see later for a basic explanation) has already been applied to the L'Aquila sequence data set by MALAGNINI *et al.* (2011). Differently than them, in the present study we finely sample the peak velocity spectra at 50 adjacent central frequencies, equally spaced in log units.

It must be clear that, unlike what was described for the L'Aquila data set, the Northridge spectral ratios are based only on spectral measurements taken on the seismic coda.

4.2. *Time-Domain Peak Values and Fourier Amplitudes*

We calculated the time domain peak "spectra" of Fig. 3, including their standard errors, from the regression of the following non-parametric equation:

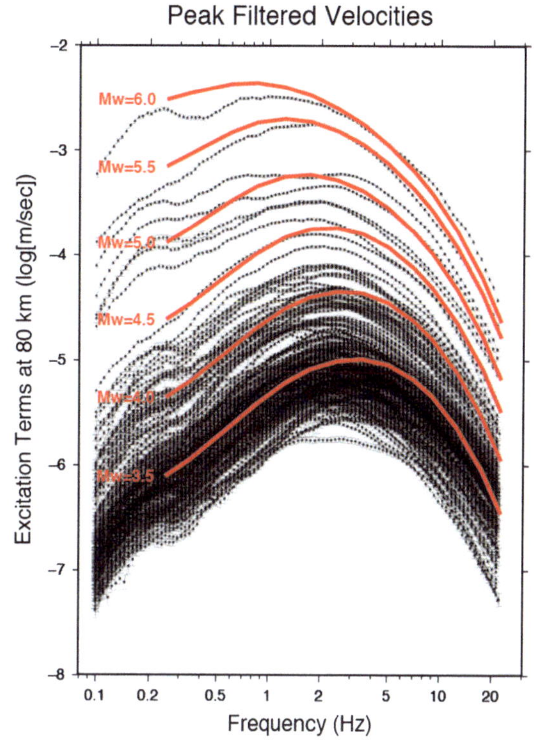

Figure 3
Black lines spectra from regressions (peak amplitudes of filtered time histories. L'Aquila sequence); *red lines* predicted peak amplitudes based on the excitation/attenuation model by Malagnini *et al.* (2011)

$$A_n(f_i, r_{jk}) = \mathrm{SRC}_j(f_i, r_0) + \mathrm{SITE}_k(f_i) + D(r_{jk}, r_0, f_i), \quad (4)$$

where: $A_n(f_i) = \log_{10} a_n(f_i)$ is the logarithm of the peak amplitude (units of m/s) observed on a narrow bandpass-filtered version of the nth time history, of central frequency f_i. In order to bracket all direct-wave arrivals carrying significant amplitudes, we use a time window defined by the S-wave onset, and a group velocity of 2 km/s. Peak values within the described time windows are relative to the jth excitation term of the data set, SRC_j (f_i, r_0), where r_{jk} is the hypocentral distance of the kth station, and SITE_k (f_i) is the corresponding kth site term. The propagation term is indicated as D (r_{jk}, r_0, f_i), and represents a contribution of the regional crustal structure that may be considered common to all recording sites. f_i is the central frequency of the bandpass filter (8-poles low-pass with corner frequency at $f_{2i} = \sqrt{2}f_i$, followed by an 8-pole high-pass with corner frequency

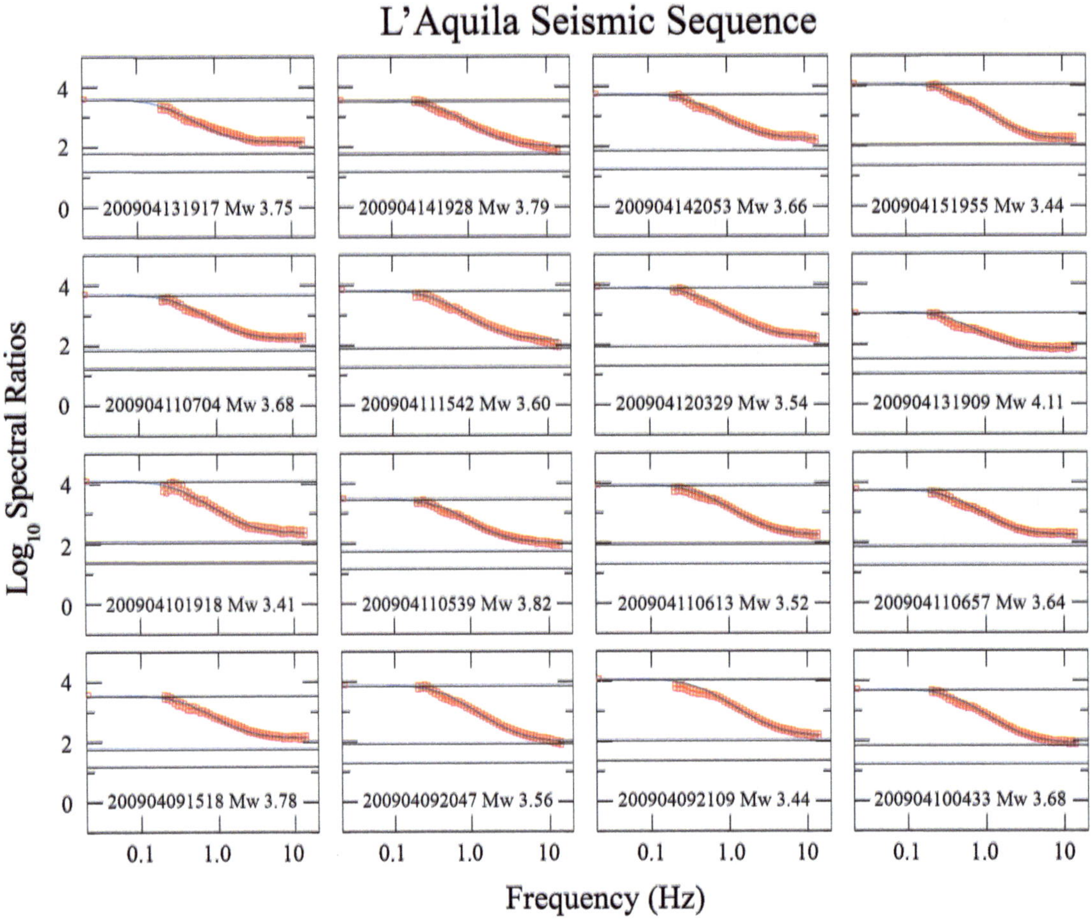

Figure 4

Sample spectral ratios, plotted in the frequency range between 0.02 and 20 Hz (L'Aquila sequence). A point equal to the ratio of the seismic moments of the two events is set at 0.02 Hz. Three *asymptotic lines* are indicated in each frame: the low-frequency asymptote (ratio of seismic moments), the self-similar, high-frequency asymptote (*lowest line* cubic root of the ratio of the moments), and the non-self-similar, high-frequency asymptote (*middle line*), as predicted by the best ψ parameter of Eqs. (13) and (14)

at $f_{1i} = f_i/\sqrt{2}$). Finally, r_0 is an arbitrary hypocentral distance at which all source terms are referred; the latter characteristic of the inverted terms is achieved by forcing the constraint $D(r_{jk} = r_0, r_0, f_i) = 0$ to the ith regression.

Specific constraints need to be forced on the site terms: generally, the summation of the logarithms of all the rock site terms equal to zero, so that the source terms resulting from the regressions (in log units) would contain (and be biased by) a site effect equal to the average site response of the rock site. If a quantitative estimate of the site effect of a specific site is available (e.g., from a 3-D numerical model), by forcing the logarithm of that site to be zero in the regression it would refer all sources to that specific site, and the other site terms would be equivalent to relative estimates of site effect. Constraints of a different nature may be forced onto the system in order to achieve specific goals (see, for example, Malagnini *et al.* 2004, 2006, who estimated absolute site terms in different regions of the world by regressing ground motion data after fixing absolute source spectra from a coda-based study, and path terms from previous inversions).

Given a stationary, random time history of length T', Random Vibration Theory (RVT, see Cartwright and Longuet-Higgins 1956) may be used to estimate its peak value:

$$\text{Peak}(a(t)) \approx \eta a_{\text{RMS}}, \tag{5}$$

where the a_{RMS} is its RMS-average calculated over T'.

In our case, RVT is applied to a filtered time history that is significantly different from zero only between $t = 0$ (the direct S-wave arrival time) and $t = T$ (the effective duration of the signal that follows the direct S-waves). Duration T can be defined on each seismogram as stated by RAOOF *et al.* (1999), and by MALAGNINI *et al.* (2000a, b): the time elapsed after 5 % of the cumulative energy, until 75 % of the cumulative energy is observed (e.g., see Fig. 4 of RAOOF *et al.* 1999).

The filtering action performed by the bandpass filter makes the Fourier amplitudes different from zero only between the two corner frequencies defined above, allowing the following equation:

$$a_{\text{RMS}}(f_i) = \sqrt{\frac{\int_0^T a^2(t, f_i)\text{d}t}{T}} = \sqrt{\frac{\int_{f_{1i}}^{f_{2i}} \hat{a}^2(f)\text{d}f}{T}}, \tag{6}$$

and $\eta = \eta(m_0, m_2, m_4)$ is a function of the specified spectral moments ($m_n = \frac{1}{\pi}\int_0^\infty \omega^n |\hat{a}(\omega)|^2 \text{d}\omega$) of the filtered time history.

The Parseval theorem, in the form of Eq. (6), and the Random Vibration Theory, in the form of Eq. (5), together with the convolution theorem, can be used to completely switch from the peak values of time histories to their Fourier spectral amplitudes. With respect to Fourier amplitudes, the great advantage of using peak values is that they are much less affected by background noise. Consequently, for what concerns source spectral ratios, peak and Fourier analyses yield totally equivalent results, but the former is characterized by much smaller standard errors.

4.3. Source Spectral Ratios

Source spectral ratios between the mainshock and all the aftershocks can easily be calculated from the quantities plotted in Fig. 3. The theoretical moment-rate spectral ratio between two generic events #1 and #2 can be written as follows:

$$\frac{\dot{M}_1(\omega)}{\dot{M}_2(\omega)} = \frac{M_{0_1}\left(1 + (\omega/\omega_{c_2})^2\right)^{p/2}}{M_{0_2}\left(1 + (\omega/\omega_{c1})^2\right)^{p/2}}, \tag{7}$$

where M_{0_i} is the seismic moment of the ith event ($i = 1, 2$); p is the high-frequency roll-off of the moment-rate spectra (the omega-square model dictates $p = 2$, and the simple consideration that the energy integral must be finite, force $p > 1.5$). Figure 4 shows example source spectral ratios between the M_W 6.13 L'Aquila main event of April 6, 2009, at 1:32 UTC, and selected aftershocks using averaged ratios obtained over regional three-component broadband stations. Two horizontal asymptotes characterize the right-hand-side of (7): at low-frequency, the ratio tends to $\frac{M_{0_1}}{M_{0_2}}$, whereas, if self-similarity holds, and $p = 2$, the high-frequency asymptote is: $\left(\frac{M_{0_1}}{M_{0_2}}\right)^{\frac{1}{3}}$ (MAYEDA *et al.* 2007). In the spectral ratios shown in Fig. 4, we indicate the two theoretical asymptotes, noting that the asymptotic value at high-frequency is almost never reached by the empirical coda-derived source ratios, a violation of self-similarity if $p = 2$.

To account for the possibility of non-self-similar scaling, we used the implementation of the source model developed by WALTER and TAYLOR (2001), which uses an omega-square source spectral model (i.e., $p = 2$) but also permits non-self-similar scaling. From their study:

$$\omega_{c_i} = 2\pi f_{c_i} = \left(\frac{k_i \sigma_{a_i}}{M_{0_i}}\right)^{1/3} \tag{8}$$

is the ith event's corner frequency (ω_{c_i} in radians/s, f_{c_i} in Hz), and k_i is for the ith earthquake, a function of the seismic velocities (α, β), corner frequencies (ω_{cP_i}, ω_{cS_i}), and average radiation patterns ($R_{\theta\varphi P}$, $R_{\theta\varphi S}$) of the appropriate wave types (P, S):

$$k_i = \frac{16\pi}{\beta^2\left(\frac{R^2_{\theta\varphi P}\omega^3_{cP_i}}{\alpha^5\omega^3_{cS_i}} + \frac{R^2_{\theta\varphi S}}{\beta^5}\right)}, \tag{9}$$

$$\sigma_{a_i} = \mu\left(\frac{E_{R_i}}{M_{0_i}}\right) \tag{10}$$

is the ith event's apparent stress. For simplicity, following WALTER and TAYLOR (2001), we use an additional parameter to link the *P*- and *S*- wave corner frequencies:

$$\omega_{cP_i} = \zeta\omega_{cS_i}, \tag{11}$$

where ζ is the scale factor. For the BRUNE (1970, 1971) omega-square model, ζ is the ratio of the wave velocities, but other studies have shown that the corner frequencies are roughly equal. In general: $1 \leq \zeta \leq \frac{\alpha}{\beta}$. However, as suggested by WALTER and TAYLOR (2001), we used $\zeta = 1$, that is:

$$\omega_{cP_i} = \omega_{cS_i}. \tag{12}$$

In any case, although the absolute values of each individual apparent stress retrieved using Eq. (7) may change depending on the choice of the scale factor ζ in (11), trends existing within the seismic sequence (the source scaling) will still be detectable, as long as the scale factor ζ will stay constant with increasing seismic moment. The spectral ratio fits in Fig. 4 are all obtained using the theoretical function given in Eq. (7), and assuming the following non-self-similar scaling for the apparent stress:

$$\sigma_{a_i} = \sigma'_a \left(\frac{M_{0_i}}{M'_0}\right)^{\psi}, \tag{13}$$

where:

$$\psi = \frac{\varepsilon}{3+\varepsilon}, \tag{14}$$

σ'_a is the apparent stress at an arbitrary reference seismic moment M'_0 (e.g., the seismic moment at $M_W = 5.0$), and the parameter ε (KANAMORI and RIVERA 2004), reflects the deviation from self-similarity in the corner frequency-seismic moment scaling relation:

$$M_0 \propto f_c^{-(3+\varepsilon)}. \tag{15}$$

Once the spectral ratios are calculated, a technique explained in MAYEDA *et al.* (2007) was used in order to get estimates of corner frequency and apparent stress of 57 events from the L'Aquila seismic sequence. We used Eqs. (7)–(15) in order to obtain best-fit spectral ratios after a grid-search on the critical parameters of the spectral model by WALTER and TAYLOR (2001) (apparent stress at the reference moment magnitude, and the ε parameter of Eq. (15)). We also decided to allow for a minimal adjustment of the moment magnitudes taken from HERRMANN *et al.* (2011) (of the order of 0.05 magnitude units, in most cases). In order to be accepted and included in the regression, our modified version of the procedure described in MAYEDA *et al.* (2007) requires each spectral ratio to stay below an arbitrary misfit threshold of 0.02 (in the L-1 norm), between the logarithm of a specific empirical spectral ratio and the logarithm of its theoretical estimate. As a result of this procedure, 57 "good" source ratios between the main event and 56 aftershocks (a total of 57 earthquakes) were selected by the best-fit procedure. For each event in the "good" list we obtained a reliable estimate of corner frequency, and a small adjustment to the original seismic moment calculated by HERRMANN *et al.* (2011). Equation (17) was then used, together with the estimates of corner frequency and adjusted seismic moment, to calculate the fracture energies of the 57 earthquakes, and Eq. (18) was used for the seismic slip. Results for the L'Aquila events are given in Table 1.

4.4. Calculation of Fracture Energies and Relationship with Stress History

Following ABERCROMBIE and RICE (2005), in the hypothesis of no undershoot or overshoot, their Eq. (4) describes the fracture energy ($G(S)$) in terms of seismic stress drop ($\Delta\sigma$), final slip (S), and apparent stress (σ_a):

$$G(S) = \frac{1}{2}(\Delta\sigma - 2\sigma_a)S. \tag{16}$$

Because our independent spectral measurements are the seismic moments and the corner frequencies of each event, we need to explicitly write the fracture energy given in (16), and the corresponding seismic slip of a circular crack, based on the results by SATO and HIRASAWA (1973), and ESHELBY (1957, 1959), and on the relationship between apparent stress and corner frequency, $f_c = \frac{1}{2\pi}\left(\frac{k\sigma_a}{M_0}\right)^{1/3}$, from the MDAC spectral model by WALTER and TAYLOR (2001):

$$G(M_0, f_c) = \frac{M_0^2 \pi^4 f_c^5}{\mu C^2 \beta^2}\left(\frac{7}{C^3\beta^3} - \frac{16}{k_i}\right). \tag{17}$$

k_i is given by Eq. (8), and, as for the crack slip S used in (16):

$$S = \frac{M_0 4\pi f_c^2}{\mu C^2 \beta^2}. \tag{18}$$

For a rupture velocity equal to $V = 0.9\beta$, $C = 1.85$ (AKI and RICHARDS 2002).

From the definition of fracture energy:

$$G(S) = \int_0^S (\sigma_F(u) - \sigma_d)\mathrm{d}u \qquad (19)$$

where $\sigma_d = \sigma_F(S)$ indicates the residual stress level of "dynamic" friction during sliding, $\sigma_F(u)$ is the shear stress on the fault and, from the differentiation of Eq. (19):

$$\frac{\mathrm{d}G(u)}{\mathrm{d}u} = -u\frac{\mathrm{d}\sigma_F(u)}{\mathrm{d}u}. \qquad (20)$$

Abercrombie and Rice (2005) used a linear fit for a compilation of data points in the fracture energy—seismic slip space:

$$\log_{10} G(S) = \log_{10} G_0 + \eta \log\left(\frac{S}{S_0}\right), \qquad (21)$$

where the parameter $S_0 = 1$ m keeps the physical dimensions of the constant G_0 in J/m^2. It follows:

$$\sigma_F(S) = \sigma_0 - \frac{\eta}{\eta - 1}\frac{G_0}{S_0}\left(\frac{S}{S_0}\right)^{\eta-1}, \qquad (22)$$

where σ_0 is the fault's prestress, and:

$$\Delta\sigma(S) = \frac{\eta}{\eta - 1}\frac{G_0}{S_0}\left(\frac{S}{S_0}\right)^{\eta-1} \qquad (23)$$

is the earthquake stress drop. We remind the reader that S is the final slip.

From a compilation of fracture energies, Abercrombie and Rice found: $\frac{\mathrm{d}G(S)}{\mathrm{d}S} = 6.72 \cdot 10^6 \left(\frac{S}{S_0}\right)^{0.28}$ J/m (S in meters). Consequently, $\sigma_F(S) = \sigma_0 - 24 \times 10^6 \left(\frac{S}{S_0}\right)^{0.28}$ Pa. In terms of stress drop: $\Delta\sigma(S) = 24 \times 10^6 \left(\frac{S}{S_0}\right)^{0.28}$ Pa.

Thc functional form (21) does not have a steady-state value. When applied to individual sequences, the exponent $(\eta - 1) \approx 1$ (see Fig. 5 for a linear fit of the fracture energies of the events of the L'Aquila seismic sequence analyzed in this study), and Cocco and Tinti (2008) found an exponent $(\eta - 1) \approx 0.8$. In the case studied in this paper, the steepness of the best-fit line of Fig. 5, for a larger crustal earthquake ($M \sim 7$), would predict an unrealistically high stress drop, of the order of hundreds of MPa for large events with several meters of slip.

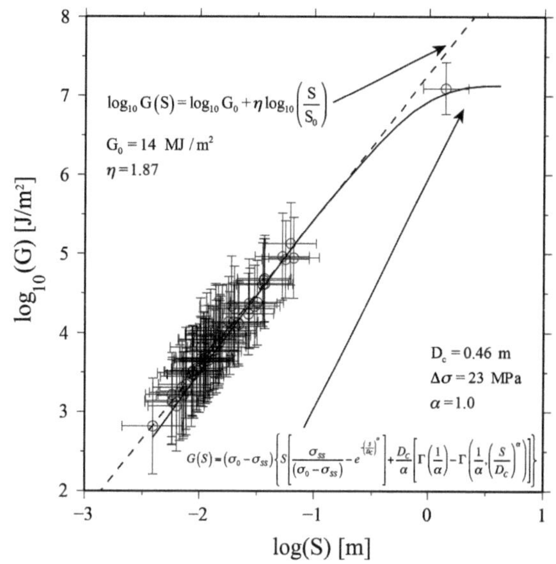

Figure 5
Fracture energy (G) vs. slip (S). L'Aquila seismic sequence. We plot the *error bars* of all the estimates of fracture energy. The *two curves* plotted in the picture indicate: (i) the *best-fit lines* that indicate the functional form of Eq. (21), and that of Eq. (25). For both *best-fit lines*, we provide their specific parameters. In the framework described in the present paper, the main event of the L'Aquila seismic sequence was characterized by a stress drop of around 23 MPa

A stress function characterized by an exponential decay of the shear stress toward a steady state (σ_{SS}) would avoid the difficulty of an unbounded stress drop as a function of seismic slip. Our preferred functional form for the shear stress on the fault plane is the following:

$$\sigma_F(S) = \sigma_{SS} + \Delta\sigma \exp\left(-\left(\frac{S}{D_C}\right)^{\alpha}\right). \qquad (24)$$

In our simple hypothesis the strength excess is null because $\sigma_y = \sigma_0$. A thorough grid-search over α yielded a best-fit parameter very close to unity; consequently, we drop the issue and use $\alpha = 1$ in all our computations. σ_{SS} in (24) is the steady stress value reached asymptotically for large values of slip.

Using Eqs. (24) and (19), we find the following general form for the fracture energy (where $\Gamma\xi$) is the Gamma function, and $\Gamma(\xi, \psi)$ is the incomplete Gamma function):

$$G(S) = (\sigma_0 - \sigma_{SS})\left\{ S\left[\frac{\sigma_{SS}}{(\sigma_0 - \sigma_{SS})} - e^{-\left(\frac{S}{D_C}\right)^{\alpha}}\right] + \frac{D_C}{\alpha}\left[\Gamma\left(\frac{1}{\alpha}\right) - \Gamma\left(\frac{1}{\alpha}, \left(\frac{S}{D_C}\right)^{\alpha}\right)\right]\right\}. \quad (25)$$

The parameters of two best-fit functional forms that go through the data points of Fig. 5 are estimated: those of the linear fit described by Eq. (21): $G_0 = 1.4 \times 10^7$ J/m^2 and $\eta = 1.87$, and those of Eqs. (24) and (25) (with $\alpha = 1$): $D_C = 0.46$ m, $\Delta\sigma = 23$ MPa (at steady-state $\sigma_{SS} \approx 26$ MPa). With a moderate pore-fluid overpressure ($\lambda_f = 0.65$), the effective shear prestress acting at nucleation depth (~ 10 km) on the L'Aquila (Paganica) normal fault (dip angle 40°) is estimated to be around 49 MPa (see the weakening curve of Fig. 1).

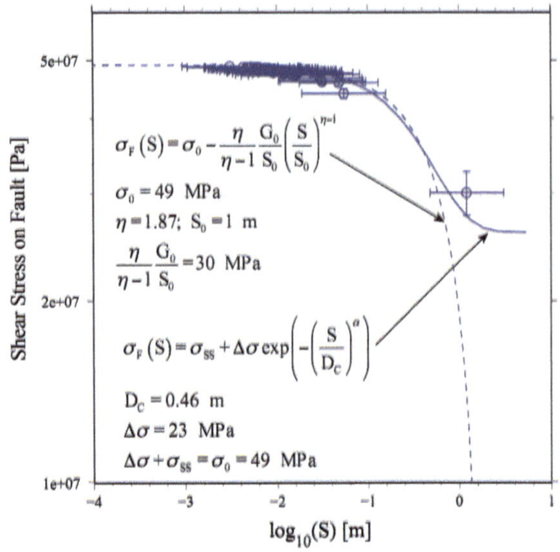

Figure 6
Shear stress on the fault plane estimated for 57 events of the L'Aquila seismic sequence. The *plotted solid curved* indicate the unbounded polynomial weakening (a la Abercrombie and Rice, 2005), and the exponential weakening used for fitting the distribution of fracture energies vs. slip that is plotted in Fig. 5. The stress drop actually estimated for the main event using Eq. (24) is around 23 MPa

4.5. *Apparent Stress vs. Dynamic Stress Drop: Estimating Shear Stress vs. Slip*

Malagnini *et al.* (2010) used a crack model in order to demonstrate that the dynamic stress drop ($\Delta\sigma_d$) of a radially growing circular crack may be quantified as:

$$\Delta\sigma_d = F\sigma_a. \quad (26)$$

In Fig. 6 we used Eqs. (3a) and (26) (obviously: $\sigma_{F_i}(S_i) = \sigma_{0_i} - \Delta\sigma_{d_i}(S_i)$, where i refers to the ith earthquake and S_i is their final slip) in order to plot the calculated average shear stresses of the 57 selected events of the L'Aquila seismic sequence, as a function of their estimated coseismic slips. Two best-fit curves are plotted (L_1-norm): the one relative to the power-law slip weakening, a la Abercrombie and Rice (2005), and the one obtained from the exponential weakening proposed in the present study. For the small events, the two curves, as expected, are equivalent. However, at large slips the exponential weakening curve is characterized by a more acceptable value of asymptotic friction.

The main physical difficulty we have with the power-law weakening is the fact that the stress drop is not bounded. The consequences of such weakening behavior are twofold: (a) unrealistic stress drops at large seismic slips, and (b) overshooting conditions for large earthquakes.

The meaning of point (b) may be explained as follows: for large-enough offsets, the power-law weakening would produce stress drops larger than the specific prestress. As a consequence, we should expect a change in the focal mechanisms of the aftershocks, in the opposite direction with respect to the mainshock.

4.6. *The Northridge Seismic Sequence*

The second data set used in this study is related to the Northridge seismic sequence (Southern California). Its mainshock struck on January 17, 1994 (M_W 6.7, Wald *et al.* 1996), and excited one of the highest ground accelerations ever instrumentally recorded in an urban area in North America (17.8 m/s^2 on the east–west direction at a site in Tarzana, California, roughly 6 km from the epicenter of the main event, Spudich *et al.* 1996). The Northridge seismic sequence is of special interest to us, because both its radiated energies and stress drops are taken from

ABERCROMBIE and RICE (2005). Specifically, the estimates of the seismic energy radiated by the main event were calculated by KANAMORI (2003, cited as personal communication in ABERCROMBIE and RICE, 2005), and by McGARR and FLETCHER (2000). For the aftershocks of the Northridge sequence we use our estimates of radiated energies, and measurements of radiated energy obtained by MORI *et al.* (2003). All energy estimates mentioned so far were used in the study by ABERCROMBIE and RICE (2005).

In Table 1, we show the information obtained for the selected earthquakes of the L'Aquila sequence. In order to validate our results, we calculate the fracture energies of a number of events (including the mainshock) of the Northridge seismic sequence of 1999. The sequence started with the main earthquake, that occurred on January 17, 1994, and was used in ABERCROMBIE and RICE (2005) in order to obtain indications for a global function of dynamic weakening based on Eqs. (19)–(23) (power law weakening). From the visual inspection of Fig. 8 by ABERCROMBIE and RICE (2005), we see that it averages sub-parallel, steeper functional forms of $\log_{10}(G)$ vs. $\log_{10}(S)$, arranged en echelon. Our analysis on an individual distribution of the same nature, calculated over fracture energies and slips of earthquakes from the same mainshock-aftershock seismic sequence, may shed light on details of the weakening function (like a reasonable steady-state friction at large values of slip) that may have been hidden in previous studies on global data sets of fracture energy vs. fault slip.

Our comparison is two-fold: first, we used Eq. (17) and obtained an estimate of the fracture energy of the Northridge mainshock ($G = 1.1 \times 10^7$ J/m^2, see Table 2), in agreement (within a factor of 2–3) with the estimates published in by ABERCROMBIE and RICE (2005) in their Table 5 (between $G = 3.3 \times 10^6$ J/m^2 and $G = 4.7 \times 10^6$ J/m^2).

Figure 7 shows the results of the analysis of fracture energies of the Northridge seismic sequence events. The dashed straight line is calculated over the aftershocks only, whereas the solid curve is obtained by grid-searching the parameters of the functional form (25) on the complete data set. Interestingly, the two curves are totally coincident at small slip values. Unfortunately, no error bars on either energies or slips are provided by ABERCROMBIE and RICE (2005).

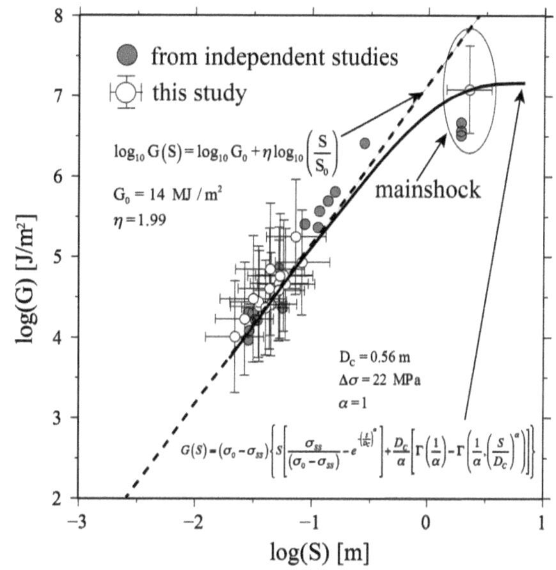

Figure 7

For comparison with the results shown in ABERCROMBIE and RICE (2005), fracture energies (16 aftershocks plus the main event, Table 2) obtained from Eq. (16) are plotted against the same estimates (14 aftershocks in common with the coda-based data set, plus the main event, Table 2), calculated using the parameters published by MORI *et al.* (2003). Data points obtained using Eq. (16) are plotted using the open circles with *error bars*. The main event (the points to the upper right end of the distribution) has four estimates of fracture energy: one with *error bars* (from this study), and the other three are taken using the parameters from the studies cited in Table 5 by ABERCROMBIE and RICE (2005). Both *best-fit curves* are obtained using only our results. For two events, our coda-based estimates do not have a corresponding value in MORI *et al.* (2003)

A weakening curve characterized by an exponential decay with slip may be a better choice for describing the stress history of the Northridge events. With a moderate pore-fluid overpressure ($\lambda_f = 0.65$), and a static friction coefficient $\mu_S = 0.7$, we find a stress drop $\Delta\sigma = 22$ MPa, and an asymptotic value of frictional stress $\sigma_{SS} \approx 203$ MPa. The shear prestress acting at the nucleation depth (14 km) on the Northridge reverse fault (dip angle of 37°) is estimated to be around 225 MPa.

4.7. Calibration of the F Parameter in (26)

The parameter F in Eq. (26) is calibrated separately for each sequence by fitting the stress curves (the power-law weakening and the exponential weakening, that overlap for small values of slip) with the aftershocks' estimates of average dynamic shear stress (Fig. 6). Specifically: $F = 12$ for Northridge, $F = 7$ for L'Aquila.

Using Eqs. (16) and (26):

$$G = H\sigma_a S, \quad (27)$$

where $H = \left(\frac{F}{2} - 1\right)$. For the L'Aquila seismic sequence: $H = 2.5$; for Northridge: $H = 5.0$. It is worth mentioning that Eq. (27) yields exactly the same results as Eq. (17).

It would be interesting to investigate the physical meaning of the constant H (of F), which may be sequence-dependent (or maybe it just depends on the tectonic environment). We here have studied only two cases, but we will provide a thorough analysis in the future.

4.8. *Comparison with Results by Abercrombie and Rice (Northridge)*

In their Fig. 8, Abercrombie and Rice (2005) plotted results from several studies in terms of fracture energy and seismic slip. Specifically, without any model limitations in terms of overshooting/undershooting, their Fig. 8 contains fracture energies and seismic slips of: (i) the Cajon Pass earthquakes; and (ii) the Northridge sequence. They used three different estimates of the radiated energy of the Northridge main event (M_W 6.7): Kanamori (pers. comm., 2003), McGarr and Fletcher (2000), and Mori *et al.* (2003). Radiated energies were estimated following Kanamori *et al.* (1993); those of the Northridge's aftershocks were taken from Mori *et al.* (2003).

Abercrombie and Rice (2005) plotted fracture energies from other major events that struck California: Hector Mine (energies from Venkataraman *et al.* 2002, and Boatwright *et al.* 2002), Landers (energies from Kanamori *et al.* 1993, and Kanamori, pers. comm. 2003), Loma Prieta (energy computed by Abercrombie and Rice 2005), as well as other smaller California shocks like that of the Sierra Madre (1991), or that of the Imperial Valley (1971). These latter energies were calculated by Abercrombie and Rice (2005) themselves, who estimated the radiated energy of the Kobe earthquake (1995, Japan), and of a smaller mainshock from California. Teleseismic estimates of energy from Choy and Boatwright (1995) were taken from the NEIC monthly PDE listings.

From a visual inspection of Fig. 8 by Abercrombie and Rice (2005), we gather that the individual subsets of data points (e.g., from Cajon Pass, or from Northridge) follow specific trends very closely. In the log–log space defined by fracture energy and seismic slip, subparallel lines that are much steeper than the one indicated by Abercrombie and Rice (2005) would give a better fit to the fracture energies of the individual seismic sequences. Cocco and Tinti (2008) found a steepness of around 1.45 for the breakdown work as a function of seismic slip.

5. *Conclusions*

We used source spectral ratios to calculate source parameters of 57 events of the 2009 L'Aquila seismic sequence, and of 17 events of the 1994 Northridge seismic sequence. In turn, these parameters were used to calculate the seismological fracture energy of all the investigated earthquakes. We used a formal relationship between fracture energy and stress parameters in order to estimate the average stress of each investigated earthquake of the sequence. Within the framework given by our working hypotheses, our results may be considered in terms of an event-averaged shear stress (i.e., a weighted average over the slipped portion of the fault plane, where the weight is given by the slip distribution). In light of this interpretation, slip estimates in Figs. 1, 5, 6, and 7 are meant to be surface averages: each one taken over the specific portion of the fault that has slipped during the specific earthquake. Finally, our results are obtained in the hypothesis of the omega-square fall-off of the source spectra $\left(A(f)_{f>>f_c} \propto f^{-2}\right)$.

From a visual inspection of Fig. 6, we see that the shear stress tends to decrease monotonically as a function of seismic slip. In the log–log space defined by fracture energy and seismic slip, the linear fit given by Eq. (21) proposed by Abercrombie and Rice (2005) would lead to an unbounded function of the seismic stress drop on the fault (see Eq. (23), and the related power-law stress history of Fig. 6).

We proposed in this study a different functional form for the fault's shear stress (described by Eq. 24) that asymptotically tends to a steady-state value of the fault's sliding stress for large, coseismic slips (and thus yielding a bounded stress drop for large earthquakes). Using the proposed functional form for stress evolution, we have derived the functional form (25), which links seismic slip and fracture energy if a generic, exponential weakening function is valid.

For the fracture energy data points of the events of the 2009 L'Aquila sequence, the exponent in Eq. (24) is equal to unity ($\alpha = 1$), the stress drop is $\Delta\sigma = 23$ MPa, and the asymptotic value of shear stress is $\sigma_{SS} \approx 26$ MPa (the steady-state value). For this fit, the normalizing weakening distance is $D_C = 0.46$ m, pointing out that it is a normalization factor and not a constitutive parameter. With a moderate pore-fluid overpressure ($\lambda_f = 0.65$), and a static friction coefficient ($\mu_S = 0.7$), the specific shear prestress acting at a 10-km depth on the L'Aquila (Paganica) normal fault (dip angle of 50°) was estimated to be around 49 MPa.

Similar results were obtained for the Northridge seismic sequence, for which our best-fit analysis on the exponential stress history (24) ($\alpha = 1$) yielded the following parameters: $D_C = 0.56$ m, $\Delta\sigma = 22$ MPa, and, with a moderate pore-fluid overpressure ($\lambda_f = 0.65$) and a static friction coefficient $\mu_S = 0.7$, a steady-state frictional stress $\sigma_{SS} \approx 203$ MPa. For the shear prestress acting at the nucleation depth (14 km) on the Northridge reverse fault (dip angle of 37°), we estimated 225 MPa.

We argue that the power-law weakening of Abercrombie and Rice (2005) would cause excessively large stress-drop estimates for large-magnitude earthquakes: a fact that was not clear from their analysis: Abercrombie and Rice (2005) regressed a heterogeneous data set of fracture energies from earthquakes belonging to different seismic sequences and different tectonic environments (from strike-slip and thrust faulting).

The power-law weakening is unable to fit the fracture energy of the main shocks analyzed in this study: not within the error bars for the L'Aquila main event (for which the error bars are provided). For the Northridge mainshock, it is very clear that a power-law weakening function would imply an excessive, physically unacceptable stress drop.

Our preferred functional form is the exponential weakening (23). A general consideration in favor of a weakening function with an asymptotic, steady-state frictional stress at large values of slip is that similar weakening slip histories are also observed in high-velocity lab experiments for large values of slip (e.g., Di Toro *et al.*, 2011).

The unbounded weakening function described by Eqs. (22) and (23) yields very large values of stress drop at large offsets (of the order of hundreds of MPa), and cannot represent the dynamics of large crustal earthquakes. Finally, an implication emerging from the results obtained in this study is the saturation of fracture energy with seismic slip at large slip values, which is a consequence of the exponential term included in Eqs. (24) and (25).

Acknowledgments

K. Mayeda was supported by the Air Force Research Laboratory under Contract No. FA8718-09-C-0014.

References

Abercrombie, R.E., and J.R. Rice (2005). *Can observations of earthquake scaling constrain slip weakening?*, Geophys. J. Int., *162*, 406–424, doi:10.1111/j.1365-246X.2005.02579.x.

Aki, K. (1967). *Scaling law of seismic spectrum*, J. Geophys. Res., *72*, 1217–1231.

Aki, K. and Chouet, B. (1975). *Origin of coda waves: Source, attenuation, and scattering effects.* Journal of Geophysical Research *80*(B23): doi:10.1029/0JGREA000080000B2300332 2000001.

Aki, K., and P.G. Richards (2002). Quantitative seismology 2-nd ed., University Science Books, 700 pp., ISBN 0-935702-96 2.

Akinci, A., Malagnini, L., Herrmann, R.B, Pino, N.A., Scognamiglio, L., Eyidogan, H. (2001). *High-Frequency Ground Motion in the Erzincan Region, Turkey: Inferences from Small Earthquakes*, Bull. Seis. Soc. Am., *91*, 1446–1455.

Akinci, A., L. Malagnini, and F. Sabetta (2010). *Characteristics of the strong ground motions from the 6 April 2009 L'Aquila earthquake, Italy*, Soil Dyn., Earth. Eng. *30*, 320–335, doi:10.1016/j.soildyn.2009.12.006.

Anderson, E.M. (1905). *The dynamics of faulting* Trans. geol. Soc. Edin., *8*, pp. 393–402.

Bizzarri, A., and M. Cocco (2006), *A thermal pressurization model for the spontaneous dynamic rupture propagation on a three-dimensional fault: 1. Methodological approach*, J. Geophys. Res., *111*, B05303, doi:10.1029/2005JB003862.

Bay F., D. Fäh, L. Malagnini and D. Giardini (2003). *Spectral shear-wave ground motion scaling for Switzerland.* Bull. Seism. Soc Am., Vol. *93*, 414–429.

Boatwright, J., G.L. Choy, and L.C. Seekins (2002). *Regional estimates of radiated seismic energy*, Bull. seism. Soc. Am., *92*, 1241–1255.

Bodin, P., L. Malagnini, and A. Akinci (2004). *Ground motion scaling in the Kachchh Basin, India, deduced from aftershocks of the 2001 Mw 7.6 Bhuj earthquake*, Bull. Seism. Soc. Am., *94*, 1658–1669; doi:10.1785/012003202.

Brodsky, E. E. and H. Kanamori (2001). *The elastohydrodynamic lubrication of faults*, J. Geophys. Res., *106*, *16*, 357–16, 374.

Brune, J. N. (1970). *Tectonic stress and the spectra of seismic shear waves from earthquakes*, J. Geophys. Res. *75*, 4997–5009.

Brune, J. N. (1971). *Correction (to Brune (1970))*, J. Geophys. Res. *76*, 5002.

Cartwright, D. E., and M. S. Longuet-Higgins (1956). *The statistical distribution of the maxima of a random function*, Proc. R. Soc. London, Ser. *A237*, 212–232.

Chiarabba, C., *et al.* (2009), *The 2009 L'Aquila (central Italy) MW6.3 earthquake: Main shock and aftershocks*, Geophys. Res. Lett., *36*, L18308, doi:10.1029/2009GL039627.

Choy, G.L., and J.L. Boatwright (1995). *Global patterns for radiated seismic energy and apparent stress*, J. geophys. Res., *100*, 18 205–18 228.

Cocco, M., E. Tinti, C. Marone, A.Piatanesi (2009). *Scaling of Slip Weakening Distance with Final Slip during Dynamic Earthquake Rupture, in: International Geophysics, Fault-Zone Properties and Earthquake Rupture Dynamics*, 94, pp. 163RLIN, doi:10.1016/S0074-6142(08)00007-7.

Cocco M., and Tinti E., (2008). *Mechanical work partitioning during dynamic earthquake propagation, Earth and Planetary Science Letters*, *273*, 123–131.

Cocco, M., P. Spudich, and E. Tinti (2006). *On the mechanical work absorbed on faults during earthquake ruptures*, on: Abercrombie, R., A. McGarr, H. Kanamori, and G. Di Toro (Eds.) (2006), Earthquakes: Radiated Energy and the Physics of Faulting, Geophys. Monogr. Ser., vol. 170, 327 pp., AGU, Washington, D. C., doi:10.1029/GM170.

D'Amico, S., A. Akinci and L. Malagnini (2012). *Prediction of High-Frequency Ground-Motion in Taiwan Based on Weak Motion Data Taiwan*, Geophys. J. Int., doi:10.1111/j.1365-246X.2012.05367.x.

Di Toro, G., R. Han, N. De Paola, S. Nielsen, K. Mizoguchi, F. Ferri, M. Cocco, and T. Shimamoto (2011). *Fault lubrication during earthquakes*, Nature.

Di Toro, G., Hirose, T. Nielsen, S., Pennacchioni, G. and T. Shimamoto (2006). *Natural and experimental evidence of melt lubrication of faults during earthquakes*, Science, 311, 647–649, doi:10.1126/science.1121012.

Di Toro, G., D.L. Goldsby, and T.E. Tullis, *Friction falls towards zero in quartz rock as slip velocity approaches seismic rates*, Nature, 427, 436–439, (2004).

Eshelby, J. D. (1957), *The determination of the elastic field of an ellipsoidal inclusion, and related problems*, Proc. R. Soc. London, Ser. A, *241*, 376–396.

Eshelby, J. D. (1959), *The elastic field outside an ellipsoidal inclusion*, Proc. R. Soc. London, Ser. A, *252*, 561–569, doi:10.1098/rspa.1959.0173.

Famin, V., S. Nakashima, A.M. Boullier, K. Fujimoto and T. Hirono (2008). *Earthquakes produce carbon dioxide in crustal faults, Earth and Planetary Science Letters 265* (3–4), pp. 487–497, doi:10.1016/j.epsl.2007.10.041.

Ferri F., G. Di Toro, T. Hirose, T. Shimamoto (2010). *Evidences of thermal pressurization in high velocity friction experiments on smectite-rich gouges*. Terra Nova, vol. *22*, pp. 347-353.

Han, R., T. Shimamoto, T. Hirose, J-H Ree, and J. Ando, (2007) *Ultralow friction of carbonate faults caused by thermal decomposition, Science*, *316*, 878–881, (2007).

Han, R., T. Hirose, and T. Shimamoto (2009). *Strong velocity weakening and powder lubrication of simulated carbonate faults at seismic slip rates*, J. Geophys. Res., in press.

Herrmann, R.B., L. Malagnini, and I. Munafo' (2011). *Regional moment tensors of the 2009 L'Aquila earthquake sequence*, Bull. Seism. Soc. Am., Vol. *101*, No. 3, doi:10.1785/0120100184.

Hirose, T., and Shimamoto, T. (2005). *Growth of molten zone as a mechanism of slip weakening of simulated faults in gabbro during frictional melting*, J. Geophys. Res. *110*, doi:10.1029/2004JB003207.

Hirose, T., Nielsen, S., Pennacchioni, G. & Shimamoto, T. (2006). *Natural and experimental evidence of melt lubrication of faults during earthquakes*. Science *311*, 647–649.

Hirose, T., and M. Bystricky (2007). *Extreme weakening of faults during dehydration by coseismic shear heating*, Geophys. Res. Lett., *34*, L14311, doi:10.1029/2007GL030049.

Ide, S., and M. Takeo (1997). *Determination of constitutive relations of fault slip based on seismic wave analysis*. J. Geophys. Res. *102*, 27379–27391.

Kanamori, H., J. Mori, E. Hauksson, T.H. Heaton, L.K. Hutton, and L.M. Jones (1993). *Determination of earthquake energy release and ML using TERRAscope*, Bull. Seism. Soc. Am. *83*, 330–346.

Kanamori, H., and Rivera, L. (2004). *Static and dynamic scaling relations for earthquakes and their implications for rupture speed and stress drop*. Bull. Seismol. Soc. Am. *94*, 314–319.

Kostrov, B.V., and S. Das, (1988). Principles of earthquake source mechanics, Cambridge University Press, ISBN 0-521-30345-1, 287 pp.

Malagnini, L., and R.B., Herrmann (2000). *Ground motion scaling in the region of the Umbria-Marche earthquake of 1997*. Bull. Seism. Soc. Am., *90*, 1041–1051.

Malagnini, L., R.B., Herrmann, and M. Di Bona (2000a). *Ground motion scaling in the Apennines (Italy)*, Bull. Seism. Soc. Am., *90*, 1062–1081.

Malagnini, L., R.B. Herrmann, and K. Koch (2000b). *Ground motion scaling in Central Europe*, Bull. Seism. Soc. Am., *90*, 1052–1061.

Malagnini, L., Akinci, A., Herrmann, R. B., Pino, N.A., Scognamiglio, L, (2002). *Characteristics of the ground motion in Northeastern Italy*, Bull. Seism. Soc. Am., *92*, 2186–2204 .

Malagnini, L., K. Mayeda, A. Akinci, and P.L. Bragato (2004). *Estimating absolute site effects*, Bull. Seism. Soc. Am., 94, 1343-1352.

Malagnini, L., P. Bodin, K. Mayeda, and A. Akinci, (2006). *Unbiased moment-rate spectra and absolute site effects in the Kachchh basin, India, from the analysis of the aftershocks of the 2001 Mw 7.6 Bhuj earthquake*, BSSA, v. *96*; no. 2; p. 456-466; doi:10.1785/0120050089.

Malagnini, L., K. Mayeda, R. Uhrhammer, A. Akinci, and R.B. Herrmann (2007). *A regional ground motion excitation/attenuation model for the San Francisco region*, Bull. Seism. Soc. Am., *97*; no. 3; 843–862; doi:10.1785/0120060101.

Malagnini, L., and K. Mayeda (2008). *High-stress strike-slip faults in the Apennines: An example from the 2002 San Giuliano earthquakes (southern Italy)*, Geophys. Res. Lett., *35*, L12302, doi:10.1029/2008GL034024.

Malagnini, L., L. Scognamiglio, A. Mercuri, A. Akinci, and K. Mayeda (2008), *Strong evidence for non-similar earthquake source scaling in central Italy*, Geophys. Res. Lett., *35*, L17303, doi:10.1029/2008GL034310.

Malagnini, L., S. Nielsen, K. Mayeda, and E. Boschi (2010). *Energy radiation from intermediate to large magnitude earthquakes: implications for dynamic fault weakening*, J. Geophys. Res., doi:10.1029/2009JB006786.

Malagnini, L., A. Akinci, K. Mayeda, I. Munafo', R.B. Herrmann, and A. Mercuri (2011). *Characterization of Earthquake-Induced Ground Motion from the L'Aquila Seismic Sequence of 2009, Italy, Geophys.* J. Int., Geophys. J. Int. *184*, 325–337, doi:10.1111/j.1365-246X.2010.04837.x.

Malagnini, L., F. Lucente, P. De Gori, A. Akinci, and I. Munafo' (2012). *Control of Pore-Fluid Pressure Diffusion on Fault Failure Mode: Insights from the 2009 L'Aquila Seismic Sequence*, J. Geophys. Res. *117*, ISSN: 0148-0227, doi:10.1029/.2011JB008911.

Malagnini, L., K. Mayeda, S. Nielsen, S.-H. Yoo, I. Munafo', C. Rawles, and E. Boschi (2014). *Scaling transition in earthquake sources: a possible link between seismic and laboratory measurements*, Pure Appl. Geophys., doi:10.1007/s00024-013-0749-8, this issue.

Mayeda, K., and W.R. Walter (1996). *Moment, energy, stress drop and source spectra of Western U.S. earthquakes from regional coda envelopes*, J. Geophys. Res., *101*, 11,195–11,208.

Mayeda, K., A. Hofstetter, J. L. O'Boyle, and W. R. Walter (2003), *Stable and transportable regional magnitudes based on coda-derived moment-rate spectra*, Bull. Seism. Soc. Am., *93*, 224–239.

Mayeda, K., L. Malagnini, and W.R. Walter (2007), *A new spectral ratio method using narrow band coda envelopes: Evidence for non-self-similarity in the Hector Mine sequence*, Geophys. Res. Lett., *34*, L11303, doi:10.1029/2007GL030041.

Mayeda, K., and L. Malagnini (2009). *An evolution of a new spectral ratio technique for the analysis of source scaling: an application to the Chi–Chi sequence*, Geophys. Res. Lett., *36*, L10308, doi:10.1029/2009GLD37421.

Mayeda, K., and L. Malagnini (2010). *Source Radiation Invariant Property of Local and Near-Regional Shear-Wave Coda: Application to Source Scaling for the Mw 5.9 Wells, Nevada Sequence*, Geophys. Res. Lett., doi:10.1029/2010GL042698.

McGarr, A. & Fletcher, J.B., 2000. *A method for mapping apparent stress and energy radiation applied to the 1994 Northridge earthquake fault zone*, Geophys. Res. Lett., *27*, 1953–1956.

Melosh, H.J (1996). *Dynamical weakening of faults by acoustic fluidization*, Nature *379*, 601–606, doi:10.1038/379601a0.

Meirova, T., R. Hofstetter, Z. Ben-Avraham, D.M. Steinberg, L. Malagnini, and A. Akinci (2008). *Weak-motion-based predictive relationships for the ground motion in Israel*, Geophys. J. Int., doi: 10.1111/j.1365-246X.2008.03953.x.

Morasca, P., Malagnini, L., A. Akinci,, D. Spallarossa, and R. B. Herrmann, (2006). *Ground-Motion Scaling in the Western Alps, Journal of Seismology*, Volume *10*, Number 3, DOI:10.1007/s10950-006-9019-X.

Mori, J., Abercrombie, R.E. & Kanamori, H. (2003). *Stress drop and radiated energies of the Northridge aftershocks*, J. geophys. Res., *108*, 10.1029/200JB000474.

Nielsen, S., G. Di Toro, T. Hirose, and T. Shimamoto (2008), *Frictional melt and seismic slip*, J. Geophys. Res., *113*, B01308, doi:10.1029/2007JB005122.

Nielsen, S., P. Mosca, G. Giberti, G. Di Toro, T. Hirose, and T. Shimamoto (2010). *On the transient behavior of frictional melt during seismic slip*, J. Geophys. Res. *115*, B10301, doi:10.1029/2009JB007020.

Niemeijer, A., C. Marone, and D. Elsworth (2010), *Frictional strength and strain weakening in simulated fault gouge: Competition between geometrical weakening and chemical strengthening*, J. Geophys. Res., *115*, B10207, doi:10.1029/2009JB000838.

Niemeijer, A., G. Di Toro, S. Nielsen, and F. Di Felice (2011), *Frictional melting of gabbro under extreme experimental conditions of normal stress, acceleration, and sliding velocity*, J. Geophys. Res., *116*, B07404, doi:10.1029/2010JB008181.

Noda H., Lapusta N. (2012). *On averaging interface response during dynamic rupture and energy partitioning diagrams for earthquakes*. J. Appl. Mech. 2012;79. doi:10.1115/1.4005964.

Noir, J., E. Jacques, S. Békri, P. M. Adler, P. Tapponnier, and G. C. P. King (1997). *Fluid flow triggered migration of events in the 1989 Dobi Earthquake sequence of central Afar*, Geophys. Res. Lett., *24*, 2335–2338, doi:10.1029/97GL02182.

Ohnaka, M. (2003). *A constitutive scaling law and a unified comprehension for frictional slip failure, shear fracture of intact rock, and earthquake rupture*, J. Geophys. Res., *108*(B2), 2080, doi:10.1029/2000JB000123.

Orowan, E., 1960. Mechanism of Seismic Faulting in Rock Deformation, Geological Society of America Memoir 79, pp. 323–345, Geological Society of America, Boulder, CO.

Raoof, M. R.B. Herrmann, and L. Malagnini (1999). *Excitation of three-component ground motion in Southern California, 89*, 888–902.

Reches, Z., and D.A. Lockner (2010). *Fault weakening and earthquake instability by powder lubrication*, Nature *467*, 452–455, doi:10.1038/nature09348.

Rempel, A. W., and J. R. Rice (2006). *Thermal pressurization and onset of melting in fault zones*, J. Geophys. Res., *111*, B09314, doi:10.1029/2006JB004314.

Rice, J. R. (2006). *Heating and weakening of faults during earthquake slip*, J. Geophys. Res., *111*, doi:10.1029/2005JB004006.

Rice, J. R., and M. Cocco (2006). *Seismic fault rheology and earthquake dynamics, Dahlem Workshop on The Dynamics of Fault Zones*, pp.99-137, edited by M. R. Handy, MIT Press, Cambridge, Mass.

Sato, T., and T. Hirasawa (1973). *Body wave spectra from propagating shear crack*, J. Phys. Earth *21*, 415–431.

Scholz, C.H. (1990). The Mechanics of Earthquakes and Faulting. Cambridge University Press, New York, 1990, 439 pp.

Scognamiglio, L., L. Malagnini, and A. Akinci (2005). *Ground motion scaling in Eastern Sicily*, Bull. Seism. Soc. Am., *95*, 2, 568–578.

Shipton, Z.K., A.M. Soden, J.D. Kirkpatrick, A.M Bright, and R.J. Lunn (2006). *How thick is a fault? Fault displacement-thickness scaling revisited, in: Earthquakes: radiated energy and the physics of faulting*, Geophysical Monograph Series *170*, doi:10.1029/170GM19.

SIBSON, R.H. (1974). *Frictional constraints on thrust, wrench and normal faults*, Nature *249*, 542–54, doi:10.1038/249542a0.

SIBSON, R.H. (1990). *Rupture nucleation on unfavorably oriented faults*, Bull Seism. Soc. Am., *80*, n. 6, pp 1580–1604.

SIBSON, R.H. (1992a). *Implications of fault-valve behaviour for rupture nucleation and recurrence*, Tectonophysics, *211*, pp. 283–293.

SIBSON, R.H. (1992b). *Fault-valve behavior and the hydrostatic-lithostatic fluid pressure interface*, Earth-Science Reviews, *32*, pp. 141-144.

SIBSON, R.H. (1994). *Crustal stress, faulting and fluid flow, Geofluids: Origin, Migration and Evolution of Fluids in Sedimentary Basins*, Parnell, J. (Ed.), Geol. Soc. Special Publication No. 78, pp. 69-84.

SPUDICH, P., M. HELLWEG, and W.H.K. LEE (1996). *Directional topographic site response at Tarzana observed in aftershocks of the 1994 Northridge, California, earthquake: Implications for mainshock motions*, Bull. Seism. Soc. Am., *86*, S193–S208.

SUPPE, J., and L. F. YUE (2008), *Pore-fluid pressures and crustal strength*, in Geophysical Research Abstracts, vol. 10, EGU2008-A-01970, 2008 SRef-ID: 1607–7962/gra/EGU2008-A-01970, EGU General Assembly 2008.

TINTI, E., P. SPUDICH, and M. COCCO, (2005), *Earthquake fracture energy inferred from kinematic rupture models on extended faults*, J. Geophys. Res., *110*, B12303, doi:10.1029/2005JB00364.

URGUHAN, B., A. AKSAN, A. AKINCI, and L. MALAGNINI (2012). *High-frequency ground motion simulations of the 2009 L'Aquila earthquake*, Italy, Bull Seism. Soc. Am., v. *102*, p. 1429–1445, doi:10.1785/0120110060.

VALOROSO, L., L. CHIARALUCE, D. PICCININI, R. DI STEFANO, D. SCHAFF, and F. WALDHAUSER (2013). *Radiography of a normal fault system by 64,000 high-precision earthquake locations: The 2009 L'Aquila (central Italy) case study*, J. Geophys. Res., *118*, 1156–1176, doi:10.1002/grb.50130.

VENKATARAMAN, A., RIVERA, L. & KANAMORI, H., 2002. *Radiated energy from the 16 October 1999 Hector Mine earthquake: regional and teleseismic estimates*, Bull. seism. Soc. Am., *92*, 1256–1265.

WALD, D.J., T.H. HEATON, and K.W. HUDNUT (1996). *The slip history of the 1994 Northridge, California, earthquake determined from strong-motion, teleseismic, GPS, and leveling data (in The Northridge, California earthquake of January 17, 1994)*, Bull. Seism. Soc. Am., *86* (1, Part B Suppl.), 49-70.

WALTER, W. R. and S. R. TAYLOR (2001). *A revised Magnitude and Distance Amplitude Correction (MDAC2) procedure for regional seismic discriminants, Lawrence Livermore National Laboratory UCRL-ID-146882*: http://www.llnl.gov/tid/lof/documents/pdf/240563.pdf

WIBBERLEY, C.A.J., and T. SHIMAMOTO (2005). *Earthquake slip weakening and asperities explained by thermal pressurization*, Nature *436*, 689–692, doi:10.1038/nature03901.

YOMOGIDA, K., and T. NAKATA (1994). *Large slip velocity of the surface rupture associated with the 1990 Luzon earthquake*, Geophys. Res. Lett., *21*, 1799–1802.

YOO, S.-H., and K. MAYEDA (2013). *Validation of non–self-similar source scaling using ground motions from the 2008 Wells, Nevada, earthquake sequence*, Bull. Seism. Soc. Am., *103*, 2508–2519; doi:10.1785/0120120327.

YUAN, F., and V. PRAKASH (2007). *Slip weakening in rocks and analog materials at co-seismic slip rates*, J. Mech. Phys. Solids *56* (2008) 542–560, doi:10.1016/j.jmps.2007.05.007.

(Received May 16, 2013, revised November 13, 2013, accepted November 23, 2013, Published online December 15, 2013)

Pure Appl. Geophys. 171 (2014), 2731–2746
© 2014 Springer Basel
DOI 10.1007/s00024-014-0856-1

Pure and Applied Geophysics

Stress Drops of the 1997–1998 Colfiorito, Central Italy Earthquakes: Hints for a Common Behaviour of Normal Faults in the Apennines

ANTONIO ROVELLI[1] and GIOVANNA CALDERONI[1]

Abstract—Stress drop estimates of moderate-magnitude earthquakes in the Umbria–Marche region, in the northern Apennines, exhibit a large scatter. For the two M_w 5.7 and 6.0 main shocks of 26 September 1997 near Colfiorito, several papers resulted in stress drop estimates of 20 MPa, but values as low as 2–3 MPa were proposed as well. Also for the largest aftershocks ($M_w > 4$), estimates spread from < 1 MPa up to values ten times larger. We have critically revisited methods and data used in the literature. We have specifically faced the trade-off between source and propagation effects, as we believe that it is responsible for a part of the large scatter. To keep this trade-off under control, we have applied a methodology that combines the best fit of both source spectra after Empirical Green's Function (EGF) deconvolution and observed ground motion spectra, finding that the results of the two different data sets converge independently at the same solution. We have used ground motions observed in the Colfiorito basin, where an accelerograph and a co-located seismological broad-band station recorded three clusters of earthquakes in a broad magnitude interval ($1.7 \leq M_w \leq 6.0$). We have found that the mainshock–aftershock sequences result in stress drops of 2–5 MPa at $M_w \geq 5.6$, with an average tendency to decrease at smaller magnitudes where stress drop variability increases. These findings confirm the source scaling recently assessed through Empirical Green's Function deconvolution for another well-monitored seismic sequence of normal-faulting earthquakes, which struck the city of L'Aquila in the central Apennines in April 2009. The similar scaling law of the two areas suggests common mechanisms of stress release for the shallow normal faults in the Apennines. The propensity of smaller earthquakes to increase in variability, with a tendency toward smaller stress drops, may reflect an effect of fault strength heterogeneities for smaller size ruptures.

1. Introduction

Estimates of stress drop ($\Delta\sigma$) show largely scattered values in the literature (e.g., see ABERCROMBIE and RICE 2005). Even for the same earthquakes, estimates by different authors can differ significantly. The 1997–1998 Umbria–Marche seismic sequence is a good example of strong disagreement between different authors. In the literature there are values as low as 1.5 and 1.9 MPa for the two M_w 5.7 and 6.0 main shocks, respectively (ZOLLO *et al.* 1999; CAPUANO *et al.* 2000; KOHRS-SANSORNY *et al.* 2005), and $1 < \Delta\sigma < 3$ MPa in the magnitude range $4.6 \leq M_L \leq 5.7$ (BINDI *et al.* 2004). In strict contrast, other authors used values as large as 20 MPa to fit amplitudes of the M_w 6.0 main shock (CASTRO *et al.* 2001) and the entire seismic sequence (MALAGNINI and HERRMANN 2000; CULTRERA *et al.* 2003). Moreover, MALAGNINI *et al.* (2008) applied an Empirical Green's Function (EGF) deconvolution procedure to coda envelopes of broad-band seismograms and found a source scaling with significant deviation from self-similarity, with stress drops increasing versus magnitude up to 20 MPa for the strongest shocks. In contrast, the trend inferred by BINDI *et al.* (2004) through a Generalized Inversion Technique (GIT) applied to strong-motion accelerograms is substantially flat.

In the present study, we analyse data that were never used in the previous papers dealing with source scaling. The goal is to yield further robust stress drop estimates of the 1997–1998 Umbria–Marche earthquakes, extending the analysis to magnitudes as small as 2. The seismic sequence was characterized by the rupture of several faults in a 50×30 km^2 area in the northern Apennines. The damaging earthquakes began in September 1997 and continued up to April 1998 (AMATO *et al.* 1998). The largest shocks of 26 September 1997 at 00:33 and 09:40 ruptured two shallow normal faults a few kilometers to the southeast and northwest of the

[1] Istituto Nazionale di Geofisica e Vulcanologia, Seismology and Tectonophysics, Via di Vigna Murata 605, 00143 Rome, Italy. E-mail: giovanna.calderoni@ingv.it

village of Colfiorito, respectively (Fig. 1). According to the DISS WORKING GROUP (2010), the rupture fault planes had an extent of 8 × 14 and 6 × 9 km^2, respectively, and a 40° dip angle; ruptures nucleated at a depth of about 5 km. On 14 October 1997 a M_w 5.6 earthquake triggered a new seismic sequence on a similarly oriented fault near the village of Sellano, about 25 km southeast of Colfiorito (Fig. 1), also showing a very similar source depth and dip and rake angles (EKSTRÖM *et al.* 1998). The seismic activity continued in the region with many other moderate-magnitude shocks, culminating with the M_w 5.6 shock near Gualdo Tadino and a new seismic sequence in April 1998 following this earthquake. During all of these shocks, a strong motion analog accelerograph (CLF) installed on the Holocene alluvial deposits of the Colfiorito plain was repeatedly activated by earthquakes exceeding its trigger threshold (it was 0.01 g on the vertical component). The local response of the basin around this station was investigated by using a local array of seismometers deployed 100 m from the accelerograph (ROVELLI *et al.* 2001) and by comparing ambient noise results with in situ down-hole measurements (DI GIULIO *et al.* 2006). Between 24 February and 19 March 1998, five broadband seismometers were installed in the Colfiorito area to study the basin resonance using aftershock waveforms (DI GIULIO *et al.* 2003). One (CLF1) of these stations was deployed about 100 m from the accelerograph, and another (CLF2) on a nearby rock edge about 600 m east of CLF (Fig. 1). These seismometers worked in continuous modality and recorded hundreds of small aftershocks with amplitudes well below the accelerograph threshold (DI GIULIO *et al.* 2002). Many of them occurred on (or close to) the fault planes ruptured by the main shocks of 26 September and 14 October 1997. The data set composed of the strong-motion accelerograms of CLF and weak-motion seismograms of aftershocks recorded at CLF1 offers a unique opportunity for a source scaling study over a broad magnitude range ($1.7 \leq M_w \leq 6.0$) at a well investigated site. The paper by ROVELLI *et al.* (2001) showed a substantial stability of the H/V spectral at the scale of hundreds of meters ratios around the accelerograph, guaranteeing that local response is not affected by strong lateral variations, and therefore the accelerograph CLF and the nearby broadband seismometer CLF1 can be used as co-located stations. We first apply the single-station spectral ratio method to infer the source scaling over a broad magnitude range (as already done by CALDERONI *et al.* 2012, 2013). This approach removes the entire-path and near-site propagation terms by deconvolving the Green's function response from the records of large shocks through smaller magnitude aftershocks, with the advantage of not assuming any propagation model and site term. However, its efficiency is guaranteed only when focal mechanism and propagation are the same, a condition that is not strictly verified at close distance. Therefore, we checked the consistency of the EGF-based source scaling with observed Fourier amplitudes when propagation and site terms are modelled on the basis of the regional crustal velocity (CHIARABBA and AMATO 2003) and local site response (DI GIULIO *et al.* 2002, 2006). Through the best fit of both spectral ratios and absolute ground acceleration spectra, we find values of 2–5 MPa for the largest shocks, confirming, within a factor of 2, the estimates by BINDI *et al.* (2004), but contradicting the higher stress drops found by other authors. We ascribe a part of this discrepancy to an overestimation due to the lack of the crust transfer function in the stress drop computation of some of the previously mentioned papers (e.g., ROVELLI *et al.* 1988; CULTRERA *et al.* 2003). Our estimates are in line with the largely accepted expectation of stress drop around 5 MPa for middle-size to large-size earthquakes at a global scale (ALLMANN and SHEARER 2009). Moreover, we demonstrate that, at moderate magnitudes, our stress drops also result in a satisfactory fit of observed ground motion peak values of the 1997–1998 seismic sequence over a broad range of distances. The low magnitudes ($M_W \approx 2$) used in this study, and which thus far had never been reached in source scaling studies of the Colfiorito earthquakes, are crucial to further constrain the stress drop trend over an extended magnitude range. We find that stress drops tend to decrease at smaller magnitudes, approximately by one decade over three decades of seismic moment, in fairly good agreement with the variation estimated for Colfiorito

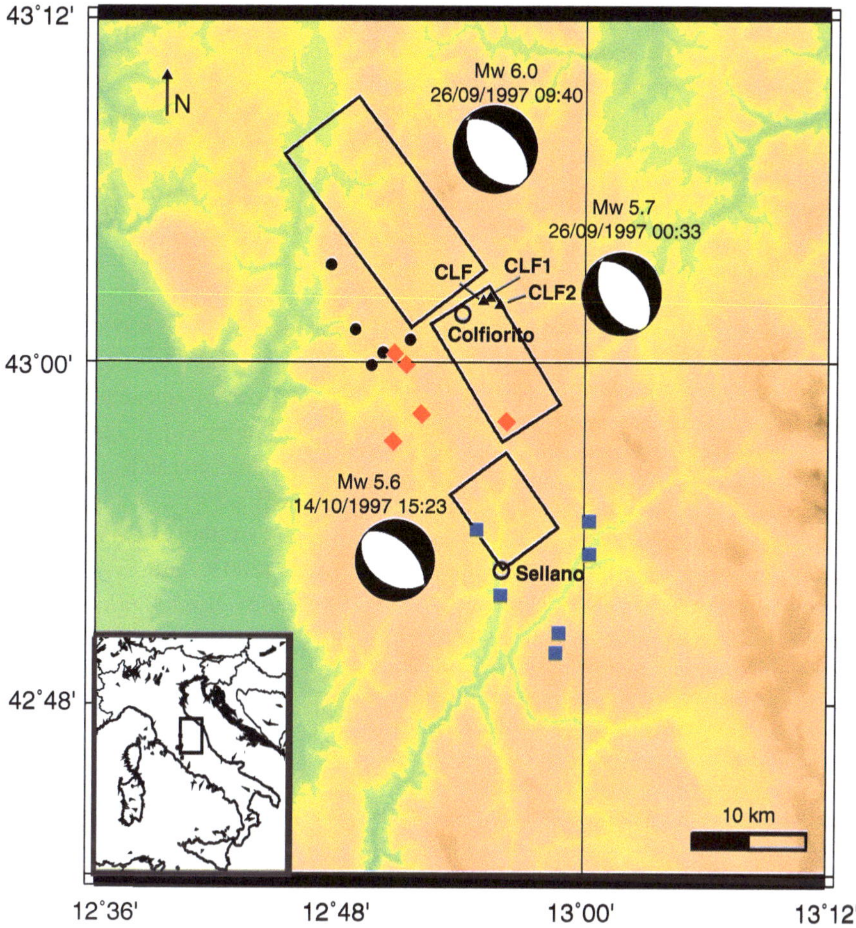

Figure 1
Map of the study area. The three *rectangles* (from northwest to southeast) represent the faults ruptured in September and October 1997 by the M_W 6.0, 5.7, and 5.6 main shocks (#2, 1, and 3, respectively, in Table 1). *Triangles* are recording stations used in this study. *Black*, *red*, and *blue dots* are the aftershocks used as empirical Green's functions (see Table 1), with different symbols for each fault. The fault *boxes* are drawn according to DISS Working Group (2010); focal mechanisms are from Ekström *et al.* (1998)

earthquakes by Malagnini *et al.* (2008). Similar values of stress drops at moderate magnitudes and a similar trend versus earthquake size were recently found for the 2009 L'Aquila seismic sequence (Malagnini *et al.* 2011; Calderoni *et al.* 2013), also generated by shallow normal faults in the Apennines, about 250 km southeast of Colfiorito. The consistent behaviour of the two seismic sequences suggests a similar strength in the crust and similar style of stress release for this type of faults in the Apennines.

An interpretation of this tendency of stress drop to decrease for smaller magnitudes and the resulting increase of variability at low magnitudes is proposed in terms of the larger sensitivity of small ruptures to fault strength heterogeneities.

2. A Critical Review of Previous Studies

The common assumption of papers dealing with models of ground motion is that Fourier amplitude spectra can be written as

$$\Omega(f) = \frac{A_0(f)}{(2\pi f)^2} \cdot P(f,R) \cdot H(f) \qquad (1)$$

where observed ground motion is the convolution of source, propagation in the crust, and near-site terms

($A_0(f)$, $P(f,R)$, and $H(f)$, respectively). According to the Brune (1970, 1971) model, the source radiated acceleration in point-source approximation is expressed as

$$A_0(f) = \frac{F_S R_{\vartheta,\phi}}{4\pi\rho\beta^3} \frac{(2\pi f)^2 M_0}{1 + (f/f_0)^2} \quad (2)$$

where F_S is a free-surface coefficient, $R_{\vartheta,\phi}$ is the station radiation pattern, ρ is the density and β the shear-wave velocity in the source volume (see also Boore 1983). M_0 and f_0 are seismic moment and corner frequency, respectively. For a circular crack with a uniform stress drop in an infinite Poisson solid, the stress drop is linked to M_0 and f_0 (Eshelby 1957; Keilis-Borok 1959; Brune 1970) through the relations

$$\Delta\sigma = \frac{7}{16} \frac{M_0}{r^3} \quad (3)$$

and

$$r = \frac{2.34\beta}{2\pi f_0} \quad (4)$$

In Eq. (1), the propagation term $P(f, R)$ represents the combined effect of geometrical spreading and crustal transfer function. The former is usually written as an inverse power of distance ($1/R^v$, where v varies for body and surface waves), the latter is controlled by the velocity and Q structure in the crust. As remarked by Calderoni *et al.* (2012), the effect of the impedance variation in the crust, first proposed by Boore (1983, 1986), is not properly taken into account when propagation is modeled only through exponential terms as a function of Q and R. In many of the above quoted papers dealing with the stress drop of the Umbria–Marche earthquakes, a correction for the crust transfer function was not applied. Some authors (e.g., Rovelli *et al.* 1988, 1991; Cultrera *et al.* 2003) used (1) and (2), assuming a model where the S-wave velocity maintained a constant value β from the source volume up to the surface simply modeling seismic attenuation through exponential terms controlled by Q and R. This assumption of a half-space oversimplification may cause an overestimation of stress drop, since it neglects the amplitude increase due to the impedance variation in the crust where shear-wave velocity decreases significantly when seismic waves approach to the surface. The amount of this overestimation is shown in Fig. 2c, where we plot the SH-wave crustal transfer function in the Umbria–Marche epicentral area as inferred from the 1-D velocity profile estimated by Chiarabba and Amato (2003) after a tomography inversion of P-wave (α) travel times (with $\alpha/\beta = 1.89$, according to their paper). The crustal transfer function shown in Fig. 2c is computed using the code *NRATTLE* (available in Boore 2005) with the assumption $Q = \infty$. For the sake of simplicity, the computation is made for vertical incidence since the result for incident SH waves is not very sensitive to the incidence angle. Note that in Fig. 2, the uppermost layer has been modified by replacing the uniform velocity $\beta = 2.1$ km/s of the original model by Chiarabba and Amato (2003) with a constant gradient layer where β varies from 2.6 km/s at the 2-km deep interface to the value of 1.2 km/s at the surface. The latter value derives from in situ down-hole measurements on the limestone outcrop where the reference station CLF2 was installed. The same value was also found at the bedrock interface beneath the soft sediments in the central part of the basin, about 500 m from CLF (Di Giulio *et al.* 2006). Thus, our final 1D crustal velocity model has, in its uppermost layer, the real value of the S-wave velocity of outcropping limestone near CLF, maintaining the travel times of the original model by Chiarabba and Amato (2003) unchanged after the linear gradient modification.

Figure 2c suggests that papers where source parameters were derived from (1) and (2) using a half-space approximation could suffer an overestimation of $\Delta\sigma$ by a factor of 2, approximately. The factor-2 amplification affects the high-frequency ($f > 0.7$ Hz) part of the spectrum that is controlled by the stress drop, whereas the effect of the vertical velocity profile is ≈ 1 at the lower frequencies where seismic moment is usually computed. Therefore, although small, stress drop overestimation is systematic when high-frequency amplitudes are not corrected for the crustal transfer function effect. Several papers using the half-space approximation (e.g., Rovelli *et al.* 1988; Cultrera *et al.* 2003) needed 20 MPa to fit the observed amplitudes of moderate magnitude earthquakes of that region.

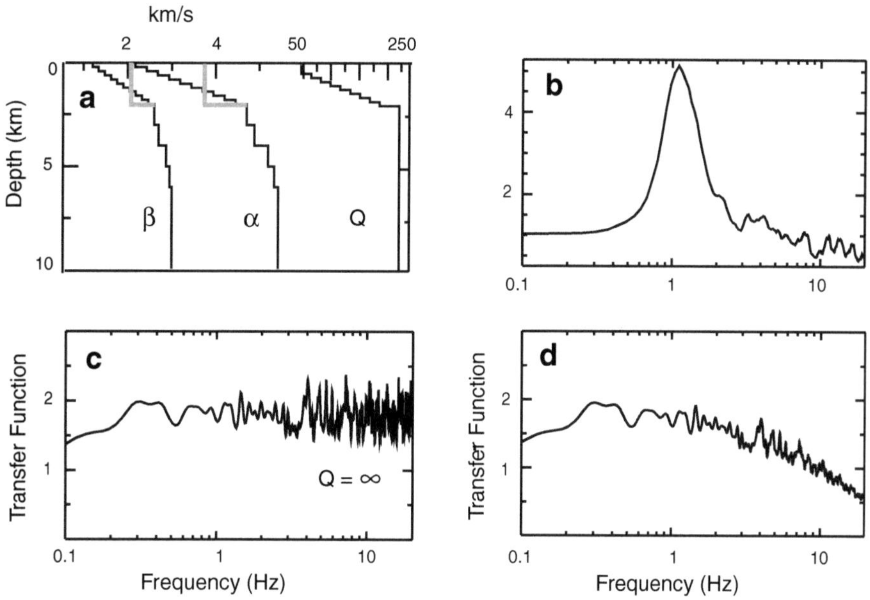

Figure 2
a 1-D velocity profile in the crust of northern Apennines as inferred by Chiarabba and Amato (2003) from tomography inversion. The original 2-km thick uppermost layer with constant velocity (*gray line*) is changed with a more realistic constant gradient layer (*black*) leaving travel times unchanged. The *Q* structure mimics the vertical velocity profiles. **b** Empirical site transfer function *H*(*f*) as estimated by Di Giulio *et al.* (2002) through spectral ratios between CLF1 (on soft sediments) and CLF2 (on limestone) (see Fig. 1). **c** Transfer function of SH waves derived for the 1-D model (**a**) with $Q = \infty$ using the code *NRATTLE* (Boore 2005). The crust structure causes an amplification effect by a factor of two, approximately, that can trade off with stress drop estimates. **d** Crustal transfer function CTF(*f*) as computed when the *Q* vertical profile of (**a**) is used

Figure 2 suggests that, if we take the crust velocity variation into account, a stress drop of 10 MPa would be more appropriate in decreasing the gap between different author estimates, this value being in the middle between the ones of Bindi *et al.* (2004) and Malagnini *et al.* (2008).

The present analysis is now performed using ground motion waveforms recorded in the Colfiorito basin (CLF and CLF1 stations), thus extending the magnitude interval ($1.7 \le M_w \le 6.0$) and offering the opportunity for further comparisons.

3. The Analysis Method

In this study, the stress drops of earthquakes of the 1997–1998 Umbria–Marche seismic sequence are computed in a two-step analysis. In the first step, spectral ratios between strong and weak motions recorded at CLF and CLF1, respectively, are used to deconvolve observed spectra from propagation and site terms and assess the source scaling following a well consolidated approach (Boore and Atkinson 1989; Rovelli *et al.* 1991; Calderoni *et al.* 2012, 2013). When two earthquakes having different magnitude M_1 and M_2 are recorded at the same station, the ratio between their spectra

$$\frac{\Omega_{M_1}(f)}{\Omega_{M_2}(f)} = \frac{A_{0,M_1}(f)P(f,R)H(f)}{A_{0,M_2}(f)P(f,R)H(f)} = \frac{M_{0,M_1}}{M_{0,M_2}} \cdot \frac{1+(f/f_{0,M_2})^2}{1+(f/f_{0,M_1})^2} \tag{5}$$

reduces to the ratio between the two source terms based on the assumption that crustal propagation is similar and the site term is the same. The method details are described by Calderoni *et al.* (2013), who demonstrated how the variation trend of $\Delta\sigma$ versus M_0 can be inferred through a best fit of the empirical spectral ratios over an extended magnitude range under the assumption of a power law such as

$$\log \Delta\sigma = p \log M_0 + q. \tag{6}$$

The misfit is given as

$$\varepsilon = \frac{1}{M}\sum_{m=1}^{M}\sqrt{\frac{\int_{f_{\min}}^{f_{\max}}(\log \mathrm{Obs}_m(f) - \log \mathrm{Theo}_m(f))^2 df}{(f_{\max} - f_{\min})}} \quad (7)$$

where $\mathrm{Obs}_m(f)$ are the empirical spectral ratios calculated from couples of events with different magnitudes and $\mathrm{Theo}_m(f)$ are the corresponding mathematical functions (5) that are written in terms of M_0, p, and q. The couple p and q that minimize (7) define the source scaling by means of (6). In (7), M is the number of the spectral ratios curves used in the fit, and $f_{\min}$ and $f_{\max}$ are the lower and upper bounds of the frequency band of analysis (0.5 and 15 Hz, respectively). Further details can be found in Calderoni *et al.* (2013), where the authors describe how method uncertainties can be assessed and noise contamination is maintained under control in this approach.

In the second step of analysis, again we use Eqs. (6) and (7) to search for the optimal values of p and q that fit the data behaviour. In this case, $\mathrm{Obs}_m(f)$ and $\mathrm{Theo}_m(f)$ are, respectively, the Fourier amplitude spectra of recorded ground acceleration and the theoretical curves of the omega-square source model (2) corrected for propagation according to (1). As first noted by Boore and Atkinson (1989) and recently exploited by Calderoni *et al.* (2012), only the simultaneous analysis of both ground motion amplitudes and spectral ratios can prevent errors deriving from an inadequate propagation model or not optimal conditions for the EGF application (propagation and focal mechanism of the target event and the aftershock must be the same). In this study, the use of both spectral ratios and Fourier amplitude spectra allows a robust consistency check between the results of the first and the second step of analysis, thus providing a double, independent confirmation for the inferred source scaling and propagation/site models. Such a check is particularly important when, like in this case, the conditions for the applicability of the EGF deconvolution are not strictly verified. One more difficulty in this study is represented by the small distance of stations from the source. In these conditions, the far-field assumption is valid only for high frequency ground motions. In the following, we will discuss these critical features in greater detail.

4. Data

The analysis is applied to strong and weak motion waveforms recorded in the Colfiorito plain, where a SMA-1 accelerometric station (CLF) belonging to the Rete Accelerometrica Nazionale (RAN, owned by Civil Defense Department of Italy) was triggered by the strongest earthquakes of the Umbria–Marche seismic sequence (Fig. 1). Table 1 lists the events used in this study. For the three strongest shocks (events #1, 2, and 3), a digitized version of the accelerograms is available (http://itaca.mi.ingv.it). The other events of Table 1 were recorded by the broadband station CLF1 installed by Di Giulio *et al.* (2003) about 100 m from the accelerometer of CLF. This station was instrumented with a three-axial Guralp CMG40T connected to a Reftek 72A07 digital recorder, and operated from 24 February to 19 March 1998. The recorded aftershocks show a satisfactory (>3) signal-to-noise ratio in the frequency band 0.5–15 Hz for magnitudes as low as ≈ 2. Aftershocks of Table 1 were selected as falling in the fault planes (according to DISS Working Group 2010) of the three main shocks, or as close as possible to them.

Ground motion waveforms were high-pass ($f > 0.3$ Hz) filtered and Fourier amplitude spectra of the two horizontal components were computed. We used a time window beginning 2–6 s before the shear wave arrival and including the most significant part of the seismograms with a duration of 10–30 s (time window increases with event magnitude). A 10 % cosine taper was applied. The squares of the two horizontal motion spectra were summed and the square-root was computed. The final result was smoothed with a running 0.5-Hz wide triangular function. The SAC software was used (Goldstein *et al.* 2003).

Seismic moments of aftershocks were computed from seismograms, whereas M_0 was taken from Ekström *et al.* (1998) for the three strongest shocks (Table 1). The Equations

$$\Omega_0 = \lim \Omega(f)_{f\to 0} \quad (8)$$

and

$$\Omega(f) = \frac{M_0 F_S < R_{\vartheta,\phi} >}{4\pi\rho\beta^3} \cdot P(f,R) \cdot H(f) \quad (8')$$

Table 1

Earthquakes analysed in this study (epicentral coordinates are taken from INGV bulletins; seismic moments of the three main shocks were computed by EKSTRÖM *et al. 1998)*

Event number	Time UTC		Lat (°)	Lon (°)	M_0 (N–m)	M_w	$\Delta\sigma$ (MPa)
1	26/09/1997	00:33	43.002	12.888	4.00×10^{17}	5.70	3.68
2	26/09/1997	09:40	43.030	12.862	1.20×10^{18}	6.00	4.91
3	14/10/1997	15:23	42.916	12.924	3.42×10^{17}	5.60	2.17
4	24/02/1998	12:32	42.835	13.028	1.12×10^{14}	3.30	0.43
5	24/02/1998	12:41	42.831	13.026	2.58×10^{14}	3.54	0.16
6	26/02/1998	14:09	43.018	12.824	2.32×10^{12}	2.18	0.05
7	26/02/1998	04:54	42.860	13.034	2.33×10^{13}	2.85	0.08
8	27/02/1998	05:32	42.949	12.924	4.11×10^{13}	3.01	0.15
9	27/02/1998	18:48	43.006	12.825	5.22×10^{12}	2.41	0.75
10	27/02/1998	22:11	42.887	13.005	3.12×10^{12}	2.27	0.02
11	27/02/1998	23:23	43.387	12.525	4.81×10^{11}	1.72	0.03
12	03/03/1998	02:32	43.017	12.821	1.49×10^{12}	2.05	0.03
13	04/03/1998	09:12	43.021	12.820	6.80×10^{12}	2.49	0.13
14	06/03/1998	03:33	42.863	12.936	1.08×10^{12}	1.96	0.04
15	07/03/1998	09:23	43.046	12.812	6.95×10^{12}	2.50	0.94
16	08/03/1998	01:26	43.833	13.026	1.28×10^{12}	2.01	0.03
17	08/03/1998	09:42	42.994	12.863	1.21×10^{12}	1.99	0.15
18	08/03/1998	13:14	42.983	12.866	1.75×10^{12}	2.10	0.56
19	09/03/1998	01:48	42.961	12.866	5.05×10^{11}	1.74	0.05

were applied to estimate M_0 of aftershocks from the low-frequency spectral plateau (Ω_0) of horizontal ground displacement, after correcting $\Omega(f)$ for propagation and site terms ($P(f, R)$ and $H(f)$, respectively). $P(f, R)$was computed as

$$P(f,R) = G(R)\mathrm{CTF}(f) \tag{9}$$

where $G(R)$ a is geometrical spreading function and CTF(f) is the crustal transfer function (as shown in Fig. 2d) that represents the effect of the vertical velocity and Q structure. A body-wave geometrical spreading ($1/R$) was assumed. The transfer function CTF(f) is computed according to vertical profiles of Fig. 2a where the velocity model is taken from CHIARABBA and AMATO (2003), and the Q structure mimics the velocity model with a constant gradient in the uppermost layer ($0 < z < 2$ km) and a constant Q for $z \geq 2$ km. The adopted Q model is thus defined by two parameters: $Q(z = 0) = 50$ and $Q(z = 2$ km$) = 250$, they were optimized to fit the observed peak ground acceleration (PGA) and velocity (PGV) versus distance of other authors (BINDI *et al.* 2006; MALAGNINI *et al.* 2008). This large-distance constraint derived from other data of different authors prevents the circularity that is typical when a unique data set is used to fix some of the model parameters (attenuation) and fit others (source).

In order to avoid complications due to the amplified site response of CLF around 1 Hz (see Fig. 2), for the computation of M_0 we used the aftershock seismograms recorded at the rock reference station CLF2, where the site term $H(f)$ can be approximated to 1 when applying Eq. (8′), since the adopted velocity model (Fig. 2) uses the in situ measured shear velocity at CLF2. The low-frequency plateau was estimated in the frequency band 0.5–2 Hz; in general, its uncertainty is <10 % due to the displacement spectrum fluctuations in its flat part. At frequencies of 0.5–2 Hz, the correction of the attenuation model has a very small role: we have checked that variations of 100 % in the Q values cause a variation of no more than 20 % in seismic moment estimates. The major source of error is the lack of a radiation pattern correction, the focal mechanisms of small events being unknown. However, we have adopted the root-mean-square value $<R_{\vartheta,\phi}>$ of S waves on the basis of its better performance in the used frequency band (according to CASTRO *et al.* 2006) as due to the predominant role of multipathing and scattering around 1 Hz, which

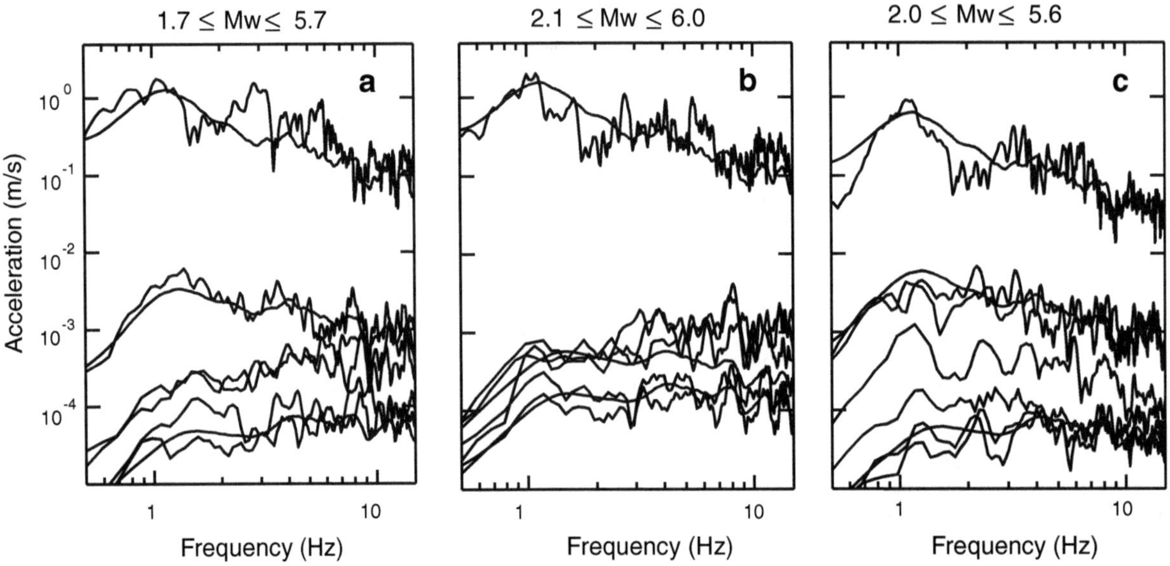

Figure 3
Fourier amplitude spectra of ground acceleration recorded at the co-located stations of CLF and CLF1 (see Fig. 1). The former (SMA1 analog accelerograph) recorded the main shocks (events #1, 2, and 3, from *left* to *right*); the latter was equipped with broad-band CMG40T seismometer and recorded the aftershocks. The overimposed theoretical curves are the omega-square spectra of Eqs. 1 and 2, scaled according to the best fit solution (14) and corrected for the propagation and site terms ($P(f, R)$ and $H(f)$, respectively) as defined in (8′)

mask azimuthal source radiation variations. BOORE and BOATWRIGHT (1984) provided the variability range of $<R_{\vartheta,\phi}>$ (a factor of 1.60): we have summed it to uncertainties deriving from the propagation model and the plateau variability, yielding a total of 1.90 that is used to assess statistical errors of M_0. Moment-magnitudes (M_w) were computed from M_0 following HANKS and KANAMORI (1979), M_0 and M_w of the selected aftershocks are listed in Table 1.

Figure 3 shows the observed horizontal ground acceleration spectra of the selected events; they are plotted separately for the three different faults investigated in this study. A source distance of 15, 10, and 18 km was used for the three main shocks (#2, 1, and 3, respectively). As mentioned above, for the largest magnitudes, the far-field assumption is not applicable at low frequencies. However, we see in Fig. 3 that a value of R can be found for each main shock, such that the observed spectra obey a far-field Brune spectral model over a broad frequency band. Strictly speaking, when R is of the order of 10 km, the far-field model is applicable only at frequencies $f > 1$ Hz. However, Fig. 1 of the paper by ICHINOSE *et al.* (2000) suggests that, for $R = 10$ km, the far-field term exceeds the near-field term above $f = 0.5$ Hz for crustal velocities comparable to those of Fig. 2. A frequency of 0.5 Hz is the lower limit of the frequency band of analysis in this study. In the following, the adequacy of far field theoretical spectra in the investigated frequency band is confirmed a posteriori by the quality of fit.

In Fig. 3, amplitudes of aftershocks are scaled to the same reference distance of the main shock of each fault (15, 10, and 18 km for events #2, 1, and 3, respectively). This distance correction was based on S–P delays instead of bulletin epicentral distances to get a higher precision.

5. *Analysis and results*

Spectral ratios of CLF were computed by dividing the spectrum of the main shock of each fault by the spectrum of the aftershocks of the same fault, all of them having been scaled to the same distance for each fault. Figure 4 shows the trend of the spectral ratios at different magnitudes: they display a regularly scaled behaviour in the spanned magnitude range.

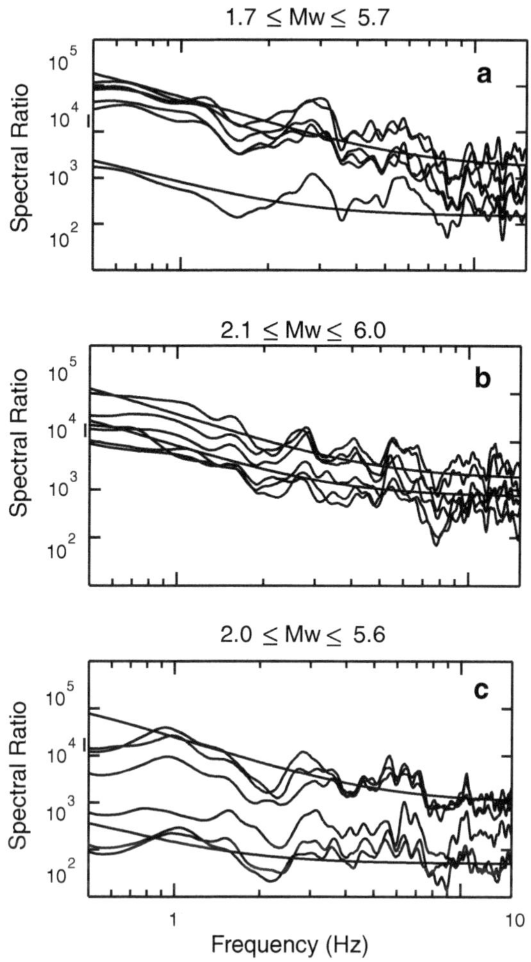

Figure 4
Spectral ratios (5) between main shock and aftershocks of the three faults of Fig. 1, using the Fourier amplitude spectra of Fig. 3. The overimposed theoretical curves correspond to the best fit solution (14)

We then used a trial-and-error approach to search for the couple (p, q) that minimizes the variance of misfit for spectral ratios and observed acceleration spectra according to (5) and (6). The grid search was performed in the intervals

$$0.00 \leq p \leq 0.60 \quad -10.0 \leq q \leq 0.0 \qquad (10)$$

at discretized steps incremented by $\Delta p = 0.01$ and $\Delta q = 0.1$. Using spectral ratios of the three faults, the least-mean-square solutions are

$$\begin{aligned} p &= 0.30 \\ q &= -5.0 \end{aligned} \qquad (11)$$

for the cluster of the M_w 6.0 rupture,

$$\begin{aligned} p &= 0.28 \\ q &= -4.2 \end{aligned} \qquad (12)$$

for the cluster of the M_w 5.7 rupture, and

$$\begin{aligned} p &= 0.35 \\ q &= -6.0 \end{aligned} \qquad (13)$$

for the cluster of the M_w 5.6 rupture. A consistent behaviour of the three faults emerges from (11), (12) and (13), and this suggests to merge all the data together; then, the resulting best-fitting parameters are

$$\begin{aligned} p &= 0.32 \\ q &= -5.3 \end{aligned} \qquad (14)$$

that fit quite satisfactorily the spectral scaling of the individual clusters, as shown in Fig. 4. Figure 5 compares the individual solutions (11), (12), (13) with the one of the merged data set (14). In this approach, the statistical uncertainty estimated through a bootstrap technique is of the order of 10 and 15 % for p and q, respectively (Calderoni *et al.* 2013). These authors found that the method uncertainty due to bandwidth limitation can increase the relative error of these parameters up to 40 % in terms of $\Delta\sigma$. To be more conservative, we apply in Fig. 5 the uncertainty band derived from this relative error in p and q. The figure shows that the individual best fit solutions (11), (12) and (13) are similar, within the uncertainty band, to the solution (14) obtained from the best-fit of the merged data set. Their equivalence is confirmed by the misfit values ε of the cost functions (7) computed for individual earthquakes. Figure 6 compares the misfit computed for the individual solutions (11), (12) and (13) with the misfit resulting from the best-fit solution (14) obtained from the entire data set: a unique scaling law is as valid as the individual ones, the improvement being negligible in terms of residuals. A T test applied to the misfit values of the individual solutions and the general one results in an equivalence probability of 0.91 to a confidence level of 95 %. However, all of the solutions unequivocally indicate a tendency toward smaller stress drop values when magnitude decreases. It is worthy to note that the parameter p as defined in (6) is equivalent to the parameter ψ used by other authors to quantify the degree of deviation from self-

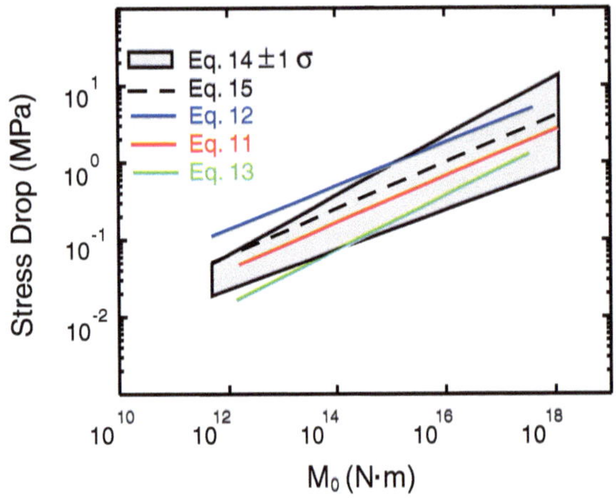

Figure 5
Best fit solutions of Eq. 6 for the three clusters (Eqs. 11, 12, and 13) as found using spectral ratios. The *shadow band* represents the best fit solution (14) of the merged data set (spectral ratios) with its statistical uncertainty; the *dashed line* correspond to the best fit solution (15) as found using Fourier amplitude spectra

similarity (MAYEDA *et al.* 2007; MALAGNINI *et al.* 2008). The value of ψ estimated for Colfiorito earthquakes by MALAGNINI *et al.* (2008) was 0.36, which differs only by a 10 % from $p = 0.32$ assessed in this study (Eq. 14).

Equation (14), which is derived from spectral ratios, yields a satisfactory fit of observed acceleration spectra of CLF and CLF1 when the propagation model $P(f, R)$ and the site term $H(f)$ are applied (Fig. 3). We have repeated the best fit (7) using the Fourier spectra of ground motion. In this case, $\mathrm{Obs}_m(f)$ and $\mathrm{Theo}_m(f)$ (both plotted in Fig. 3) are, respectively, the amplitude spectra of recorded accelerations and the theoretical curves of the omega-square source model corrected for propagation according to (8′). The grid search was made over all the earthquakes of Table 1, with the same intervals of p and q as in (10) and the same grid. The least-mean-square solution for ground acceleration Fourier amplitudes is found when

$$\begin{aligned} p &= 0.30 \\ q &= -4.8 \end{aligned} \qquad (15)$$

confirming the trend of spectral ratios (14) (Fig. 5).

The minimum misfit solutions were also sought for constant stress drop models by fixing $p = 0$ in Eq. 6 and using a denser grid for q. In this case, the value of misfit increases by more than 100 % for

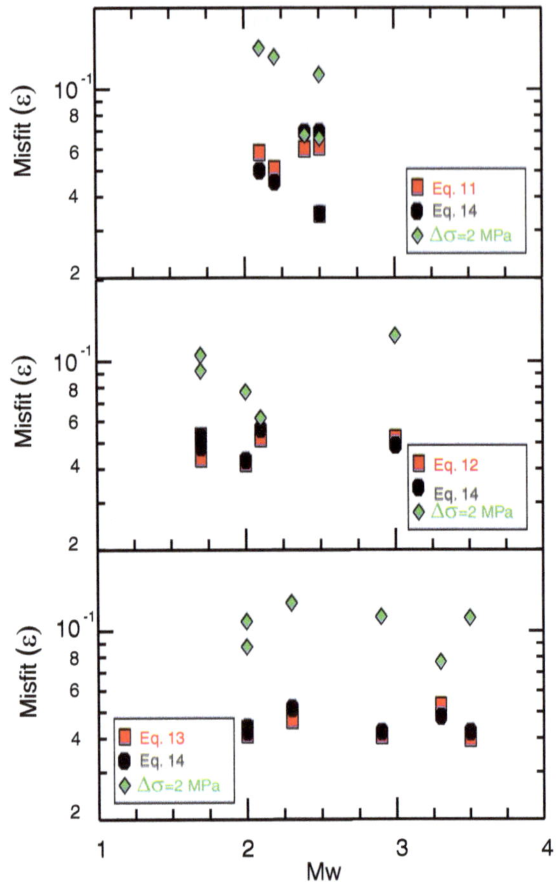

Figure 6
Misfit of individual earthquakes resulting from the cost functions (7). The best-fit solutions of each cluster (Eqs. 11, 12, and 13) are statistically consistent with the best fit solution of the merged data set (Eq. 14) in terms of residuals. When the best-fit is made with the constraint of a constant stress drop scaling, residuals increase by more than 100 %

individual earthquakes (Fig. 6). Moreover, the merged data sets of spectral ratios and Fourier amplitudes yield inconsistent values of the best-fit stress drops (17 and 0.06 MPa, respectively), whereas the varying stress drop models converge to two very similar solutions (compare 14 with 15) and realistic values of $\Delta\sigma$. These results confirm the superiority of the varying stress drop model as already found for the Apennines in previous papers using the same approach (CALDERONI *et al.* 2012, 2013).

The combined use of spectral ratios and Fourier amplitude spectra of records appears to be a very efficient tool in source scaling studies (BOORE and ATKINSON 1989; CALDERONI *et al.* 2012). The finding

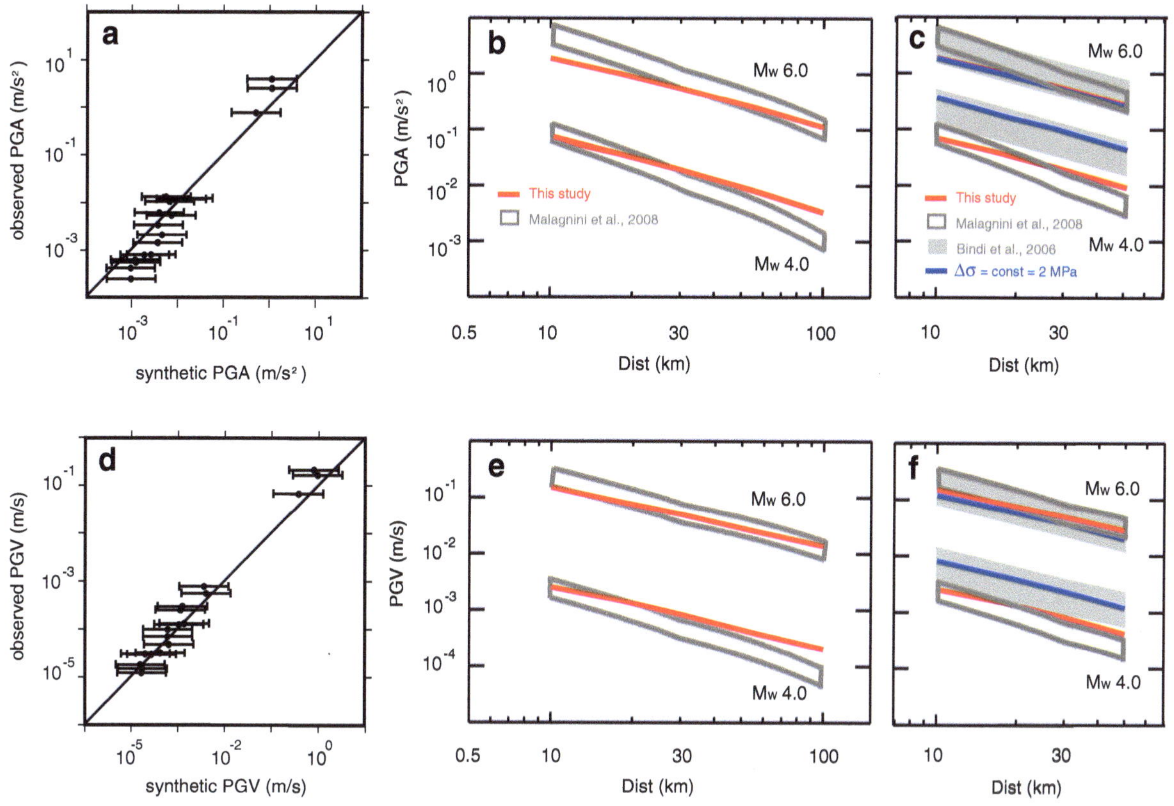

Figure 7

Comparison of observed PGA and PGV trends with those estimated through stochastic simulations (SMSIM code, see BOORE 2005) for different scaling laws. **a**, **d** compare observations at CLF with PGA and PGV, respectively, of synthetics resulting from stress drops of Eq. 14 and the propagation model given by (8′) and (9). The one standard deviation uncertainties of expected PGA and PGV are those of BINDI *et al.* (2006). In **b**, **e**, PGA and PGV of synthetics generated with the present study model are compared to the ±1 standard deviation band of MALAGNINI *et al.* (2008). In **c**, **f**, the PGA and PGV predictions by BINDI *et al.* (2006) are compared with the trend of synthetics obeying the source scaling (14), the constant (2 MPa) stress drop scaling, and the ±1 standard deviation band of MALAGNINI *et al.* (2008)

of a very similar source scaling in (14) and (15) is a double, a posteriori guarantee that (1) the spectral ratio method is applicable to the Colfiorito data set in spite of the poor consistency of propagation paths of target events and subevents, and (2) propagation is correctly modeled, at least at a zero order approximation, since it reproduces the same stress drops found from spectral ratios where no propagation correction is applied.

A final check of the scaling laws assessed in this and in other studies is performed by computing the PGA and PGV trend of synthetics obeying three different source scaling models: the one assessed in this study, the one assessed by MALAGNINI *et al.* (2008), and a constant (2 MPa according to BINDI *et al.* 2004) stress drop model with the propagation model of Eq. 9. Synthetics were generated through stochastic simulations using the SMSIM code (BOORE 2005). Figure 7a, d compares PGA and PGV, respectively, of the records used in this study with expectations of the spectral model (8′) at the soft site CLF, whereas Fig. 7b, e show the trend of rock sites for the different spectral models in the magnitude interval $4 < M_W < 6$ and the distance range $10 < R < 100$ km. Figure 7b, e show that, despite the different values, the stress drop trends inferred in this study and the one in MALAGNINI *et al.* (2008) yield acceptably similar trends of PGA and PGV, within uncertainties. Moreover, in the distance range $10 < R < 50$ km (Fig. 7c, f) that is satisfactorily constrained by data in the BINDI *et al.* (2006) analysis, all of the three scaling laws show a substantial consistency with observed PGA and PGV, falling within (or very close to) the ±1 standard deviation

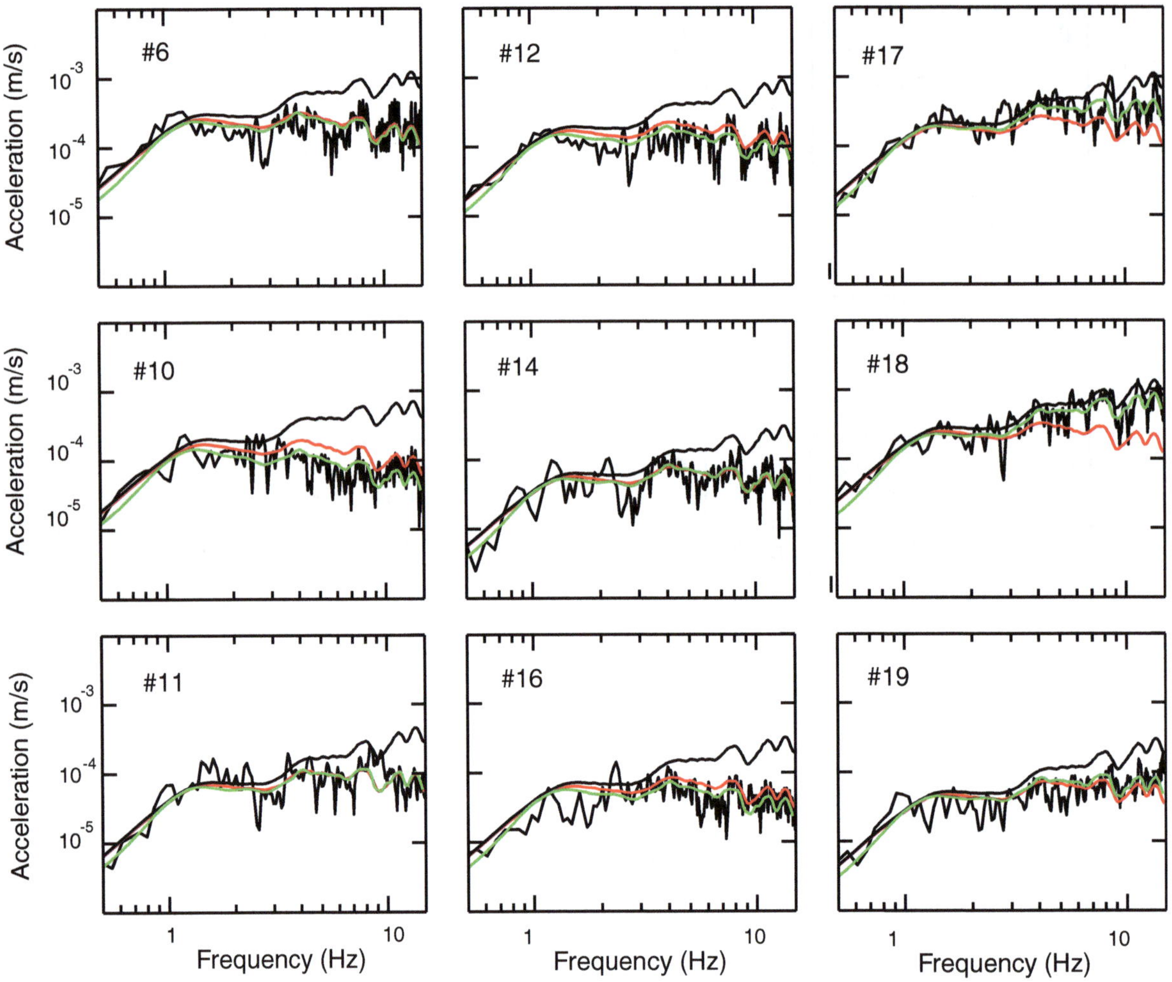

Figure 8
Acceleration spectra observed at CLF1 for small magnitudes ($M_W \approx 2$) are compared with expectations of three different models according to (1) this study, Eq. (14) (*red curve*), (2) constant stress drop scaling (*black curve*), and (3) individual-event best-fit (*green curve*)

uncertainty of the Bindi *et al.* (2006) predictive equations for moderate magnitudes. At smaller magnitudes, predictions based on this study and those by Malagnini *et al.* (2008) do not match the Bindi *et al.* (2006) theoretical trend; however, they do fit observations. The results of Fig. 7 suggest that (1) both the higher and lower stress drop EGF-based trends reproduce satisfactorily observed peak ground motions, and therefore there is a compensation in a respectively lower and higher contribution in each propagation model; (2) possible nonlinearity effects at the highest magnitudes due to the shallow soft soils of the Colfiorito basin are not significant, since the same values used for CLF also fit hard rock observations; and (3) the magnitude interval $4 < M_W < 6$ is not large enough to discern the superiority of the varying or constant stress drop scaling. We can show that, in this study, the availability of records of co-located stations with magnitudes in the range $1.7 < M_W < 6$ is crucial to constrain the slope of Eq. 14. Figure 8 compares the fit of high-frequency ground motions at small magnitudes with stress drops inferred from Eq. 14, a constant stress drop of 2 MPa, and the best fit value of each event when the propagation model (9) is used. The values of the individual-event best fit are listed in Table 1. This comparison remarks the sensitivity of the used frequency band ($f < 15$ Hz) to stress drop variations and the better performance of a decreasing stress drop model at smaller magnitudes.

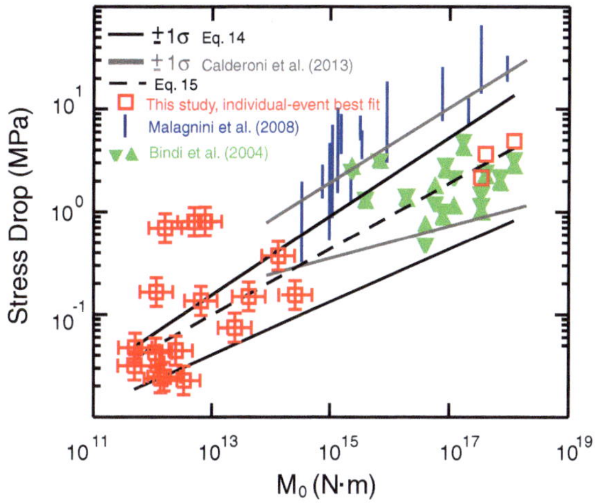

Figure 9
Scaling laws of Colfiorito earthquakes (this study and previous papers) are compared to the scaling law of the 2009 L'Aquila earthquakes by CALDERONI *et al.* (2013). A substantial similarity emerges in the behaviour of the two seismic areas that share the same extensional tectonic regime and are only 250 km apart

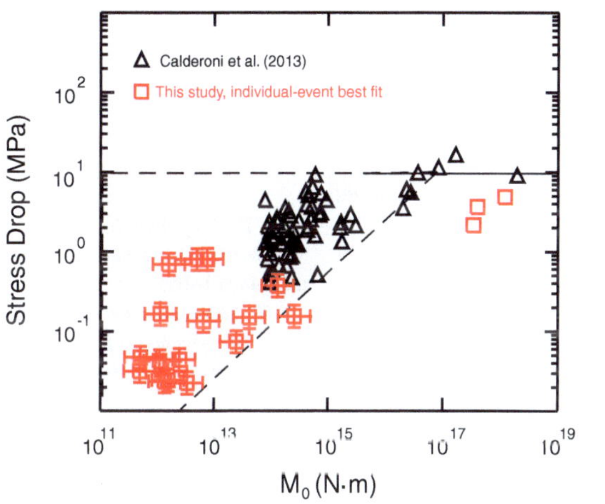

Figure 10
Stress drop estimates of the Colfiorito earthquakes (individual best-fits, this study) are overlapped with those of the 2009 L'Aquila earthquakes computed with the same approach (redrawn from CALDERONI *et al.* 2013). A similarity emerges, both in terms of value consistency for the main shocks and increase of variability for decreasing magnitudes

6. Discussion

Although it is based on only 19 records, the scaling law described by (14) and (15) is robust, since it consistently fits both spectral ratios and absolute amplitudes of the investigated data set and satisfactorily reproduces the expected trends of PGA and PGV of that seismic sequence up to large distances. Figure 9 shows that stress drops of this study agree, within statistical uncertainties, with the BINDI *et al.* (2004) estimates, although the resulting trend over the seismic moment interval of analysis is much more similar to the one proposed by MALAGNINI *et al.* (2008). Moreover, it is worthy to note that (14) and (15) strictly match the result recently found by CALDERONI *et al.* (2013) for a much larger number of earthquakes (64) occurring in the normal faults near L'Aquila (Fig. 9). The finding of such a similar trend for these two regions likely suggests a common pattern of stress release along the Apennine chain, where the present-day extension (D'AGOSTINO *et al.* 2011) is responsible for a frequent moderate-magnitude seismicity due to active normal faults (see also GHISETTI and VEZZANI 2002).

Based on the similarity between the trend of the 1997–1998 Umbria–Marche and the 2009 L'Aquila stress drops (both are shown in Fig. 9), the interpretation of the stress drop variation versus magnitude already proposed by CALDERONI *et al.* (2013) seems to be applicable to the seismic sequence of Colfiorito as well. The large number of records available for L'Aquila earthquakes allowed CALDERONI *et al.* (2013) to observe that the trend of stress drop decreasing for smaller earthquakes is probably caused by an increase of variability at small magnitudes (Fig. 10). This is consistent with a higher sensitivity of smaller size ruptures to stress and strength heterogeneities of the focal volume. As noted by BEN-ZION and ZHU (2002) and simulated by BEN-ZION *et al.* (2003), small earthquakes are likely to propagate in a strongly heterogeneous stress field that prevents them from growing to a larger size. During earthquake failure, the amount of slip depends on the degree of heterogeneity of the stress field (FISHER *et al.* 1997; BEN-ZION 2008), implying a larger variability of stress drops at smaller magnitudes. Figure 10 suggests that it is not necessary to invoke breakdown of self-similarity in the rupture process to explain observations: for many of the smaller earthquakes self-similarity holds, but the probability of less cohesive patches increases for smaller ruptures, mostly during the aftershock sequence, increasing the number of small earthquakes with lower stress drops.

Figure 10 shows that the model proposed for L'Aquila earthquakes by CALDERONI *et al.* (2013), with a stress drop substantially constant above $M_0 \approx 10^{17}$ N m, is consistent with the results of this study. For both Colfiorito and L'Aquila earthquakes, the variability increases below the threshold $M_0 \approx 10^{17}$ N m and both large (up to 10 MPa) and small ($\approx$0.1 MPa) values of stress drop are possible, depending on the larger heterogeneity of the fault strength for small patches.

7. Concluding Remarks

We have used ground motions recorded in the basin of Colfiorito to assess a scaling law over an extended magnitude range ($1.7 \leq M_W \leq 6.0$). This analysis is based on an independent fit of spectral ratios and amplitude spectra that converge to values of 2–5 MPa for the stress drop of these earthquakes at moderate magnitude. Each of these fit methods, alone, would be affected by significant limitations: lack of the ideal conditions in the EGF approach and oversimplification of the used attenuation model in the amplitude fit. However, we think that their convergence to the same solution is straightforward, suggesting that intrinsic limitations have a not-crucial role on the estimate reliability. At moderate magnitudes, our stress drops are strictly consistent with those by Bindi *et al.* (2004), which were estimated by a completely different approach and using different data.

In this study, we have also found a tendency toward smaller stress drops for smaller shocks, confirming the findings of Calderoni *et al.* (2013) and Malagnini *et al.* (2011) for other normal faults in the Apennines, in a similar tectonic regime about 250 km southeast of Colfiorito. The tendency of stress drops to decrease for smaller magnitudes, which is a matter of debate among many authors, is interpreted here in terms of the high heterogeneity of the fault structure at small scale, without invoking a breakdown of self-similarity in the rupture process.

Acknowledgments

Conversations with Dave Boore, about the trade-off between source and propagation and methods to reduce it as much as possible, were the basis of this paper. Special thanks are due to Dino Bindi and Luca Malagnini for their critical comments, which were decisive in correcting and improving several previous versions of the manuscript. Suggestions by Yehuda Ben-Zion are gratefully acknowledged. We thank Adrien Oth for his encouragement to contribute to this Special Issue, and Kevin Mayeda and an anonymous reviewer for useful comments in the review process.

References

Abercrombie, R., and Rice, J. R. (2005), *Can observations of earthquake scaling constrain slip weakening?*, Geophys. J. Int., *162*, 406–424. doi:10.1111/j.1365-246X.2005.02579.x.

Allmann, B. P., and Shearer, P. M. (2009), *Global variations of stress drop for moderate to large earthquakes*, J. Geophys. Res., *114*, B01310. doi:10.1029/2008JB005821.

Amato, A., Azzara, R., Chiarabba, C., Cimini, G. B., Cocco, M., Di Bona, M., Margheriti, L., Mazza, S., Mele, F., Selvaggi, G., Basili, A., Boschi, E., Courboulex, F., Deschamps, A., Gaffet, S., Bittarelli, G., Chiaraluce, L., Piccinini, D., and Ripepe, M. (1998), *The 1997 Umbria-Marche, Italy, earthquake sequence: A first look at the main shocks and aftershocks*, Geophys. Res. Lett., *25*(15), 2861–2864.

Ben-Zion, Y. (2008), *Collective behavior of earthquakes and faults: Continuum-discrete transitions, evolutionary changes and corresponding dynamic regimes*, Rev. Geophysics, *46*, RG4006. doi:10.1029/2008RG000260.

Ben-Zion, Y., and Zhu, L. (2002), *Potency-magnitude scaling relations for Southern California earthquakes with $1.0 < M_L < 7.0$*, Geophys. J. Int., *148*, F1–F5.

Ben-Zion, Y., Eneva, M. and Liu, Y. (2003), *Large earthquake cycles and intermittent criticality on heterogeneous faults due to evolving stress and seismicity*, J. Geophys. Res., *108*, No. B6, 2307. doi:10.1029/2002JB002121.

Bindi, D., Castro, R. R., Franceschina, G., Luzi, L., and Pacor, F. (2004), *The 1997-1998 Umbria-Marche sequence (central Italy): Source, path, and site effects estimated from strong motion data recorded in the epicentral area*, J. Geophys. Res., *109*, B04312. doi:10.1029/2003JB002857.

Bindi, D., Luzi, L., Pacor, F., Franceschina, G., and Castro, R. R. (2006), *Ground-motion predictions from empirical attenuation relationships versus recorded data: the case of 1997-1998 Umbria-Marche, central Italy, strong motion data set*, Bull. Seism. Soc. Am., *96*, 984–1002, doi:10.1785/0120050102.

Boore, D. M. (1983), *Stochastic simulation of high-frequency ground motions based on seismological models of the radiated spectra*, Bull. Seism. Soc. Am., *73*, 1865–1894.

Boore, D. M. (1986), *Short-period P- and S-wave radiation from large earthquakes: implications for spectral scaling relations*, Bull. Seism. Soc. Am., *76*, 43–64.

Boore, D. M. (2005), *SMSIM–Fortran Programs for Simulating Ground Motions from Earthquakes: Version 2.3 – A Revision of OFR 96-80-A*, U. S. Geological Survey Open-File Report 00-509, revised 15 August 2005, 55 pp. Available from the online publications link on http://www.daveboore.com.

BOORE, D. M., and ATKINSON, G. M. (1989), *Spectral scaling of the 1985 to 1988 Nahanni, Northwest Territories, earthquakes,* Bull. Seism. Soc. Am., *79*, 1736–1761.

BOORE, D. M., and BOATWRIGHT, J. (1984), *Average body-wave radiation coefficients,* Bull. Seism. Soc. Am., *74*, 1615–1621.

BRUNE, J. N. (1970), *Tectonic stress and the spectra of seismic shear waves from earthquakes*, J. Geophys. Res., *75*, 4997–5009.

BRUNE, J.N. (1971), *Correction (to Brune 1970),* J. Geophys. Res., *76*, 5002.

CALDERONI, G., ROVELLI, A., MILANA, G., and VALENSISE, G. (2012), *Do strike-slip faults of Molise, central southern Italy, really release a high stress?* Bull. Seismol. Soc. Am. *100*, 307–2324, doi:10.1785/0120090046.

CALDERONI, G., ROVELLI, A., and SINGH, S. K. (2013), *Stress drop and source scaling of the 2009 April L'Aquila earthquakes,* Geophys. J. Int., 1–15, doi:10.1093/gji/ggs011.

CAPUANO, P., ZOLLO, A., EMOLO, A., MARCUCCI, S., and MILANA, G. (2000), *Rupture mechanism and source parameters of Umbria-Marche mainshocks from strong motion data,* Journal of Seismology, *4*, 463–478.

CASTRO, R. R., ROVELLI, A., COCCO, M., PACOR, F., and DI BONA, M. (2001), *Stochastic simulation of strong-motion records from the September 26, 1997 (Mw = 6) Umbria-Marche (central Italy) earthquake,* Bull. Seismol. Soc. Am., *91*, 27–39.

CASTRO, R. R., FRANCESCHINA, G., PACOR, F., BINDI, D., and LUZI, L. (2006), *Analysis of the frequency dependence of the S-wave radiation pattern from local earthquakes in central Italy*, Bull. Seismol. Soc. Am., *96*, 415–426.

CHIARABBA, C., and AMATO, A. (2003), *Vp and Vp/Vs images in the M_w 6.0 Colfiorito fault region (central Italy): A contribution to the understanding of seismotectonic and seismogenic processes,* J. Geophys. Res., *108*, no. B5, 2248, doi:10.1029/2001JB001665.

CULTRERA, G., ROVELLI, A., MELE, G., AZZARA, R., CASERTA, A., and MARRA, F. (2003), *Azimuth-dependent amplification of weak and strong ground motions within a fault zone (Nocera Umbra, central Italy),* J. Geophys. Res., *108*, no. B3, 2156, doi:10.1029/2002JB001929.

D'AGOSTINO, N., MANTENUTO, S., D'ANASTASIO, E., GIULIANI, R., MATTONE, M., CALCATERRA, S., GAMBINO, P., and BONCI, L. (2011), *Evidence for localized active extension in the central Apennines (Italy) from global positioning system observations,* Geology *39*, 291–294, doi:10.1130/G31796.1.

DI GIULIO, G., CARA, F., AZZARA, R. M., ROVELLI, A., and LENTI, L. (2002), *Seismic hazard increase in intermontane basin: the case study of Colfiorito, Central Italy,* EOS Trans. AGU, *83*(47), Fall Meeting Suppl., abstract #S71B-1076, 2002.

DI GIULIO, G., ROVELLI, A., CARA, F., AZZARA, R. M., MARRA, F., BASILI, R., and CASERTA, A. (2003), *Long-duration asynchronous ground motions in the Colfiorito plain, central Italy, observed on a two-dimensional dense array,* J. Geophys. Res., *108*, no. B10, 2486, doi:10.1029/2002JB002367.

DI GIULIO, G., CORNOU, C., OHRNBERGER, M., WATHELET, M., and ROVELLI, A. (2006), *Deriving wavefield characteristics and shear-velocity profiles from 2-D small-aperture arrays analysis of ambient vibrations in a small-size alluvial basin, Colfiorito, Italy,* Bull. Seism. Soc. Am., *96*, 1915–1933, doi:10.1785/0120060119.

DISS WORKING GROUP (2010), *Database of Individual Seismogenic Sources (DISS), Version 3.1.1: A compilation of potential sources for earthquakes larger than M 5.5 in Italy and surrounding areas,* http://diss.rm.ingv.it/diss/, © INGV 2010 - Istituto Nazionale di Geofisica e Vulcanologia, Rome, Italy.

EKSTRÖM, G., MORELLI, A., BOSCHI, E., and DZIEWONSKI, A. M. (1998), *Moment tensor analysis of the central Italy earthquake sequence of September–October 1997,* Geophys. Res. Lett., *25*, 1971–1974.

ESHELBY, J. (1957), *The elastic field of an ellipsoid inclusion and related problems,* Proc. R. Soc. Lond., *241*(1226), 376–396.

FISHER, D. S., DAHMEN, K., RAMANATHAN, S. and BEN-ZION, Y. (1997), *Statistics of Earthquakes in Simple Models of Heterogeneous Faults,* Phys. Rev. Lett., *78*, 4885–4888.

GHISETTI, F., and VEZZANI, L. (2002), *Normal faulting, transcrustal permeability and seismogenesis in the Apennines (Italy)*, Tectonophysics *348*, 155–168, doi:10.1016/S0040-1951(01)00254-2.

GOLDSTEIN, P., DODGE, D., FIRPO, M., and MINNER, L. (2003), *SAC2000: Signal Processing and Analysis Tools for Seismologists and Engineers,* Invited contribution to *"The IASPEI International Handbook of Earthquake and Engineering Seismology",* W. K. Lee, H. Kanamori, P. Jennings, and C. Kisslinger Edts., Academic Press, London.

ICHINOSE, G. A., GOLDSTEIN, P., and RODGERS, A. J. (2000), *Relative importance of near-, intermediate- and far-field displacement terms in layered Earth synthetic seismograms,* Bull. Seismol. Soc. Am., *90*, 531–536.

HANKS, T. C., and KANAMORI, H. (1979), *A moment magnitude scale,* J. Geophys. Res., *84*, B5, 2348–2350.

KEILIS-BOROK, V. (1959), *On estimation of the displacement in an earthquake source and of source dimensions,* Annali di Geofisica, *12*, 205–214.

KOHRS-SANSORNY, C., COURBOULEX, F., BOUR, M., and DESCHAMPS, A. (2005), *A two-stage method for ground-motion simulation using stochastic summation of small earthquakes,* Bull. Seism. Soc. Am., *95*, 1387–1400.

MALAGNINI, L., and HERRMANN, R. (2000), *Ground-motion scaling in the region of the 1997 Umbria-Marche earthquake (Italy),* Bull. Seismol Soc Am, *90*, 1041–1051, doi:10.1785/0119990150.

MALAGNINI, L., SCOGNAMIGLIO, L., MERCURI, A., AKINCI, A., and MAYEDA, K. (2008), *Strong evidence for non-similar earthquake source scaling in central Italy*, Geophys. Res. Lett., *35*, L17303, doi:10.1029/2008GL034310.

MALAGNINI, L., AKINCI, A., MAYEDA, K., MUNAFÒ, I., HERRMANN, R. B., and MERCURI, A. (2011), *Characterization of earthquake-induced ground motion from the L'Aquila seismic sequence of 2009, Italy,* Geophys. J. Int., *184*, 325–337, doi:10.1111/j.1365-246X.2010.04837.x.

MAYEDA, K., MALAGNINI, L., and WALTER, W.R. (2007), *A new spectral ratio method using narrow band coda envelopes: evidence for nonselfsimilarity in the Hector Mine sequence,* Geophys. Res. Lett., *35*, L11303, doi:10.1029/2007GL030041.

ROVELLI, A., BONAMASSA, O., COCCO, M., DI BONA, M., and MAZZA, S. (1988), *Scaling laws and spectral parameters of ground motion in active extensional areas in Italy,* Bull. Seismol. Soc. Am., *78*, 530–560.

ROVELLI, A., COCCO, M., CONSOLE, R., ALESSANDRINI, B., and MAZZA, S. (1991), *Ground motion waveforms and source spectral scaling from close-distance accelerograms in a compressional regime area (Friuli, northeastern Italy),* Bull. Seism. Soc. Am., *81*, 57–80, 1991.

ROVELLI, A., SCOGNAMIGLIO, L., MARRA, F., and CASERTA, A. (2001), *Edge-diffracted 1-s surface waves observed in a small-size intermontane basin (Colfiorito, central Italy)*, Bull. Seismol. Soc. Am., *91*, 1851–1, 866.

ZOLLO, A., MARCUCCI, S., MILANA, G., and CAPUANO, P. (1999), *The 1997 Umbria-Marche (central Italy) earthquake sequence: insights on the mainshocks ruptures from near source strong motion records*, Geophys. Res. Lett. *26*, 3165–3168.

(Received April 29, 2013, revised April 15, 2014, accepted April 25, 2014, Published online May 14, 2014)

Pure Appl. Geophys. 171 (2014), 2747–2766

DOI 10.1007/s00024-014-0774-2

Pure and Applied Geophysics

Source Scaling of Inland Crustal Earthquake Sequences in Japan Using the S-Wave Coda Spectral Ratio Method

Kazuhiro Somei,[1] Kimiyuki Asano,[2] Tomotaka Iwata,[2] and Ken Miyakoshi[1]

Abstract—We estimate corner frequencies and stress drops for 298 events ranging from M_w 3.2–7.0 in 17 inland crustal earthquake sequences in Japan to investigate the source scaling and variation in stress drops. We obtain the source spectral ratio from observed records by the S-wave coda spectral ratio method. The advantage of using the S-wave coda is in obtaining much more stable source spectral ratios than using direct S-waves. We carefully examine the common shape of the decay of coda envelopes between event pair records. The corner frequency and stress drop are estimated by modeling the observed source spectral ratio with the omega-square source spectral model. We investigate the dependences of stress drops on some tectonic effects such as regionality, focal mechanism, and source depth. The principal findings are as follows: (1) a break in self-similar source scaling is found in our dataset. Events larger than M_w 4.5 show larger stress drops than those of smaller events. (2) Stress drops of aftershocks are mostly smaller than those of mainshocks in each sequence. (3) There are no systematic differences between stress drops of events occurring inside and outside the Niigata-Kobe Tectonic Zone in Japan. (4) Clear dependence of the faulting type on stress drops cannot be seen. (5) Stress drops of aftershocks depend on their source depth. (6) The crack size obtained from the corner frequency corresponds to the total rupture area of heterogeneous slip models for large events.

Key words: Source scaling, corner frequency, stress drop, inland crustal earthquake sequences, S-wave coda spectral ratio method.

1. Introduction

Source scaling is an important issue for earthquake source physics. Since Aki (1967) proposed the seismic scaling law of source spectrum, it has been widely accepted that the stress drop has a constant value independent of magnitude. Seismic moment M_0 is inversely proportional to the cube of corner frequency f_c ($M_0 \sim f_c^{-3}$) (e.g., Kanamori and Anderson, 1975). This self-similarity in the source scaling has been supported by many studies (e.g., McGarr, 1999; Prejean and Ellsworth, 2001; Ide *et al.*, 2003; Imanishi *et al.*, 2004; Prieto *et al.*, 2004; Oth *et al.*, 2010; Baltay *et al.*, 2011). On the other hand, evidence for non-self-similarity has been shown for more than a decade (e.g., Iio, 1986; Mayeda and Walter, 1996; Jost *et al.*, 1998; Izutani and Kanamori, 2001; Takahashi *et al.*, 2005; Tajima and Tajima, 2007; Yoo *et al.*, 2010).

Another issue is whether variations in stress drops can be related to differences in tectonic conditions. From many studies on stress drops (e.g., Kanamori and Anderson, 1975; Allmann and Shearer, 2009), it is generally accepted that stress drops have been variable over more than three orders of magnitude (0.1–100 MPa), independent of seismic moment, even though the absolute stress drop depends on its assumed crack model. Kanamori and Anderson (1975) showed that stress drops for large earthquakes (M_0 10^{18}–10^{23} Nm) scatter within a range of 1–10 MPa, and also showed intra-plate earthquakes have larger stress drops than inter-plate earthquakes worldwide. Cocco and Rovelli (1989) estimated and compiled the stress drops of reverse and normal faulting earthquakes in Italy. They found that stress drops are approximately three times larger for reverse faulting events than those for normal faulting events. Lindley (1994) calculated stress drops for the 1992 Joshua Tree, California earthquake sequence. Stress drops of aftershocks and a foreshock varied from 0.4 to 4.1 MPa. The stress drop for the mainshock of 20.3 MPa is significantly larger than any of the aftershocks and the foreshock. Shearer *et al.* (2006) showed that median stress drop increased with depth

[1] Geo-Research Institute, 4-3-2 Itachibori, Nishi, Osaka 550-0012, Japan. E-mail: somei@geor.or.jp
[2] Disaster Prevention Research Institute, Kyoto University, Gokasho, Uji, Kyoto 611-0011, Japan.

from about 0.6 MPa at the surface to about 2.2 MPa at 8 km. ALLMANN and SHEARER (2009) estimated and compiled stress drops for global earthquakes over a wide range of seismic moment M_0 (10^9–10^{22} Nm). They showed that the variation in stress drops is typically between 0.1 and 100 MPa, independent of seismic moment. They also pointed out that stress drops for intra-plate earthquakes are larger than those for inter-plate earthquakes, and also found a dependence of stress drop on focal mechanism with larger stress drops for strike-slip earthquakes from global variation of stress drops for moderate to large earthquakes.

Investigating the source scaling and variation in stress drops contributes not only to better understanding of source physics, but also to further development of strong ground motion predictions for future earthquakes (COTTON *et al.*, 2013). Source scaling, or the relationship between seismic moment and rupture area, is one of the essential factors for strong ground motion prediction. As stress drop controls high-frequency ground motion generation, the obtained source scaling relationships in the target area are quite useful for strong ground motion prediction. If we could find non-self-similar scaling relationships and discover that they were caused by stress drop variations on tectonic effects such as regionality, focal mechanism and source depth, these characteristics would also be useful for source modeling in strong ground motion prediction. For example, ASANO and IWATA (2011a) showed that the stress drop on the asperities clearly depended on its depth from the results of characterization of heterogeneous stress drop distributions of large inland crustal earthquakes in Japan, and they proposed an empirical scaling relationship between the stress drop on the asperity and its depth.

We estimate the source spectral ratios for 298 events ranging from M_w 3.2–7.0 in 17 inland crustal earthquake sequences in Japan (Fig. 1) and estimate the corner frequencies and stress drops to investigate the source scaling and variation in stress drops. In order to obtain stable and reliable source spectra, we use the spectral ratio method for S-wave coda. Based on the estimated stress drops, we discuss their dependences on some tectonic effects such as regionality, focal mechanism, and source depth. We also

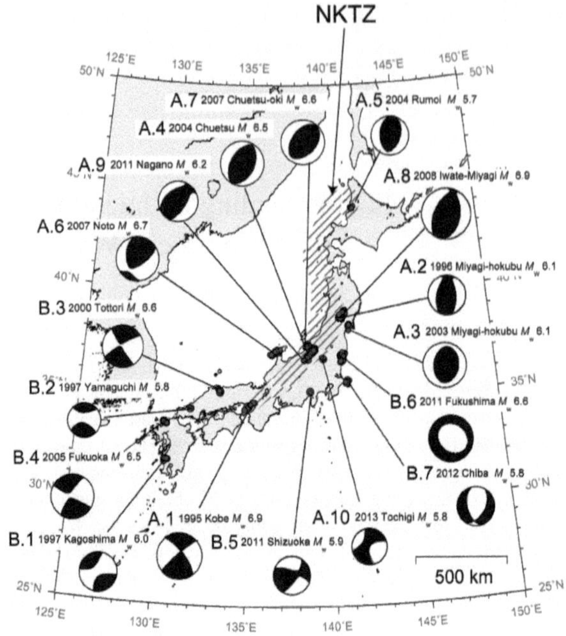

Figure 1
Distribution of 298 events analyzed in this study (*small solid circles*). The 17 mainshocks are indicated with their focal mechanisms in the lower hemisphere projection. The shaded belt-like area shows the outline of the NKTZ (SAGIYA *et al.*, 2000)

examine the relationship between the crack sizes calculated from estimated corner frequencies and the total rupture area and the asperity area of the heterogeneous slip models for large events derived by kinematic waveform inversions to discuss the source dimensions obtained from the source spectrum and the heterogeneous slip models.

In this study, we focus on the Niigata-Kobe Tectonic Zone (NKTZ) (SAGIYA *et al.*, 2000) as one of the important regional tectonic conditions from both the geodetical and the geological point of view in Japan. In terms of geodesy, SAGIYA *et al.* (2000) found the significant strain accumulation in the northern Chubu and Kinki districts as well as plate boundary regions by Global Positioning System (GPS) observations. In terms of geology, OKAMURA (2010) pointed out the dense distribution of active faults and fold structures along the eastern margin of the Japan Sea. SAGIYA *et al.* (2000) also implied the existence of this tectonic zone as the northern extension of NKTZ. Hereafter, we collectively refer to these belt-like zones along the Japan Sea coast of northeast and central Japan shown in Fig. 1 as the

NKTZ. It should be noted that many disastrous inland crustal earthquakes have occurred in this tectonic zone in history such as the 1891 Nobi (*M* 8.0), 1964 Niigata (*M* 7.5), 1983 Japan Sea (*M* 7.7), 1993 Hokkaido Nansei-oki (*M* 7.8), and earthquakes in group A used in this study (see Table 1). Therefore, it is also important for strong ground motion prediction for future earthquakes in NKTZ to discuss the source scaling and variation in stress drops.

2. Dataset

The epicenters of analyzed events in 17 earthquake sequences are shown in Fig. 1 with moment tensor solutions of the mainshocks. These seismic moment tensors were routinely determined by the moment tensor inversion of the F-net operated by the National Research Institute for Earth Science and Disaster Prevention, Japan (NIED) (FUKUYAMA *et al.*, 1998), except for the 1995 Kobe earthquake and the 1996 Miyagi-hokubu earthquake. For the focal mechanisms of the 1995 Kobe earthquake and the 1996 Miyagi-hokubu earthquake, reference was made to KATAO *et al.* (1997) and the Japan Meteorological Agency (JMA), respectively.

Of the events analyzed in this study, 166 events in ten sequences occurred inside the NKTZ. In order to compare source characteristics among events occurring inside and outside the NKTZ, we used the characters "A" and "B" in the event numbers shown in Fig. 1 and Table 1 to designate events occurring inside and outside the NKTZ, respectively.

We employ strong motion triggered records with finite time length observed by the dense K-NET (KINOSHITA, 1998) and KiK-net (AOI *et al.*, 2000) networks, both operated by NIED. For KiK-net stations, the data from the downhole sensor is used. In addition, we also use the broadband continuous records provided by the F-net broadband seismograph network of NIED. The number of F-net stations is smaller than that of K-NET and KiK-net stations. However, as these F-net records have high signal to noise (SN) ratios in the lower frequency range, we can obtain the source spectral ratios more widely in the frequency range. For the sequence of #A.1, we use the strong motion records observed by the Committee of Earthquake Observation and Research in the Kansai Area (CEORKA) (KAGAWA *et al.*, 2004) and also use the strong motion seismograph network around Lake Biwa (MATSUNAMI and HACHIRI, 1995) installed by the Disaster Prevention Research Institute (DPRI), Kyoto University and Shiga prefectural government. The datasets used for each sequence are listed in Table 1.

3. S-Wave Coda Spectral Ratio Method

The observed seismic waveform can be expressed as a convolution of the source, propagation path, and site amplification effects in the time domain. There are some approaches to separate the source effect (i.e., the source spectrum) from the observed waveform. For example, the deconvolution method using an empirical Green's function (EGF) has been applied to this kind of study (e.g., HARTZELL, 1978; MUELLER, 1985; HOUGH, 1997). In the frequency domain, the spectral separation method using inversion techniques (e.g., IWATA and IRIKURA, 1988; KAWASE and MATSUO, 2004; OTH *et al.*, 2011) is widely used. The other approach to isolate the source spectrum in the frequency domain is the spectral ratio method (e.g., IZUTANI and KANAMORI, 2001; MIYAKE *et al.*, 2003; TAJIMA and TAJIMA, 2007).

The spectral ratio method using the direct P-waves or S waves requires the records from an event pair which is co-located and has a different magnitude with the similar source mechanism at common stations to empirically remove the path and site effects. Unfortunately, the spectral ratios are often scattered from station to station because it is not always easy to select an event pair that is exactly co-located and has the similar source mechanism. Additionally, MIYAKE *et al.* (2001) pointed out the azimuth and distance dependences of observed source spectra in the near source area, i.e., rupture directivity effects. In order to reduce these effects, the averaged spectral ratio using many stations is typically used (IZUTANI and KANAMORI, 2001; TAJIMA and TAJIMA, 2007). However, in practice, it is also difficult to realize an ideal distribution of stations covering the

Table 1

List of earthquake sequences used in this study with their dataset information

No.	Org. time of mainshock (JST = UT + 9 h)	Hypocenter information	Number of events	M_w range	Data[a] A	Data[a] B	Seismic moment[b] Main.	Seismic moment[b] After.	Fault type[c]	β^d range (km/s)	Depth range (km)	Name of earthquake
Inside NKTZ												
A.1	1995/01/17,05:46	Katao *et al.* (1997)	5	4.1–7.0	○	–	R1	R2	S	3.60–3.77	10.6–22.6	1995 Kobe
A.2	1996/08/11,03:12	JMA	4	4.7–6.0	○	–	JC	JC	R	3.45–3.60	8.6–13.1	1996 Miyagi-hokubu
A.3	2003/07/26,07:13	JMA	11	3.4–6.1	○	–	F	F	R	3.51–3.58	11.5–13.7	2003 Miyagi-hokubu
A.4	2004/10/23,17:56	Kato *et al.* (2006)	48	3.2–6.6	○	○	F	F	R	2.90–3.76	1.2–15.7	2004 Chuetsu
A.5	2004/12/14,14:56	JMA	3	3.8–5.7	○	○	F	F	R	3.36–3.61	8.1–15.3	2004 Rumoi
A.6	2007/03/25,09:41	Kato *et al.* (2008)	18	3.4–6.7	○	○	F	F	R	3.46–3.50	0.5–11.9	2007 Noto
A.7	2007/07/16,10:13	JMA	13	3.5–6.6	○	○	F	F	R	3.54–3.84	13.6–27.4	2007 Chuetsu-oki
A.8	2008/06/14,08:43	JMA	49	3.5–6.9	○	○	F	F	R	3.20–3.61	2.1–12.3	2008 Iwate-Miyagi
A.9	2011/03/12,03:59	JMA	11	3.6–6.2	○	○	F	F	R	3.31–3.50	0.0–9.4	2011 Nagano
A.10	2013/02/25,16:23	JMA	4	3.4–5.8	○	○	F	F	S	3.33–3.39	2.8–4.0	2013 Tochigi
Outside NKTZ												
B.1	1997/03/26,17:31	JMA	8	3.6–6.1	○	–	F	F	S	3.48–3.62	8.1–14.3	1997 Kagoshima
B.2	1997/06/25,18:50	JMA	2	3.9–5.8	○	–	F	F	S	3.51–3.57	8.3–10.8	1997 Yamaguchi
B.3	2000/10/06,13:30	Fukuyama *et al.* (2003)	7	3.6–6.6	○	–	F	F	S	3.44–3.57	7.3–12.4	2000 Tottori
B.4	2005/03/20,10:53	Uehira *et al.* (2006)	34	3.3–6.6	○	○	F	F	S	3.37–3.52	2.7–15.0	2005 Fukuoka
B.5	2011/03/15,22:31	JMA	2	3.7–5.9	○	○	F	SE	S	3.61–3.65	11.7–14.3	2011 Shizuoka
B.6	2011/04/11,17:16	JMA	67	3.2–6.6	○	○	F	F	N	3.01–3.81	3.8–19.5	2011 Fukushima
B.7	2012/03/14,21:05	JMA	12	3.6–5.7	○	○	F	F	N	3.40–3.77	9.0–17.5	2012 Chiba

A strong motion records from K-NET, KiK-net and F-net. For Event #A.1, CEORKA and strong motion seismograph network around Lake Biwa (Matsunami and Hachiri, 1995) are used, *B* broadband seismograph records from F-net, *F* moment tensor solution determined by F-net, *JC* moment tensor solution determined by JMA, *R1* Sekiguchi *et al.* (2002), *R2* Nakagawa (1997), *SE* estimated by source spectrum in this study, *R* reverse-slip fault type, *N* normal-slip fault type, *S* strike-slip fault type

[a] Dataset used in this study

[b] Seismic moment used in this study

[c] Fault mechanism type of mainshock

[d] The shear wave velocity in the source from Matsubara *et al.* (2008)

epicenters of the event pair. Moreover, the spectrum from the P-wave portion cannot exclude the near-field term, which raises doubts concerning the operation of fitting the ω^{-2} source model for the far-field term to the observed P-wave source spectrum for strong ground motion data.

MAYEDA *et al.* (2007) introduced and tested a new coda spectral ratio method on the 1999 Hector Mine sequence to avoid these effects. They showed that the scatter of the S-wave coda source spectral ratios by station to station is significantly less than those of direct P-wave or S-wave source spectral ratios. This method has been subsequently applied to some earthquake sequences or large events over the world such as the 1999 Chi-Chi, Taiwan (MAYEDA and MALAGNINI, 2009), the 2002 San Giuliano, Italy (MALAGNINI and MAYEDA, 2008), and the 2005 Fukuoka, Japan (YOO *et al.*, 2010).

In this study, we also use the S-wave coda spectral ratio method to reduce the scatter of spectral ratios by station to station. We briefly describe our data processing as follows by showing an event pair (the mainshock and an aftershock) of the 2005 Fukuoka (event #B.4) for an example.

First, we select the event pairs occurring close to each other. Figure 2a shows the epicenters of an event pair with their moment tensor solutions and the distribution of stations used for this event pair. According to AKI and CHOUET (1975), the narrow-band RMS envelope of S-wave coda from different earthquakes decays as a function of the lapse time in the same manner independent of the magnitude, the distance and the nature of the direct path from the source to the station. RAUTIAN and KHALTURIN (1978) showed that S-wave coda are established after twice the S-wave travel time. We define the beginning of time window for the S-wave coda as 2 $T_s + T_d$, where T_s and T_d are the S-wave travel time and the source duration of a large event. We use the empirical relationship, $T_d = 2.10 \times 10^{-8} M_0^{1/3}$, proposed by EKSTRÖM *et al.* (2005). For calculating the amplitude spectra, we use the 20.48 s for strong motion records or 40.96 s for broadband records of S-wave coda parts from the same lapse time between the event pair as the analyzed time window T ($T = \{t_1, t_2, \ldots, t_N\}$). Figure 2b shows examples of waveforms recorded at the F-net INN station with the analyzed S-wave coda windows for this event pair.

In Fig. 2c, we examine the commonality of the decay of coda envelopes between the event pair records for five frequency bands (0.25–0.5, 0.5–1.0, 1.0–2.0, 2.0–4.0, and 4.0–8.0 Hz). For each frequency band, firstly, the RMS envelope is calculated by the root mean square of three components of whole waveforms. A moving average of ± 10 s for each time point is applied to estimate the RMS envelope. Then, we take the RMS envelope of the analyzed S-wave coda window to be defined in the same way as Fig. 2b. The RMS envelopes of large and small events and the normalized envelope ratio are defined E_L (t_i), E_S (t_i) and R (t_i), respectively.

$$R(t_i) = \frac{E_L(t_i)/E_S(t_i)}{E_L(t_1)/E_S(t_1)}, \quad (1)$$

where t_1 means the first time point in the analyzed S-wave coda window and E_L (t_1)/E_S (t_1) is the normalized factor considering the differences of the ratios for each event pair and/or each frequency band. Then the log average and the log standard deviation of R (t_i) for all over the window are obtained. If the shapes of the decay curve of coda envelopes for the event pair are exactly same, both of log average and the log standard deviation of R (t_i) for all over the window are obtained as 0.0. In this study, we first carefully remove the records contaminated by the signals from small earthquakes and any other noise that distorts the shape of the decay curves of coda envelopes in the analyzed windows by visual inspection. Then, we also remove the records of event pairs whose the log standard deviation of R (t_i) for all over the window is larger than 0.1 for any frequency band. The envelopes in Fig. 2c are the examples which satisfy the above criterion.

Next, we calculate the Fourier amplitude spectra for analyzed S-wave coda windows of the event pair. The vectorial summation of the three components is used in this analysis. A moving average of ± 10 % window for each frequency point is applied to smooth the amplitude spectrum. We also calculate the amplitude spectrum for the noise part of the small event in the event pair to examine the signal to noise ratio (S/N). We use the windows of 10.24 s for

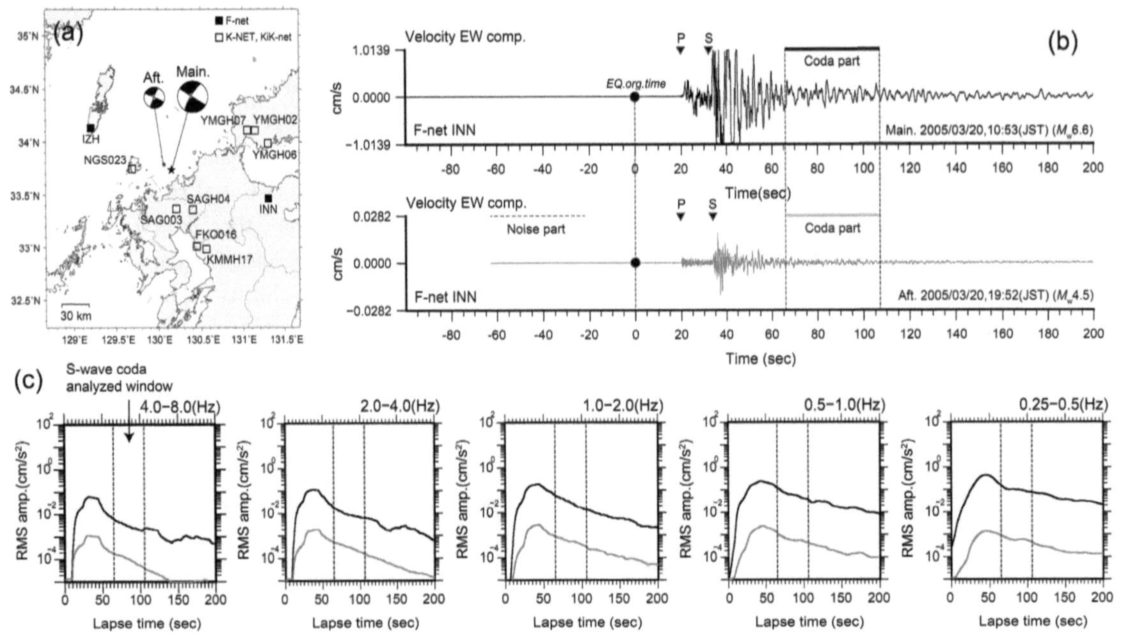

Figure 2

a Map showing the epicenters and stations used for sequence #B.4. The *large* and *small stars* represent the epicenters of a mainshock and an aftershock for an event pair with their focal mechanisms determined by F-net. *Open* and *solid squares* indicate the strong motion stations and F-net broadband stations used for this event pair. **b** Velocity waveforms (EW component) of mainshock (*black line*) and aftershock (*gray line*) at F-net INN station. The time windows of coda parts and the noise part of the small event used to calculate the Fourier amplitude spectra are shown with the *bold* and the *broken bars* above each seismogram, respectively. For the mainshock (*black line*), the waveform is saturated in the S-wave part due to the instrumental response; however, this is not an issue for using the S-wave coda. **c** Comparisons of decay rate curves in coda parts for five frequency bands between mainshock (*black line*) and aftershock (*gray line*) by RMS amplitude envelopes

strong motion records and 40.96 s for broadband records before the arrivals of the P-waves as the noise parts.

The source spectral ratio for one station is obtained by the amplitude spectral ratio of large and small events. Then the log average and the log standard deviation of the source spectral ratios for all stations whose spectra of the small event satisfy the S/N more than 2.0 are defined as $S_{\mathrm{obs}}(f_i)$ and S.D.(f_i) ($i = 1$ to m), respectively. The model function $S_{\mathrm{syn}}(f_i)$ based on the omega-squared (ω^{-2}) source model by Brune (1970, 1971), is given by

$$S_{\mathrm{syn}}(f_i) = \frac{M_0}{m_0} \cdot \frac{1 + (f_i/f_{\mathrm{cS}})^2}{1 + (f_i/f_{\mathrm{cL}})^2}, \quad (2)$$

where M_0/m_0 indicates the seismic moment ratio between the large and small events, and f_{cL} and f_{cS} are the corner frequencies for the large and small events, respectively. We search the minimum value of the weighted least squares (Eq. 3) to estimate the f_{cL} and f_{cS}.

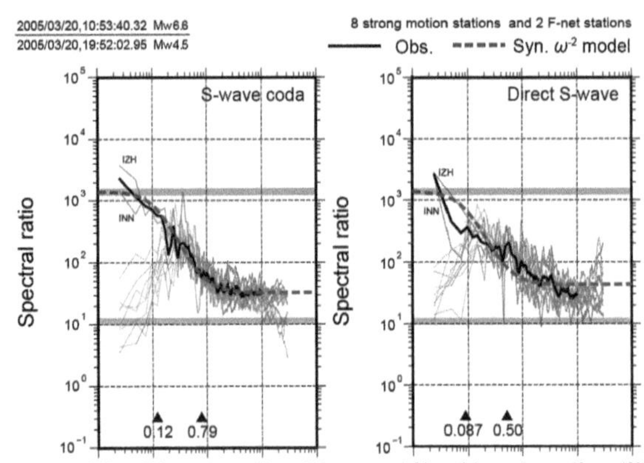

Figure 3

Example of the observed source spectral ratios for each station (*thin gray lines*), the average observed source spectral ratio (*thick black line*), and the ω^{-2} source spectral ratio model (*broken gray line*) fitted to the observation. The *triangles* indicate the corner frequencies for the mainshock and the aftershock. *Upper* and *lower gray horizontal lines* represent low and high frequency asymptotes predicted by the ratio of seismic moments of the event pair and self-similar ω^{-2} source spectral model, respectively. The source spectral ratios by using S-wave coda (*left*) and those by using direct S-wave (*right*) are shown

$$\sum_{i=1}^{m}\left\{\frac{\log[S_{obs}(f_i)]-\log[S_{syn}(f_i)]}{\log \mathrm{S.D.}(f_i)}\right\}^2 \times \Delta f_i \rightarrow \min,$$
$$\Delta f_i = \frac{\log(f_{i+1})-\log(f_{i-1})}{2} \quad (3)$$

The M_0/m_0 is fixed by the seismic moments which are determined by previous studies listed in Table 1. The target frequency range for modeling is defined by a criterion which the Fourier spectrum of a small event satisfies the S/N more than 2.0 as mentioned above (e.g., 0.1–10 Hz). The search windows are from 0.01 to 1 Hz for f_{cL} and from 0.1 to 20 Hz for f_{cS} with a search interval of 0.001 Hz in 0.01–0.1 and 0.01 Hz in 0.1–20 Hz. For some of the large earthquakes such as mainshocks, we can estimate the f_{cL} from more than two event pairs. In this case, we adopt the f_{cL} which is calculated by log averaging the f_{cL}'s derived from those event pairs.

For comparison, the observed source spectral ratio for the direct S-wave is obtained by correcting the propagation path and site effects from the observed record. The propagation path effect is given by the geometrical spreading for the body wave and by a frequency-dependent attenuation factor $Q_s(f)$ for the S-wave. The examples of source spectral ratios

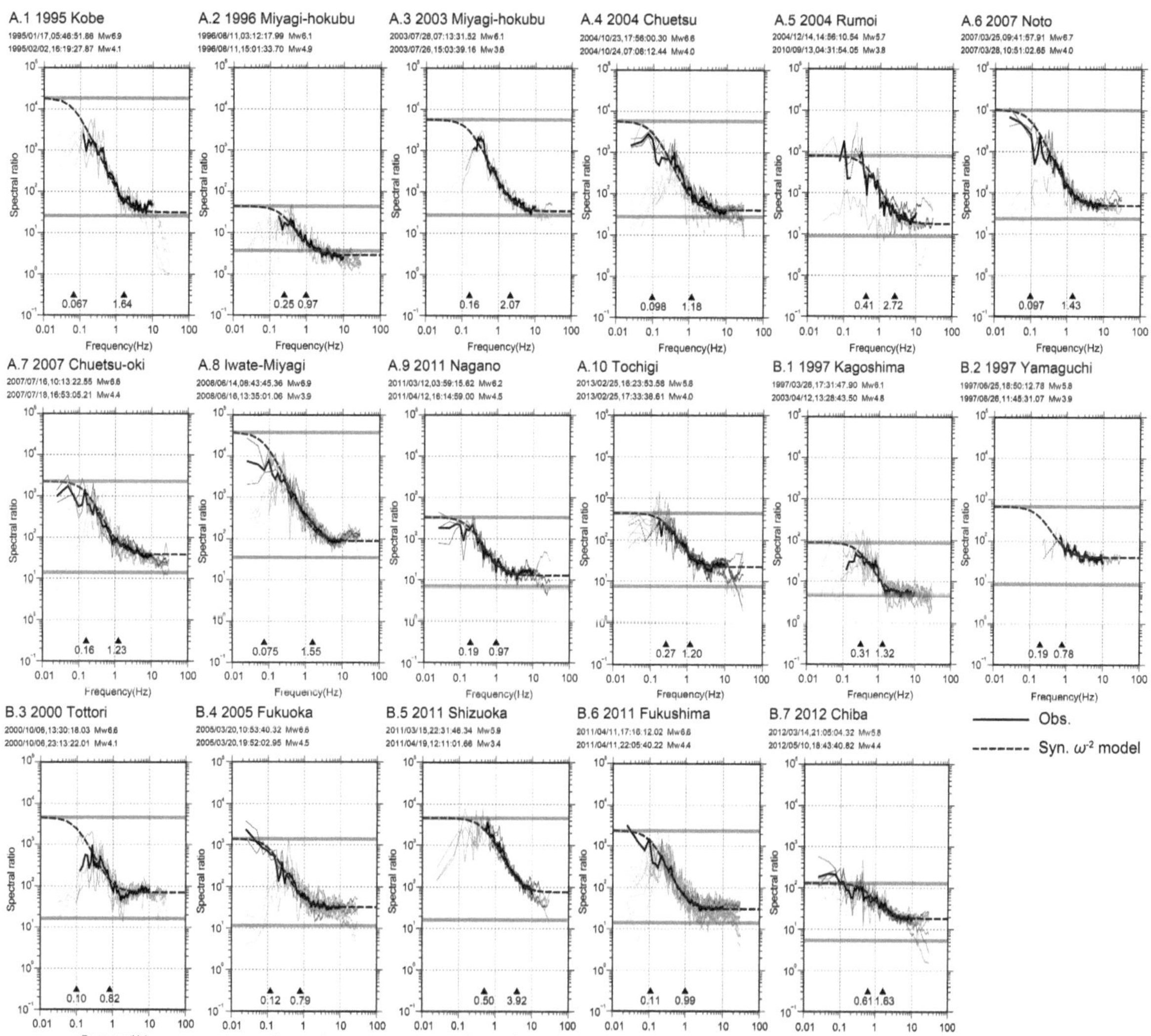

Figure 4
Examples of source spectral ratio fitting for each earthquake sequence. The notations are the same as in Fig. 3

obtained from the S-wave coda and direct S-wave are shown in Fig. 3. We can clearly see that the source spectral ratios by using S-wave coda are much more stable than those by using direct S-waves.

Another point to note for Fig. 3 is that the observed source spectral ratio has high S/N in the lower frequency range by using the F-net broadband records. Because the source spectra of large earthquakes have the lower corner frequencies, wide frequency range spectral ratios are necessary to estimate the corner frequencies of those earthquakes.

Figure 4 shows 17 source spectral ratios including the mainshocks for each sequence. The observed source spectral ratios are fitted by the ω^{-2} source spectral ratio models well, and we could estimate the reliable corner frequencies for large and small events. For example, the corner frequency of the 2005 Fukuoka mainshock is estimated as 0.15 Hz in this study. Yoo *et al.* (2010) also estimated the corner frequencies for the 2005 Fukuoka events by their coda spectral ratio method. They used the waveforms recorded at four F-net stations around the events. The event pairs used in their study are not the same as those used in this study. They estimated a corner frequency value of 0.1534 Hz for the 2005 Fukuoka mainshock. This similarity of estimated f_c's indicates that the coda spectral ratio method is useful to estimate the reliable and stable corner frequency independent of the stations and event pairs.

Figure 6 ▶

The relationships between corner frequency f_c and seismic moment M_0 (*lower scale*) and moment magnitude M_w (*upper scale*) for each sequence (*solid symbols*) and all events (*open symbols*). The mainshocks for each sequence are represented by *red symbols*. The *dashed red line* for each sequence indicates the stress drop of mainshock

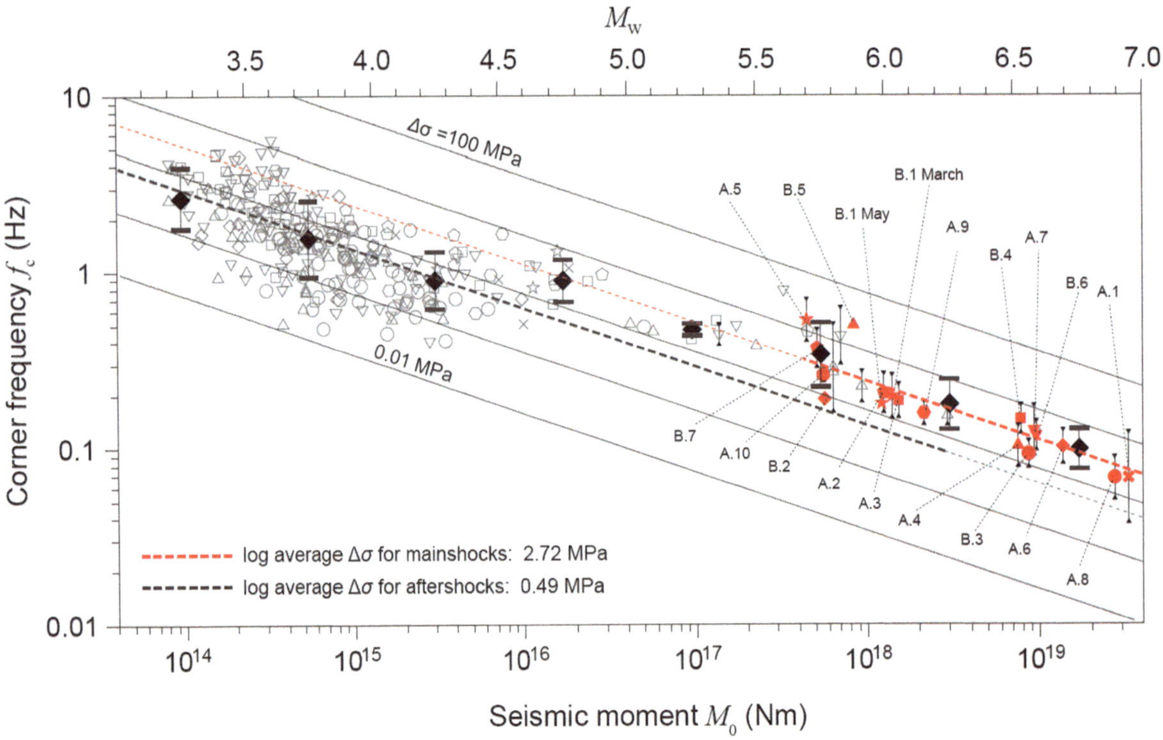

Figure 5

The relationships between corner frequency f_c and seismic moment M_0 (*lower scale*) and moment magnitude M_w (*upper scale*) for all events. The log average corner frequencies for bins of 0.5 in M_w are represented by *solid diamonds* with the *standard deviation bars*. The mainshocks and the aftershocks are shown by *red* and *gray symbols*, respectively. The *dashed red* and *gray lines* indicate the log average stress drops of mainshocks and aftershocks. The log standard deviations of f_c for earthquakes which are estimated from more than two event pairs are also shown. The *solid lines* indicate constant stress drops of 0.01, 0.1, 1, 10 and 100 MPa

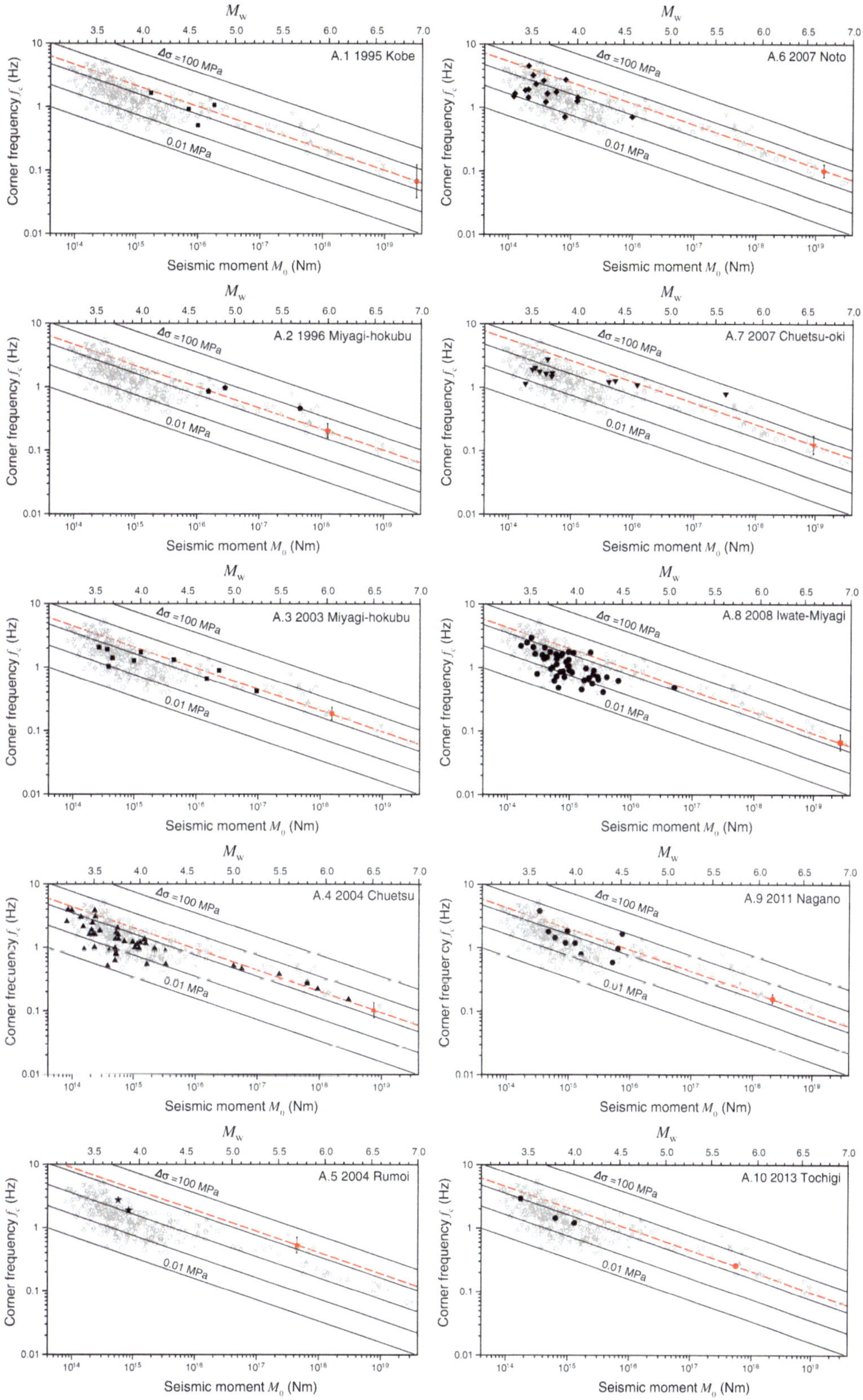
A.1 1995 Kobe
A.2 1996 Miyagi-hokubu
A.3 2003 Miyagi-hokubu
A.4 2004 Chuetsu
A.5 2004 Rumoi
A.6 2007 Noto
A.7 2007 Chuetsu-oki
A.8 2008 Iwate-Miyagi
A.9 2011 Nagano
A.10 2013 Tochigi
$\Delta\sigma$ =100 MPa
0.01 MPa
M_w
Corner frequency f_c (Hz)
Seismic moment M_0 (Nm)

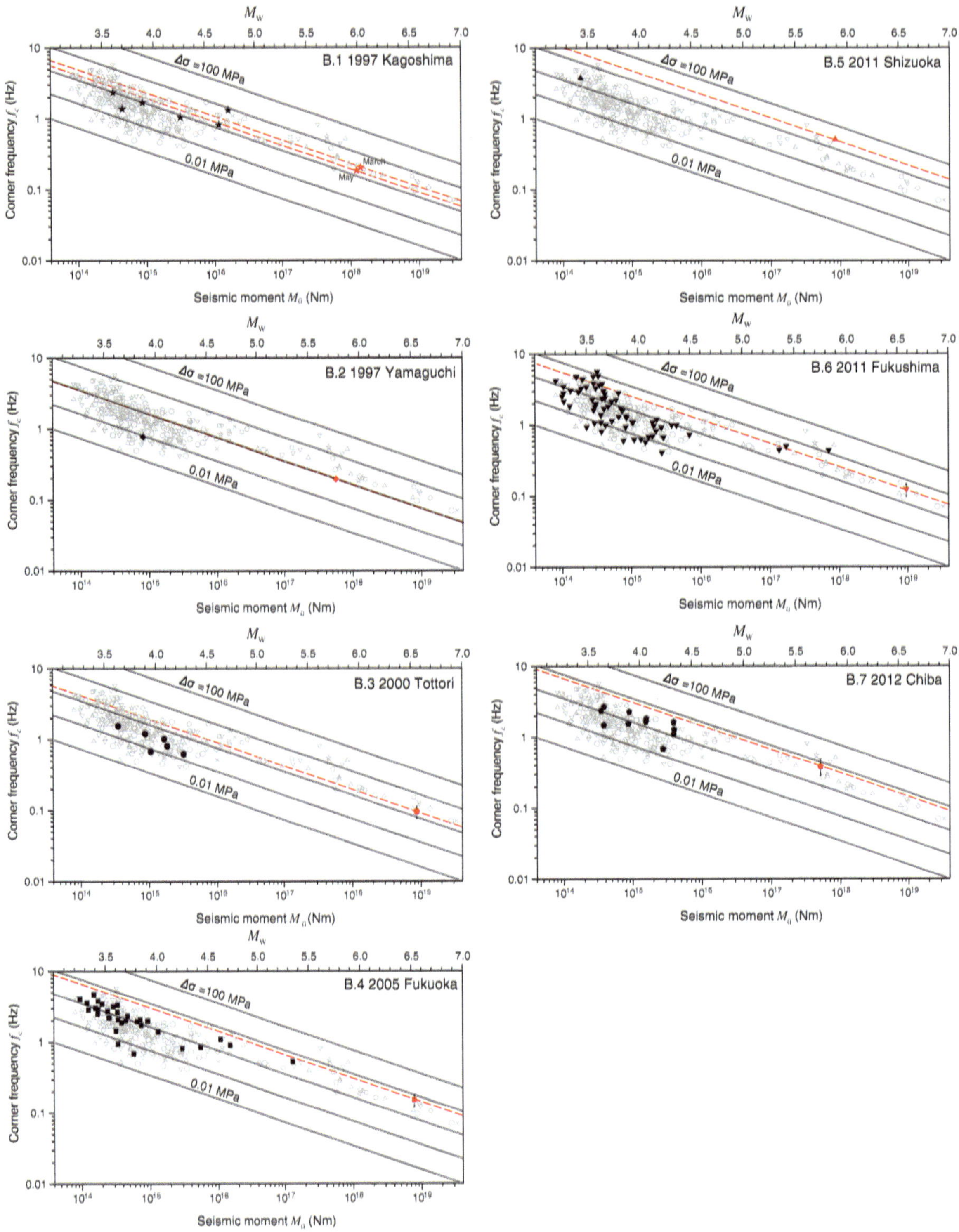

Figure 6
continued

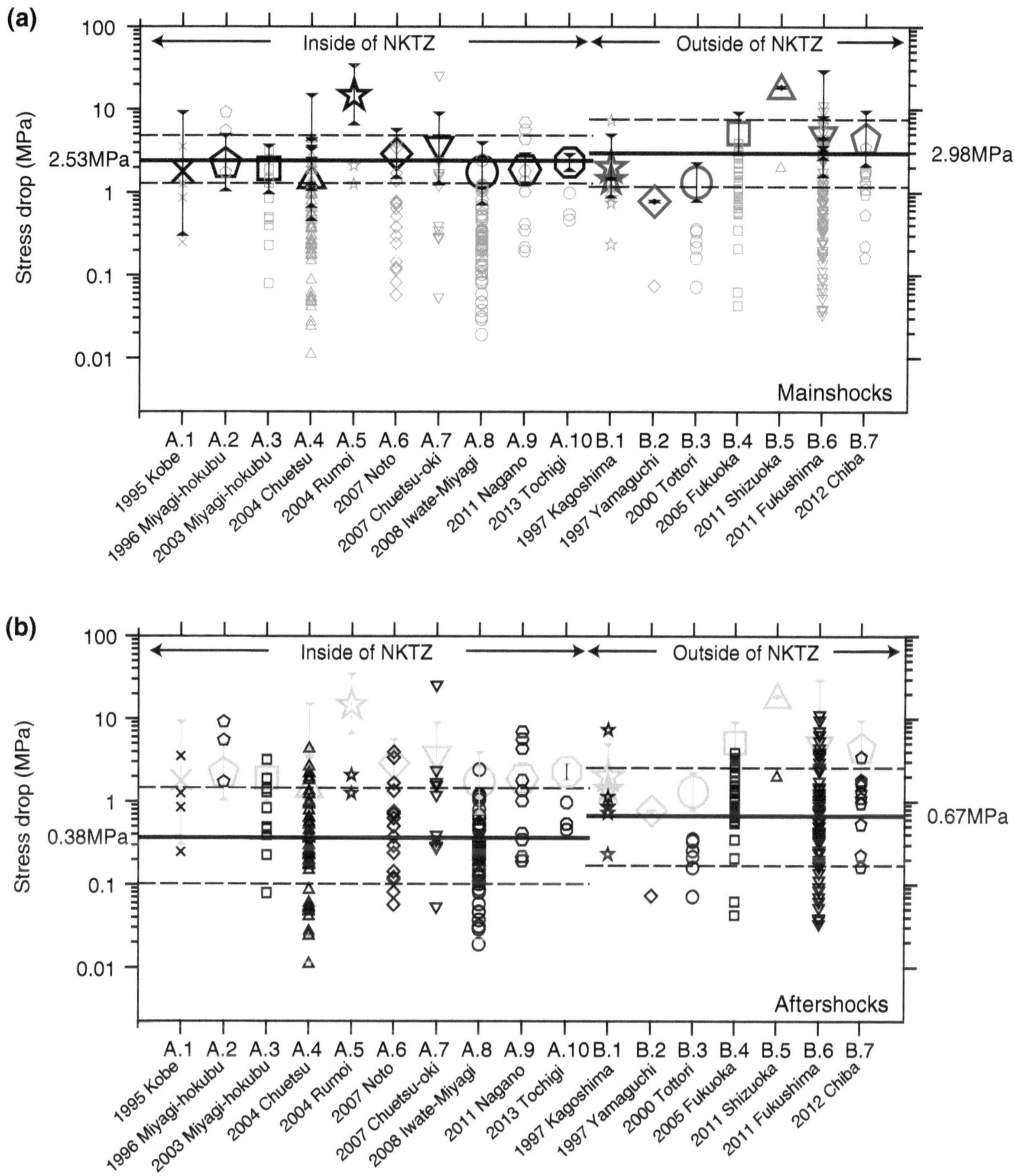

Figure 7

Stress drops of events in each earthquake sequence. Stress drops of mainshocks and aftershocks are represented by *large* and *small symbols*, respectively. The log average stress drops of mainshocks (**a**) and aftershocks (**b**) for events occurring inside and outside the NKTZ are separately indicated by *bold lines* with their log standard deviations (*dashed lines*)

4. Source Scaling

Assuming a circular crack fault, the Brune stress drop (hereafter "stress drop") $\Delta\sigma$ can be estimated from M_0 and f_c using the following relations (ESHELBY, 1957; BRUNE, 1970, 1971):

$$\Delta\sigma = \frac{7}{16}\left(\frac{M_0}{r^3}\right),\ f_c = 0.37\frac{\beta}{r}, \qquad (4)$$

where r is the source radius and β is the shear wave velocity at the source. We used the β extracted from the three-dimensional velocity structure model beneath the Japanese islands by MATSUBARA *et al.* (2008). The ranges of β used in this study for each sequence are listed in Table 1.

Our result of the relationship between the corner frequency f_c and seismic moment M_0 from all 17 sequences is shown in Fig. 5 with constant $\Delta\sigma$'s of 0.01, 0.1, 1, 10, and 100 MPa (solid lines). The log average corner frequencies for bins of 0.5 in M_w (solid diamonds in Fig. 5) indicate the difference between the stress drops of large and small events. The average stress drop of larger events ($4.5 \leq M_w \leq 7.0$) is 2.45 MPa, whereas that of smaller events ($3.2 \leq M_w < 4.5$) is 0.41 MPa. The stress drops of the mainshocks ($5.7 \leq M_w \leq 7.0$) are from 0.8 to 19 MPa over two orders of magnitude. In contrast, the stress drops of aftershocks ($3.2 \leq M_w \leq 6.2$) are scattering from 0.02 to 26 MPa over three orders of magnitude. The log average stress drops of mainshocks and aftershocks are 2.72 and 0.49 MPa, respectively. In these variations, the stress drops of most of the aftershocks do not exceed the log average stress drop of the mainshocks. Figure 6 also shows the source scaling for each sequence. The stress drops of aftershocks are mostly smaller than those of mainshocks for each sequence in Fig. 6. But occasionally, we can see aftershocks that have larger stress drops than the mainshock in Fig. 6.

5. Discussion

In the previous section, we showed the variations in stress drops are over two orders of magnitude for mainshocks ($5.7 \leq M_w \leq 7.0$) and three orders of magnitude for aftershocks ($3.2 \leq M_w \leq 6.2$). The average stress drops for each group are 2.72 and 0.49 MPa, respectively, and the stress drops depend on the seismic moment, which means that the self-similar source scaling is broken. This result is similar to source scalings for the 1999 Hector Mine earthquake sequence (MAYEDA *et al.*, 2007) and the 1999 Chi-Chi, Taiwan, earthquake sequence (MAYEDA and MALAGNINI, 2009). In Fig. 6, this kind of non-self-similar source scaling is shown in several event sequences, such as the 2004 Chuetsu, the 2005 Fukuoka, the 2007 Noto, the 2007 Chuetsu-oki, the 2008 Iwate-Miyagi, and the 2011 Fukushima earthquake sequences. On the contrary, self-similar source scalings can be found for the 1995 Kobe, the 1996 Miyagi-hokubu, the 1997 Kagoshima, and the 2011 Nagano earthquake sequences. TAJIMA and TAJIMA (2007, 2008) estimated the source scaling for three inland crustal earthquake sequences in Japan (the 2004 Chuetsu, the 2005 Fukuoka, and the 2007 Chuetsu-oki, which are the same as the sequences used in this study), by using the spectral ratio method for P-waves. IZUTANI (2005) also estimated the source scaling for the 2004 Chuetsu events by using the spectral ratio method for S-waves. YOO *et al.* (2010) estimated it for the 2005 Fukuoka events by using the coda spectral ratio method. They also showed the breaking self-similarity in each source scaling to be the same as our results for these earthquake sequences even though they used a different dataset and method of processing to estimate the source scaling.

We analyze 17 earthquake sequences. Although each earthquake sequence has either non-self or self-similar source scaling relationship, a break in self-similar source scaling can be found for the whole dataset as shown in Fig. 5. We show the average stress drop of larger events ($4.5 \leq M_w \leq 7.0$) is 2.45 MPa, whereas that of smaller events ($3.2 \leq M_w < 4.5$) is 0.41 MPa. There is a possibility for breaking self-similar source scaling about M_w 4.5. On the other, the average stress drops of 17 mainshocks and their aftershocks are 2.72 and 0.49 MPa, respectively, and those values are similar to the average stress drops of larger and smaller event groups. Therefore, the breakage of self-similar source scaling would be caused by the difference in source characteristics between the mainshocks and aftershocks. Stress drop analysis of mainshock events about M_w 4.0–5.0 is needed for the further discrimination.

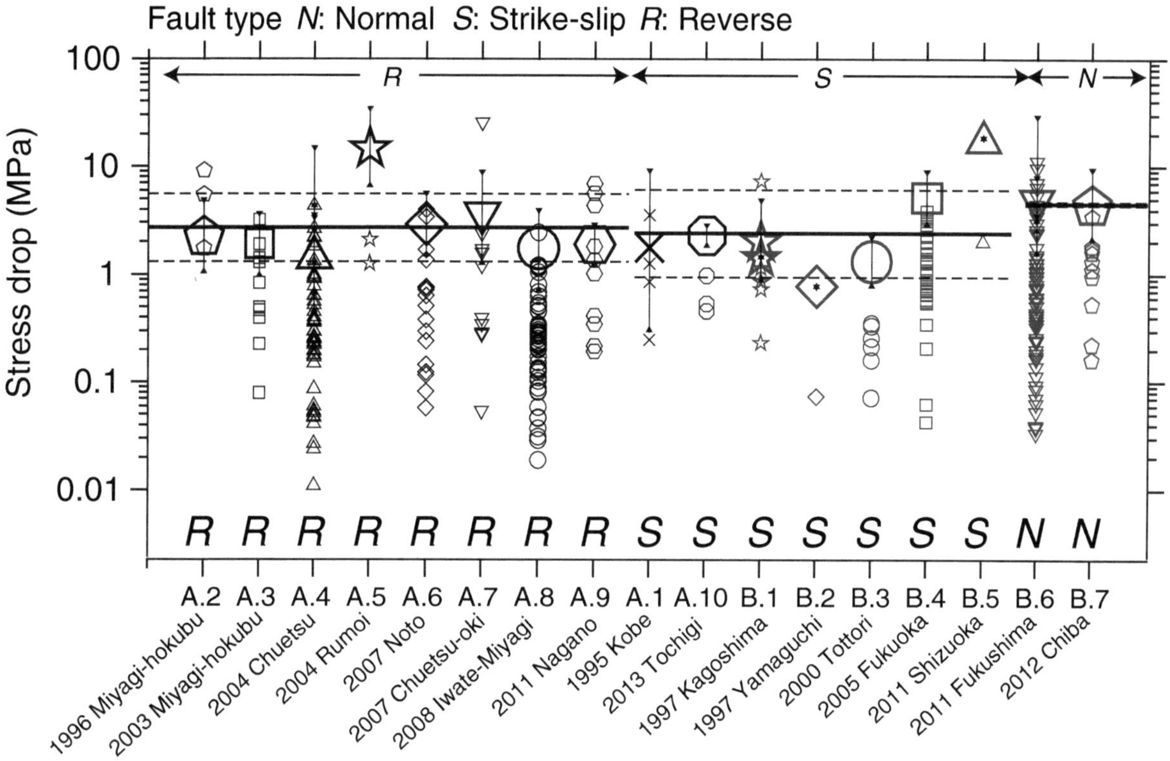

Figure 8
The log average stress drops of mainshocks for reverse, strike-slip, and normal fault types are separately indicated by *bold line* with their log standard deviations (*dashed lines*). The characters of "*R*", "*S*", and "*N*" indicate the reverse fault type, strike-slip fault type, and normal fault type. The notations are the same as in Fig. 7

From another point of view, we can also see the non-self-similar source scaling in source spectral ratios. In Figs. 3 and 4, the theoretical low and high frequency asymptote lines of the source spectral ratios are shown. The low frequency asymptote line is defined by the seismic moment ratio. The high frequency asymptote line is predicted by the ω^{-2} source spectral model with self-similar source scaling (i.e., the cubic root of the ratio of the seismic moments). If both ω^{-2} source spectral model and self-similar source scaling are valid in observation, the flat level at high frequencies for both the observed and modeled spectral ratios would be equal to the predicted high frequency asymptote line. From the results in Figs. 3 and 4, the predicted high frequency asymptote lines for most event pairs are not consistent with the observed and modeled spectral ratios. These results also give the information of non-self-similar source scaling in our dataset.

To investigate the factors of variations in stress drops independent of seismic moment, we examined

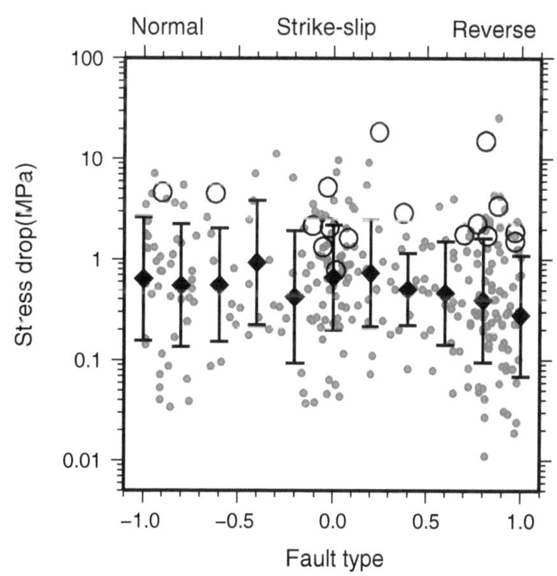

Figure 9
The relationship between the stress drop and fault type. The log average stress drops (*solid diamonds*) in 0.2 of Fault type bins with the *standard error bars* are calculated by aftershocks (*gray solid circles*). *Open circles* represent the mainshocks

possible correlations between stress drop and some tectonic effects such as regionality, focal mechanism, and source depth. First, we focus on a regional tectonic condition where the earthquake sequences have occurred. The 17 inland crustal earthquake sequences analyzed in this study are widely distributed in and around Japanese Islands (Fig. 1). As stated before, these sequences can be divided into two groups according to where a sequence has occurred inside or outside of the NKTZ. Figure 7 shows stress drops of all events with respect to each sequence. In Fig. 7a, the stress drops of mainshocks are represented by larger symbols with the log average stress drops of mainshocks and their log standard deviations occurring inside (i.e., group A) and outside (i.e., group B) the NKTZ. The average stress drops of mainshocks are 2.53 MPa for group A and 2.98 MPa for group B, respectively. For the mainshocks of the 2004 Rumoi (event #A.5) and the 2011 Shizuoka (event #B.5), we estimated larger stress drops than the average values. These mainshocks occurred in blind faults, which have not been identified by detailed geological surveys. However, for the other mainshocks which could be also categorized as events occurring in blind faults (e.g., the 1997 Kagoshima; the 1997 Yamaguchi; the 2013 Tochigi), their stress drops are not as large as those of the 2004 Rumoi and the 2011 Shizuoka events.

In Fig. 7b, the log average stress drops of aftershocks are shown with their log standard deviations. The average stress drops of aftershocks are 0.38 MPa for group A and 0.67 MPa for group B, respectively. The variation in stress drops of aftershocks is over three orders of magnitude. Although the number of aftershocks depends on each earthquake sequence, we could see that some earthquake sequences (e.g., the 2004 Chuetsu; the 2007 Chuetsu-oki; the 2011 Fukushima) have large variation in stress drop of aftershocks.

For both mainshocks and aftershocks, the log average stress drops are slightly different between the events occurring inside and outside the NKTZ. However, there is no obvious difference from each other considering their log standard deviations. These results suggest the possibility of other factors which are more effective to variation in stress drops.

To examine the dependence of our estimated stress drops on the fault mechanism types, we categorize the mainshocks into three fault mechanism types (i.e., reverse faults, strike-slip faults, and normal faults). Satoh (2006) estimated stress drops of small and moderate crustal earthquakes in Japan and found that the stress drops for reverse earthquakes are two times larger than those for normal earthquakes. For global earthquakes, Allmann and Shearer (2009) showed that stress drops for strike-slip earthquakes are larger than those for the other earthquakes for moderate to large earthquakes. The comparisons of the stress drops of mainshocks among three fault mechanism types are shown in Fig. 8 with their log average stress drops and log standard deviations. The log average stress drops of mainshocks with reverse, strike-slip, and normal fault types are 2.72, 2.39 and 4.56 MPa, respectively. Among the mainshocks we have analyzed, the average stress drop of the normal fault type is slightly larger than those of the strike and reverse fault types. This result is different from that of Satoh (2006). As we have only two events for the normal fault type, it is hard to say that our result demonstrates common characteristics for the fault type. Both log standard deviations of stress drops for reverse and strike-slip faulting mainshocks are larger than those for normal faulting mainshocks. Considering these log standard deviations, there might be other dependence of stress drops of mainshocks for reverse and strike-slip fault types.

We also examine the stress drops for all events including aftershocks with respect to fault type by parameterizing the moment tensor solution with scalar value from -1 to 1. The fault type is calculated by using the two rake angles described by Shearer *et al.* (2006). As shown in Fig. 9, there are no obvious differences of stress drop for aftershocks in fault type from the average stress drops in 0.2 of fault type bins excluding the mainshocks. Therefore, we conclude that a clear dependence of the faulting type of the mainshocks and aftershocks on stress drops cannot be seen in these results.

Next, we consider the depth dependence of stress drop. The stress drops of all the events excluding the mainshocks are plotted against their depth in Fig. 10. In Fig. 10, the stress drops for each event occurring inside the NKTZ (group A) and outside the NKTZ

Figure 10 ▶
The relationship between the stress drop and source depth for events occurring inside the NKTZ (Group A) (*top*), events occurring outside the NKTZ (Group B) (*middle*), and all events (*bottom*). The log average stress drops (*solid diamonds*) in 2 km depth bins with the *standard error bars* are calculated by events (*thick gray squares*) that we used from the relocated hypocenter information listed in Table 1. The other events are represented by *thin gray squares*. The *broken lines* represent the empirical relationships between the stress drop and source depth for all relocated events

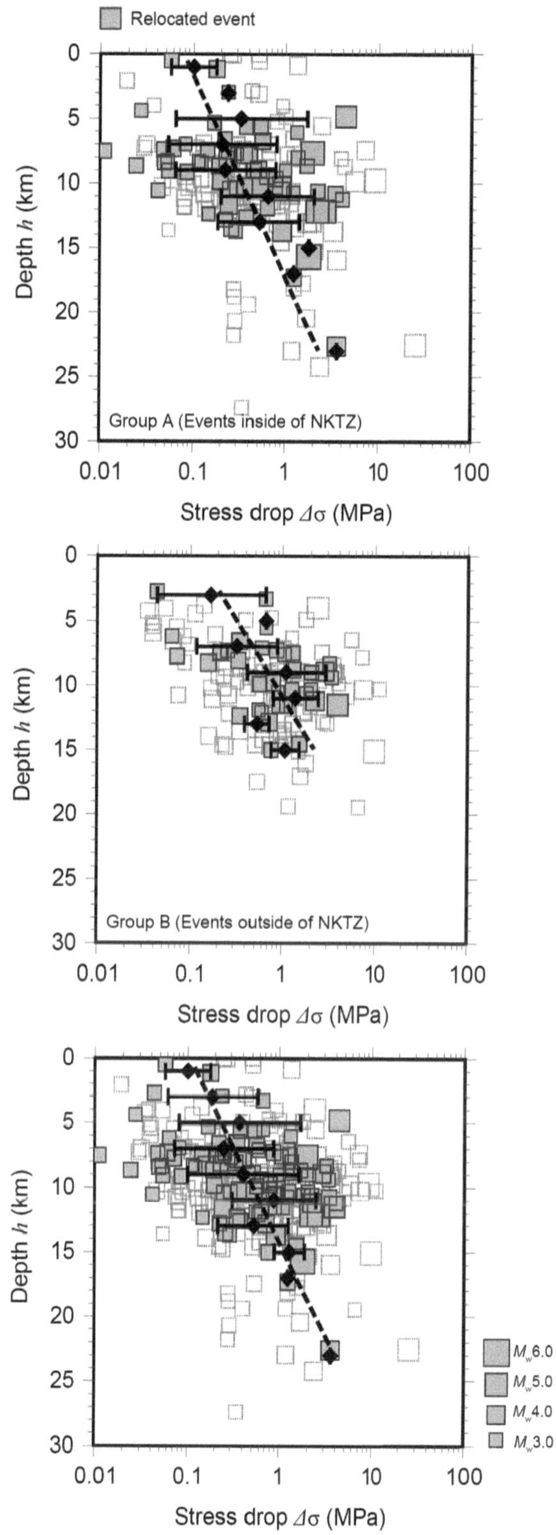

(group B) are also plotted. The average stress drops in 2 km depth bins with their standard deviations are calculated only for events for which we have the relocated hypocenter information listed in Table 1. In order to check the depth-dependency of the stress drop, an empirical relationship between the stress drop and source depth is estimated by using a dataset of all relocated events. As seen in Fig. 10, the stress drops of aftershocks clearly depend on their source depth. That is, a deeper event has a relatively larger stress drop than a shallower one. The difference in source characteristics between the events occurring inside NKTZ (group A) and outside NKTZ (group B) is not clear. A growth in stress drop with source depth was previously noted for various regions (e.g., JONES and HELMBERGER, 1996 for events in the Eastern Transverse Ranges, California; MORI *et al.*, 2003 for the 1994 Northridge, California, aftershocks; ASANO and IWATA, 2011a for the asperities of inland crustal earthquakes). Because the shear strength of the lithosphere increases with depth (SCHOLZ, 1990), the depth dependence of stress drops is reasonable.

Finally, we compare the crack size calculated from the corner frequency in this study with the total rupture area and the combined area of asperity which were characterized from the heterogeneous slip model obtained by waveform inversion listed in Table 2 to discuss the source dimensions obtained from the source spectrum and the heterogeneous slip models. The total rupture area and the combined area of asperity were extracted according to the definitions of SOMERVILLE *et al.* (1999). For three earthquakes (the 2011 Nagano, the 2011 Shizuoka, and the largest aftershock of the 2005 Fukuoka), total rupture areas are also estimated by distributions of the relocated aftershocks. We use the distributions of relocated

Table 2

The total rupture areas and the combined areas of asperities characterized from heterogeneous slip models

No.	Name of earthquake	S_{total} (km^2)	S_{asp} (km^2)	Slip model	Slip characterization
Inside NKTZ					
A.1	1995 Kobe	1,303.0	244.0	Sekiguchi *et al.* (2002)	Miyakoshi *et al.* (2008)
A.2	1996 Miyagi-hokubu	–	–	–	–
A.3	2003 Miyagi-hokubu	180.0	20.0	Hikima and Koketsu (2004)	This study
A.4	2004 Chuetsu	504.0	84.0	Asano and Iwata (2009)	Asano and Iwata (2011a)
A.5	2004 Rumoi	–	–	–	–
A.6	2007 Noto	480.0	96.0	Asano and Iwata (2011b)	Asano and Iwata (2011a)
A.7	2007 Chuetsu-oki	720.0	64.0	Aoi *et al.* (2008)	Miyakoshi *et al.* (2008)
A.8	2008 Iwate-Miyagi	684.0	124.0	Asano and Iwata (2011a)	Asano and Iwata (2011a)
A.9	2011 Nagano	308.0[a]	72.0	JMA (2011)	This study
A.10	2013 Tochigi	–	–	–	–
Outside NKTZ					
B.1	1997 Kagoshima (Mar.)	120.0	18.0	Miyakoshi *et al.* (2004)	Miyakoshi *et al.* (2008)
	1997 Kagoshima (May)	–	–	–	–
B.2	1997 Yamaguchi	112.0	18.0	Miyakoshi *et al.* (2004)	Miyakoshi *et al.* (2008)
B.3	2000 Tottori	598.0	101.0	Iwata and Sekiguchi (2002)	Miyakoshi *et al.* (2008)
B.4	2005 Fukuoka	468.0	64.0	Asano and Iwata (2006)	Asano and Iwata (2011a)
	2005 Fukuoka aft.	64.0[b]	9.0	Asano and Iwata (2006)	Asano and Iwata (2011a)
B.5	2011 Shizuoka	96.0[c]	16.0	JMA (2011)	This study
B.6	2011 Fukushima	640.0	144.0	Hikima (2011)	This study
B.7	2012 Chiba	–		–	–

S_{total} and S_{asp} are the total rupture area and combined area of asperity, respectively

Slip characterizations are carried out following the criteria of Somerville *et al.* (1999)

Rupture area based on the relocated distribution of aftershocks

[a] 122.0 km^2 (from the distribution of relocated aftershocks by Takeda, 2011a)

[b] 25.0 km^2 (from the distribution of relocated aftershocks by Uehira *et al.*, 2006)

[c] 41.8 km^2 (from the distribution of relocated aftershocks by Takeda, 2011b)

aftershocks whose periods are half a day after the 2011 Nagano mainshock, two days after the 2011 Shizuoka mainshock, and one day after the largest aftershock of the 2005 Fukuoka, respectively. As shown in Fig. 11 the crack size estimated from the corner frequency seems to correspond to the total rupture area of the heterogeneous slip model. Thus, we can treat the stress drops estimated in this analysis as the stress drops of total rupture area for large events.

6. Conclusions

We estimated corner frequencies and stress drops of 298 inland crustal earthquakes ($3.2 \leq M_w \leq 7.0$) including 17 mainshocks in Japan, to investigate the source scaling and variation in stress drops. We used the S-wave coda spectral ratio method to obtain the source spectral ratio from the observed record. The observed source spectral ratios from S-wave coda were more stable than those from direct S-waves. We obtained the corner frequency and the stress drop from the observed source spectral ratio by assuming the ω^{-2} source spectral model.

We could see a break in self-similar source scaling for the whole dataset from 17 earthquake sequences. The average stress drop for events larger than M_w 4.5 was larger than that for events smaller than M_w 4.5. In terms of the source scaling for each sequence, we could see either non-self or self-similar source scaling. However, the stress drops for most aftershocks did not exceed the stress drop of the mainshock. The breakage of self-similar source scaling about M_w 4.5 would be caused by the difference in source characteristics between the mainshocks and aftershocks. In our dataset, all small events ($M_w < 4.5$) are aftershocks. It is an important

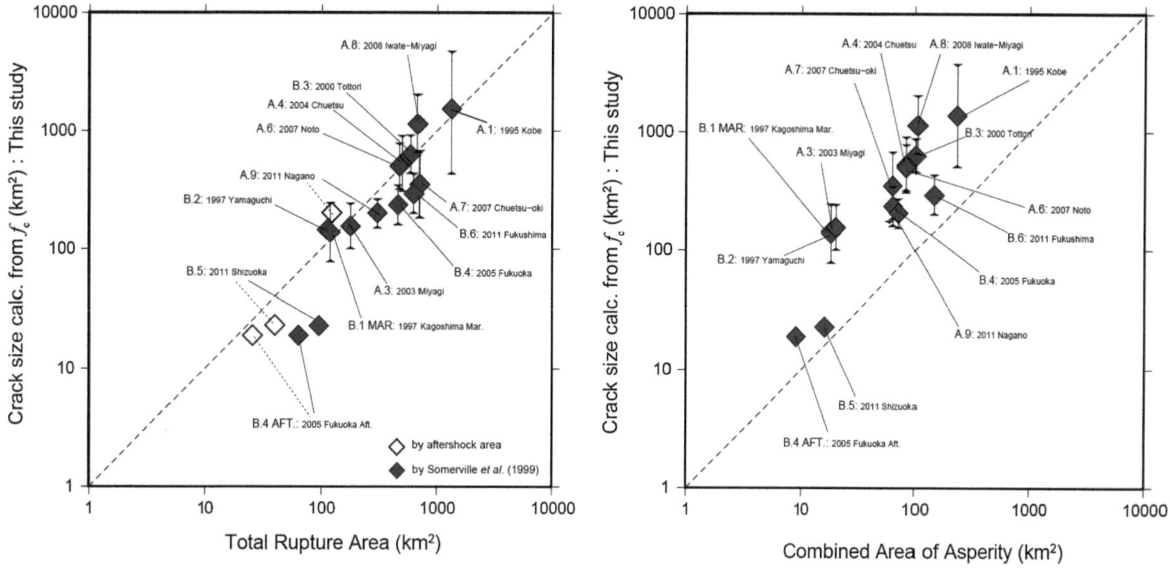

Figure 11

Comparison between the crack sizes calculated from the corner frequencies in this study with total rupture area (*left*) and combined area of asperity (*right*) characterized from the heterogeneous slip models by kinematic waveform inversions listed in Table 2. For three earthquakes (the 2011 Nagano, the 2011 Shizuoka, and the largest aftershock of the 2005 Fukuoka), *open symbols* indicate that the total rupture area was estimated from the relocated distribution of aftershocks

future task for the discrimination to analyze the stress drop of mainshocks about M_w 4.0–5.0. Occasionally, we could see aftershocks that have larger stress drops than the mainshock. Examining the characteristics of these aftershocks is also a task for the future. One of the possible reasons is that these aftershocks have occurred independent of the fault plane of the mainshock, that is, the effects of the mainshock are less for these aftershocks in the sequence.

Overall, the variations in stress drops were over two orders of magnitude for mainshocks ($5.7 \leq M_w \leq 7.0$) and three orders of magnitude for aftershocks ($3.2 \leq M_w \leq 6.2$). For these variations in stress drops independent of seismic moment, we examined possible relationships between stress drops and other properties. There were no obvious differences between stress drops of events occurring inside and outside the NKTZ in Japan. We could not see a clear dependence of the faulting type on stress drops. The stress drop clearly depended on depth of aftershocks. By comparing the crack size calculated from the estimated corner frequency of some large earthquakes in this study with the total rupture area and the asperity area characterized from the heterogeneous slip model, the crack size seemed to correspond to total rupture area. It could mean that we estimated the stress drop of total rupture area for the large earthquake in this analysis. We could suggest that the average source scaling relationships for inland crustal earthquakes could be effective to predict the strong ground motion from an earthquake scenario occurring in the NKTZ.

Acknowledgments

We deeply appreciate Aitaro Kato and Kenji Uehira for allowing us to use their relocated hypocenter catalogs. We would like to also thank Kazuhito Hikima and Kazuhiro Iwakiri for providing their slip models. The comments from two anonymous reviewers and a guest editor were quite helpful in improving the manuscript. The waveform data were provided by NIED (K-NET, KiK-net, and F-net), CEORKA, and DPRI, Kyoto University. A hypocenter catalog provided by JMA and moment tensor solutions determined by F-net of NIED were used in this study. Most of figures were drawn by using the Generic Mapping Tools (Wessel and Smith, 1998). This study was supported by the Research Project for

Intensive Survey and Study on the Concentrated Strain Zone of the Ministry of Education, Culture, Sports, Science and Technology (MEXT), Japan. Finally, we thank Adrien Oth, corresponding editor of this special issue, for his kind encouragement to submit this work.

REFERENCES

AKI, K. (1967), Scaling Law of Seismic Spectrum, J. Geophys. Res. *72*(4), 1217–1231, doi:10.1029/JZ072i004p01217.

AKI, K., and CHOUET, B. (1975), Origin of Coda Waves: Source, Attenuation, and Scattering Effects, J. Geophys. Res. *80*, 3322–3342.

ALLMANN, B.P., and SHEARER, P.M. (2009), Global Variations of Stress Drop for Moderate to Large Earthquakes, J. Geophys. Res. *114*, B01310, doi:10.1029/2008JB005821.

AOI, S., OBARA, K., HORI, S., KASAHARA, K., and OKADA, Y. (2000), New Japanese Uphole/Downhole Strong-Motion Observation Network: KiK-net, Seismol. Res. Lett. *72*, 239.

AOI, S., SEKIGUCHI, H., MORIKAWA, N., and KUNUGI, T. (2008), Source Process of the 2007 Niigata-ken Chuetsu-oki Earthquake Derived from Near-Fault Strong Motion Data, Earth Planets Space *60*, 1131–1135.

ASANO, K., and IWATA, T. (2009), Source Rupture Process of the 2004 Chuetsu, Mid-Niigata Prefecture, Japan, Earthquake Inferred from Waveform Inversion with Dense Strong-Motion Data, Bull. Seismol. Soc. Am. *99*, 123–140.

ASANO, K., and IWATA, T. (2006), Source Process and Near-Source Ground Motions of the 2005 West Off Fukuoka Prefecture Earthquake, Earth Planets Space *58*, 93–98.

ASANO, K., and IWATA, T. (2011a), Characterization of Stress Drops on Asperities Estimated from the Heterogeneous Kinematic Slip Model for Strong Motion Prediction for Inland Crustal Earthquakes in Japan, Pure Appl. Geophys. *168*, 105–116.

ASANO, K., and IWATA, T. (2011b), Source-Rupture Process of the 2007 Noto Hanto, Japan, Earthquake Estimated by the Joint Inversion of Strong Motion and GPS Data, Bull. Seismol. Soc. Am. *101*, 2467–2480.

BALTAY, A., IDE, S., PRIETO, G., and BEROZA, G.C. (2011), Variability in Earthquake Stress Drop and Apparent Stress, Geophys. Res. Lett. *38*, L06303, doi:10.1029/2011GL046698.

BRUNE, J.N. (1970), Tectonic Stress and Spectra of Seismic Shear Waves from Earthquakes, J. Geophys. Res. *75*, 4997–5009.

BRUNE, J.N. (1971), Tectonic Stress and Seismic Shear Waves from Earthquakes, Correction, J. Geophys. Res. *76*, 5002.

COCCO, M., and ROVELLI, A. (1989), Evidence for the Variation of Stress Drop Between Normal and Thrust Faulting Earthquakes in Italy, J. Geophys. Res. *94*, 9399–9416.

COTTON, F., ARCHULETA, R., and CAUSSE, M. (2013), What is Sigma of the Stress Drop? Seismol. Res. Lett. *84*, 42–48, doi:10.1785/0220120087.

EKSTRÖM, G., DZIEWONSKI, A.M., MATERNOVSKAYA, N.N., and NETTLES, M. (2005), Global Seismicity of 2003: Centroid-Moment Tensor Solutions for 1087 Earthquakes, Phys. Earth Planet. Inter. *148*, 327–351.

ESHELBY, J. (1957), The Determination of the Elastic Field of an Ellipsoidal Inclusion, and Related Problems, Proc. R. Soc. London, Ser. A *241*, 376–396.

FUKUYAMA, E., ELLSWORTH, W.L., WALDHAUSER, F., and KUBO, A. (2003), Detailed Fault Structure of the 2000 Western Tottori, Japan, Earthquake Sequence, Bull. Seismol. Soc. Am. *93*, 1468–1478.

FUKUYAMA, E., ISHIDA, M., DREGER, D.S., and KAWAI, H. (1998), Automated Seismic Moment Tensor Determination Using On-Line Broadband Seismic Waveforms, Zisin 2 (J. Seismol. Soc. Jpn.) *51*, 149–156 (in Japanese with English abstract).

HARTZELL, S. (1978), Earthquake Aftershocks as Green's Functions, Geophys. Res. Lett. *5*, 1–14.

HIKIMA, K. (2011), Rupture Process of the April 11, 2011 Fukushima Hamadori Earthquake (Mj7.0) Inferred from Strong Motion Data, Zisin 2 (J. Seismol. Soc. Jpn.) *64*, 243–256 (in Japanese with English abstract).

HIKIMA, K., and KOKETSU, K. (2004), Source Processes of the Foreshock, Mainshock and Largest Aftershock in the 2003 Miyagi-ken Hokubu, Japan, Earthquake Sequence, Earth Planets Space *56*, 87–93, 2004.

HOUGH, S. (1997), Empirical Green's Function Analysis: Taking the Next Step, J. Geophys. Res. *102*, 5369–5380.

IDE, S., BEROZA, G.C., PREJEAN, S.G., and ELLSWORTH, W.L. (2003), Apparent Break in Earthquake Scaling Due to Path and Site Effects in Deep Borehole Recordings, J. Geophys. Res. *108*(B5), 2271, doi:10.1029/2001JB001617.

IIO, Y. (1986), Scaling Relation Between Earthquake Size and Duration of Faulting for Shallow Earthquakes in Seismic Moment Between 10^{10} and 10^{25} dyne·cm, J. Phys. Earth *34*, 127–169.

IMANISHI, K., TAKEO, M., ELLSWORTH, W.L., ITO, H., MATSUZAWA, T., KUWAHARA, Y., IIO, Y., HORIUCHI, S., and OHMI, S. (2004), Source Parameters and Rupture Velocity of Microearthquakes in Western Nagano, Japan, Determined using Stopping Phases, Bull. Seismol. Soc. Am. *94*, 1762–1780.

IWATA, T., and IRIKURA, K. (1988), Source Parameters of the 1983 Japan Sea Earthquake Sequence, J. Phys. Earth *36*, 155–184.

IWATA, T., and SEKIGUCHI, H. (2002), Rupture Process and Near-Source Ground Motions of the 2000 Tottoriken-Seibu Earthquake, Proc. 11th Japan Symp. Earthq. Eng. 121–125 (in Japanese with English abstract).

IZUTANI, Y. (2005), Radiated Energy from the Mid Niigata, Japan, Earthquake of October 23, 2004, and its Aftershocks, Geophys. Res. Lett. *32*, L21313, doi:10.1029/2005GL024116.

IZUTANI, Y., and KANAMORI, H. (2001), Scale-Dependence of Seismic Energy-to-Moment Ratio for Strike-Slip Earthquakes in Japan, Geophys. Res. Lett. *28*(20), 4007–4010, doi:10.1029/2001GL013402.

Japan Meteorological Agency (2011), http://www.seisvol.kishou.go.jp/eq/sourceprocess/.

JONES, L.E., and HELMBERGER, D.V. (1996), Seismicity and Stress-Drop in the Eastern Transverse Ranges, Southern California, Geophys. Res. Lett. *23*, 233–236.

JOST, M.L., BÜßELBERG, T., JOST, Ö., and HARJES, H.-P. (1998), Source Parameters of Injection-Induced Microearthquakes at 9 km Depth at the KTB Deep Drilling Site, Germany, Bull. Seism. Soc. Am. *88*, 815–832.

KAGAWA, T., IEMURA, H., IRIKURA, K., and TOKI, K. (2004), Strong Ground Motion Observation by the Committee of Earthquake Observation and Research in the Kansai Area (CEORKA), J. Japan Assoc. Earthq. Eng. *4*(3):128–133.

KANAMORI, H., and ANDERSON, D.L. (1975), Theoretical Basis of Some Empirical Relations in Seismology, Bull. Seismol. Soc. Am. *65*, 1073–1095.

KATAO, H., MAEDA, N., HIRAMATSU, Y., IIO, Y., and NAKAO, S. (1997), Detailed Mapping of Focal Mechanisms in/around the Hyogo-ken Nanbu Earthquake Rupture Zone, J. Phys. Earth *45*, 105–119.

KATO, A., SAKAI, S., HIRATA, N., KURASHIMO, E., IIDAKA, T., IWASAKI, T., and KANAZAWA, T. (2006), Imaging the Seismic Structure and Stress Field in the Source Region of the 2004 Mid-Niigata Prefecture Earthquake: Structural Zones of Weakness and Seismogenic Stress Concentration by Ductile Flow, J. Geophys. Res. *111*, B08308, doi:10.1029/2005JB004016.

KATO, A., SAKAI, S., IIDAKA, T., IWASAKI, T., KURASHIMO, E., IGARASHI, T., HIRATA, N., KANAZAWA, T., and Group for the Aftershock Observations of the 2007 Noto Hanto Earthquake (2008), Three-Dimensional Velocity Structure in the Source Region of the Noto Hanto Earthquake in 2007 Imaged by a Dense Seismic Observation, Earth Planets Space *60*, 105–110.

LINDLEY, G.T. (1994), Source Parameters of the 23 April 1992 Joshua Tree, California, Earthquake, Its largest Foreshock, and Aftershocks, Bull. Seismol. Soc. Am. *84*, 1051–1057.

KAWASE, H., and MATSUO, H. (2004), Amplification Characteristics of K-NET, KiK-net, and JMA Shindokei Network Sites Based on the Spectral Inversion Technique, Proc. 13th World Conf. Earthq. Eng. paper no. 454.

KINOSHITA, S. (1998), Kyoshin-net (K-NET), Seismol. Res. Lett. *69*, 309–332.

MALAGNINI, L., and MAYEDA, K. (2008), High-Stress Strike-Slip Faults in the Apennines: An Example from the 2002 San Giuliano Earthquakes (Southern Italy), Geophys. Res. Lett. *35*, L12302, doi:10.1029/2008GL034024.

MATSUBARA, M., OBARA, K., and KASAHARA, K. (2008), Three-Dimensional P-and S-Wave Velocity Structures Beneath the Japan Islands Obtained by High-Density Seismic Stations by Seismic Tomography, Tectonophysics *454*, 86–103, doi:10.1016/j.tecto.2008.04.016.

MATSUNAMI, K., and HACHIRI, T. (1995), Seismic Observations for Strong Ground Motions around Lake Biwa, Annuals Disas. Prev. Res. Inst., Kyoto Univ. 38B-1, 1–12 (in Japanese with English abstract).

MAYEDA, K., MALAGNINI, L., and WALTER, W.R. (2007), A New Spectral Ratio Method Using Narrow Band Coda Envelopes: Evidence for Non-Self-Similarity in the Hector Mine Sequence, Geophys. Res. Lett. *34*, L11303, doi:10.1029/2007GL030041.

MAYEDA, K., and MALAGNINI, L. (2009), Apparent Stress and Corner Frequency Variations in the 1999 Taiwan (Chi–Chi) Sequence: Evidence for a Step-Wise Increase at Mw-5.5, Geophys. Res. Lett. *36*, L10308, doi:10.1029/2009GL037421.

MAYEDA, K., and WALTER, W.R. (1996), Moment, Energy, Stress Drop, and Source Spectra of Western United States Earthquakes from Regional Coda Envelopes, J. Geophys. Res. *101*(B5), 11,195-11,208, doi:10.1029/96JB00112.

MCGARR, A. (1999), On Relating Apparent Stress to the Stress Causing Earthquake Fault Slip, J. Geophys. Res. *104*(B2), 3003–3011, doi:10.1029/1998JB900083.

MIYAKE, H., IWATA, T., and IRIKURA, K. (2001), Estimation of Rupture Propagation Direction and Strong Motion Generation Area from Azimuth and Distance Dependence of Source Amplitude Spectra, Geophys. Res. Lett. *28*, 2727–2730.

MIYAKE, H., IWATA, T., and IRIKURA, K. (2003), Source Characterization for Broadband Ground-Motion Simulation: Kinematic Heterogeneous Source Model and Strong Motion Generation Area, Bull. Seismol. Soc. Am. *93*, 2531–2545.

MIYAKOSHI, K., KURAHASHI, S., and IRIKURA, K. (2008), Review of Source Models for the 2007 Niigata ken Chuetsu Oki Earthquake, Abstr. Seismol. Soc. Jpn. Fall Meet. C11-01, 2008 (in Japanese).

MIYAKOSHI, K., PETUKHIN, A., and KAGAWA, T. (2004), Source Modeling of Inland Earthquakes for the Intermediate Period Range Case study of 1997 Kagoshima Ken Hokuseibu (March) and 1997 Yamaguchi ken Hokubu earthquakes, Abstr. Seismol. Soc. Jpn. Fall Meet. P-065, 2004 (in Japanese).

MORI, J., ABERCROMBIE, R.E., and KANAMORI, H. (2003), Stress Drops and Radiated Energies of Aftershocks of the 1994 Northridge, California, Earthquake, J. Geophys. Res. *108*(B11), 2545, doi:10.1029/2001JB000474.

MUELLER, C. (1985), Source Pulse Enhancement by Deconvolution of an Empirical Green's Function, Geophys. Res. Lett. *12*, 33–36.

NAKAGAWA, H. (1997), Source Parameters of the 1995 Hyogo-ken Nanbu Earthquake Sequences, M. Thesis, Kyoto University.

OKAMURA, Y. (2010), Relationships between Geological Structure and Earthquake Source Faults along the Eastern Margin of the Japan Sea, J. Geol. Soc. Japan, *116*(11), 582–591 (In Japanese with English abstract).

OTH, A., BINDI, D., PAROLAI, S., and DI GIACOMO, D. (2010), Earthquake Scaling Characteristics and the Scale-(in)dependence of Seismic Energy-to-Moment Ratio: Insights from KiK-net Data in Japan. Geophys. Res. Lett. *37*, L19304, doi:10.1029/2010GL044572.

OTH, A., BINDI, D., PAROLAI, S., and DI GIACOMO, D. (2011), Spectral Analysis of K-NET and KiK-net Data in Japan, Part II: On Attenuation Characteristics, Source Spectra, and Site Response of Borehole and Surface Stations, Bull. Seismol. Soc. Am. *101*(2), 667–687.

PREJEAN, S.G., and ELLSWORTH, W.L. (2001), Observations of Earthquake Source Parameters and Attenuation at 2 km Depth in the Long Valley Caldera, Eastern California, Bull. Seismol. Soc. Am., *91*, 165–177.

PRIETO, G.A., SHEARER, P.M., VERNON, F.L., and KILB, D. (2004), Earthquake Source Scaling and Self-Similarity Estimation from Stacking P and S Spectra, J. Geophys. Res., *109*, B08310, doi:10.1029/2004JB003084.

RAUTIAN, T.G., and KHALTURIN, V.I. (1978), The Use of the Coda for Determination of the Earthquake Source Spectrum, Bull. Seismol. Soc. Am. *68*(4), 923–948.

SAGIYA, T., MIYAZAKI, S., and TADA, T. (2000), Continuous GPS Array and Present-Day Crustal Deformation of Japan, Pure Appl. Geophys. *157*, 2302–2232.

SATOH, T. (2006), Influence of Fault Mechanism, Depth, and Region on Stress Drops of Small and Moderate Earthquakes in Japan, Struct. Eng./Earthq. Eng., Jpn. Soc. Civil Eng. *23-1*, 125–134.

SCHOLZ, C.H., The Mechanics of Earthquakes and Faulting (Cambridge University Press, New York, 1990).

SEKIGUCHI, H., IWATA, T., and IRIKURA, K. (2002), Source Inversion for Estimating Continuous Slip Distribution on the Fault – Introduction of Green's Functions Convolved with a Correction Function to give Moving Dislocation Effects in Subfaults, Geophys. J. Int. *150*, 377–391.

SHEARER, P., PRIETO, G., and HAUKSSON, E. (2006), Comprehensive Analysis of Earthquake Source Spectra in Southern California, J. Geophys. Res. *111*, B06303, doi:10.1029/2005JB003979.

SOMERVILLE, P., IRIKURA, K., GRAVES, G., SAWADA, S., WALD, D., ABRAHAMSON, N., IWASAKI, Y., KAGAWA, T., SMITH, N., and KOWADA, A. (1999), Characterizing Crustal Earthquake Slip Models for the Prediction of Strong Ground Motion. Seism. Res. Lett. *70*, 59–80.

TAJIMA, R., and TAJIMA, F. (2007), Seismic Scaling Relations and Aftershock Activity from the Sequences of the 2004 Mid Niigata and the 2005 West Off Fukuoka Earthquakes (Mw 6.6) in Japan, J. Geophys. Res. *112*, B10302, doi:10.1029/2007JB004941.

TAJIMA, R., and TAJIMA, F. (2008), Seismic Scaling Relation of the 2007 Off Mid Niigata, Japan, Earthquake (M_w 6.6) Sequence in Comparison with Two Other Earthquake (M_w 6.6) Sequences, Earth Planets Space *60*, 1137–1141.

TAKAHASHI, T., SATO, H., OHTAKE, M., and OBARA, K. (2005), Scale Dependence of Apparent Stress for Earthquakes along the Subducting Pacific Plate in Northeastern Honshu, Japan, Bull. Seismol. Soc. Am., *95*, 1334–1345.

TAKEDA, T. (2011), The Earthquake in the Vicinity of Nagano and Niigata Prefecture Boundary on Mar. 12, 2011, Report of the Coordinating Committee for Earthquake Prediction, Japan *86*, 7–1 (in Japanese).

TAKEDA, T. (2011), The Earthquake in the East Area of Shizuoka Prefecture on Mar. 15, 2011, Report of the Coordinating Committee for Earthquake Prediction, Japan *86*, 6–3 (in Japanese).

UEHIRA, K., YAMADA, T., SHINOHARA, M., NAKAHIGASHI, K., MIYAMACHI, H., IIO, Y., OKADA, T., TAKAHASHI, H., MATSUWO, N., UCHIDA, K., KANAZAWA, T., and SHIMIZU, H. (2006), Precise Aftershock Distribution of the 2005 West Off Fukuoka Prefecture Earthquake (Mj = 7.0) using a Dense Onshore and Offshore Seismic Network, Earth Planets Space *58*, 1605–1610.

WESSEL, P., and SMITH, W.H.F. (1998), New, Improved Version of Generic Mapping Tools Released, Eos Trans. AGU *79*, 579.

YOO S.H., RHIE, J., CHOI, H., and MAYEDA, K. (2010), Evidence for Non-Self-Similarity and Transitional Increment of Scaled Energy in the 2005 West Off Fukuoka Seismic Sequence, J. Geophys. Res. *115*, B08308, doi:10.1029/2009JB007169.

(Received May 21, 2013, revised December 28, 2013, accepted January 2, 2014, Published online February 28, 2014)

Pure Appl. Geophys. 171 (2014), 2767–2782

DOI 10.1007/s00024-013-0751-1

Pure and Applied Geophysics

Stress Release and Source Scaling of the 2010–2011 Canterbury, New Zealand Earthquake Sequence from Spectral Inversion of Ground Motion Data

Adrien Oth[1] and Anna E. Kaiser[2]

Abstract—The Canterbury earthquake sequence beginning with the 2010 M_W 7.2 Darfield earthquake is one of the most notable and well-recorded crustal earthquake sequences in a low-strain-rate region worldwide and as such provides a unique opportunity to better understand earthquake source physics and ground motion generation in such a tectonic setting. Ground motions during this sequence ranged up to extreme values of 2.2 g, recorded during the February 2011 M_W 6.2 event beneath the city of Christchurch. A better understanding of the seismic source signature of this sequence, in particular the stress release and its scaling with earthquake size, is crucial for future ground motion prediction and hazard assessment in Canterbury, but also of high interest for other low-to-moderate seismicity regions where high-quality records of large earthquakes are lacking. Here we present a source parameter study of more than 200 events of the Canterbury sequence, covering the magnitude range M_W 3–7.2. Source spectra were derived using a generalized spectral inversion technique and found to be well characterized by the ω^{-2} source model. We find that stress drops range between 1 and 20 MPa with a median value of 5 MPa, which is a factor of 5 larger than the median stress drop previously estimated with the same method for crustal earthquakes in much more seismically active Japan. Stress drop scaling with earthquake size is nearly self-similar, and we identify lateral variations throughout Canterbury, in particular high stress drops at the fault edges of the two major events, the M_W 7.2 Darfield and M_W 6.2 Christchurch earthquakes.

1. Introduction

On 4 September 2010, the M_W 7.2 Darfield earthquake struck the Canterbury region on the South Island of New Zealand, giving rise to the beginning of the remarkable Canterbury earthquake sequence (Bannister and Gledhill 2012). The Darfield event occurred on the previously unmapped Greendale fault in the Canterbury Plains (Quigley *et al.* 2010), an area of comparatively low seismicity prior to this event. The area is located ~100 km from the major plate boundary through the South Island (Alpine Fault; Fig. 1) and deformation rates over the ~120 km wide Canterbury Plains region are estimated to be ~2 mm/year based on GPS-derived strain rates (Wallace *et al.* 2007). For these reasons, the source region of this sequence can be viewed as an intra-plate tectonic setting.

On 22 February 2011, the Darfield event was followed by a devastating aftershock of moment magnitude M_W 6.2, located almost immediately below the city of Christchurch, and two further events nearby on 13 June 2011 (M_W 6.0) and 23 December 2011 (M_W 5.9). In particular, the former event had tragic consequences, claiming 185 lives and causing large-scale destruction throughout the city, with over 9 billion US$ in damage (Bannister and Gledhill 2012).

The Canterbury earthquakes have produced remarkably intense ground shaking (Fry *et al.* 2011) and liquefaction phenomena have been widespread throughout the region (Orense *et al.* 2011; Kaiser *et al.* 2012). During the M_W 6.2 Christchurch event, ground motions of up to 2.2 g (vertical, 1.7 g on the horizontal components) were recorded close to the epicenter (Kaiser *et al.* 2012), and the records of this event are characterized by a very rich high frequency content evident on the vertical component of recordings and at near-source rock site LPCC. Clearly, site response phenomena play an important role in the generation of strong ground motion (e.g.,

Electronic supplementary material The online version of this article (doi:10.1007/s00024-013-0751-1) contains supplementary material, which is available to authorized users.

[1] European Center for Geodynamics and Seismology, 19, rue Josy Welter, 7256 Walferdange, Grand-Duchy of Luxembourg. E-mail: adrien.oth@ecgs.lu

[2] GNS Science, 1 Fairway Drive, Avalon, 5010, PO Box 30-368, Lower Hutt, New Zealand. E-mail: a.kaiser@gns.cri.nz

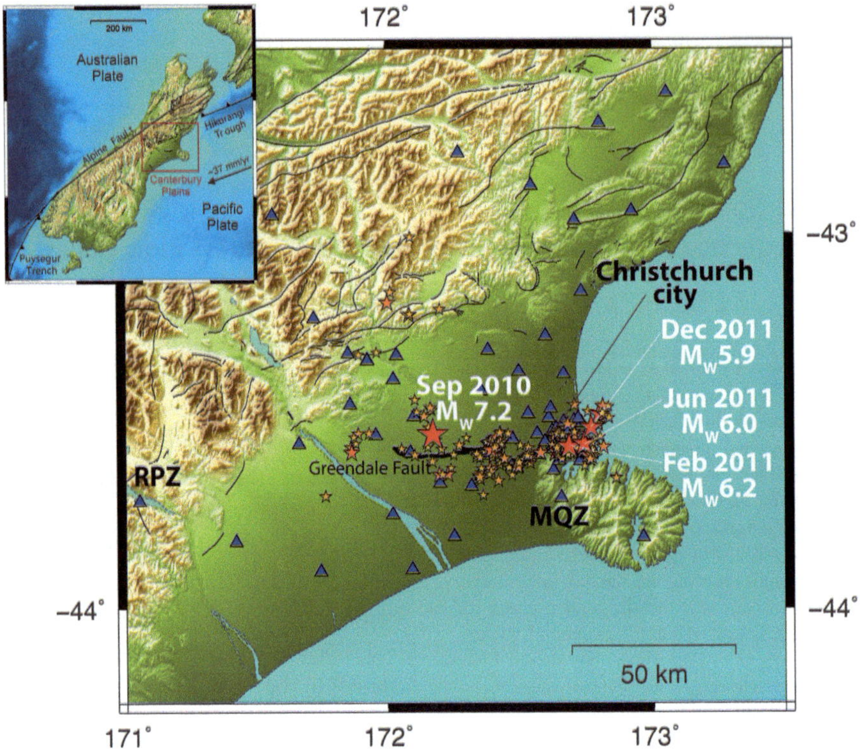

Figure 1

Epicenters of earthquakes (*stars*) and stations (*blue triangles*) used in this study. *Red stars* indicate the September 2010 (M_W 7.2, Darfield earthquake), February 2011 (M_W 6.2, Christchurch earthquake), June 2011 (M_W 6.0) and December 2011 (M_W 5.9) shocks. The Greendale fault (main fault involved in the 2010 Darfield event) is indicated as a *black line*. Locations for the most significant events are derived from additional double difference relocation analysis (e.g., Bannister and Gledhill 2012)

Fry *et al.* 2011; Bradley 2012). However, previous studies came to the conclusion that these features observed for the largest Canterbury events are also linked to particularly strong seismic energy radiation relative to their seismic moment, and thus high apparent stress (Fry and Gerstenberger 2011; Holden 2011).

The Canterbury earthquake sequence provides a unique opportunity to learn more about the source characteristics of major earthquakes in intra-plate tectonic settings, where events such as the September 2010 Darfield or February 2011 Christchurch earthquakes represent extreme events with very low probability of occurrence. As a consequence, very few high-quality near-source recordings of such events exist worldwide, and as Fry and Gerstenberger (2011) point out, it may well be the case that such extreme ground motions could be a common feature for these types of events. In this context, the source characteristics of the large magnitude events are clearly of interest, but in particular also the scaling of these parameters between small and large events. Upscaling techniques using recordings of small earthquakes as empirical Green's functions (e.g., Irikura 1986) are a key element in deterministic seismic hazard assessment in regions where only very limited data from past large magnitude events are available.

The aim of this study is therefore to make use of the rich database of strong motion recordings within the New Zealand GeoNet monitoring network (Petersen *et al.* 2011) in order to provide an in-depth analysis of the source spectral characteristics of the Canterbury earthquakes. To this end, we use a spectral inversion technique that allows for a separation of source, path and site contributions within the recorded ground motion spectra, and focus on the source term in this article. We first provide a brief overview of the dataset and the methods used, followed by a discussion of the source spectra resulting from the inversion and the calculated stress drops, with particular emphasis on their scaling behavior and lateral variations.

2. *Dataset and Spectra Calculation*

The dataset under investigation consists of 2,415 accelerograms from 205 earthquakes that were recorded at 64 stations throughout the Canterbury Plains, with a particular density of recording sites in the Christchurch urban area (Fig. 1). Of these 64 stations, five belong to the New Zealand National Seismograph Network (NZNSN, Petersen *et al.* 2011), where broadband and strong-motion instruments are collocated. Strong motion data are acquired with Kinemetrics episensors and 24-bit Quanterra Q330/Q4120 data loggers, sampled at 100 Hz. The remaining stations are part of the New Zealand National Strong Motion Network and are equipped either with Kinemetrics episensors or Canterbury Seismic Instruments (CSI) CUSP units, coupled with 24-bit Quanterra Q330/Q4120 or Kinemetrics Basalt data loggers, most of these providing acceleration records with 18-bit resolution. The sampling rate at the strong motion stations is 200 Hz, and the full recording range is ±4 g. Detailed information on the seismic networks in New Zealand can be found on the GeoNet website (www.geonet.org.nz).

All records are sourced from the GeoNet strong motion catalogue (ftp://ftp.geonet.org.nz/strong/) and have signal-to-noise ratio larger than 3 in the frequency band of analysis, 0.5–20 Hz. During the data selection process, only stations and earthquakes with at least three recordings were considered, and records with peak ground accelerations higher than 0.15 g were excluded in order to avoid bias in the source spectra due to the presence of strong non-linear site effects (the applied spectral inversion scheme is based on the assumption of linear soil response and the occurrence of widespread liquefaction clearly indicates the existence of non-linear soil behavior). While seismic moments for all events except the largest ones are determined through the analysis (see following section), it is desirable to have access to independent, robust magnitude estimates serving as benchmarks. For this purpose, where available, we utilize moment magnitude (M_W) derived from GeoNet regional moment tensor analysis (Ristau 2008). For the remaining events where only GeoNet local magnitudes (M_L) (e.g., Haines 1981) are available, we estimate M_W based on these values. In New Zealand M_L has been observed to be systematically biased upwards with respect to M_W (Ristau 2013). To account for this, we adjusted the M_L estimates using the following empirical relationship derived for Canterbury earthquakes: $M_W = M_L - 0.34$ (J. Ristau, personal communication). These magnitude values (hereinafter referred to $M_{W,GeoNet}$) serve as a benchmark for crosschecking the moment magnitudes derived from the spectral fitting procedure detailed below.

Figure 2 shows the distribution of data points in terms of magnitude with respect to hypocentral distance and focal depth. The M_W-distance coverage is excellent with very large numbers of ray path crossings in the source region of the Canterbury earthquake sequence, thereby fulfilling the fundamental prerequisites for the application of the non-parametric spectral inversion scheme outlined below. Most events have focal depths ranging between 5 and 12 km.

For the spectral analysis, we calculated the Fourier amplitude spectra of horizontal component S-wave windows starting 0.5 s before the S-wave onset and ending when 80 % of record's energy has been reached (Oth *et al.* 2011a, b), with a 5 % cosine taper applied to the selected windows and a minimum duration of 5 s in order to ensure enough spectral resolution at lowest frequencies. With this approach, the vast majority of records have window lengths between 5 and 10 s. For the M_W 7.2 Darfield earthquake that involved complex rupture on multiple fault planes (Beavan *et al.* 2012), we allowed for a maximum window length of 20 s, as a compromise between ensuring the inclusion of the entire S-wave window while avoiding too much contamination by surface waves. The obtained spectra were smoothed around 40 frequency points equidistant on log scale between 0.5 and 20 Hz using the Konno and Ohmachi (1998) windowing function with $b = 30$ and combined into their root-mean-square average. Two examples for this data processing procedure are shown in Fig. 3.

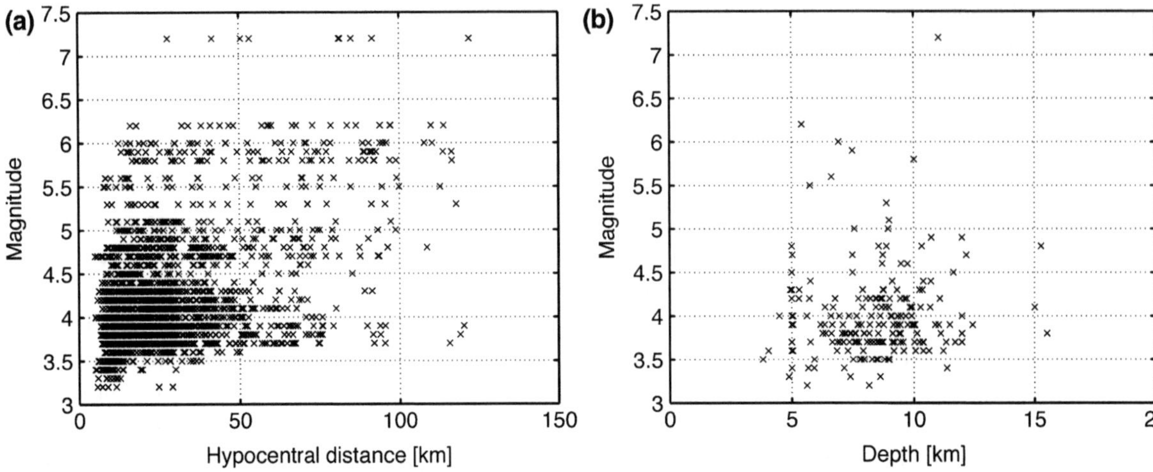

Figure 2
Dataset characteristics. **a** GeoNet moment magnitude ($M_{W,GeoNet}$) versus hypocentral distance. **b** Magnitude versus event depth

3. *Spectral Inversion Approach*

In order to isolate the earthquake source spectra from the observed S-wave amplitude spectra, we follow the one-step non-parametric generalized inversion technique as outlined by Oth *et al.* (2011b). We only provide a brief summary of the method here and refer the reader to the above-mentioned article and the references therein for further details.

Under the assumption of a convolutional ground motion model, the amplitude spectrum of earthquake ground motions can be written as:

$$U_{ij}(f, M_i, R_{ij}) = S_i(f, M_i) \cdot A(f, R_{ij}) \cdot G_j(f), \quad (1)$$

where $U_{ij}(f, M_i, R_{ij})$ represents the observed spectral amplitude at frequency f (in our case acceleration) obtained at the jth station resulting from the ith earthquake with magnitude M_i, R_{ij} is the hypocentral distance, $S_i(f, M_i)$ represents the source spectrum of the ith earthquake, $A(f, R_{ij})$ accounts for the path effects and $G_j(f)$ is the site response function of the jth station (assuming here that the instrument response is corrected). We linearize Eq. (1) by taking the logarithm:

$$\log_{10} U_{ij}(f, M_i, R_{ij}) = \log_{10} S_i(f, M_i) + \log_{10} A(f, R_{ij}) + \log_{10} G_j(f), \quad (2)$$

and obtain this way a linear system of the form $\boldsymbol{Ax} = \boldsymbol{b}$, which can be solved using appropriate algorithms (e.g., Menke 1989).

If, as in the present case, a dataset with good distance coverage is available, a non-parametric inversion approach can be used rather than pre-defining a functional form for $A(f, R_{ij})$. In this case, the non-parametric function $A(f, R_{ij})$ implicitly includes all attenuation effects along the travel path (geometrical spreading, anelastic and scattering attenuation, refracted arrivals, etc.) in a 1D model, and based on the idea that these properties vary slowly with distance, $A(f, R_{ij})$ is only constrained to be a smooth function of distance [see for instance Castro *et al.* (1990), for the implementation of this condition] and to take the value $A(f, R_0) = 1$ at some reference distance R_0, which is set to 5 km in this work.

Finally, in order to resolve a remaining degree of freedom (Andrews 1986), a reference condition either for the source or site part needs to be set. A common such reference is to fix the site response of a single rock site or the average of a set of rock sites to be equal to unity, independent of frequency. However, if not only the relative site terms (i.e., amplification relative to the given reference condition), but also the absolute source spectra derived in the inversion are of interest, extreme care must be taken that this reference condition is reasonably chosen, in order to avoid bias in the source spectral shape. Particularly in Canterbury this is a difficult issue since most rock stations exhibit their own site response (van Houtte *et al.* 2012). In order to choose

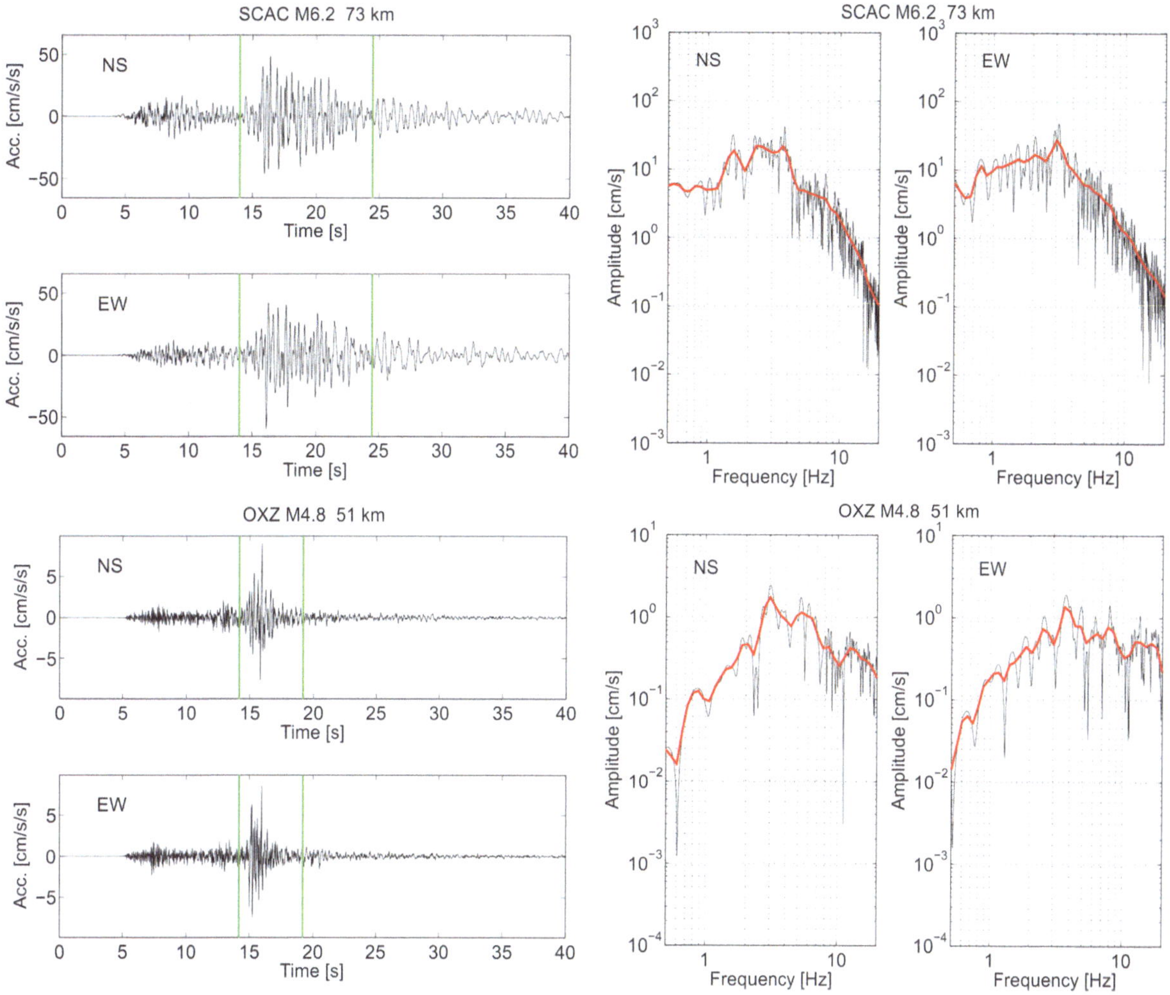

Figure 3

Examples of the event recordings. *Top left* recording of the February 2011 Christchurch event ($M_W = 6.2$) at station SCAC (NS and EW components, hypocentral distance 73 km). The *green lines* indicate the selected time window. *Top right* corresponding Fourier amplitude spectra, *red lines* indicate the smoothed spectra using the Konno and Ohmachi (1998) filter with $b = 30$. *Bottom* same as *top* row for a moderate ($M_W = 4.8$) earthquake located at a distance of 51 km from station OXZ

an appropriate reference condition, we, therefore, took into consideration the geological setting of the sites and the shape of the *H*/*V* ratios (calculated directly from the amplitude spectra used in the inversions), as well as the results from a range of trial runs using various reference conditions. Since (a) the NZNSN stations MQZ and RPZ (Fig. 1) are located on rock, (b) the H/V spectral ratios for these stations are reasonably flat over the entire frequency range of analysis and (c) trial runs with these stations provided results with reasonable spectral shapes of the source and site terms in the inversion, we came to the conclusion that imposing the average of the site response at these two stations to be equal to one is the most appropriate choice for our purposes. It should be noted at this point that imposing a well-founded reference condition is of crucial importance, since the site response functions of all other sites as well as the source spectra resulting from the inversion will be relative to the imposed constraint. Therefore, if an inappropriate reference condition were to be chosen, all the source spectra would show a systematic bias due to site amplification peaks or troughs of the reference site(s) not taken into account in the reference

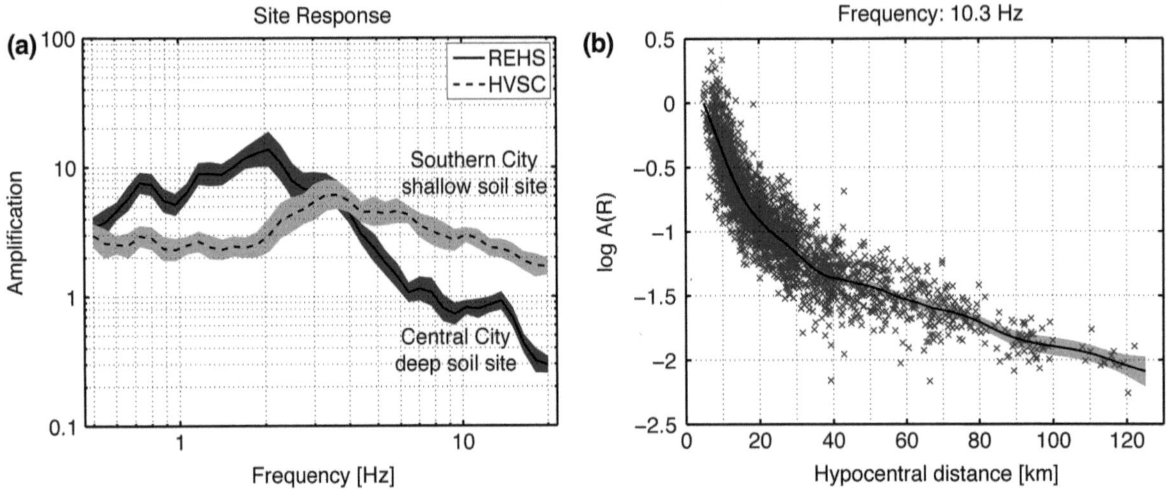

Figure 4
a Two examples of site response functions resulting from the spectral inversion. REHS is a deep soil site in the Christchurch central city, while HVSC is a shallow soil site in southern Christchurch. Note the lower predominant frequency for REHS as compared with the response at HVSC. **b** Example for the non-parametric attenuation curves at frequency 10.3 Hz (*black line, gray shaded area* denotes one standard deviation of the bootstrap analysis) and *individual data points* (corrected for source and site terms)

condition. The influence of potential high-frequency diminution effects, which are commonly parameterized by an exponential term $\exp(-\pi\kappa f)$ (e.g., ANDERSON and HOUGH 1984), will be discussed in the following section. Finally, uncertainty estimates on the various non-parametric components are derived by running a set of 100 bootstrap inversions (OTH *et al.* 2011b).

Figure 4 shows two examples for the site response functions derived under this reference condition assumption as well as an example for the 1D attenuation models obtained. As one would expect, the deep soil site shows a lower fundamental frequency ($\sim$2 Hz) than the shallow soil site ($\sim$3–4 Hz) as well as high-frequency de-amplification. In terms of attenuation, at each analyzed frequency, the spectral data points corrected for their source and site contributions follow the 1D-attenuation curves very well, showing that these are well-constrained by the data. The site response functions and non-linear soil behavior issues, as well as the attenuation characteristics will be discussed in detail in a dedicated article, and in the following, we will concentrate on the source components.

4. Source Spectra and Stress Drop Calculation

The isolated non-parametric $S_i(f, M_i)$ terms represent the earthquake acceleration source spectra at the reference distance R_0. These are not subject to any specific assumption about functional form in terms of either attenuation or source spectral shape, and can be interpreted in the framework of an appropriate earthquake source model. However, as mentioned previously, their spectral shape is dependent on the reliability of the constraint assumed for the site response of the reference station(s).

Figure 5 shows four examples of these source spectra (black lines, gray shaded area denotes standard deviation from bootstrap analysis). While for instance the source spectrum of the M_W 7.2 Darfield event is nearly entirely flat over the bandwidth of analysis, those of the smaller events (Fig. 5c, d) show an increase at low frequencies approximately proportional to f^2 and a plateau at high frequencies, consistent with the ω^{-2} source model (AKI 1967). Indeed, fitting the ω^{-2}-model (BRUNE 1970, 1971) to the inverted source spectra using the following equation (OTH *et al.* 2010):

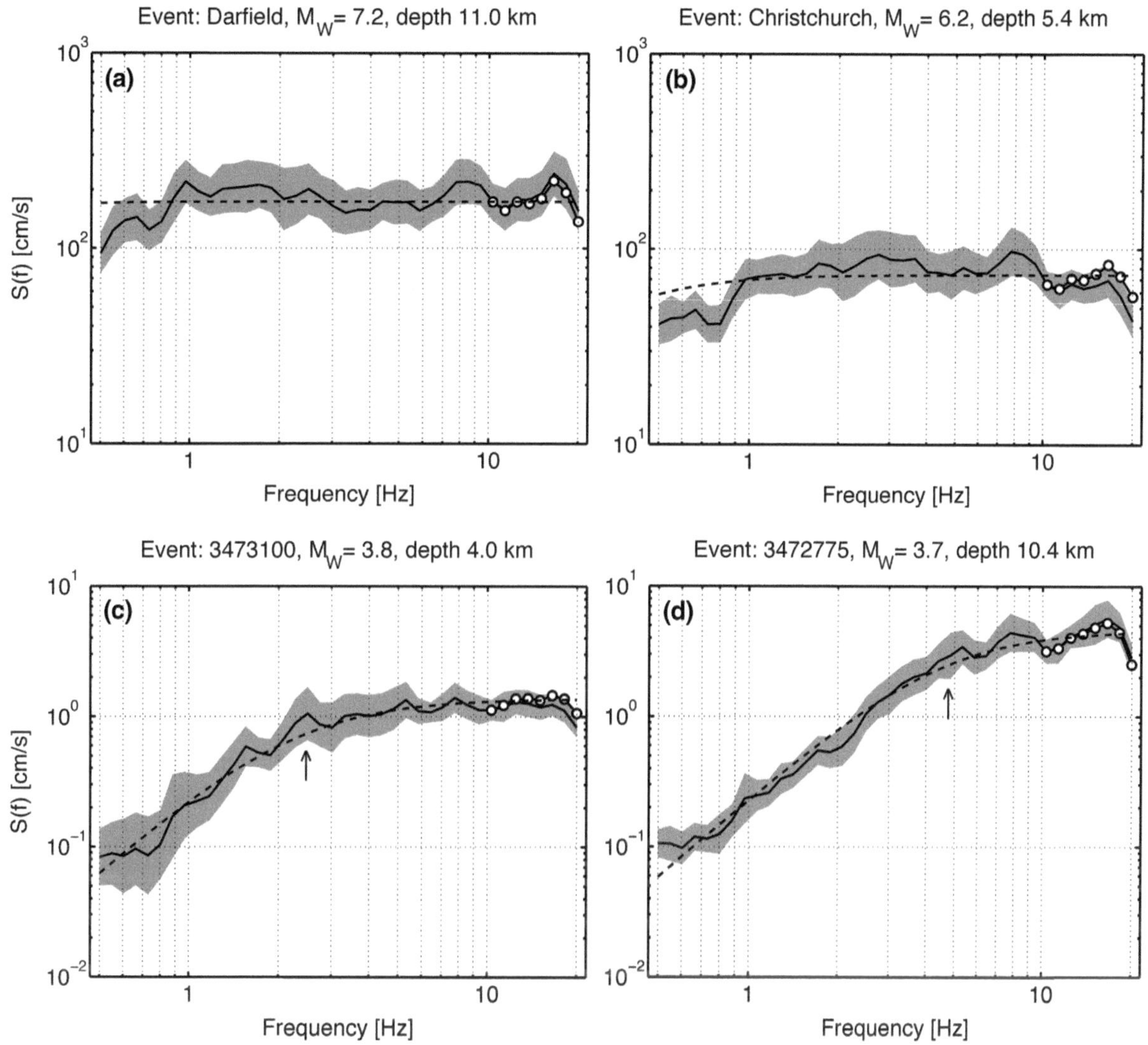

Figure 5
Four examples of acceleration source spectra derived from GIT inversion (*black line* average; *gray shaded area* standard deviation from bootstrap analysis) for four events. *Dashed lines* denote the best ω^{-2} source spectral fit. *White dots* at frequencies larger than 10 Hz show the high-frequency κ correction (Oth *et al.* 2011b) (see also text and Fig. 6), which only has a negligibly small effect in this dataset. **a** Source spectrum of the September 2010 Darfield earthquake, M_W 7.2. Note that the source spectrum is practically flat over the entire bandwidth of analysis, characteristic of an ω^{-2} source spectrum of a large earthquake with corner frequency lower than the lower bandwidth limit. **b** February 2011 Christchurch event, M_W 6.2. **c, d** Two examples for lower magnitude events. The events show an excellent fit with the ω^2-model and have nearly the same M_W, but very different corner frequencies (indicated by the *arrows*) and, consequently, stress drops

$$S(f) = (2\pi f)^2 \frac{R^{\theta\varphi}VF}{4\pi\rho v_S^3 R_0} M(f), \text{ with } M(f) = \frac{M_0}{1 + (f/f_C)^2}, \quad (3)$$

we can see that the ω^{-2} source model provides an excellent fit to the source spectra (Fig. 5, dashed black lines). In Eq. (3), *M*(*f*) denotes the moment rate spectrum, $R^{\theta\varphi}$ represents the average radiation pattern of S-waves set to 0.55 (Boore and Boatwright 1984), $R_0 = 5$ km is the reference distance, $V = 1/\sqrt{2}$ accounts for the separation of S-wave energy onto two horizontal components, $F = 2$ is the free surface factor and ρ and v_S are density and shear wave velocity estimates in the source region, for which we adopted standard upper crustal density of $\rho = 2.7$ g/cm^3 and $v_S = 3.3$ km/s (derived from the

3D velocity model of Fry *et al.* 2013). With this fitting procedure (using non-linear least squares), we determine the seismic moment M_0 (respectively, moment magnitude M_W, Hanks and Kanamori 1979) and corner frequency f_C for each earthquake. One restriction in this procedure applies to the larger events of $M_{W,GeoNet} \geq 5.5$: for such events, f_C is likely to be smaller than the lowest frequency of our analysis bandwidth, and in this case, it is impossible to reliably determine M_0 and f_C simultaneously (Oth *et al.* 2010). For this reason, we constrained M_0 to the value given in the GeoNet database for those events, which is well constrained by regional moment tensor inversion. Because of the necessity of constraining the moments for the largest events, it is important to ensure that the M_W values derived from the spectral fits are consistent with the $M_{W,GeoNet}$ values, since otherwise the scaling trend discussed below might be biased due to inconsistencies between the magnitudes of large and small events (Oth 2013, see also Fig. 9).

A further noteworthy observation is that the source spectra do not show any significant decay at high frequencies. It is well known that the Fourier amplitude spectra of ground motion records usually show a decay at high frequencies that can be approximated with an exponential of the form $\exp(-\pi\kappa f)$. This κ term is often associated with near-surface attenuation at the observation sites (Anderson and Hough 1984; Boore 2003) and, therefore, is often considered a site effect, even though some path and source contributions to this high-frequency diminution effect are also under debate (e.g., Hanks 1982; Papageorgiou and Aki 1983). Oth *et al.* (2011b) showed that a site-related κ effect at the reference site, if not taken into account when imposing the reference constraint, can be systematically moved into the source spectra and may need to be taken into account during their interpretation. Path-related high-frequency diminution effects, in contrast, will flow into the general attenuation effects that lead to the non-parametric attenuation curves $A(f, R_{ij})$.

Since indeed the reference condition in our inversion does not include such a κ decay, one might expect that the high-frequency diminution effect related to the reference site(s) is moved into the source spectral estimates, as was previously observed in Japan (Oth *et al.* 2011b). This effect, if present, needs to be corrected in the framework of a source spectral interpretation within the ω^{-2} model. In order to quantify how strongly this issue might affect our source spectral estimates at high frequencies, we systematically calculate κ for our source spectra by fitting the following relation to the high-frequency part:

$$\log_{10} S_i(f > f_E) = S_{0,i} - \log_{10} e \cdot \pi\kappa_i f, \qquad (4)$$

with f_E being the lower frequency bound considered (Anderson and Hough 1984). In view of the magnitude range of the events considered in this study, we set $f_E = 10$ Hz in order to be well beyond the corner frequency range of the events (Parolai and Bindi 2004), but strictly speaking, f_E would have to be appropriately chosen for each individual spectrum.

The distribution of source κ values is shown in Fig. 6. The obtained values are very small, with an average of 0.006 s, and roughly normally distributed. These values are in contrast to the rock site value of 0.03 s found in the source spectra obtained by Oth *et al.* (2011b) in Japan and thus mean that the reference site condition of unity imposed on the average of the sites MQZ and RPZ does not introduce a significant high-frequency diminution effect in the source spectra. Instead this implies that this effect, if present, must have been completely taken up in the attenuation operator $A(f, R_{ij})$. In Fig. 5, the white circles at high frequencies represent the high-frequency source spectra (for frequencies higher than 10 Hz) following the automatic κ-correction, showing that the effect is overall negligible in the case of the Canterbury dataset.

Stress drop estimates $\Delta\sigma$ are computed based on the determined M_0 and f_C values following Hanks and Thatcher (1972):

$$\Delta\sigma = 8.5 M_0 \left(\frac{f_C}{v_S}\right)^3 \qquad (5)$$

These stress drop values relate to the model introduced by Brune (1970, 1971), and it should be noted here that using a different model, such as the well-known Madariaga (1976) one, will result in different stress drop values (higher by a factor of 5.5 in the case of the Madariaga model). Therefore, before comparing stress drop estimates from different studies, consistency in terms of underlying model

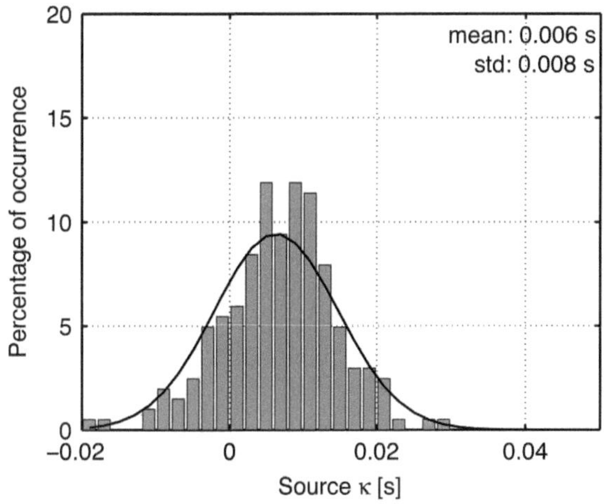

Figure 6
Distribution of residual high-frequency decay parameter κ in the source spectra, estimated from the high-frequency decay for $f > 10$ Hz. Most values are well below 0.02 s, indicating that also at highest frequencies the source spectra are approximately flat, as expected for ω^{-2} acceleration source spectra. The *black line* indicates the fit of a normal distribution function to the data

assumptions should imperatively be verified. Also note that the link between stress drop and the so-called apparent stress, which is commonly determined through integration of the earthquake source spectrum and is directly related to an estimate of the radiated energy of the earthquake, is given via the assumed source model. For source spectra that are well characterized by the ω^{-2}-model, proportionality between these two quantities is expected (see for instance Singh and Ordaz 1994). However, this is not necessarily always the case, and the terms *apparent stress* and *stress drop* are not simply interchangeable.

In order to provide an uncertainty estimate of seismic moment M_0, corner frequency f_C and stress drop $\Delta\sigma$, we follow the approach of Viegas *et al.* (2010). This involves determining the bounds of the range of values for M_0 and f_C where the variance of the best fit between the observed and theoretical source spectrum increases by 5 %.

One final concern that might be raised lies in the fact that the attenuation model considered in this work is only 1D. If strong 2D/3D attenuation heterogeneities were present in the vicinity of the source zones, they could in principle be mapped into the source spectra and lead to biased stress drop estimates, such that any lateral stress drop variations would rather represent lateral attenuation variations. While such a trade-off between source spectral level and attenuation characteristics can never be fully excluded, we can test whether or not the data points from events with different stress drop ranges show any systematic differences relative to the 1D attenuation model derived from the entirety of the dataset. At a given frequency, source spectral level is not distant-dependent whereas attenuation definitely is, and therefore, differing trends in the distance-dependence of data points from earthquakes with very low or very high stress drops might be expected if such a trade-off was to bias the source spectra. Figure 7 shows that source- and site-corrected data points from events with different stress drop levels follow the attenuation model overall equally well, both at low and high frequency. It is only at the largest distances that data points from the lowest and highest stress drop events (lowest and highest 5 % of stress drop distribution) show some slight deviation. However, these data points are only very few and are thus statistically not relevant. Therefore, this test does not provide any indication that 2D/3D attenuation effects not taken into account would significantly bias the stress drops determined in this work.

5. *Results and Discussion*

Stress drop is the source parameter governing high-frequency energy radiation during earthquakes and, as such, is a key parameter for ground motion prediction and seismic hazard assessment. In this context, its variability in particular is of great importance (Cotton *et al.* 2013). Studies involving large datasets on global scales or covering large regions usually come to the conclusion that stress drop varies over at least three orders of magnitude, if not more (e.g., Shearer *et al.* 2006; Allmann and Shearer 2007; Oth *et al.* 2010).

Figure 5c and d provide an excellent illustration of the effect of varying stress drop of two earthquakes with nearly the same seismic moment. While the low-frequency asymptote is nearly the same for these two events, showing that they have approximately identical seismic moments, their corner frequencies and high-frequency plateau levels vary considerably. The

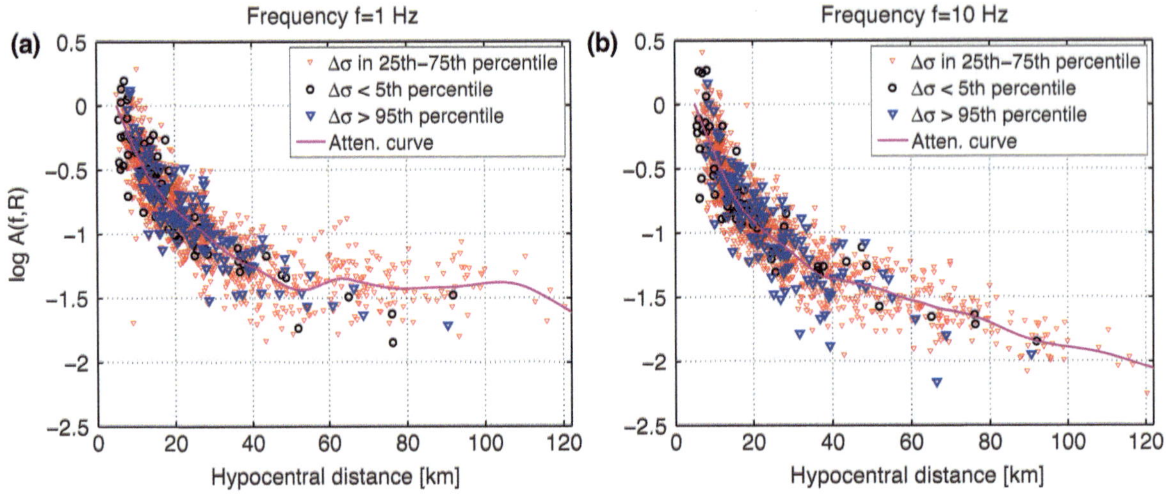

Figure 7

Comparison of source and site-corrected spectral amplitude decay with distance from earthquakes with different stress drop ranges and the inverted attenuation model at frequencies **a** 1 and **b** 10 Hz. *Red triangles* data from moderate stress drop events within the interquartile range. *Black circles* data from 5 % lowest stress drop events. *Blue triangles* data from 5 % highest stress drop events. *Magenta line* 1D attenuation model. Spectral amplitudes from events with different stress drop ranges follow the 1D attenuation model equally well. It is only at the largest distances that data points from the highest stress drop events could be slightly biased relative to the attenuation curve, but these data points are very few and thus have only very limited influence on the estimated stress drops

event in Fig. 5d has almost one order of magnitude higher stress release ($\sim$9.5 MPa) than the one in Fig. 5c ($\sim$1.5 MPa) and is hence significantly more energetic.

Stress drops of the Canterbury earthquakes vary between about 1–20 MPa (Fig. 8), thus covering a little more than one order of magnitude variability. This is considerably less than the above-mentioned overall scatter of at least three orders of magnitude observed on large-scale datasets. This observation of lower stress release variability in individual earthquake sequences has also been made in Japan (Oth 2013) and provides indications that on local (regional) scales, stress release variability is only of about one order of magnitude. However, significant variations in the average stress drop level between different areas exist. In Canterbury, the median of the stress drop distribution is about 5 MPa (Fig. 8b), which is high compared to other regions where stress drops of crustal earthquakes have been derived using the methodology applied in this study. For crustal earthquakes in Japan for instance, a median stress drop of only $\sim$1 MPa was obtained (Oth *et al.* 2010), whereas for the L'Aquila sequence in Italy, the average stress drop ranged around 3 MPa (Ameri *et al.* 2011, all calculated using the same source model). The high stress drops of Canterbury earthquakes obtained here are in good agreement with previous findings of high apparent stress or implied stress drop (Fry and Gerstenberger 2011; Holden 2011) and most likely reflect the fact that these earthquakes take place in an immature intra-plate setting, with little seismic activity prior to this sequence. As noted by Fry and Gerstenberger (2011), the Canterbury events are most likely taking place on strong faults characterized by high friction. In contrast, Japan is overall characterized by much higher seismicity rates and, therefore, also by more mature crustal features. However, in contrast to the apparent stress results of Fry and Gerstenberger (2011) that are based on teleseismically estimated radiated energy calculations, we obtain larger stress drop for the M_W 6.2 Christchurch event than for the M_W 7.2 Darfield one. A potential explanation for this fact could lie in the focal mechanism correction applied in the methodology of Choy and Boatwright (1995) for calculating radiated energy. As noted by Di Giacomo *et al.* (2010), teleseismic energy

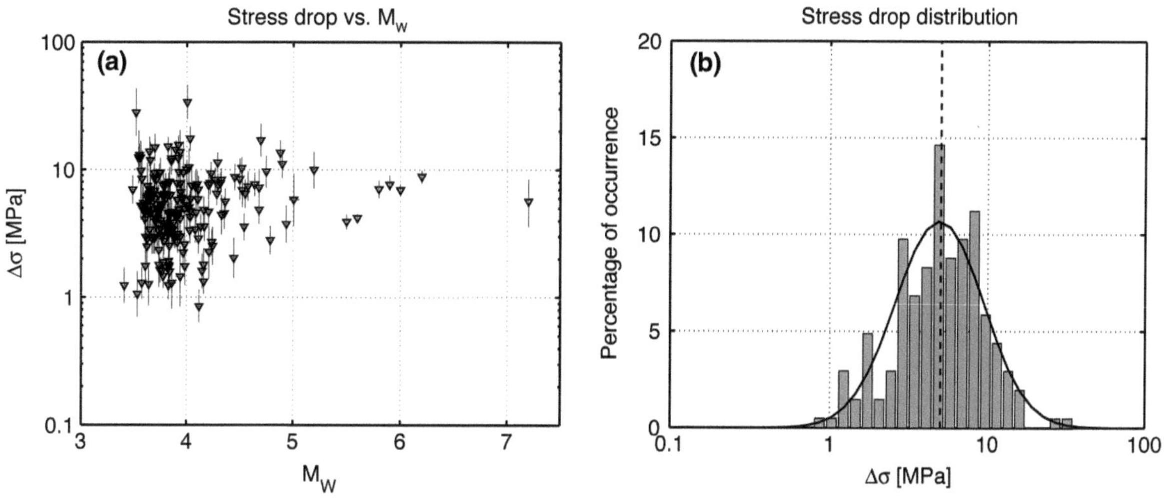

Figure 8
Stress drops of the 2010–2011 Canterbury sequence. **a** Stress drop versus magnitude. Uncertainty ranges are determined following the approach of VIEGAS *et al.* (2010). Values range from 1 to 20 MPa, with two outliers between 30 and 40 MPa at low magnitude. Note that the 2010 Darfield earthquake (largest event) shows a comparatively low stress drop relative to the 2011 February, June and December events (magnitude range 5.9–6.2). **b** Stress drop distribution as a histogram plot. Tentative log-normal fit to the distribution is overlain as *black line*. The median of the distribution (indicated by the *dashed line*) is ~5 MPa

magnitude estimates of strike-slip earthquakes including this correction are systematically higher by about 0.2–0.3 magnitude units than those without, and this is only the case for strike-slip events. This issue notwithstanding, this study clearly corroborates that stress drops in Canterbury are indeed higher than in other, higher strain-rate, regions.

As mentioned earlier, stress drop scaling with earthquake size is a key aspect in seismic hazard assessment. In particular, if earthquakes scale self-similarly with seismic moment, stress drop is constant and $M_0 \propto f_C^{-3}$ (AKI 1967). Indeed, datasets involving large earthquake populations usually do not find a significant deviation from this scaling rule (e.g., SHEARER *et al.* 2006; ALLMANN and SHEARER 2007; OTH *et al.* 2010). However, the findings of some studies on individual earthquake sequences indicate that significant breaks in self-similarity may occur (e.g., MAYEDA and MALAGNINI 2010), even though contradictory results have been obtained for the same earthquake sequences (BALTAY *et al.* 2010). KANAMORI and RIVERA (2004) proposed to quantify deviations from the self-similarity principle through the so-called ε parameter, such that $M_0 \propto f_C^{-(3+\varepsilon)}$. With this parameterization, $\varepsilon = 0$ in the case of self-similar scaling, while $\varepsilon > 0$ indicates increasing stress drop with earthquake size and $\varepsilon < 0$ results in the opposite trend.

The scaling characteristics of the Canterbury earthquake sequence are shown in Fig. 9a in the form of an $M_0 - f_C$ plot. In the framework of the uncertainty estimates calculated following the approach of VIEGAS *et al.* (2010), the corner frequencies and seismic moments are robustly constrained for almost all events. The determination of ε results in the value $\varepsilon = 0.16 \pm 0.17$, where the uncertainty range given is the standard deviation resulting from a bootstrap analysis of 100 resampled datasets. This result indicates that, if any, there is only a very weak increase of stress drop with seismic moment, and self-similarity is certainly an option within the uncertainty bounds. As indicated in the previous section, it is important that the moment magnitudes of the small/moderate events derived from the spectral fitting procedure are consistent with the ones fixed for the larger shocks. Figure 9b shows the comparison of $M_{W,\text{GeoNet}}$ and $M_{W,\text{spectral fit}}$. In the M_W range 4–5, the $M_{W,\text{GeoNet}}$ values tend to be slightly larger than the values derived from the spectral fit, while for the smallest events, this trend reverses. However, the differences

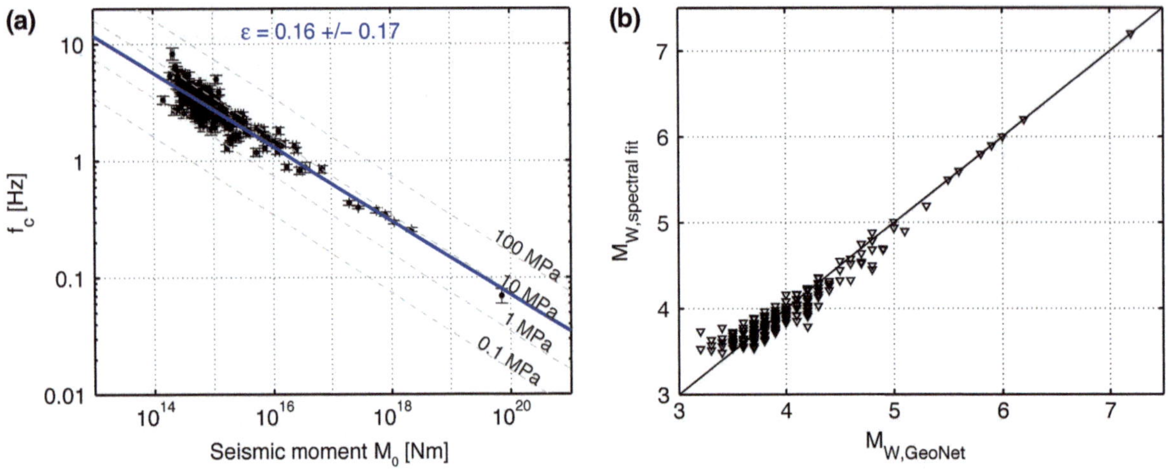

Figure 9

Scaling characteristics of the 2010–2011 Canterbury earthquake sequence. **a** $f_C - M_0$ plot, with constant stress drop values indicated by *dashed lines*. E is determined from the slope of the linear fit in log f_C − log M_0 (*blue line*), and the standard deviation is obtained from bootstrap resampling. **b** Comparison of M_W values derived from the spectral fitting procedure with the estimated GeoNet moment magnitudes, $M_{W,GeoNet}$ (see text for explanations). Note that for events with $M_{W,GeoNet} \geq 5.5$, the moment for the spectral fits is set according to this value (see text)

are well within 0.1 magnitude units in most cases, and thus we conclude that overall, the seismic moments determined for the small/moderate events are compatible with the values set for the large ones. Therefore, we do not expect any significant bias of the scaling results in that respect.

Finally, an important aspect of stress drop variability lies in the question of whether there are systematic lateral variations. Figure 10 shows the stress drop measurements as given in Fig. 8a in map view. There are several noteworthy clusters appearing in the evolution of the Canterbury sequence (Bannister and Gledhill 2012). The M_W 7.2 complex Darfield mainshock rupture began on a subsidiary thrust fault, with a hypocenter about 4 km north of the near-vertical strike-slip Greendale Fault that represents the main part of the rupture (Gledhill *et al.* 2011). Aftershock activity decayed relatively rapidly in the zone extending to the west and north of the Greendale fault, whereas aftershocks clustered persistently at the eastern edge of the Greendale fault, including a significant number of moderate events immediately following the Darfield rupture (Bannister and Gledhill 2012). On 22 February 2011, the M_W 6.2 oblique-reverse Christchurch event occurred. Aftershock activity was triggered throughout the entire Canterbury region, but particularly pronounced in the vicinity of the M_W 6.2 epicenter. The M_W 6.0 June 2011 event, involving right-lateral strike-slip motion, occurred on a fault plane thought to intersect the eastern edge of the fault plane associated with the February event, and was in turn followed by the reverse faulting M_W 5.9 December 2011 event offshore. Overall, strike-slip faulting dominated along the trace of the Greendale Fault and its eastern edge, while some reverse faulting mechanisms are apparent to the west and north of the aftershock zone, as well as offshore following the M_W 5.9 December earthquake (Ristau *et al.* 2013). Oblique mechanisms are dominant throughout much of the vicinity of Christchurch city (Gledhill *et al.* 2011).

In terms of stress drop variations, these clusters show interesting features (Fig. 10). The M_W 7.2 Darfield mainshock shows a slightly elevated stress drop compared to the average, which, in view of the significant uncertainties that are generally associated with stress drop estimates, should, however, not be over-interpreted. We also note at this point that the stress drop estimate for the Darfield event showed a non-negligible dependence on the maximum window length allowed in the calculation of the spectra

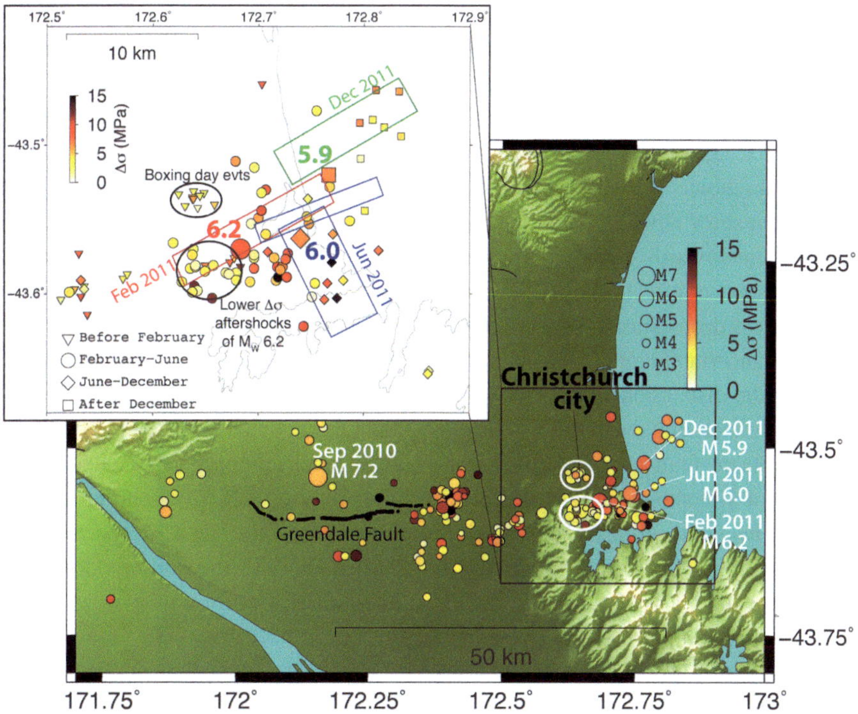

Figure 10

Lateral stress drop variations during the 2010–2011 Canterbury sequence (see also Fig. 1 for event locations). *Dotted black lines* in the center indicate the location of the Greendale fault. Note that high stress drop events tend to cluster at the eastern edge of the Greendale fault and east of the 2011 February (M_W 6.2) hypocenter. The two *white ellipses* highlight two clusters of comparatively low stress drops, which are the cluster of the Boxing Day events (northern ellipse, see inset and text) and the aftershocks of the M_W 6.2 Christchurch events located southwest of the event's epicenter. Inset close-up of the Christchurch area, with fault rupture surface projections as estimated from geodetic inversion (Beavan *et al.* 2012). The two *black ellipses* correspond to the *white* ones in the main figure. *Red rectangle* represents 2011 February (M_W 6.2), *blue* the 2011 June (M_W 6.0) and *green* the 2011 December (M_W 5.9) event. The epicenters of these three events are denoted by a *larger symbol* than the remaining events. Note that the fault geometries of the June and December events are not very well constrained. Events are represented by *different symbols* depending on their occurrence time (legend of inset)

(increasing with increasing window length up to about 20 s). We, therefore, allowed for a maximum of 20-second window length, since allowing windows longer than this did not significantly increase the stress drop further. However, a non-negligible portion of surface wave energy in addition to S-wave energy may therefore be included in the spectra, and, therefore, the stress drop estimate for the Darfield event should most probably be considered as an upper bound estimate. As a general rule, stress drop estimates determined from the spectra of such large events as the Darfield one, involving complex ruptures, are to be interpreted with caution. For aftershocks of the Darfield event, stress drops were particularly high in the cluster at the eastern edge of the Greendale Fault.

Before moving on to the M_W 6.2 Christchurch event and its aftershocks, we also remind the reader once more of the uncertainties that stress drop calculations always involve. Indeed, besides the error estimates derived from the regression analysis of the inverted source spectra (which indicate that the source spectra can robustly be modeled as ω^{-2} sources, see Fig. 8a), other effects that are more difficult to quantify should not be forgotten, such as potential trade-offs between source terms and near-source attenuation, or between source and site terms. As discussed in the previous section, near-source attenuation variations could, in principle, bias the observed stress drop variations. We investigated this possibility (Fig. 7) and could not find any significant hints for such a bias. Because each station recorded multiple events and each event was recorded at multiple stations, the most significant danger for a trade-off between source and site terms comes from the choice of an inappropriate reference condition, and as discussed in Sect. 3, we are confident for several reasons that our choice is reasonable. The source spectra also do not show any significant peaks or troughs that would hint towards a contamination with a residual site response effect. Nevertheless, spectrally determined stress drop estimates inherently involve many factors of uncertainty, and these should be kept in mind during the discussion.

While the M_W 6.2 Christchurch event shows very high stress release, a notable feature is the comparatively low stress drops of aftershocks in two clusters in its vicinity (marked by white ellipses and black ellipses in the inset in Fig. 10). The cluster to the north (<2 km away from the Christchurch city center) actually occurred well before the M_W 6.2 event in February 2011, starting with a M_W 4.7 event on 26 December 2010 that caused significant damage. This earthquake was followed by a sequence including two more events of magnitude larger than 4 within a couple of hours (this sequence was termed the *Boxing Day earthquakes*) and a series of events in the vicinity within the following 3–4 weeks. Similarly, the aftershocks of the M_W 6.2 February 2011 Christchurch event southwest of its epicenter (partially within the southwestern part of the estimated fault plane projection for the M_W 6.2 event, see inset of Fig. 10) are also consistently lower stress drop, and are clearly distinguished from the higher stress drops to the east as well as at the eastern edge of the Greendale Fault.

A cluster of high stress drop events also followed the 2011 February M_W 6.2 event and is located between the estimated source areas of the 2011 February and the 2011 June (M_W 6.0) events (inset of Fig. 10). This cluster stands in contrast to the lower stress drop aftershocks of the 2011 February event to the southwest of the rupture plane mentioned previously. This apparent clustering of high stress drop aftershocks both east of the Greendale fault and east of the M_W 6.2 Christchurch event may point to stress concentration effects at the rupture edges of these events as a result of loading due to their rupture processes. Such effects are not easily discernible for the 2011 June event, where some high stress drop aftershocks occur well within the assumed rupture plane projection (Fig. 10, inset). However, it should also be noted that the extent of the 2011 June event fault plane is not well constrained (Beavan *et al.* 2012). In order to shed light on the lateral stress drop pattern of the 2011 June event's aftershocks and to better constrain the evolution of the stress drop offshore following the M_W 5.9 December 2011 event, the inclusion of more events in this area is necessary in the future. Apart from these apparent clusters of high and low stress drop mentioned above, we could not find an evident temporal change of average stress drop level among earthquakes occurring before and after the M_W 6.2 Christchurch and the 2011 M_W 6.0 June event in their respective surroundings, in agreement with the findings of Shearer *et al.* (2006) in southern California.

6. Conclusions

In this study, we investigated the stress release of the Canterbury earthquake sequence within the magnitude range M_W 3–7.2, taking advantage of the availability of a very dense set of recordings from the GeoNet strong motion database. This earthquake sequence provides a unique opportunity to improve our understanding of the stress release mechanisms of earthquakes in low-seismicity regions.

We showed that the source spectra of the Canterbury sequence are very well characterized by the ω^{-2} model, with on average close to self-similar scaling of stress drop with earthquake size. Furthermore, the Canterbury earthquakes are characterized by high stress drops, in the range 1–20 MPa, with a median of $\sim$5 MPa. These values are high compared to more seismically active regions such as Japan, as a result of the Canterbury events taking place on more immature crustal features (a similar analysis to the one presented in this article using the same methods determined a median stress drop for crustal earthquakes in Japan of the order of 1 MPa, Oth *et al.* 2010; Oth 2013). Of the larger earthquakes in the sequence, the M_W 6.2 Christchurch event that occurred in February 2011 was particularly energetic, which in addition to other factors such as directivity and site effects, certainly played a key role in its destructiveness. Lateral variations in stress drops during the aftershock sequence indicate that higher stress events seem to occur at the edges of major fault planes.

In contrast to large-scale earthquake populations (global or covering large regions) that show stress drop variations over three orders of magnitude, the Canterbury sequence stress drops show variation over little more than one order of magnitude. This variability range is in good agreement with the results derived for individual earthquake sequences in Japan (Oth 2013) and implies that the stress release variability on local/regional scales (such as in Canterbury) is much smaller than the three orders of magnitude obtained from global/large area datasets. This means that narrower variability ranges, centered on appropriate regional average stress drop values, can be used for ground motion prediction and hazard assessment.

Acknowledgments

We wish to thank the staff involved in the GeoNet project for making the databases available, as well as Martin Reyners for providing additional phase picks used in window selection. GeoNet (www.geonet.org.nz) is sponsored by the New Zealand Government through the Earthquake Commission (EQC), GNS Science and Land Information New Zealand (LINZ). This research was supported by the New Zealand Natural Hazards Platform. We furthermore thank Rachel Abercrombie and an anonymous reviewer for constructive comments that helped to improve the paper, and Bill Fry for fruitful discussions.

References

Aki, K. (1967). *Scaling law of seismic spectrum*. J. Geophys. Res. *72*, 1217–1231, doi:10.1029/JZ072i004p01217.

Allmann, B.P., and Shearer, P.M. (2007). *Spatial and temporal stress drop variations in small earthquakes near Parkfield*, California. J. Geophys. Res. *112*, B04305, doi:10.1029/2006JB004395.

Ameri, G., Oth, A., Pilz, M., Bindi, D., Parolai, S., Luzi, L., Mucciarelli, M., and Cultrera, G. (2011). *Separation of source and site effects by generalized inversion technique using the aftershock recordings of the 2009 L'Aquila earthquake*. Bull. Earthq. Eng. *9*, 717–739, doi:10.1007/s10518-011-9248-4.

Anderson, J.G., and Hough, S.E. (1984). *A model for the shape of the Fourier amplitude spectrum of acceleration at high frequencies*. Bull. Seismol. Soc. Am. *74*, 1969–1993.

Andrews, D.J. (1986). Objective determination of source parameters and similarity of earthquakes of different size. In Earthquake Source Mechanics, S. Das, J. Boatwright, and C.H. Scholz, eds. (Washington DC: American Geophysical Union).

Baltay, A., Prieto, G., and Beroza, G.C. (2010). *Radiated seismic energy from coda measurements and no scaling in apparent stress with seismic moment*. J. Geophys. Res. *115*, B08314, doi:10.1029/2009JB006736.

Bannister, S., and Gledhill, K. (2012). *Evolution of the 2010–2012 Canterbury earthquake sequence*. New Zealand J. Geol. Geophys. *55*, 295–304, doi:10.1080/00288306.2012.680475.

Beavan, J., Motagh, M., Fielding, E.J., Donnelly, N., and Collett, D. (2012). *Fault slip models of the 2010–2011 Canterbury, New Zealand, earthquakes from geodetic data and observations of postseismic ground deformation*. New Zealand J. Geol. and Geophys., *55*, 207–221.

Boore, D.M. (2003). *Simulation of ground motion using the stochastic method*. Pure Appl. Geophys. *160*, 635–676.

Boore, D.M., and Boatwright, J. (1984). *Average body-wave radiation coefficients*. Bull. Seismol. Soc. Am. *74*, 1615–1621.

Bradley, B.A. (2012). *Strong ground motion characteristics observed in the 4 September 2010 Darfield, New Zealand earthquake*. Soil Dyn. Earthq. Eng. *42*, 32–46, doi:10.1016/j.soildyn.2012.06.004.

Brune, J.N. (1970). *Tectonic stress and the spectra of seismic shear waves from earthquakes*. J. Geophys. Res. *75*, 4997–5009.

Brune, J.N. (1971). *Correction*. J. Geophys. Res. *76*, 5002.

Castro, R.R., Anderson, J.G., and Singh, S.K. (1990). *Site response, attenuation and source spectra of S waves along the Guerrero, Mexico, subduction zone*. Bull. Seismol. Soc. Am. *80*, 1481–1503.

Choy, G.L., and Boatwright, J.L. (1995). *Global patterns of radiated seismic energy and apparent stress*. J. Geophys. Res. *100*, 18205–18228.

Cotton, F., Archuleta, R., and Causse, M. (2013). *What is Sigma of the Stress Drop?* Seismol. Res. Lett. *84*, 42–48, doi:10.1785/0220120087.

Di Giacomo, D., Parolai, S., Bormann, P., Grosser, H., Saul, J., Wang, R., and Zschau, J. (2010). *Suitability of rapid energy magnitude determinations for emergency response purposes*. Geophys. J. Int. *180*, 361–374, doi:10.1111/j.1365-246X.2009.04416.x.

Fry, B., and Gerstenberger, M.C. (2011). *Large Apparent Stresses from the Canterbury Earthquakes of 2010 and 2011*. Seismol. Res. Lett. *82*, 833–838, doi:10.1785/gssrl.82.6.833.

Fry, B., Benites, R., and Kaiser, A. (2011). *The Character of Accelerations in the Mw 6.2 Christchurch Earthquake*. Seismol. Res. Lett. *82*, 846–852, doi:10.1785/gssrl.82.6.846.

Fry, B., Eberhart-Phillips, D., and Davey, F. (2013). Mantle accommodation of lithospheric shortening as seen by combined interferometry and body-wave imaging in the South Island, New Zealand, submitted to Geophys. J. Int.

Gledhill, K., Ristau, J., Reyners, M., Fry, B., and Holden, C. (2011). *The Darfield (Canterbury, New Zealand) Mw 7.1 Earthquake of September 2010: A Preliminary Seismological*

Report. Seismol. Res. Lett. *82*, 378–386, doi:10.1785/gssrl.82.3.378.

HAINES, A.J. (1981). *A local magnitude scale for New Zealand earthquakes*. Bull. Seismol. Soc. Am. *71*, 275–294.

HANKS, T.C. (1982). *fmax*. Bull. Seismol. Soc. Am. *72*, 1867–1879.

HANKS, T.C., and KANAMORI, H. (1979). *A moment magnitude scale*. J. Geophys. Res. *84*, 2348–2350.

HANKS, T.C., and THATCHER, W. (1972). *A graphical representation of seismic source parameters*. J. Geophys. Res. *77*, 4393–4405.

HOLDEN, C. (2011). *Kinematic Source Model of the 22 February 2011 Mw 6.2 Christchurch Earthquake Using Strong Motion Data*. Seismol. Res. Lett. *82*, 783–788, doi:10.1785/gssrl.82.6.783.

IRIKURA, K. (1986). Prediction of strong acceleration motion using empirical Green's function. Proceedings of 7th Japan Earthquake Engineering Symposium, 151–156.

KAISER, A., HOLDEN, C., BEAVAN, J., BEETHAM, D., BENITES, R., CELENTANO, A., COLLETT, D., COUSINS, J., CUBRINOVSKI, M., DELLOW, G., DENYS, P., FICLDING, E., FRY, B., GERSTENBERGER, M., LANGRIDGE, R., MASSEY, C., MOTAGH, M., PONDARD, N., McVERRY, G., RISTAU, J., STIRLING, M., THOMAS, J., UMA, S.R., and ZHAO, J. (2012). *The Mw 6.2 Christchurch earthquake of February 2011: Preliminary report*. New Zealand J. Geol. Geophys. *55*, 67–90, doi:10.1080/00288306.2011.641182.

KANAMORI, H., and RIVERA, L. (2004). *Static and dynamic scaling relations for earthquakes and their implications for rupture speed and stress drop*. Bull. Seismol. Soc. Am. *94*, 314–319.

KONNO, K., and OHMACHI, T. (1998). *Ground-motion characteristics estimated from spectral ratio between horizontal and vertical components of microtremor*. Bull. Seismol. Soc. Am. *88*, 228–241.

MADARIAGA, R. (1976). *Dynamics of an expanding circular fault*. Bull. Seismol. Soc. Am. *66*, 639–666.

MAYEDA, K., and MALAGNINI, L. (2010). *Source radiation invariant property of local and near-regional shear-wave coda: Application to source scaling for the Mw 5.9 Wells, Nevada sequence*. Geophys. Res. Lett. *37*, L07306, doi:10.1029/2009GL042148.

MENKE, W. (1989). Geophysical Data Analysis: Discrete Inverse Theory (New York: Academic Press).

ORENSE, R.P., KIYOTA, T., YAMADA, S., CUBRINOVSKI, M., HOSONO, Y., OKAMURA, M., and YASUDA, S. (2011). *Comparison of liquefaction features observed during the 2010 and 2011 Canterbury earthquakes*. Seismol. Res. Lett. *82*, 905–918, doi:10.1785/gssrl.82.6.905.

OTH, A., BINDI, D., PAROLAI, S., and DI GIACOMO, D. (2010). *Earthquake scaling characteristics and the scale-(in)dependence of seismic energy-to-moment ratio: Insights from KiK-net data in Japan*. Geophys. Res. Lett. *37*, L19304, doi:10.1029/2010GL044572.

OTH, A., PAROLAI, S., and BINDI, D. (2011a). *Spectral Analysis of K-NET and KiK-net Data in Japan, Part I: Database Compilation and Peculiarities*. Bull. Seismol. Soc. Am. *101*, 652–666, doi:10.1785/0120100134.

OTH, A., BINDI, D., PAROLAI, S., and DI GIACOMO, D. (2011b). *Spectral Analysis of K-NET and KiK-net Data in Japan, Part II: On Attenuation Characteristics, Source Spectra, and Site Response of Borehole and Surface Stations*. Bull. Seismol. Soc. Am. *101*, 667–687, doi:10.1785/0120100135.

OTH, A. (2013). *On the characteristics of earthquake stress release variations in Japan*. Earth Planet. Sci. Lett., 377–378, 132–141, doi:10.1016/j.epsl.2013.06.037.

PAPAGEORGIOU, A.S., and AKI, K. (1983). *A specific barrier model for the quantitative description of inhomogeneous faulting and the prediction of strong ground motion. Part I. Description of the model*. Bull. Seismol. Soc. Am. *73*, 693–722.

PAROLAI, S., and BINDI, D. (2004). *Influence of soil-layer properties on k evaluation*. Bull. Seismol. Soc. Am. *94*, 349–356.

PETERSEN, T., GLEDHILL, K., CHADWICK, M., GALE, N.H., and RISTAU, J. (2011). *The New Zealand National Seismograph Network*. Seismol. Res. Lett. *82*, 9–20, doi:10.1785/gssrl.82.1.9.

QUIGLEY, M., VILLAMOR, P., FURLONG, K., BEAVAN, J., VAN DISSEN, R., LITCHFIELD, N., STAHL, T., DUFFY, B., BILDERBACK, E., NOBLE, D., BARRELL, D., JONGENS, R., and COX, S. (2010). *Previously Unknown Fault Shakes New Zealand's South Island*. Eos Trans. AGU *91*, 49.

RISTAU, J. (2008). *Implementation of Routine Regional Moment Tensor Analysis in New Zealand*. Seismol. Res. Lett. *79*, 400–415, doi:10.1785/gssrl.79.3.400.

RISTAU, J. (2013). Update of Regional Moment Tensor Analysis for Earthquakes in New Zealand and Adjacent Offshore Regions, *Bull. Seismol. Soc. Am., 103,* in press.

RISTAU, J., HOLDEN, C., KAISER, A., WILLIAMS, C., BANNISTER, S., and FRY, B. (2013). The Pegasus Bay aftershock sequence of the Mw 7.1 Darfield (Canterbury), New Zealand earthquake. Submitted to Geophys. J. Int.

SHEARER, P.M., PRIETO, G.A., and HAUKSSON, E. (2006). *Comprehensive analysis of earthquake source spectra in southern California*. J. Geophys. Res. *111*, B06303, doi:10.1029/2005JB003979.

SINGH, S.K., and ORDAZ, M. (1994). *Seismic energy release in Mexican subduction zone earthquakes*. Bull. Seismol. Soc. Am. *84*, 1533–1550.

VAN HOUTTE, C., KTENIDOU, O.-J., LARKIN, T., and KAISER, A. (2012). *Reference stations for Christchurch*. Bull. New Zealand Soc. Earthq. Eng. *45*, 184–195.

VIEGAS, G., ABERCROMBIE, R.E., and KIM, W.-Y. (2010). *The 2002 M5 Au Sable Forks, NY, earthquake sequence: Source scaling relationships and energy budget*. J. Geophys. Res. *115*, B07310, doi:10.1029/2009JB006799.

WALLACE, L.M., BEAVAN, J., MCCAFFREY, R., BERRYMAN, K., DENYS, P. (2007). *Balancing the plate motion budget in the South Island, New Zealand using GPS, geological and seismological data*. Geophys. J. Int. *168:* 332–352, doi:10.1111/j.1365-246X.2006.03183.x

(Received May 7, 2013, revised November 4, 2013, accepted November 23, 2013, Published online December 5, 2013)

Pure Appl. Geophys. 171 (2014), 2783–2804

DOI 10.1007/s00024-013-0750-2

Pure and Applied Geophysics

Small Scale Earthquake Mechanisms Induced by Fluid Injection at the Enhanced Geothermal System Reservoir Soultz (Alsace) in 2003 using Alternative Source Models

JAN ŠÍLENÝ,[1] ZUZANA JECHUMTÁLOVÁ,[1] and CATHERINE DORBATH[2]

Abstract—The geothermal reservoir at Soultz-sous-Forêts is a valuable natural laboratory for understanding the mechanisms of microearthquakes generated during stimulations and circulation tests. An ongoing effort currently exists regarding the retrieval of mechanisms aimed to indicate the type of fracturing of the rock massif. As a default, a moment tensor description has been applied. Nevertheless, the retrieval of the mode of fracturing still remains ambiguous. Recent studies indicate a prevailing shear slip but, rarely, a non-shear pattern has also been observed. The moment tensor, used today as a universal tool for descriptions of the mechanism, captures general balanced dipole sources. However, in the case of small-scale earthquakes, the moment tensor need not always be reliably determined. In an effort to fit the data, there may be notable non-shear components caused by the low quality of input data. Constraining the source model to directly determine a simpler one is convenient for describing the physical phenomena expected for a particular focus. An opening of new fractures can be described, to a first approximation, by a tensile crack, optionally combined with a shear slip. Such an alternative model is called a shear-tensile crack (STC) source model. The combination is practical, and can be used to both identify events that reflect purely mode-I (tensile) failure and to determine the dilation angle of the fracture undergoing shear. The latter is particularly important in enhanced geothermal system reservoirs such as Soultz, where shear-related dilation is believed to be the primary mechanism underpinning permeability creation during stimulation injections. We performed a synthetic case study by simulating seismic data as recorded by the actual seismic array installed at Soultz-sous-Forêts. Synthetic P and S amplitudes for several shear-tensile source models were inverted for several types of station coverage. The analysis explored how results were influenced by mislocation, mismodeling, and noise contamination of data. In most cases, the orientation of the mechanism was well resolved. Determination of shear vs. non-shear content within the mechanism was more difficult. From all of the factors influencing resolution that we explored, the quality of the monitoring system (the number of stations and their distribution with respect to the focus) and noise contamination were of the highest impact. The STC source model yielded considerably less spurious non-shear fracture components than the moment tensor. From the bulk of the seismicity recorded during the stimulation in 2003, we concentrated on the first phase of the injection when only a single borehole at the site was stimulated. We processed thirteen small earthquakes with magnitudes larger than 1.4 that were not treated in previous studies. We determined that their source mechanisms were dominantly pure shear slips on pre-existing faults, just as the earthquakes investigated earlier were. The results were also in agreement with the stress pattern from in situ measurements.

[1] Institute of Geophysics, Academy of Sciences, Prague, Czech Republic. E-mail: jsi@ig.cas.cz
[2] EOST, University of Strasbourg, Strasbourg, France.

1. Introduction

It is widely accepted that the injection of a fluid into crystalline or firmly consolidated sedimentary rocks may cause a shear slip on existing faults and fractures, inducing microseismic events. During shear, displacement acting toward the opening of existing fractures or originating new ones may occur, and can be caused by a roughness of the fault surface, its bending, or the creation of tensile fractures nearby, yielding the mesh-like structure proposed by HILL (1977). As proposed by JULIAN *et al.* (2010) to explain the prevailing tensile mechanisms of injection-induced events within the Coso geothermal field in California during 2005, a sequence of 'tensile crack and subsequent shear slip along wing faults inclined to the tensile crack' is also feasible. Since it results in the enhancement of permeability within the reservoir, the process of opening faults just undergoing shear slip or creating tensile cracks in the vicinity is particularly important at geothermal sites. In Soultz, EVANS *et al.* (2005) analyzed microseismic data due to the massive injection of fluid at depths of 3–3.5 km during 1993, and, as a result of the precise location of thousands of microearthquakes, revealed a narrow sub-vertical structure for the fractured zone. They concluded that shear slip occurs in combination with tensile displacement on the jogs of a fault, and that a set of shear and tensile fractures [according to HILL

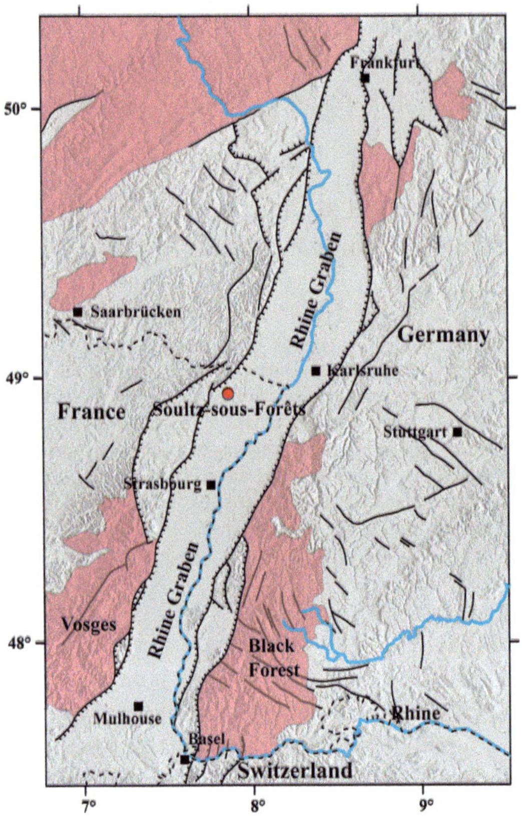

Figure 1
The location of the Soultz-sous-Forêts EGS test site in the Rhine Graben. Outcropping crystalline rocks are shown in pink. (Redrawn from AHORNER and SCHNEIDER 1974).

(1977)] may be created. As a result of the opening of voids due to the tensile process, permeability of the zone is enhanced.

In the Enhanced Geothermal System (EGS) facility at Soultz-sous-Forêts (Alsace, France) (Fig. 1), several hydraulic stimulation experiments were performed during 2000, 2003, and 2004, with the aim of improving the connectivity of drilled boreholes within the natural fracture system. Stimulations were accompanied by abundant seismic activity (CUÉNOT *et al.* 2008; DORBATH *et al.* 2009; GENTER *et al.* 2009). The 2000 stimulation induced more than 700 earthquakes with magnitudes greater than 1.0. Seismicity depicted a dense, homogeneous cloud without any apparent structure. The stimulation reactivated a 3-D dense network of fractures. The 2003 stimulation induced approximately 250 events with magnitudes greater than 1.0, but with a greater proportion of large events up to $M = 2.9$. The hypocenters formed clear structures identified as faults. The 2004 stimulation produced even fewer induced events, a situation that made interpretations difficult. The difference between the seismic response during 2000 and 2003 has been interpreted as due to faults cut by GPK3 or within its vicinity, and reached by injected water (DORBATH *et al.* 2009).

The source mechanisms of induced microearthquakes can be used to determine whether or not events occurred on pre-existing faults or opened due to new cracks. CUÉNOT *et al.* (2006) determined fault plane solutions on the basis of the first-motion polarities of thousands of microearthquakes induced during the 2000 and 2003 injections, and determined prevailing normal-faulting movements with a small strike slip component for both experiments. Additionally, these authors estimated the moment tensors (MT) of selected events from the 2003 injection by applying a common least-squares method of P-wave amplitude inversion, and determined significant non-double-couple (DC) components within the source mechanisms for some events, particularly those located near the injected borehole. Since increasing pore fluid pressure is likely to reduce the effective normal stress on fractures, DORBATH *et al.* (2010) assumed that large water injections induce shearing on pre-existing fractures favorably oriented within the regional stress field. The authors automatically determined focal mechanism solutions constrained by the P-wave first motion polarities recorded by both surface and downhole instruments, and refined them manually to avoid an occasional polarity mismatch.

HORÁLEK *et al.* (2010) retrieved the MT of a statistically representative number of events throughout the 2003 massive fluid injection experiment and investigated their temporal and spatial variations. They analyzed the 45 largest earthquakes with magnitudes between $M = 1.6$ and 2.9 occurring during and after an injection which had a high signal-to-noise ratio. The unconstrained moment tensor expression of the mechanism was applied, allowing both DC and non-DC sources, including tensile fractures. The MT of the 45 events were well constrained and stable in the presence of noise contamination and structure mismodeling. All but one was found to be predominantly pure shear slip, with at most, a few percent of non-shear components. The focal mechanisms varied from normal dip-slip, through oblique normal to strike-slip, indicating that the direction of the maximum stress axis, σ_1, and the intermediate stress axis, σ_2, were interchanged. Injected fluids activated two pre-existing fault segments that displayed different source mechanism patterns. HORÁLEK *et al.* (2010) inferred that injected fluids decreased the Coulomb friction at strength and triggered intense seismicity. All larger ($M \geq 1.6$) induced earthquakes had DC mechanisms, just as the prevailing majority of natural tectonic earthquakes. However, the result did not, a priori, exclude the possibility that some smaller events may reflect tensile fracturing. For this reason, they should be investigated, as tensile fracturing due to injection may be more relevant due to the small-scale heterogeneity of the stress field, including the tensional component.

The analysis of small earthquakes is more troublesome for several reasons. In general, a smaller magnitude tends to result in lower quality input data for an inversion. Fewer seismic stations with relevant data exist. Therefore, the inversion may become unstable. The signal-to-noise ratio is generally lower, and the amplitudes picked may be biased. As a result, the MT of such events are not always well constrained, and may contain spurious non-DC components. Additional problems arise from inaccurate locations and the use of an improper velocity model (ŠÍLENÝ 2009). BEAN *et al.* (2008) demonstrated that structure mismodeling impacts the source solution by simulating the absence of information on near-surface structures during the inversion of the long-period seismic signals that are observed in volcanoes, and stressed the importance of employing a priori source information to reduce the number of free parameters in the inversion. Taking this into account, adopting a simple source model that provides an adequate description of the anticipated physical phenomena seems reasonable. A simple combination of a shear slip with a tensile crack may be a good model both for natural earthquakes and for the induced events that occur during injection operations. A slip along a fault with an off-plane component of the relative displacement can be described using four angles that define the orientation of the fault plane normal and the slip vector (not necessarily in-plane), and by the magnitude of the slip vector (i.e. using five model parameters, in total, instead of the six for the unconstrained MT). Such is helpful for solving the inverse problem in the presence of poor data due to, for example, inaccurate locations, simplified structural models, noisy data, and poor coverage of the focal sphere by stations. We refer to this source model as a shear-tensile crack (STC). As discussed by JULIAN *et al.* (1998), who proposed the superposition of a shear fault and tensile cracks that are mutually oblique and that span both shear faults that open due to, for example, the roughness of the surface during shear slip and the shear fault accompanied by wing tensile cracks at the tips originated by the tensile stress concentration due to shear slip, the STC model can be understood as a special case for the combination of tensile and shear faulting. Configurations of oblique shear and tensile fractures with a reverse initiation were also included: tensile cracks are accompanied by wing shear faults (JULIAN *et al.* 2010) and Hill's mesh (HILL 1977).

In practice, we suggest that it is reasonable to evaluate the mechanism using an inversion of seismic data using both descriptions (i.e. the MT and STC). A comparison of the retrieved parameters for both models in terms of the orientation and the content of non-shear components can help us assess the reliability of the retrieved mechanisms.

In the first part of this work, we compared the resolution of the STC source model and the MT model using synthetic data that simulated the data recorded during injections at Soultz-sous-Forêts. A similar synthetic study, dealing only with MTs and ignoring noise contamination, was described by ŠÍLENÝ (2009). In this work, we investigated the impact of hypocenter mislocation, errors in the velocity model, noise contamination, and a reduction in the coverage of surface stations that recorded the event. Several types of shear-tensile sources ranging from pure shear slips to tensile cracks were used in the experiments.

In the second part of our work, the amplitudes of 13 weak earthquakes recorded during the early stages of the 2003 fluid injection at the Soultz-sous-Forêts EGS site were processed in terms of the full moment tensor and the STC models. Here, the reliability of the estimated parameters of both models is discussed; in particular, the certainty of the determination of the non-shear component of the slip vector indicating the type of fracturing of the rock mass. To discern the mode of fracturing, it is important to properly estimate the error associated with the obtained parameters. We constructed confidence regions by mapping the L_2 misfit function and evaluated the posterior probability density (PPD). The confidence region was then constrained as the portion of the model space that possessed an a-priori specified probability level.

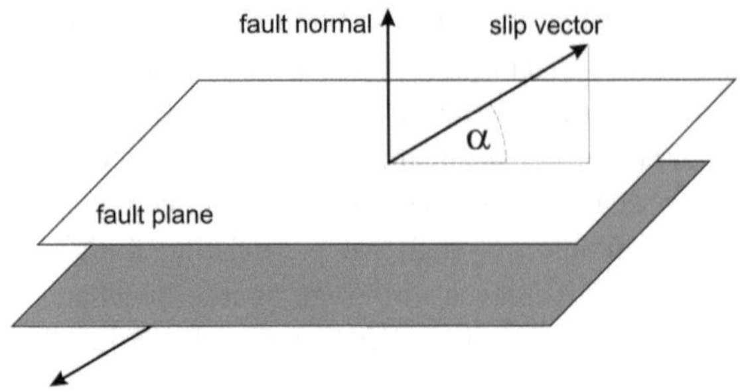

Figure 2
The STC source model: a combination of shear-slip and tensile opening represented by the slip vector directed from the fault plane by the slope angle α ($\alpha = 0°$ is the pure shear slip, and $\alpha = 90°$ is a tensile crack)

2. Source Models

The most comprehensive description of both shear and non-shear sources is the unconstrained MT. The MT is a general dipole source, but its generality may sometimes be inconvenient, since it also intrinsically includes non-physical sources.[1] Advantageously, the MT inversion is linear, allowing the fast retrieval of six independent components—M_{11}, M_{22}, M_{33}, M_{12}, M_{13}, and M_{23}. However, for cases of noisy or poor data, spurious non-DC components may result (the DC component is then generally distorted, as well, of course). For this reason, we also inverted an alternative source model that can simulate a tensile crack, a shear crack, or a combination of the two. We call the model a STC source model (Fig. 2). Such a model was originally proposed by Kozák and Šílený (1985), who were concerned with representing a tensile crack originating at the tip of a fault undergoing a shear slip. Slip along a fault creates a stress concentration at its tip that results in a tensile fracturing that radiates seismic energy almost synchronously with the shear slip. The reverse sequence of a slip on the fault, initiated by the jog collapse of a mined-out slot nearby, represents the shear-implosive source model (Teisseyre 1980; Rudajev and Šílený 1985). A model that allowed for a general angle between the fault and the direction of the tensile crack running out of its tip was mentioned by Julian *et al.* (1998). Dufumier and Rivera (1997) suggested that shear slip along a fault was accompanied by a simultaneous opening/closing. Such a source model, characterized by a slip vector that is not within the plane of the fault, was revisited and theoretically explored by Vavryčuk (2001, 2011). The source model, a priori, avoids combinations of moment tensor components that do not correspond to simple, although not necessarily a fault-parallel, slip vector. Such a source model can be described by four angles that define the direction of the fault plane normal, as well as the slip vector (strike, dip, rake, and slope angle) and the slip magnitude. Therefore, the number of model parameters is reduced from six to five, making the inversion robust even for fewer input parameters. On the other hand, the inversion becomes non-linear, slowing down the calculation.

Among constrained models of the mechanism, the STC is advantageous from a conceptual view, and can be considered as the simplest generalization of the traditional DC corresponding to a pure shear slip. If the inversion is not well posed and a constraint is necessary, the STC is capable of resolving the shear slip and, additionally, possibly providing information on the volume change within the focus.

To estimate the reliability of the five inverted STC parameters, we computed confidence zones describing the distributions of model parameters that yielded a good match to the data. The size and shape of the

[1] The compensated linear-vector dipole (CLVD), a part of the traditional decomposition of a general MT, if considered as a single phenomenon, lacks a clear physical explanation. If its major axis is tensional it is commonly described as a penny-shaped opening tensile crack accompanied by the influx of a fluid into an opened space, and vice versa, if the major couple is along the pressure axis. However, such an intuitive view is incorrect because CLVD is a system of forces, not the physical fracture itself.

confidence zones indicated how certain the parameters were determined, as follows: if the confidence zone of a parameter was large, it was poorly resolved, whereas if it was small and compact the parameter was well constrained. We considered the confidence zone of the model parameters, $\boldsymbol{m}$, with the probability content p to be a set of points, $\boldsymbol{m}$, satisfying the following:

$$\chi^2(\boldsymbol{m}) < \chi_p^2$$

where $\chi^2(\boldsymbol{m})$ is the likelihood function constructed from the residual least squares of data and synthetics, and the dispersion, χ_p^2, is determined from the following condition:

$$\frac{\int_{\chi^2(m) < \chi_p^2} \text{PPD d}m}{\int_m \text{PPD d}m} = p$$

that specifies the ratio of the cumulative probability in the region to the integral of the PPD (e.g., TARANTOLA 2005) across the entire model space. The PPD is defined as follows:

$$\text{PPD} = \exp\left(-\frac{1}{2}\chi^2\right).$$

Using the function $\chi^2(\boldsymbol{m})$, we determined particular confidence zones as regions limited by the contour χ_p^2 and possessing the probability content requested (ŠÍLENÝ 1998). In particular, we constructed confidence zones for the P, T, and N axes of the DC portion of the mechanism, which provided us with information on the uncertainty for determining the orientation, and on the confidence zone for the slope angle α, which yielded information on the uncertainty of shear vs. non-shear content. A high content of a non-shear slip component was indicated by an increased uncertainty in the P and N axes in cases of the fault opening, and the T and N axes if the fault closed. For a pure tensile crack, mode I of the fracturing, the P and N (or T, N) axes were undetermined, and their confidence zones merged into a single belt within the plane perpendicular to the crack opening (or closing). For the case of the MT source model, we investigated the confidence zones of the P, T, and N axes again, as well as the MT decomposition in terms of the percentage of the DC, the volumetric component (V), and the compensated linear vector dipole (CLVD).

3. *Synthetic Tests*

3.1. *Configuration*

We performed a synthetic case study by simulating the configuration of the surface seismic network and monitoring the seismicity at the European EGS site at Soultz-sous-Forêts, Alsace, deployed by the École et Observatoire des Sciences de la Terre (EOST), University Louis Pasteur of Strasbourg, France (CUÉNOT *et al.* 2006). We located the hypocenter of our hypothetical source near the center of the array and close to the bottom of the seismic zone, at a depth of 5.25 km. To examine the effects of hypocenter mislocation, velocity structure mismodeling, and station coverage on the recovered source parameters, the synthetic P and S amplitudes for several shear-tensile source models were inverted.

The theoretical source used to generate synthetic data was a vertical plane striking W that slipped within the strike-slip model and that was optionally combined with an off-plane component. In total, the following six types of STC sources were investigated: (i) a pure shear-slip (dip 90°, strike 270°, rake 0°), and STC sources with increasing fault openings defined by the slope angle α values of (ii) 10°, (iii) 20°, (iv) 30°, (v) 45°, and (vi) 90°, the last one representing a pure tensile crack. The increasing component of the fault opening yielded a progressive change in the P-radiation pattern from the four-quadrant form of the shear-slip (the DC in terms of body-force equivalents) to the axially-symmetric, equal-sign radiation patterns of the tensile crack (Fig. 3). Only a small change in the slope, α, produced a large increase in the non-DC portion of the equivalent MT (Table 1).

For the crust in the area surrounding the Soultz EGS facility, a 1-D, P-velocity model consisting of eight horizontal homogeneous layers was derived by EOST (L. DORBATH, personal comm., 2007; Table 2a). S-wave velocities for the eight layers were evaluated by assuming a constant ratio $v_P/v_S = \sqrt{3}$. Here, we refer to this model as a *reference model*. To investigate the effects of errors in the velocity model on the inversion of amplitudes, two additional models with different degrees of simplification were introduced (Table 2). In the first

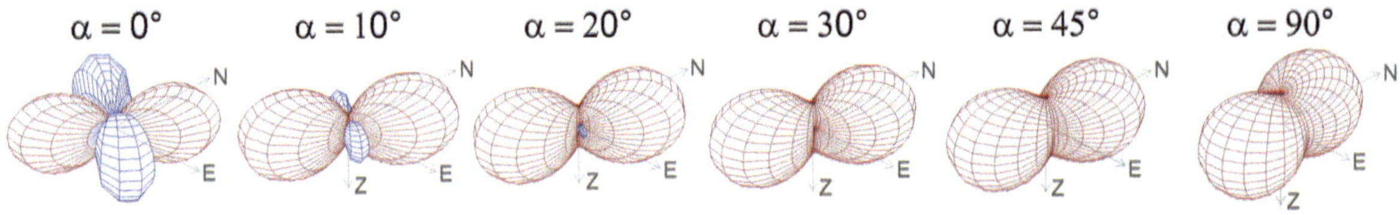

Figure 3
The radiation pattern of the STC source with a varying slope angle, α. Compressions—*red*, dilatations—*blue*

Table 1

The percentage of the non-DC component in the MT equivalent to the STC source for various slope angles, α [decomposition from Vavryčuk (2001)]

α (°)	0	10	20	30	45	90
Non-DC (%)	0	38.7	60.9	75	87.9	100

Table 2

(a) The layered 1-D velocity model of the Soultz area EOST (L. Dorbath, personal comm., 2007) used as a reference model, (b) a moderate simplification obtained by decreasing the number of layers, and (c) a strong simplification obtained by replacing the stack of layers with a homogeneous half-space overlain by a thin low-velocity layer

Depth (km)	0	0.25	0.45	0.65	0.85	1.05	1.53	1.8
(a) Reference model								
v_P (km/s)	1.85	2.36	2.63	3.15	3.71	4.54	5.48	5.89
v_S (km/s)	1.07	1.36	1.52	1.82	2.14	2.62	3.16	3.40
Depth (km)	0	0.25				1.05		
(b) Moderate simplification								
v_P (km/s)	1.85	2.96				5.89		
v_S (km/s)	1.07	1.71				3.40		
Depth (km)	0	0.25						
(c) Strong simplification								
v_P (km/s)	1.85	5.89						
v_S (km/s)	1.07	3.40						

The depth of the upper boundary of each layer is indicated

model, eight layers were reduced to three. Here, the model is referred to as a *moderate simplification.* In the second model, eight layers were replaced with a homogeneous half-space overlain by a thin surficial layer in order to rectify the dip angles of the straight-line rays to the observed values. The model is referred to here as the *strong simplification*, and was intended to mimic a situation where there was a complete lack of information regarding the medium. Green's functions for the three velocity models listed in Table 2 and configurations of the hypocenter and the stations were evaluated with a ray method using the program package CRT of Červený *et al.* (1988).

We inverted the P-wave and S-wave amplitudes in order to obtain parameters for the STC model, and to determine the resolution of the mechanisms for five types of station configurations. The first station configuration, 'stats1', included all 23 stations within the network; the second, 'stats2', included a subset of ten regularly distributed stations that kept the aperture of the network unchanged. The third, 'stats3', was a subset of ten stations with an irregular distribution that reduced the focal sphere coverage; the fourth, 'stats4', and fifth, 'stats5', were subsets of the five stations from the 'stats3' with regular and irregular configurations, respectively (Fig. 4).

3.2. *The Results for Noise-Free Data*

Common problems in retrieving earthquake focal mechanisms from the inversion of seismic data are posed by uncertainties in hypocenter locations and the velocity/attenuation model of a medium in the region under study. Both effects impact the resolution of the mechanism, as they introduce errors into the Green's functions used in the inversion. To estimate the effect of mislocation error, we evaluated the GF obtained by shifting the hypocenter 330 m upward and 350 m horizontally from its true location. Inversion of the data with an improper structure was simulated by employing "moderate" and "strong" simplifications from the eight-layer velocity model (Table 2b, c, respectively). The impact of these perturbations on the source location and the velocity model was explored for five station distributions that varied from very good ('stats1') to poor ('stats5') (Fig. 4). Results describing the resolution of the non-DC component and the fault plane orientation obtained from the full-MT and STC model inversions are listed in Tables 3, 4, 5, 6. Error in the estimate for the non-DC

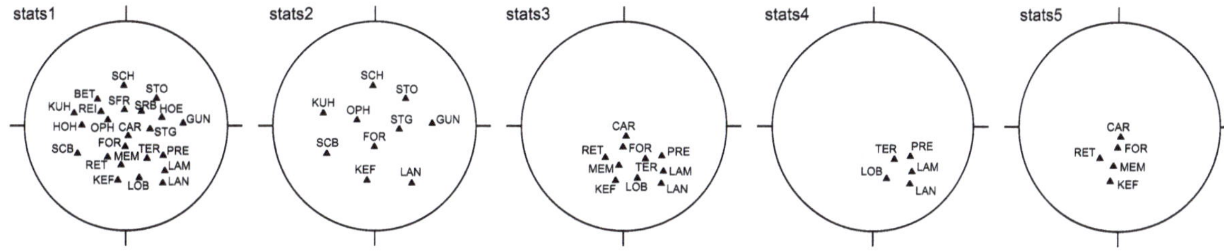

Figure 4
The configuration of the surface seismic stations at Soultz (the equal-area projection of the lower focal hemisphere), simulating the deteriorating quality of the coverage of the focal sphere as viewed from the hypocenter of the theoretical source used to generate the synthetic data

Table 3

Synthetic experiments showing the deviation of the non-DC component in the resolved source from that in the theoretical source (in percent), MT source model

non DC	mislocation						strong simplification of the velocity model					
deviation	$\alpha = 0°$	$\alpha = 10°$	$\alpha = 20°$	$\alpha = 30°$	$\alpha = 45°$	$\alpha = 90°$	$\alpha = 0°$	$\alpha = 10°$	$\alpha = 20°$	$\alpha = 30°$	$\alpha = 45°$	$\alpha = 90°$
stats1	0.3	0.8	1.1	1.1	1.1	-1.6	2.1	0.7	0.0	0.0	3.2	-3.1
stats2	1.2	1.5	1.5	1.3	1.0	-1.5	0.9	3.4	4.3	4.6	4.5	-3.8
stats3	2.6	-0.3	4.7	2.7	3.2	-8.7	8.3	0.5	4.9	-1.3	7.7	-9.0
stats4	5.0	-1.9	0.2	1.1	0.9	-4.8	36.3	-15.2	2.3	-0.7	-17.5	-20.5
stats5	1.8	4.3	15.2	5.4	5.2	-3.9	7.9	9.8	9.1	7.5	4.9	-7.2

The inversion of synthetic data assuming a mislocated hypocenter (left) and an improper velocity model (right; strong simplification of the reference velocity model) in the station configurations from 'stats1' to 'stats5'. Synthetic data generated by the vertical strike-slip with an increasing off-plane component of the slip ranging from $\alpha = 0°$ (shear slip) to $\alpha = 90°$ (tensile crack). The color displays the amplitude of the deviation (red—positive values, blue—negative values)

component is expressed as the percent deviation of the estimated component from that inherent in the source. For the MT model, estimation of the non-DC component was straightforward from the MT decomposition. For the STC model, the equivalent MT was computed and the corresponding non-DC percentage evaluated. The error in the fault plane orientation for both models was calculated as the sum of the angular differences of the deviations of the P axis and T axis of retrieved and true sources. In the case of a pure tensile crack ($\alpha = 90°$), deviation of the T axis was only recorded because the position of the P axis was ambiguous. To point out the difference in the definition, the corresponding columns in Tables 4, 6, and 7 are framed in gray.

The orientation of the resolved mechanism was largely insensitive to errors in location and velocity and for the range of network geometries considered (Tables 4, 6, 7). Not much difference was determined in the performance of the MT and STC source models. Mismodeling seemed to yield smaller errors than mislocation. In comparing individual station configurations, those with a larger number of stations and regular distributions worked better, as expected. An exception was 'stats4', the performance of which was markedly worse than 'stats5', perhaps because it was more compact (also see the comment in the "Results" section for a noisy data inversion).

Resolution of the non-DC component within the mechanism was somewhat better for the STC model,

Table 4

Synthetic experiments showing the deviation of the orientation of the resolved source from the theoretical one (in degrees), the MT source model

orientation deviation	mislocation						strong simplification of the velocity model					
	$\alpha = 0°$	$\alpha = 10°$	$\alpha = 20°$	$\alpha = 30°$	$\alpha = 45°$	$\alpha = 90°$	$\alpha = 0°$	$\alpha = 10°$	$\alpha = 20°$	$\alpha = 30°$	$\alpha = 45°$	$\alpha = 90°$
stats1	5.3	5.2	6.4	9.0	13.9	7.3	0.7	1.1	0.2	1.5	1.1	0.8
stats2	4.8	5.0	5.8	8.8	13.4	7.2	0.4	0.2	0.2	0.9	0.8	0.1
stats3	5.1	5.6	6.0	7.2	9.0	6.4	2.6	1.3	1.6	6.3	23.2	4.7
stats4	5.1	5.6	5.6	6.3	10.5	5.0	17.3	10.6	5.2	34.2	63.8	25.6
stats5	6.3	6.3	5.4	6.8	9.7	7.0	2.5	5.6	8.8	12.7	22.0	2.9

For $\alpha < 90°$, the summary deviation angle of the P and T axes is depicted; for $\alpha = 90°$ (framed in gray), the deviation of the T axis is shown. For details see the caption in Table 3. The red color indicates the amplitude of the deviation of the orientation

in some cases markedly so (Tables 3 and 5). Errors in the velocity model tended to have a greater impact on the resolution of the non-DC component than the hypocenter mislocation, which was opposite to the tendency seen for resolution of the fault plane orientation.

3.3. Results of Experiments with Noisy Data

In addition to introducing errors within the hypocenter location and the velocity models, we also modeled the deterioration of data quality with noise. To add to each data point a random value constrained by an a priori percentage of the amplitude, we simulated the noise using a pseudorandom generator. Two levels of noise contamination were selected, namely 10 % and 20 %, and combined with the following three types of STC source: (i) the slope angle $\alpha = 0°$ (i.e. a pure shear), (ii) $\alpha = 10°$, and (iii) $\alpha = 20°$. These six data sets were inverted in terms of both the MT and STC source models using the Green's functions constructed for the following cases, as follows: (a) no hypocenter error and an eight-layer reference velocity model; (b) the hypocenter mislocated 330 m upward and 350 m horizontally for the eight-layer velocity model; (c) no hypocenter error and a moderately simplified velocity model (Table 2b); (d) no hypocenter error and a strongly simplified velocity model (Table 2c); and (e) the mislocated hypocenter and the strongly simplified velocity model [i.e., combining the effects of (b) and (d)]. The impact of these combinations of noise-free/noisy data and correct/improper GFs were evaluated for the five network geometries 'stats1'–'stats5'. In this manner, 300 synthetic inversion experiments were performed.

The orientation of the source inverted in terms of the MT was resolved surprisingly well, even if the hypocenter was mislocated and the velocity model was inaccurate (Šílený 2009). The resolution began to fail only when the station distribution became extremely irregular. Inversion into the STC yielded a similar pattern. Contamination of the data by noise resulted in the same behavior as the inversion of noise-free data. Even with a noise level as high as 20 % of the data amplitude, the resolution of the T, P, and N axes in both the MT (Fig. 5) and STC (Fig. 6) descriptions was excellent from the 'stats1' to the 'stats3' station distributions, for all of the inconsistencies simulated (vertical axis of Figs. 5, 6). The network geometries 'stats4' and 'stats5' yielded reasonably good insight regarding the source orientation, although they were less certain than the better configurations.

Similar to the insight from noise-free tests, it is worth noting that the resolution of the orientation was

Table 5

Synthetic experiments indicating the deviation of the non-DC component in the resolved source from that in the theoretical source (in percent), for the STC source model

non DC deviation	mislocation						moderate simplification of the velocity model						strong simplification of the velocity model					
	α = 0°	α = 10°	α = 20°	α = 30°	α = 45°	α = 90°	α = 0°	α = 10°	α = 20°	α = 30°	α = 45°	α = 90°	α = 0°	α = 10°	α = 20°	α = 30°	α = 45°	α = 90°
stats1	1.3	0.6	0.5	1.0	1.0	-1.1	0.0	0.6	1.8	1.7	1.6	-1.2	0.0	0.6	1.8	2.4	2.6	-1.9
stats2	1.3	2.0	0.9	0.8	0.6	-1.1	2.5	2.7	2.5	2.4	1.8	-1.3	1.3	3.4	3.7	3.8	3.6	-2.2
stats3	-3.8	0.0	-0.1	1.7	2.7	-1.5	-2.5	0.0	2.5	3.2	3.7	-3.9	-2.5	0.6	4.1	5.9	5.9	-4.4
stats4	-6.3	-2.2	-0.4	0.7	1.3	-1.2	-10.9	-3.8	0.9	1.7	2.0	-3.3	-25.0	-12.8	1.3	4.7	2.2	-12.5
stats5	0.0	2.0	-0.1	1.8	1.0	-1.0	1.3	2.0	3.0	1.8	4.9	-1.6	1.3	7.2	4.5	3.4	5.3	-2.5

The inversion of synthetic data assuming a mislocated hypocenter (left) and an improper velocity model (middle—moderate simplification of the reference velocity model; right—strong simplification) in the station configurations from 'stats1' to 'stats5'. Synthetic data generated by the vertical strike-slip where the increasing off-plane component of the slip ranged from $\alpha = 0°$ (shear slip) to $\alpha = 90°$ (tensile crack). The color indicates the amplitude of the deviation (red—positive values, blue—negative values)

Table 6

Synthetic experiments showing the absolute values of the deviation of the slope angle α of the resolved source from the theoretical one (in degrees), STC source model

slope deviation	mislocation						moderate simplification of the velocity model						strong simplification of the velocity model					
	α = 0°	α = 10°	α = 20°	α = 30°	α = 45°	α = 90°	α = 0°	α = 10°	α = 20°	α = 30°	α = 45°	α = 90°	α = 0°	α = 10°	α = 20°	α = 30°	α = 45°	α = 90°
stats1	0.25	0.25	0.25	1	1.5	14.25	0	0.25	1	1.5	2.75	15	0	0.25	1	2.25	4.5	19.25
stats2	0.25	0.75	0.5	0.75	1	14.5	0.5	1	1.5	2.25	3	15.75	0.25	1.25	2.25	3.5	6.5	20.5
stats3	0.75	0	0	1.5	4.75	17.5	0.5	0	1.5	3	6.75	21.75	0.5	0.25	2.5	6	11.5	28.5
stats4	1.25	0.75	0.25	0.5	2.25	15.5	2.25	1.25	0.5	1.5	3.5	25	5.75	4	0.75	4.5	3.75	45.5
stats5	0	0.75	0	1.75	1.5	13.75	0.25	0.75	1.75	1.75	9.25	18	0.25	2.75	2.75	3.25	10	21.5

For details see the caption in Table 5

Table 7

Synthetic experiments showing the deviation of the orientation of the resolved source from the theoretical one (in degrees), STC source model

orientation	mislocation						moderate simplification of the velocity model						strong simplification of the velocity model					
deviation	α = 0°	α = 10°	α = 20°	α = 30°	α = 45°	α = 90°	α = 0°	α = 10°	α = 20°	α = 30°	α = 45°	α = 90°	α = 0°	α = 10°	α = 20°	α = 30°	α = 45°	α = 90°
stats1	5.4	5.0	8.4	8.9	14.6	10.3	0.2	0.2	1.1	1.0	2.7	7.7	0.2	0.6	2.0	2.2	3.0	10.1
stats2	4.9	4.6	7.7	9.0	14.7	10.2	1.0	1.0	2.0	2.6	4.0	7.7	0.5	0.5	1.9	3.0	5.6	10.7
stats3	4.6	4.6	7.7	8.2	8.9	12.4	1.0	0.2	2.6	4.9	20.4	15.1	1.0	0.4	3.3	8.0	40.7	17.4
stats4	6.4	6.6	7.4	8.1	5.7	11.6	10.8	2.1	4.6	5.1	26.3	20.3	17.4	9.9	2.1	19.5	44.7	44.6
stats5	5.9	6.0	8.2	7.6	12.1	9.4	0.6	0.5	4.7	6.6	34.0	9.3	0.6	4.7	6.9	10.5	45.0	10.3

For $\alpha < 90°$, a summary deviation angle of the P and T axes was depicted; for $\alpha = 90°$ (framed in gray), the deviation of the T axis is shown. For details see the caption in Table 5. Red color indicates the amplitude of the deviation of the orientation

worse for the 'stats4' than for the 'stats5' configuration; see the discussion section. For cases of inversion with a strongly incorrect velocity model, see Fig. 5d and e; the additional feature of the 'stats4' configuration was a systematic shift of clusters corresponding to the P and N axes. The corresponding STC solutions were similarly biased, but to a slighter extent; the clusters of the N axes were closer to the center of the focal sphere projections than for the case of the MT solutions, Figs. 5 and 6.

Regardless of the details for the 'stats4' configuration, the individual rows in Figs. 5 and 6 appeared to be fairly similar, indicating that 20 % of the noise in the data impacted the resolution of the orientation of the source mechanism much more than hypocenter mislocation and an incorrect velocity model.

Better performance of the STC as compared to the MT source model was obvious for the decomposition of the mechanisms retrieved by inverting noisy data. The MT solutions were clusters surrounding the proper point within the Hudson plot—the center for the case of a shear-slip source in Fig. 7. Clusters enlarged with the deteriorating quality of monitoring—decreasing the number of stations and departing from the regularity of their distribution with respect to focus. As a result of inherent constraint within the STC model that ruptured into a STC, STC solutions within the Hudson plot were located along the line (−Crack, +Crack). Therefore, by definition, the size of the cloud in the direction perpendicular to this line shrunk. In most cases, the other dimension, namely the length of the STC line, remained similar to the corresponding size of MT clouds. However, there were exceptions, inversions with strongly improper velocity models where the STC line did not intersect the MT cloud but passed nearby (Fig. 7d, e, 'stats4'). For these cases, the STC solution was not biased by a systematic shift from the Hudson plot center such as the MT solution; and the length of the STC line was shorter than the corresponding size of the MT cluster.

The STC line within the Hudson plot in Fig. 7 is a 1-D record of the distribution of values of slope angle α. The red dot density should indicate the multitude of the values of α. However, we displayed it directly, and assessed the most frequent value of α for individual cases (Fig. 8). The histograms were fairly narrow and peaked correctly at zero up to the station configuration 'stats3', demonstrating very good resolution of the slope angle. Configurations 'stats4' and 'stats5' (both irregularly distributed sets of five stations) yielded notably different patterns of the α distribution—broader than for 'stats1'–'stats3' and centered off the zero value. In the configuration 'stats5', the peak was slightly positive; in 'stats4', it was negative. For 'stats4', if the GF was constructed for the strongly different velocity model, the histogram peak was notably father from zero. However, it is worth noting that similar to the clouds of the P, T, and N axes specifying uncertainty within the retrieval of the orientation of the source in Fig. 6, α-histograms indicated that mislocation and both types of mismodeling were only impacted in comparison to noise contamination.

An interesting feature of the orientation of the MT and STC solutions for sources with a zero (pure shear) and non-zero slope was demonstrated. In our synthetic experiments, there was a vertical strike-slip striking W (i.e. the N axis vertical), and T and P were determined in the horizontal plane to the NE and NW, respectively, of the N–S and E–W nodal lines. The display, in terms of the clouds of the P, T, and N axes, and bunches of the nodal/source lines were the same for MT and STC, Fig. 9, $\alpha = 0°$. For sources with an off-plane component of the slip vector (i.e., a slip accompanied by an opening of the fault), the P and T axes and the nodal lines did not indicate the directions of the fault and slip, but were rotated. Proper orientation of the fault and slip were indicated by source lines of the STC (Fig. 9, $\alpha = 10°$, right), which were no more perpendicular (Vavryčuk 2011).

4. Application

We analyzed small earthquakes from the beginning of the 2003 fluid injection at the Soultz-sous-Forêts EGS site. We selected thirteen events from the first phase of the injection when only the borehole GPK3 was stimulated (Table 8), not processed by Horálek *et al.* (2010). The records had a low signal-to-noise ratio and were not detected by all 15 stations of the Soultz surface network consisting of nine three-component, and six one-component stations, indicating that the MTs need not always be well constrained and that noise contamination and structure mismodeling may produce spurious non-shear components. Inversion of the STC source model, consisting exclusively of the physical components, can reveal this fact.

The identification of selected events such as the origin time and magnitude, together with a comparison of the minimum normalized residual sums for the MT and STC solutions, are presented in Table 8. All but one of the NRMS values for the MT inversion were lower than the NRMS values for the STC model. The STC was designed to have less parameters than the MT model (five vs. six). Decomposition of the MTs into the percentage of shear (DC) and non-shear [V and compensated linear-vector dipole (CLVD)] components can be compared with the slope angle α of the STC in Table 8; for small values of α the non-DC component should be smaller, and vice versa. Non-shear components of the MT often have opposite signs that are not easy to interpret within the physical source model of a crack. Additionally, even when they are in the sign of conformity, the ratio of V to CLVD components cannot be explained using tensile fracturing. In view of the fact that the values of the slope angle are all very small, the non-DC components of the MT were likely due to high noise within the input data and to a small number of seismic stations that detected the event. For an easier comparison of the MT and STC solutions, the traditional fault-plane solution plots of the source mechanisms are displayed in Fig. 10.

Application of the STC model to this data set appeared to be appropriate. Detailed results are shown in Fig. 11. The source lines of the best STC are displayed together with the directions of the fault normal and the slip vector. The source lines (Vavryčuk 2011) in the STC model are analogous to the nodal lines in the fault-plane solution of the MT. However, contrary to the nodal lines, they need not be perpendicular. If the mechanism deviates from a pure shear, this occurs. The confidence zones of the fault normal/slip vector and the slope angle histograms allowed us to assess the reliability and the accuracy of the obtained solutions. The uncertainty in determination of the principal axes T, P, and N in the form of confidence zones is also displayed. For most of the events investigated, the confidence zones of the

Table 8

A comparison of the moment tensor (MT) solutions and the shear-tensile crack (STC) solutions for selected earthquakes from Soultz

Event	Date (yy/mm/dd)	Time (h:min)	*M*	MT				STC	
				NRMS	DC (%)	V (%)	CLVD (%)	NRMS	α (°)
2	03/05/28	01:42	1.9	0.184	59	−6	−35	0.207	0.25
3	03/05/29	03:44	2.3	0.222	96	−3	−1	0.228	−1.25
5	03/05/29	04:28	1.5	0.374	85	3	12	0.382	1.25
6	03/05/29	14:25	1.4	0.443	74	11	15	0.426	3.75
8	03/05/30	18:46	1.4	0.321	45	10	−45	0.350	1.00
9	03/05/30	21:11	1.5	0.344	79	9	−12	0.354	2.25
10	03/05/31	02:23	1.4	0.333	74	13	−13	0.353	4.50
12	03/05/31	03:29	1.6	0.381	52	6	42	0.433	0.25
14	03/05/31	11:33	1.4	0.137	70	−5	−25	0.165	−0.25
19	03/06/01	05:42	1.5	0.280	32	−4	−64	0.355	−3.25
21	03/06/01	11:57	1.5	0.283	83	9	8	0.289	1.00
23	03/06/01	14:27	1.5	0.469	67	5	−28	0.477	−1.25
25	03/06/02	07:44	1.5	0.196	53	−2	−45	0.257	1.75

For a comparison of data fit, the normalized residual sums (NRMS) for both solutions are listed. The decomposition of the MTs of the events examined into percentages of the volumetric component (V), the double-couple (DC), and the compensated linear-vector dipole (CLVD) can be compared with the slope angle of the STC. A negative value of V indicates an implosive mechanism; a negative value of the CLVD indicates the principal dipole in the CLVD aligned with the P axis

T, P, and N axes (or of the source plane poles) were small, indicating a well constrained orientation for the source. Exceptions were events 10 and 19, for which the zones of the T and N axes were stretched in the plane perpendicular to the P axis that was well constrained. However, the histograms of the slope angle were all fairly narrow and located at or very close to the zero value, suggesting that the off-plane slip component is negligible. In other words, the events represented pure shear slips.

A detailed comparison of two alternative source models is shown for three events (8, 12, and 19) that had the largest non-DC component for the MT solution. The MT solutions of these events are displayed in Fig. 12. For each event, the traditional fault-plane solution and the principal T, P, and N axes (red, blue and green, respectively) are presented in an equal-area, lower-hemisphere projection. Histograms of the individual components of the decomposed MT, namely the DC, V, and CLVD, are also visible. The actual percentage of these components for the best solution was also determined. Colors represent the confidence zones within different probabilities (90, 95, and 99 %). STC solutions of the same three events are shown in Fig. 13. Plots of the source lines (Vavryčuk 2011) and the directions of fault normal/slip vectors, as well as plots of the principal T, P, and N axes (in turn, red, blue, and green), are presented for the same projections mentioned above. The slope angle α histograms are also displayed. To be allowed to immediately compare the confidence of the MT and STC solutions, histograms of DC, V, and CLVD corresponding to the MTs evaluated from the resolved STC solutions are also presented. For visualizing the confidence, the same percentages as for the MT confidence assessment were applied.

Agreement in the retrieval of the orientation of the source using both models, Figs. 12 and 13, was determined as compared to the principal axes and the nodal lines in Fig. 12, using the principal axes and the source lines in Fig. 13. However, the confidence zones of the T, P, and N axes were considerably smaller for the STC solutions, indicating that the STC results were more reliable. Contrary to the similarity of the orientations retrieved, the shear vs. non-shear content for both solutions completely differed. At the same time, their NRMS values were nearly the same (i.e. both solutions fit the data nearly equally). However, the confidence zones of the STC solutions were considerably smaller than the confidence zones of the MT solutions. Thus, the STC results should be

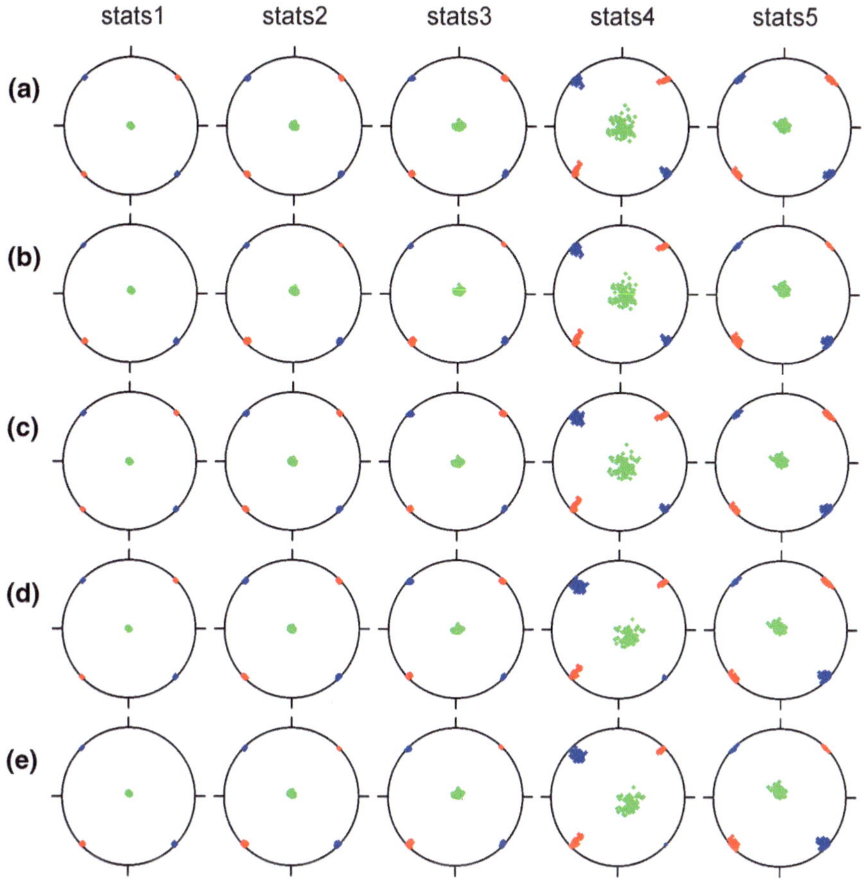

Figure 5
The inversion of noisy data in terms of the MT. Theoretical source: shear-slip (Fig. 3—$\alpha = 0°$). T axes—*red*, P axes—*blue*, and N axes—*green*. One hundred trials were performed for randomly generated noise superimposed on the data. At most, the noise amplitude reached 20 % of the data amplitude. Station configurations from 'stats1' to 'stats5' from Fig. 4. **a** Exact data and model, **b** mislocation, **c** moderate simplification of the velocity model, **d** strong simplification, and **e** the combination of mislocation and strong simplification of the velocity model

considered more reliable than the MT solutions. As a result, despite a rather large non-DC in the MT solution of events 8, 12, and 19, all of the events treated can be regarded as shear slips.

5. Discussion

A comparison of the source mechanisms resulting from the alternative approaches, MT and STC, offered a clue for estimating the reliability of shear vs. non-shear source components, and helped to discern fracture types. The performance of an inverse task for retrieving the parameters of the seismic source depends, in general, on the monitoring configuration, the quality of the data in terms of the frequency band recorded and noise contamination, and success in properly constructing the Green's function (i.e. for a proper hypocenter and velocity/attenuation model of the medium within the region under study). To assess their influence on resolution of the mechanism, we performed several series of numerical experiments in which either noise-free or noisy data were inverted using GF mimicking mislocation, mismodeling, and by taking into account various types of seismic station configuration. Each test was performed by assuming both moment tensor and STC source models.

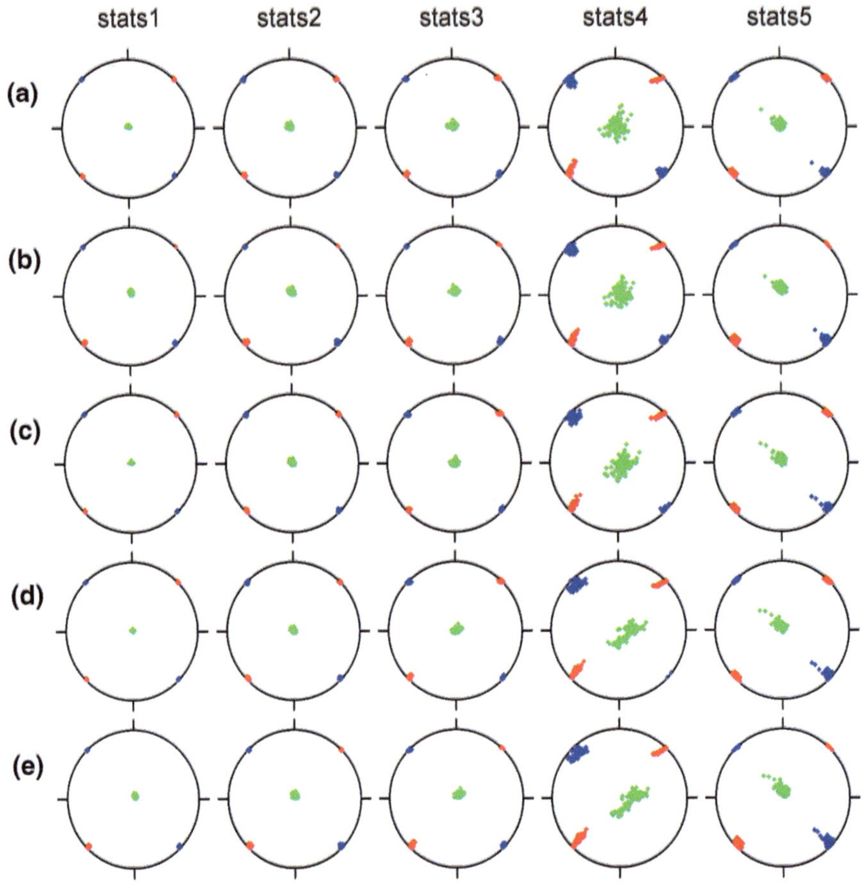

Figure 6
The inversion of noisy data in terms of the STC. For details see the caption in Fig. 5

5.1. Synthetic Experiments—Noise-Free Data

According to previous results (e.g., Šílený 2009), the orientation of the mechanism is a remarkably robust feature of a source, resistant even against rather large departures from the consistent set-up of inverse tasks (Tables 4 and 7). Our experiments clearly demonstrated this fact. Errors in the orientation retrieval were generally low, remaining mostly below 10° due to the effect of mislocation, and even less for cases of velocity mismodeling. The error tended to grow with an increasing slope angle, α; this was not a surprise when taking into account the fact that with an increased tensile component, the radiation pattern becomes less variable, transforming from a quadrant pattern with lobes of changing sign for the shear slip (α = 0°) to an everywhere compressive pattern with a rotational axis for the tensile crack (α = 90°). Mismodeling, even a strong one, yielded a smaller error than mislocation. The finding is somewhat surprising, because a mislocation hundreds of meters away may be well within the common location uncertainty for networks of Soultz apertures. However, a homogeneous half-space is undoubtedly a very rough approximation, and more detailed, and more realistic models are largely available. Few 'outliers' exist within our Tables. When faced with the mismodeling configuration, 'stats4' was markedly worse than the others. A comparison should be made with 'stats5' (both comprised of five stations), where we anticipated poor performance as a result of the linear distribution of the stations. Close to single-azimuth monitoring (with the only exception of the station RET), this was the case when the complete MT was not sufficiently constrained (Vavryčuk

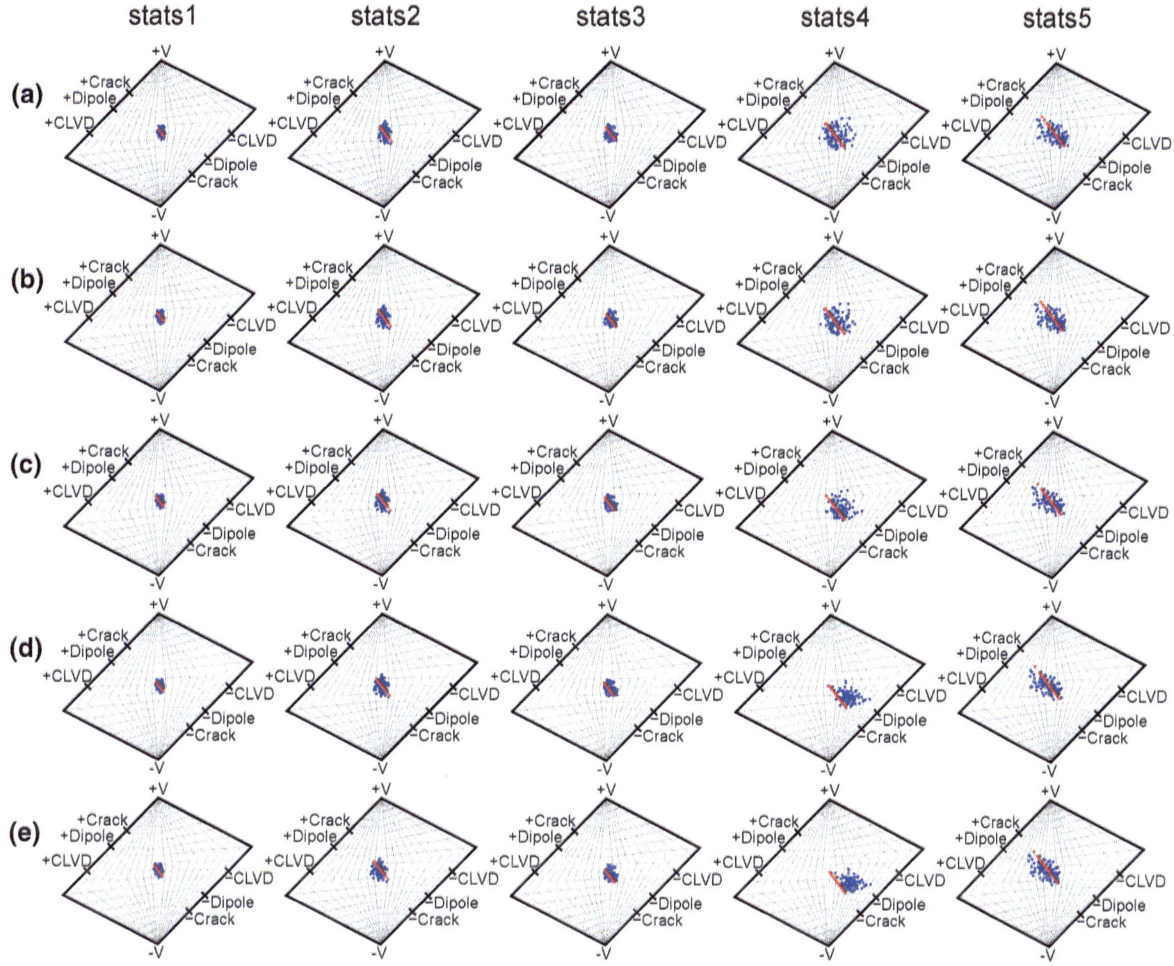

Figure 7
Hudson plots of the MT (*blue*) and STC (*red*) solutions of the noisy data inversion. Theoretical source: shear-slip (Fig. 3—$\alpha = 0°$). One hundred trials were performed with randomly generated noise superimposed on the data. At most, the noise amplitude reached 20 % of the data amplitude. The station configurations from 'stats1' to 'stats5' from Fig. 4. **a** The exact data and the model, **b** mislocation, **c** the moderate simplification of velocities, **d** strong simplification, and **e** the combination of mislocation and strong simplification

2007). Although the STC still worked (Šílený 2012). The configuration 'stats5' performed better than 'stats4', suggesting that the aperture of 'stats5' was greater than that of 'stats4' and was more important than the 'stats5' quasi-linearity, cf. 'stats4' and 'stats5' in Fig. 4.

Unlike the orientation of the retrieved source, the content of the non-DC component differed with respect to the source model applied for the inversion. Table 5 is paler in color than Table 3, indicating that the STC provided us with a more accurate estimate of the share of DC vs. non-DC components within the mechanism than the MT. Obviously, this is the consequence of a higher robustness of the STC inversion based in less parameters needed to describe the model than in case of the MT (five parameters vs. six).

5.2. Synthetic Experiments—Noisy Data

A noise superimposed on the data basically represents perturbed data. Inverting the perturbed data allowed us to assess the resolution of the source parameters. By repeating the procedure many times, we drew statistically reliable conclusions regarding the mechanism and its uncertainty due to the various

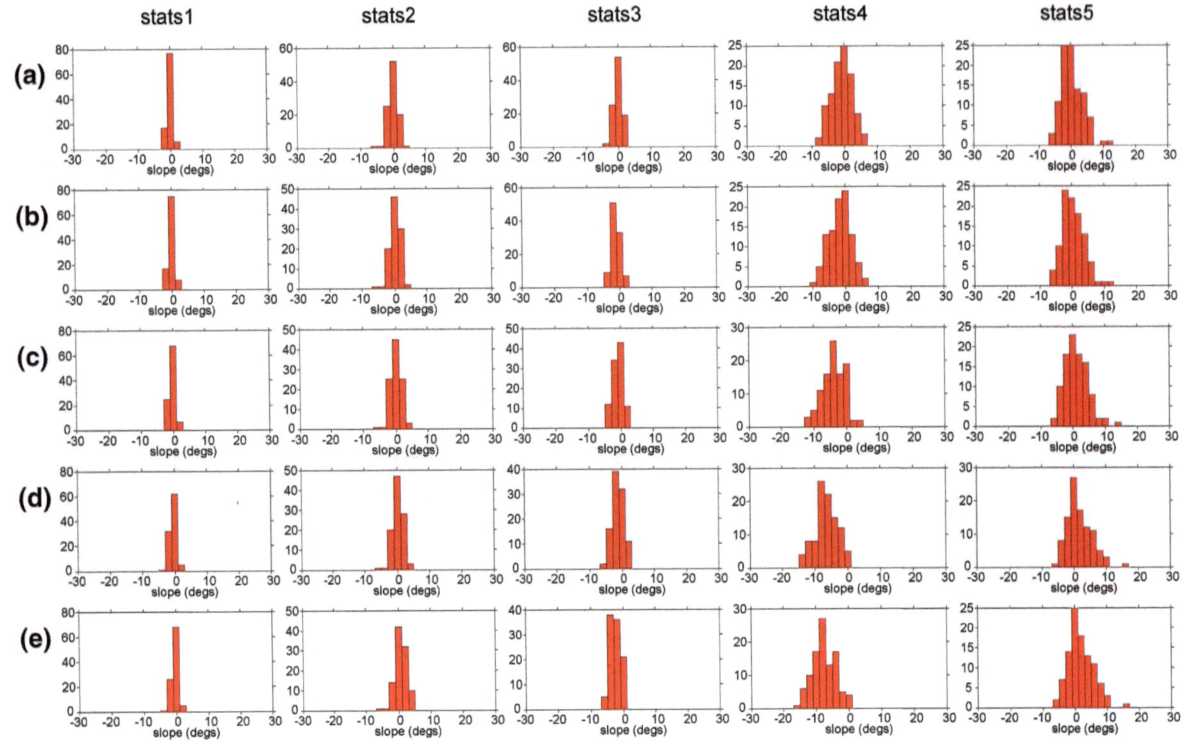

Figure 8
Histograms of the slope angle, α, for the STC solutions of the noisy data inversion. Theoretical source: shear-slip (Fig. 3—$\alpha = 0°$). One hundred trials were performed for randomly generated noise superimposed on the data. At most, the noise amplitude reached 20 % of the data amplitude. The station configurations from 'stats1' to 'stats5' from Fig. 4. **a** The exact data and the model, **b** mislocation, **c** moderate simplification of the velocities, **d** strong simplification, and **e** the combination of mislocation and strong simplification

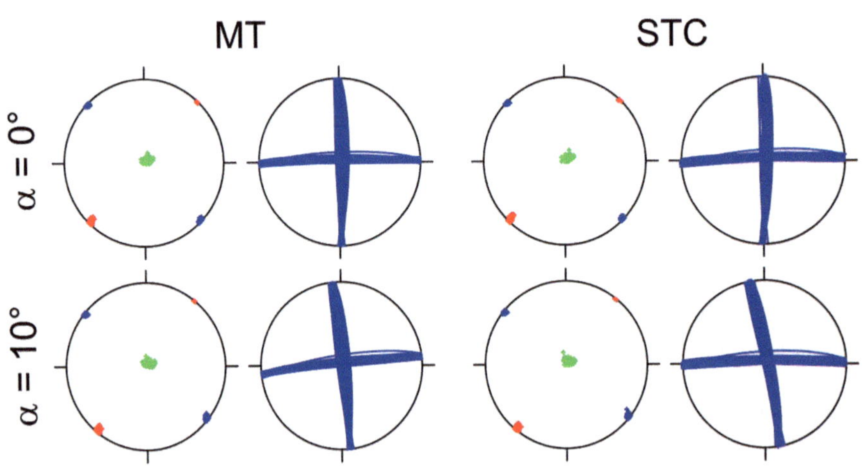

Figure 9
The MT vs. STC solution of the inversion of noisy data generated by a shear slip ($\alpha = 0°$), and the shear slip accompanied by an opening of the fault ($\alpha = 10°$). One hundred of trials were performed with randomly generated noise superimposed on the data. The noise amplitude reached 20 % of the data amplitude at most. T axes—*red*, P axes—*blue*, N axes—*green*, *blue lines* in the MT solution are nodal lines, and *blue lines* in the STC are source lines

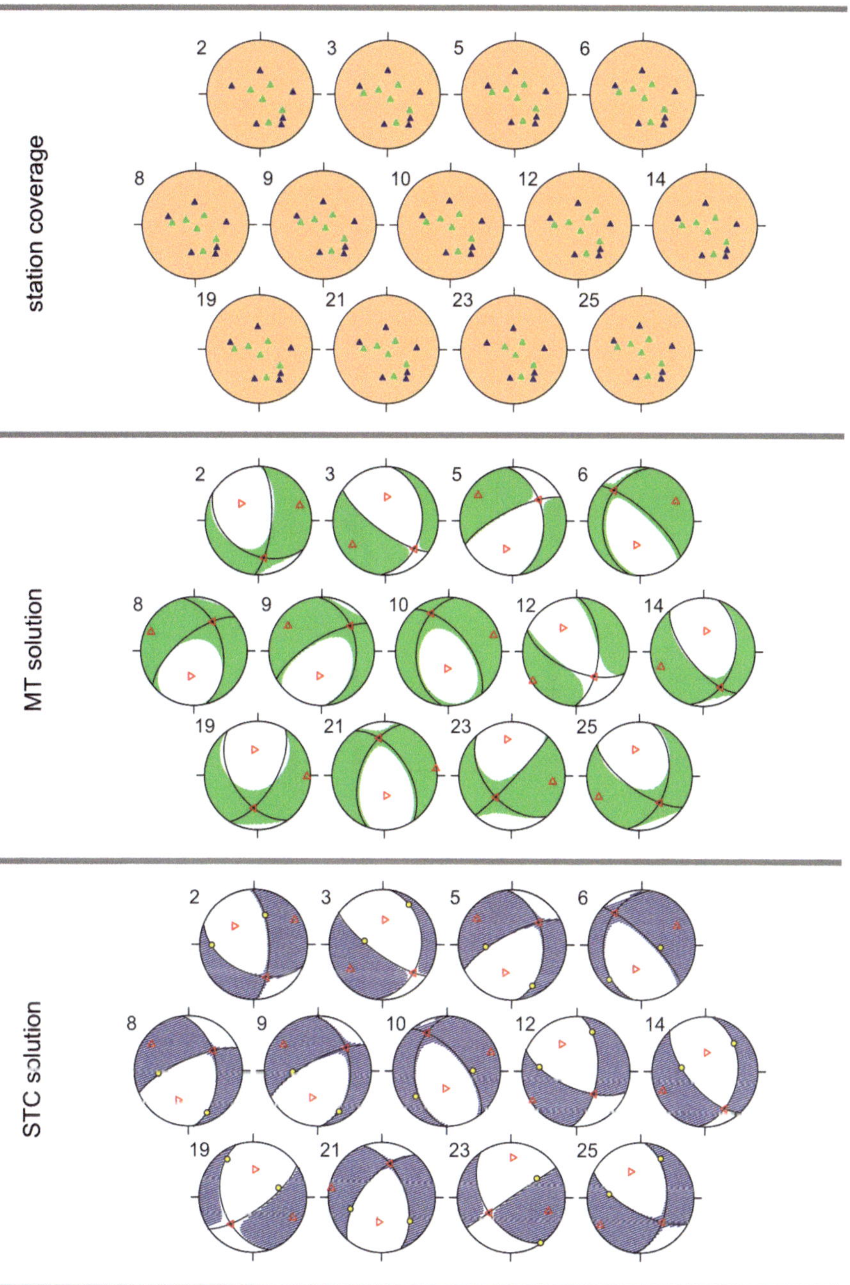

Figure 10
A comparison of moment tensor (*MT*) solutions and shear-tensile crack (*STC*) solutions. The source mechanisms are represented within the traditional fault-plane solution plots. The nodal lines of the DC portion of the retrieved MTs and the source lines of the STC are displayed using *gray lines*. The corresponding principal axes T, P, and N are displayed using a *triangle apex up*, *right*, and *left*, respectively, within an equal-area, lower-hemisphere projection. Top—station coverage (lower hemisphere equal area plot): three-component stations—*green triangles*, one-component stations—in *blue*

factors involved in the inversion. From the extent of the volume comprising solutions of the inversion of perturbed data sets, we could estimate the sensitivity to the mechanism and the factors impacting the inversion, namely the quality of the distribution of stations recording an event, the accuracy of the

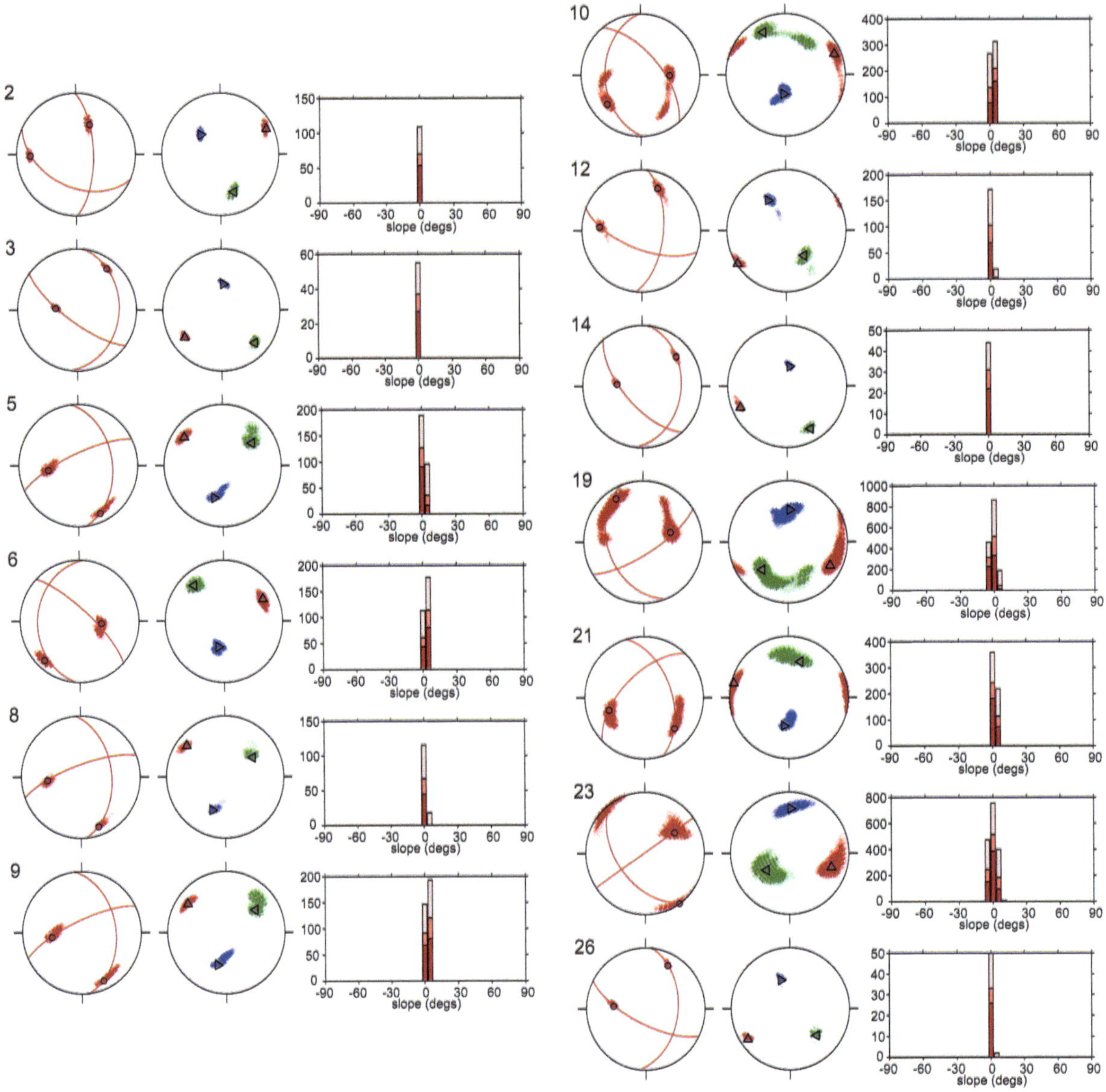

Figure 11
Shear-tensile crack (STC) solutions. The first column displays source lines (*red lines*) and the direction of the fault normal/slip vector (*circle*). In the second column, there are principal axes T, P, and N (a *triangle apex*, in turn, *up*, *right*, and *left*) in the equal-area, lower-hemisphere projections. In the last column, there are slope angle α histograms. *Different colors* represent confidence zones with different probabilities. *Dark*, *medium*, and *light colors* correspond, in turn, to probabilities of 90, 95, and 99 %

localization of the hypocenter, and the relevance of the velocity model. In particular, we were interested in how sensitivity is impacted by employing alternative models within the source, the MT, and the STC.

Tests with noisy data confirmed the finding from noise-free experiments, namely the advantageous insensitivity of the orientation of the retrieved mechanism to mislocation and velocity mismodeling. The orientation remained well constrained (the clouds of the P, T, and N axes corresponding to 100 inversion trials with noisy data were fairly small) down to the rather irregular distribution of ten monitoring stations, leaving three quadrants of the focal sphere projection vacant. The orientation was estimated well even further, using no more than five stations. However, its determination was less

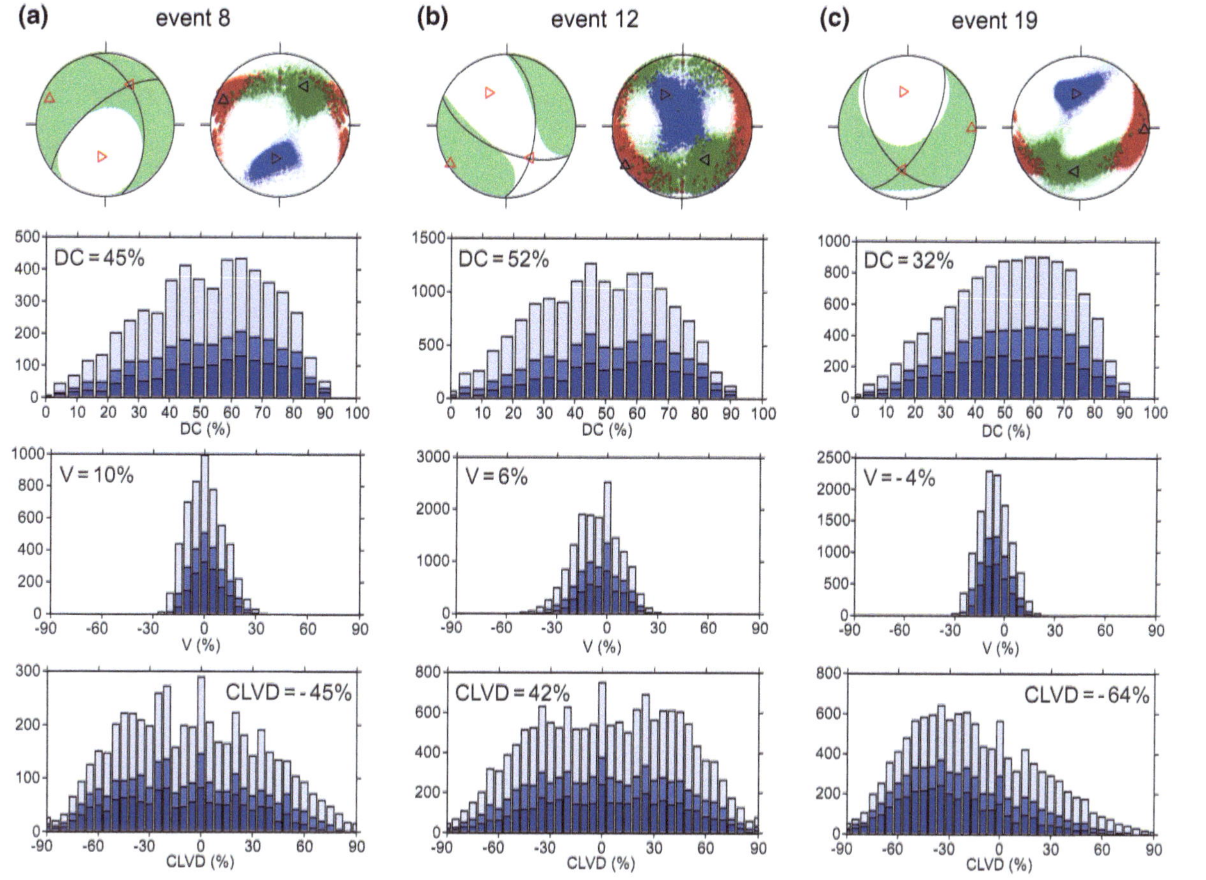

Figure 12
The source mechanisms of events 8, 12, and 19 retrieved using a general moment tensor. The traditional fault-plane solution and the principal T, P, and N axes are displayed together with the histograms of the individual components of the decomposed MT (DC, V, and CLVD). For details, see the caption in Fig. 11

constrained. Concerning the correctness of the determination, similar to noise-free experiments, there was trouble with the 'stats4' station configuration that provided a biased position for P and N clusters if the inversion was performed using the strongly simplified velocity model. The reason the 'stats4' configuration lacked resolution remains unknown. However, failure was weaker if the inversion was performed in the STC model instead of the MT.

Resolution of the decomposition mechanism, in particular the content of the DC vs. non-DC components, depended on the choice of the source model more than the resolution of the orientation. Repeated inversions of noisy data onto the MT yielded clouds within the Hudson plot that became larger when the number of monitoring stations dropped and their distribution with respect to the focus became irregular. However, the STC, thanks to its inherent constraint in involving just shear and tensile slips, displayed a line instead of a cloud, excluding, a priori, V and CLVD combinations that did not comply with the STC. For this reason, the STC solutions were a priori constrained better than the MT, and were especially beneficial for poor configurations, 'stats4' and 'stats5', in our notation. For 'stats4', where the MT solutions were biased when obtained using a strongly simplified velocity model, the STC provided us with a better estimate of the DC vs. non-DC content within the mechanism.

An interesting feature of the inversion of noisy data is that the noise, with the exception of the station number and distribution, was a dominant factor controlling the retrieval of the mechanism. Noise was more important than the accuracy of the location

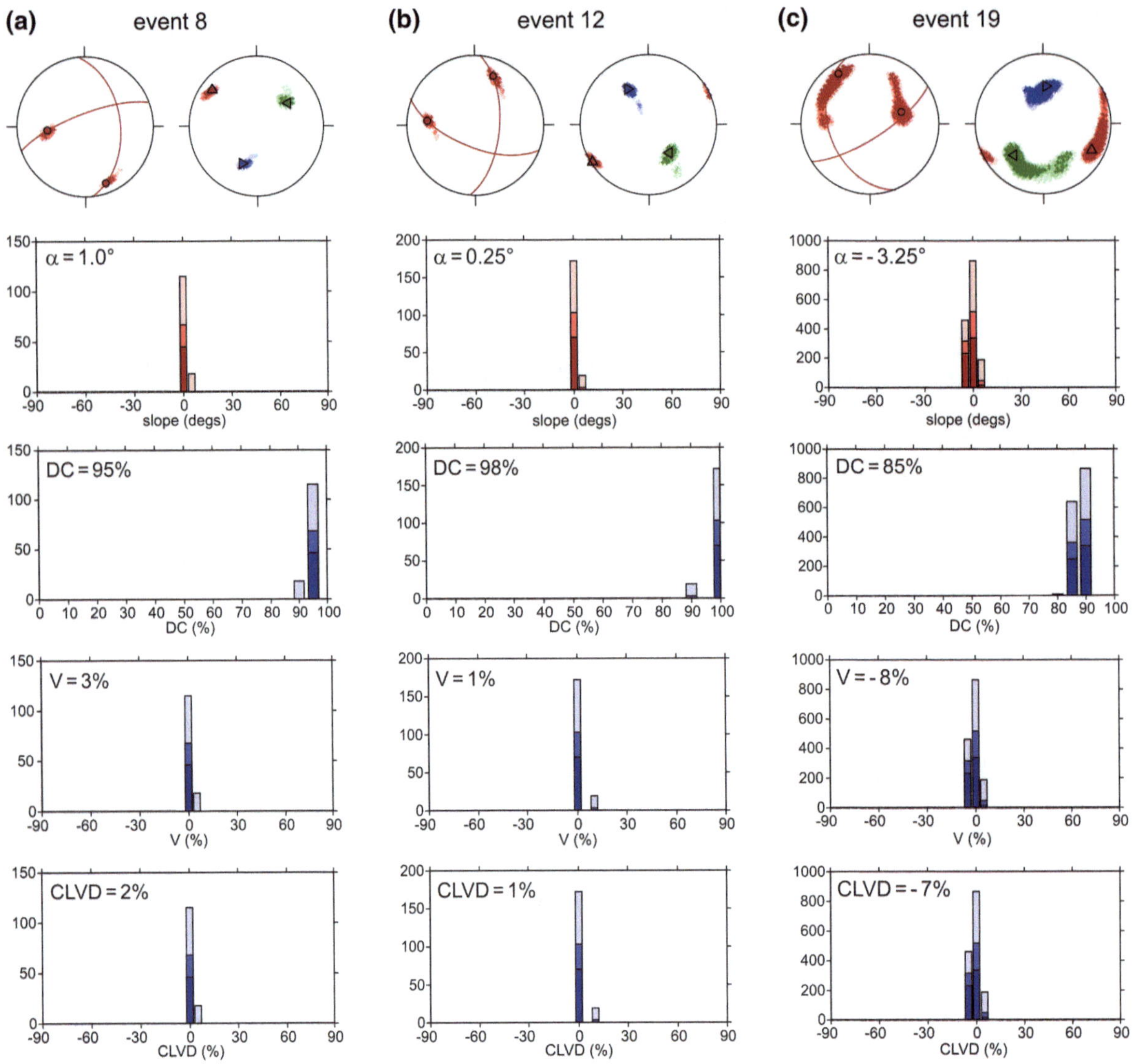

Figure 13
The source mechanisms of events 8, 12, and 19, retrieved using a STC model. Source lines, as well as the directions of fault normal/slip vectors and the principal T, P, and N axes, were displayed together with histograms of the slope angle and the DC, V, and CLVD components of the MTs determined from the STC solutions. For details, see the caption in Fig. 11

and the velocity model used throughout the inversion (see the individual rows in Figs. 5, 6, 7). Taking into account the fact that 10 % and 20 % of the noise considered in synthetic experiments did not contain a high level of contamination, with an assumption of strong mismodeling, we replaced the true 1-D layered velocity model with a homogeneous half-space (with an overlying thin low-velocity layer) that appeared to be an extreme case of the velocity model simplification. In practice, more information for velocities is usually available. The advantageous small sensitivity to the velocity model and the hypocenter location was obviously a benefit of using amplitudes in the inversion instead of waveforms when we ignored the kinematics of seismic waves, which is largely impacted by the model.

5.3. The Inversion of Soultz Events

We applied the MT and STC inversion to 13 small-scale earthquakes from the beginning of the 2003 injection that ranged in magnitude from 1.4 to

2.3. The orientations of the DC portion of the MT were in good agreement with the source lines of the STC model. The MT possessed non-DC portions between 4 and 68 %. The STC model resulted in slope angles between $-3.5°$ and $4.5°$, and the pattern of their distribution indicated that non-zero values were not significant, demonstrating that the non-DC components of the MT were spurious and caused by noise contamination, structure mismodeling, and a lack of input data. Therefore, investigation of all 13 induced events with magnitudes larger than 1.4, and not previously processed, suggested that their source mechanisms were dominantly pure shears (dip-slip and oblique normal), just as the earthquakes processed by HORÁLEK *et al.* (2010) were. The directions of the T axes were stable and sub-horizontal in the E-W direction. On the other hand, the directions of the P axes varied from the vertical to the horizontal in the N–S direction. All of the results were very similar to previous analyses (HORÁLEK *et al.* 2010) and in agreement with the stress pattern obtained from in situ measurements (EVANS 2005; VALLEY and EVANS 2006). Even small earthquakes from the first phase of the injection in 2003, with magnitudes larger than 1.4, were pure shear slips on pre-existing faults.

6. Conclusions

Synthetic experiments exploring the performance of the MT and STC source models in the P and S amplitude inversion on a local scale simulate (i) the different types of station distribution from very good to very poor, (ii) the inaccurate location of the hypocenter, (iii) improper modeling of the velocities, and (iv) the noise contamination of the data yield of the following:

- The number of stations and the regularity of their distribution are of crucial importance for the retrieval of both the orientation of source mechanisms and the DC vs. non-DC content;
- Noise in the data harms the retrieval of the mechanism more than the uncertainty in the location and velocity modeling;
- Determination of the orientation of the mechanism is fairly robust with respect to all of the factors (i–iv), excluding very poor station distributions, regardless of the source model employed;
- Determination of the DC vs. non-DC content is more sensitive than the orientation, poor station distributions may be largely biased, and the STC model works better than the MT during its retrieval.

Based on the parallel determination of the mechanism together with its uncertainty, in terms of the MT and STC for 13 small earthquakes from the 2003 Soultz injection, all of the events can be classified as pure shear, regardless of the rather large non-DC component in the MTs of the three.

Acknowledgments

We are indebted to Gillian Foulger and an anonymous reviewer for comments that helped improve the manuscript. Green's functions were kindly provided by Petr Bulant, Department of Geophysics, Charles University, Prague. Our research was supported by grants of the Czech Republic Grant Agency 'Non-double-couple mechanisms: through induced seismicity to fluid-driven earthquakes', Grant Agreement No. P210/10/1728, and 'Constrained models of seismic source: in between a double-couple and moment tensor', Grant Agreement No. P210/12/2235.

REFERENCES

AHORNER, L. and SCHNEIDER, G. (1974). Herdmechanismen von Erdbeben im Oberrheingraben und in seinen Randgebirgen. In: K. Fuchs and J.H. Illies (Editors), *Approaches to Taphrogenesis*, Stuttgart, pp. 104–117.

BEAN, CH. and O'BRIEN, G. (2008). *Influence of near-surface volcanic structure on long-period seismic signals and on moment tensor inversions: Simulated examples from Mount Etna*, J. Geophys. Res. *113*, B08308, doi:10.1029/2007JB005468.

CUÉNOT, N., CHARLÉTY, J., DORBATH, L. and HAESSLER, H. (2006). *Faulting mechanisms and stress regime at the European HDR site of Soultz-sous-Forêts, France. Geothermics*, *35*, 561–575.

CUÉNOT, N., DORBATH, C. and DORBATH, L. (2008). *Analysis of the microseismicity induced by fluid injection in the Hot Dry Rock site of Soultz-sous-Forêts (Alsace, France): Implications for the characterization of the geothermal reservoir properties.* Pure appl. Geophys., *165*, 797–828, doi:10.1007/s00024-008-0335-7.

ČERVENÝ, V., KLIMEŠ, L. and PŠENČÍK, I. (1988). Complete seismic-ray tracing in three-dimensional structures. In: Doornbos, D.J. (ed.), *Seismological Algorithms*, pp. 89–168, Academic Press, New York.

Dorbath, L., Cuénot, N., Genter, A. and Frogneux, M. (2009). *Seismic response of the fractured and faulted granite of Soutz-sous-Forêts (France) to 5 km deep massive water injections.* Geophys. J. Int., *177*, 653–675, doi:10.1111/j.1365-246X.2009.04030.x.

Dorbath, L., Evans, K., Cuénot, N., Valley, B., Charléty, J. and Frogneux, M. (2010). *The stress field at Soultz-sous-Forêts from focal mechanisms of induced seismic events: Cases of the wells GPK2 and GPK3.* Comptes Rendus Geoscience, doi:10.1016/j.crte.2009.12.003.

Dufumier, H. and Rivera, L. (1997). *On the resolution of the isotropic component in moment tensor inversion.* Geophys. J. Int., *131*, 595–606, doi:10.1111/j.1365-246X.1997.tb06601.x.

Evans, K. F. (2005). *Permeability creation and damage due to massive fluid injections into granite at 3.5 km at Soultz: Part 2 - Critical stress and fracture strength.* J. Geophys. Res., *110*, B04204, doi:10.1029/2004JB003169.

Evans, K.F., Moriya, H., Niitsuma, H., Jones, R.H., Phillips, W.S., Genter, A., Sausse, J., Jung, R., Baria, R., 2005. *Microseismicity and permeability enhancement of hydro-geologic structures during massive fluid injections into granite at 3 km depth at the Soultz HDR site.* Geophys. J. Int., *160*, 388–412.

Genter, A., Fritsch, D., Cuénot, N. Baumgärtner, J. and Graff, J.-J. (2009). Overview of the current activities of the european EGS Soultz project: From exploration to electricity production. *Proceedings Thirty-Fourth Workshop on Geothermal Reservoir Engineering*, Stanford, California, February 9–11, 2009, SGP-TR-187.

Hill, D. P., 1977. *A model for earthquake swarms.* J. Geophys. Res., *82*, 1347–1352, doi:10.1029/JB082i008p01347.

Horálek, J., Jechumtálová, Z., Dorbath, L. and Šílený, J. (2010). *Source mechanisms of micro-earthquakes induced in a fluid injection experiment at the HDR site Soultz-sous-Forêts (Alsace) in 2003 and their temporal and spatial variations.* Geophys. J. Int., *181*, 1547–1565, doi:10.1111/j.1365-246X.2010.04506.x.

Julian, B. R., A. D. Miller, and G. R. Foulger (1998), *Non-double-couple earthquakes 1. Theory.* Rev. Geophys., *36*, 525–549.

Julian, B.R., Foulger, G.R., Monastero, F.C., and Bjornstad, S., 2010. *Imaging hydraulic fractures in a geothermal reservoir.* Geophys. Res. Lett., *37*, L07305, doi:10.1029/2009GL040933.

Kozák, J. and Šílený, J. (1985). *Seismic events with non-shear component I. Shallow earthquakes with tensile source component.* Pure Appl. Geophys., *123*, 1–15.

Rudajev, V. and Šílený, J. (1985). *Seismic events with non-shear component II. Rock bursts with implosive source component.* Pure Appl. Geophys., *123*, 17–25.

Šílený, J., 1998. *Earthquake source parameters and their confidence regions by a genetic algorithm with a "memory".* Geophys. J. Int., *134*, 228–242.

Šílený, J. (2009). *Resolution of non-double-couple-mechanisms: simulation of hypocenter mislocation and velocity structure mismodeling.* Bull. Seismol. Soc. Am., *99*, doi:10.1785/0120080335.

Šílený, J. (2012). Shear-tensile/implosion source model vs. moment tensor: benefit in single-azimuth monitoring. Cotton Valley set-up. 74th EAGE conference & exhibition incorporating SPE EUROPEC 2012. Houten : EAGE, 2012, s. 1-5. ISBN 978-90-73834-27-9.

Tarantola, A. (2005). Inverse Problem Theory and Methods for Model Parameter Estimation, SIAM, Philadelphia, Second Ed.

Teisseyre, R. (1980). Some remarks on the source mechanism of rockbursts in mines and on the possible source extension. *Acta Montana CSAV Praha*, *58*, 7–13.

Valley, B. and Evans, K. F. (2006). Stress state at Soultz to 5 km depth from wellbore failure and hydraulic observations, *Proceedings EHDRA Scientific Conference*, Soultz-sous-Forêts, 15/16 June 2006, 15 pp.

Vavryčuk, V. (2001). *Inversion for parameters of tensile earthquakes.* J. Geophys. Res., *106*, 16339–16355, doi:10.1029/2001JB000372.

Vavryčuk, V. (2007). *On the retrieval of moment tensors from borehole data.* Geoph. Prospecting *55*(3):381–391.

Vavryčuk, V. (2011). *Tensile earthquakes: Theory, modeling and inversion.* J. Geophys. Res., *116*, B12320, doi:10.1029/2011JB008770.

(Received May 2, 2013, revised November 7, 2013, accepted November 23, 2013, Published online January 9, 2014)

Pure Appl. Geophys. 171 (2014), 2805–2817

DOI 10.1007/s00024-013-0758-7

Pure and Applied Geophysics

Full Moment Tensor Variations and Isotropic Characteristics of Earthquakes in the Gulf of California Transform Fault System

ROBERTO ORTEGA,[1] LUIS QUINTANAR,[2] and LUIS RIVERA[3]

Abstract—The full moment tensor is a mathematical expression of six independent variables; however, on a routine basis, it is a common practice to reduce them to five assuming that the isotropic component is zero. This constraint is valid in most tectonic regimes where slip occurs entirely at the fault surface (e.g. subduction zones); however, we found that full moment tensors are best represented in transform fault systems. Here we present a method to analyze source complexity of earthquakes of different sizes using a simple formulation that relates the elastic constants obtained from independent studies with the angle between the slip and the fault normal vector, referred to as angle θ; this angle is obtained from the full moment tensors. The angle θ, the proportion of volume change (k) and the constant volume (shear) component (T) are numerical indicators of complexity of the source; earthquakes are more complex as θ deviates from $\pi/2$ or as T and k deviate from zero as well. These parameters are obtained from the eigensolution of the full moment tensor. We analyzed earthquakes in the Gulf of California that exhibit a clear isotropic component and we observed that the constant volume parameter T is independent of scalar moments, suggesting that big and small earthquakes are equally complex. In addition, simple models of one single fault are not sufficient to describe physically all the combinations of θ in a source type plot. We also found that the principal direction of the strike of the Transform Fault System in the Gulf of California is following the first order approximation of the normal surface of the full moment tensor solution, whereas for deviatoric moment tensors the principal direction does not coincide with the strike of the Transform Fault System. Our observations that small and large earthquakes are equally complex are in agreement with recent studies of strike-slip earthquakes.

Key words: Earthquake source observations, Isotropic component, transform faults, Gulf of California.

[1] Unidad La Paz, Centro de Investigación Científica y de Educación Superior de Ensenada, Miraflores 334, Col. Bellavista, 23050 La Paz, BCS, Mexico. E-mail: ortega@cicese.mx

[2] Instituto de Geofísica, Universidad Nacional Autónoma de México, Circuito de la Investigación Científica, 04510 Coyoacán, D.F., Mexico.

[3] Institute de Physique du Globe de Strasbourg, 5 rue Descartes, 67084 Strasbourg Cedex, France.

1. Introduction

The isotropic component of the moment tensor has been a topic of research for decades (HARA 1996; KAWAKATSU 1996). Although there have been many attempts to identify the isotropic component from the seismic waveforms (HARA 1996), researchers have not succeeded in explaining the nature of either shallow or deep earthquakes with isotropic component. Recent studies, based on many seismic observations, suggest that earthquake ruptures different than pure double couple dislocation occur in nature. Frequently, a non-double couple earthquake (NDC) was considered to be the result of modeling errors (e.g., ADAMOVA and SILENY 2010) but there are many cases that suggest that NDC earthquakes actually occur. Among such examples are transform faults, volcanic fields and calderas, ridges, mining collapses, and well-induced earthquakes, ranging from negative magnitude (mining and well-induced) to moderate to large earthquakes (transform faults). As the quantity of seismic data has increased and the new advances in seismic instrumentation have brought seismograms with unprecedented quality, modeling methods have improved significantly. In addition it has been shown that, for some earthquakes, the non-double couple solutions reflect the complexity of the source mechanism. Some well-documented cases of non-double couple sources are caused by volumetric changes in geothermal sites, volcanic and ridge opening (AKI 1984; FOULGER and LONG 1984; JULIAN and SPIKIN 1985; DREGER *et al.* 2000, ORTEGA and QUINTANAR 2010) and some seismic events may have important non-double couple components based on laboratory observations (BRUNE *et al.* 1993) or changes in the elastic moduli at the source region (Knopoff and Randall 1970; Vavryčuk 2004; Ben-Zion and Ampuero 2009; Shi and Ben-Zion 2009). Some recent earthquakes with important non-double couple components are: the 2000 Sumatra earthquake (Abercrombie and Antolik 2003); the 2010 Samoa Islands earthquake (Lay *et al.* 2010); the 2010 Cucapah—El Mayor, Mexicali Mexico earthquake (Hauksson *et al.* 2011).

The purpose of this article is to characterize some earthquake source parameters, namely the constant volume parameter T, the proportion of volume change parameter k (Hudson *et al.* 1989), the angle between the resultant vector of slip and the fault system normal vector θ, $\hat{n}$ and $\hat{\upsilon}$. The last three parameters are described in the method section. These values are obtained from different earthquake sizes by inverting the full moment tensor, and then all these values are compared.

2. Methods

The definition of a moment tensor as a function of the isotropic fourth order tensor c_{ijkl} and the vectors $n_k v_l$ represented in Fig. 1 is:

$$M_{ij} = c_{ijkl} n_k v_l SD \tag{1}$$

where n_k is the unit vector related to the fault and v_l is the unit vector related to the slip. In the general case v_l may not coincide with the fault plane. S is the scalar that defines the fault area and D is the average slip.

Therefore, the full moment tensor is (Ben-Menahem and Singh 1981; Aki and Richards 2002):

$$M_{ij} = SD\left[\delta_{ij}\lambda n_k v_k + \mu\left(\upsilon_i n_j + v_j n_i\right)\right] = SD\begin{bmatrix} \lambda(\hat{n}\cdot\hat{v}) + 2\mu n_1 v_1 & \mu(n_1v_2 + n_2v_1) & \mu(n_1v_3 + n_3v_2) \\ \mu(n_1v_2 + n_2v_1) & \lambda(\hat{n}\cdot\hat{v}) + 2\mu n_2 v_2 & \mu(n_2v_3 + n_3v_2) \\ \mu(n_1v_3 + n_3v_2) & \mu(n_2v_3 + n_3v_2) & \lambda(\hat{n}\cdot\hat{v}) + 2\mu n_3 v_3 \end{bmatrix} \tag{2}$$

where λ and μ are the Lamé constants.

The trace of $\mathbf{M}$ is $[3\lambda(\hat{n}\cdot\hat{v}) + 2\mu(\hat{n}\cdot\hat{v})]SD$. The characteristic solution of $\mathbf{M}$ is:

$$\begin{aligned} &\text{Eigenvector } s_1 = \hat{n} + \hat{v} \text{ eigenvalue } e_1 = \lambda(\hat{n}\cdot\hat{v})SD + \mu[1 + (\hat{n}\cdot\hat{v})],\ SD, \\ &\text{Eigenvector } s_2 = \hat{n} \times \hat{v},\ \text{eigenvalue } e_2 = \lambda(\hat{n}\cdot\hat{v})SD, \\ &\text{Eigenvector } s_3 = \hat{n} - \hat{v},\ \text{eigenvalue } e_3 = \lambda(\hat{n}\cdot\hat{v})SD - \mu[1 - (\hat{n}\cdot\hat{v})],\ SD. \end{aligned} \tag{3}$$

Then if $(\hat{n}\cdot\hat{v}) = cos\theta$ we have that

$$\theta = \text{arcos}\ \left[(\lambda/\mu)^{-1}(2e_2/(e_1 - e_3))\right] \tag{4}$$

It follows that

$$\begin{aligned} \hat{n} &= 1/2(s_1 + s_3) = 1/2[(\hat{n} + \hat{\upsilon}) + (\hat{n} - \hat{\upsilon})] \\ \hat{\upsilon} &= 1/2(s_1 - s_3) = 1/2[(\hat{n} + \hat{\upsilon}) - (\hat{n} + \hat{\upsilon})] \end{aligned} \tag{5}$$

The full moment tensor is obtained using a waveform modeling as described by Minson and Dreger (2008) and Ford *et al.* (2009). The inversion was carried out in two steps, first the observed and synthetic waveforms were aligned in such a way to have the S arrival at the same time; the synthetic S arrival is obtained directly from the output of the program whereas the observed arrival is picked visually using the origin time as a guide. Then, the moment tensor inversion is carried out constraining the correlation factor with a time shift <50 s which corresponds to two times the period of the analyzed band-pass. In this way, we were able to align the synthetic and observed waveforms trying to avoid misalignments of further phases. However, we found that in some cases the best waveform fitting was low and we discarded those events. In any case, full moment tensor inversion requires a good velocity model in order to obtain robust results (Křížová 2013).

After we computed the moment tensor, we solved the characteristic equation computing the eigenvalues and eigenvectors with an iterative Jacobi method. The eigenvalue–eigenvector pairs correspond to Eq. (3). Then θ, $\hat{n}$ and $\hat{\upsilon}$ are obtained from (4) and (5). The elastic constant ratio λ/μ is required to compute θ and it is better obtained by independent studies in order to avoid unrealistic rheological values (Dufumier and Rivera 1997).

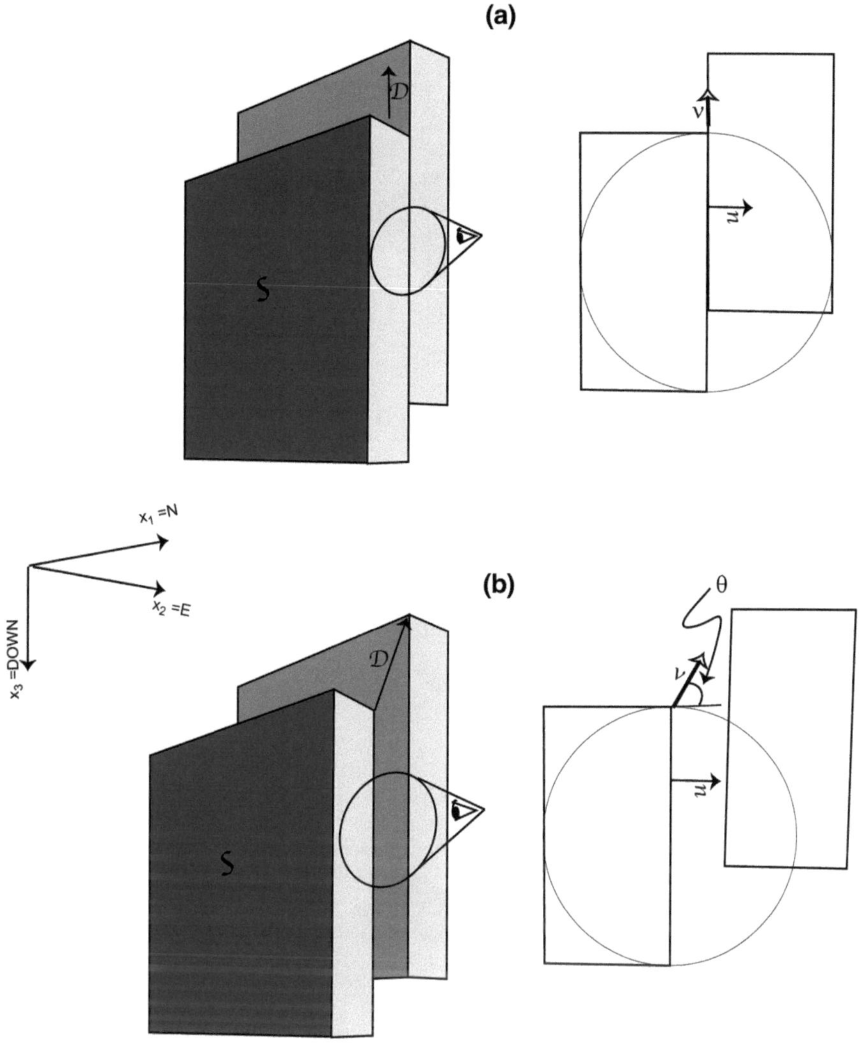

Figure 1
a Shear dislocation model without isotropic component, **b** general displacement model with isotropic component; vectors $\hat{n}$ and $\hat{v}$ are the unit vector perpendicular to the surface *S* and the vector parallel to the displacement *D*, respectively. As expected $\hat{n}$ and $\hat{v}$ are orthogonal for the shear dislocation model whereas in the general model both vectors are related by a θ angle. The Cartesian system used in this article is represented at the *left* of the figure

A full moment tensor represents simple mechanical models of double couples and linear dipoles. On a general view, the moment tensor of six components is not referenced to a particular source model, each eigenvalue–eigenvector pair represents a linear dipole, so the full moment tensor is in fact the combination of three orthogonally oriented pairs of forces acting in opposite directions. However, on the restricted deviatoric form, it is generally parameterized by the scalar seismic moment and three Euler angles for the fault plane. These angles are defined as the dip, strike and rake of the fault; this case is only represented by pure double couple model with $m_1 = -m_3 = \mu SD$ and $m_2 = 0$. In any other case, separation into main double couple (DC) or compensated linear vector dipole (CLVD) is generally ambiguous (Jost and Herrmann 1989).

There are many ways of representing a full moment tensor with different types of parameters. Each of these methods requires a specific approach so that a fundamental characteristic is observed; the source type plot (Hudson *et al.* 1989) is a simple

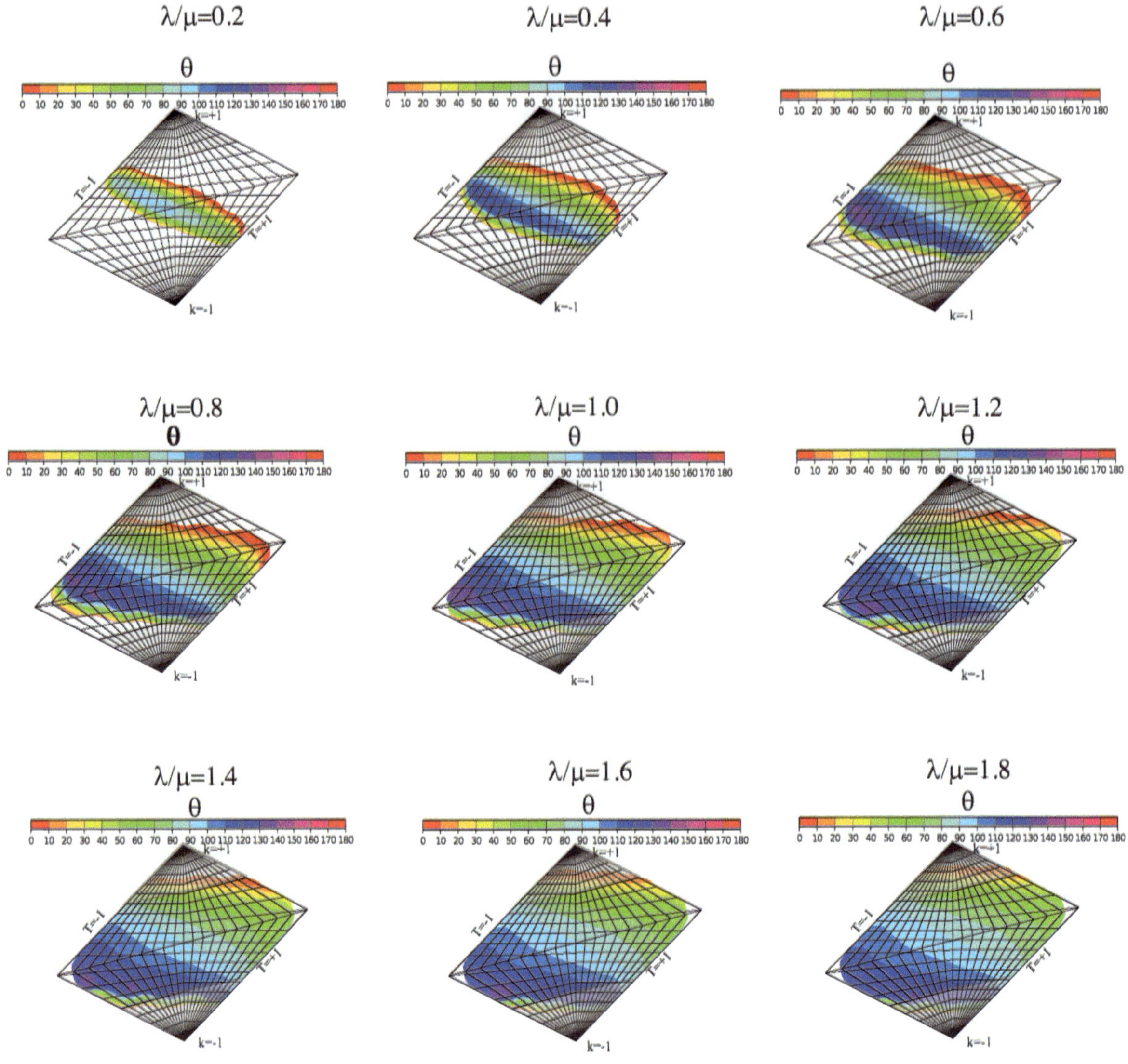

Figure 2
Source type plots for different elastic constant ratios. Each *color* represents different values of θ. *Shaded regions* represent the solution space where nodal planes are not present (*solid circles* as focal mechanisms). As the principal moments are approaching to pure explosion or pure implosion, the argument $(\lambda/\mu)^{-1}(2e2/(e1-e3))$ has no real values for θ. The middle column $(0.8 < \lambda/\mu < 1.2)$ is where most of the crustal earthquakes occur

representation of the source for general moment tensors that expresses the moment tensor as function of two basic parameters; constant volume (shear) component T and the proportion of volume change k. This plot is capable of interpretation in terms of one or a combination of physical processes of failure resulting from an accumulated change in the stress field. The processes may take any form such as shear or tensile fracture, explosion or implosion.

Based on the general assumption that all the earthquakes have a maximum magnitude M, if we suppose that any moment tensor has the same probability of occurrence, we can plot in an equal area a source type plot. This plot has a diamond shape and is symmetrical about the axis that corresponds to when the intermediate eigenvalue = 0. The source type plot does not depend on the fault orientation since it is only described by the eigenvalues. Therefore, a pure DC, regardless of

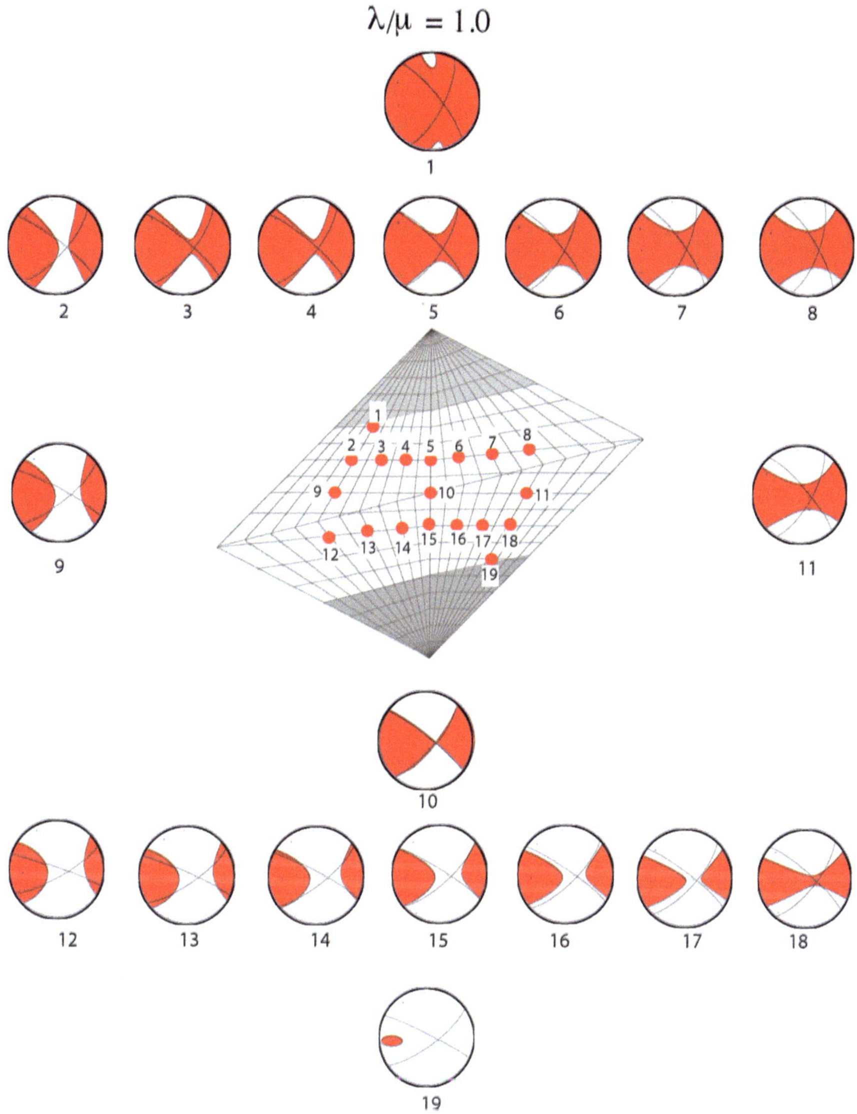

Figure 3
Source type plot and focal mechanisms for a fixed P, T, NULL direction and $\lambda/\mu = 1$. The *great circles* in the focal mechanisms are the planes that are represented by the normal vectors $\hat{n}$ and $\hat{v}$

whether it corresponds to a strike-slip, normal or inverse fault, will map at the origin (0, 0) of the source type plot. Any characteristic, which depends only on the eigenvalues and describes the relation of explosion, compensated linear vector dipole (CLVD), double couple (DC) and cracks, is suitable to be represented by the source type plot.

In Fig. 2, all the possible eigenvalue combinations of full moment tensors in the $T-k$ space is obtained by fixing the eigenvectors and perturbing the eigenvalues in order to have different percentages of DC, CLVD, cracks and explosions (Bowers and Hudson 1999), where:

$$T = 2\frac{e_1'}{|e_3'|} \text{ and } k = \frac{M_{\mathrm{ISO}}}{|M_{\mathrm{ISO}}| + |m_3'|}$$

e_1', e_2' and e_3' are the deviatoric eigenvalues for the N, P and T axes, respectively; the deviatoric eigenvalues are ordered in such a way that $|e_1'| \leq |e_2'| \leq |e_3'|$ and $M_{\mathrm{ISO}} = 1/3\mathrm{Trace}(M_{ij})$. Then the value of θ is computed (Eq. 4) and we plot it in a source type plot for different combinations of λ/μ.

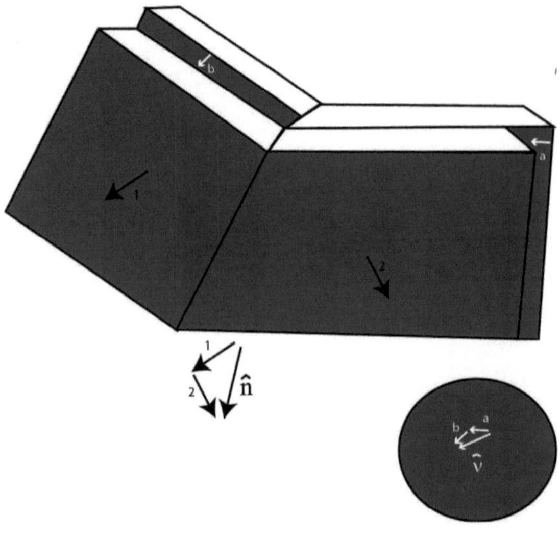

Figure 4
Schematic representation of normal vectors and slip directions of a typical transform fault system

As expected, in the case of pure DC ($\kappa = 0$ and $T = 0$), the value of θ is equal to $\pi/2$. However, the implosion and explosion sides do not agree with the $\theta < \pi/2$ and $\theta > \pi/2$ regions. In contrast, the region where the e'_2 eigenvalue is equal to zero [$T = -2k/(1 - |k|)$] corresponds to $\theta = \pi/2$. In this case the moment tensor has only two and not three degrees of freedom, since one eigenvalue is zero. At $\theta = \pi/2$, the trend of $\hat{n}$ coincides with the resultant value of the slip of the orthogonal plane of $\hat{\upsilon}$. In addition, the nodal planes intersect in a point and the radiation pattern has nodal planes that show radial symmetry for positive and negative polarities. The $\theta = \pi/2$, is a mechanism that has four radiation pattern lobes in a ribbon shape that shift from major compressive (or dilatational, as the case may be) for constant P and T axis. The radiation patterns at $\theta = \pi/2$ show two nodal planes and the intermediate axis N is equal to zero. Therefore, as θ deviates from $\pi/2$ the radiation patterns are more complex.

In addition, in the shaded area of Fig. 2, nodal planes are no longer present and the focal mechanism is a solid or open circle for explosion or implosion respectively. Moreover, as the moment tensor is approaching to pure explosion or pure implosion, the argument $(\lambda/\mu)^{-1}$ $(2e2/(e1 - e3))$ has no real values for θ (Eq. 4) and for this reason those regions are not filled with θ values (Fig. 2). It is also clear that the elastic constant ratio λ/μ controls the space solution of θ. Earthquakes that occur in the crust have a λ/μ close to unity. Therefore, the three cases that are in the middle row of Fig. 2 are the most common rheological scenarios; however, it is worth noting that some other cases are not included, for instance, introducing a seismic source in bimaterial interfaces or other heterogeneous media. In such cases it is possible to have strong variations of λ/μ, which may vary over the rupture zone.

If we fix the P, T and NULL directions (Fig. 3), and we construct moment tensors with different percentages of DC, CLVD, cracks and explosions, then each point in the source type plot represents a focal mechanism with different nodal surfaces and different values of θ, $\hat{n}$ and $\hat{\nu}$. We will explain later a physical interpretation of these values.

In Fig. 3, the great circles in the focal mechanisms are the planes that are represented by the normal vectors $\hat{n}$ and $\hat{\nu}$. Therefore, the planes may not necessarily be orthogonal. At $T = 0$ (5 and 15 of Fig. 3), the nodal surfaces are flanked by the planes of the normal vectors $\hat{n}$ and $\hat{\nu}$. This is the case where the pure isotropic component is present and no other contribution of linear dipoles affects the forces at the source. As described by HUDSON *et al.* (1989), we construct the source type plot by projecting the full moment tensor into its deviatoric component and then decomposing the isotropic part. Since the simple deviatoric model of a single planar surface represented by $\hat{n}$ and $\hat{\nu}$ will lead only to orthogonal angles, then a better model than just one simple planar fault is required. Thus, a new model than the one sketched in Fig. 1 is needed; the new model of a fault system in a transform fault is sketched in Fig. 4. For this new model, at least two faults are required in order to construct arbitrary directions of $\hat{n}$ and $\hat{\upsilon}$.In addition, the possible use of a non-planar rupture seemed particularly promising for the study of deep earthquakes (KIRBY *et al.* 1996; OKAL 1996; RUSSAKOFF *et al.* 1997; ESTABROOK 1999), which were discovered by Wadati as early as the 1920s. Their origin of a non-planar surface is still vague, although various models, including a phase transition model, were proposed. In any case, our new model allows the decomposition of the moment tensor with simple fault systems. The deviatoric case is also considered

Table 1

Full moment tensor solutions

Date mm/dd/yyyy	Time hh:mm:ss.s	Latitude (°)	Longitude (°)	Depth (km)	M_0 (N–m)	m11	m12	m13	m22	m23	m33
03/10/2002	16:08:29.6	23.32	−108.53	18	2.40E + 18	1.44E + 25	−9.44E + 24	−4.71E + 24	−1.37E + 25	-3.01E + 24	9.47E + 24
12/03/2003	23:41:32.9	26.56	−110.59	10	3.57E + 18	−2.66E + 25	1.16E + 25	−1.32E + 24	3.15E + 25	1.58E + 24	5.21E + 24
12/11/2003	04:54:56.4	28.97	−113.22	8	4.74E + 17	−1.93E + 24	−1.37E + 23	2.03E + 23	2.22E + 24	2.64E + 23	−4.05E + 24
18/02/2004	10:59:19.4	23.64	−108.82	6	8.97E + 17	−5.18E + 24	6.05E + 23	1.59E + 24	1.67E + 24	−1.92E + 24	−6.81E + 24
24/09/2004	14:43:11.4	28.57	−112.72	12	9.15E + 17	−6.09E + 24	4.19E + 24	−3.51E + 23	7.51E + 24	−1.38E + 24	1.40E + 24
22/02/2005	19:15:49.9	25.67	−109.97	16	4.64E + 17	−3.29E + 24	7.15E + 23	−4.22E + 23	2.15E + 24	−7.41E + 22	−2.44E + 24
05/06/2005	08:28:50.9	23.67	−108.37	8	4.94E + 17	−3.22E + 24	1.06E + 24	4.87E + 23	1.94E + 24	−2.53E + 23	−2.89E + 24
04/01/2006	01:05:08.8	27.96	−112.14	8	1.26E + 17	−4.13E + 23	−5.61E + 23	7.35E + 22	7.88E + 23	−2.01E + 23	−7.16E + 23
04/01/2006	08:32:32.4	28.16	−112.12	8	1.06E + 19	−7.48E + 25	2.90E + 25	−2.73E + 24	7.89E + 25	−1.23E + 25	3.52E + 25
28/05/2006	14:18:02.7	26.7	−111.23	18	8.38E + 16	−4.09E + 23	6.83E + 22	−1.77E + 23	7.97E + 23	1.18E + 23	1.68E + 23
06/12/2006	14:09:58.8	27.59	−111.29	6	4.06E + 16	−3.24E + 23	1.29E + 23	−1.34E + 22	3.13E + 23	−1.13E + 23	−4.40E + 22
30/07/2006	01:20:59.2	26.86	−111.21	10	7.88E + 17	−5.40E + 24	1.10E + 24	−9.99E + 23	7.64E + 24	6.84E + 23	1.09E + 24
23/02/2007	04:16:15.2	25.97	−110.35	10	1.58E + 16	−1.28E + 23	2.71E + 22	2.81E + 22	1.79E + 20	−2.64E + 22	−9.44E + 22
23/02/2007	07:12:18.2	25.95	−110.52	10	8.35E + 16	−3.80E + 23	5.75E + 22	1.30E + 23	1.72E + 23	−1.17E + 23	−7.60E + 23
23/02/2007	07:30:12.1	25.99	−110.5	10	1.06E + 17	−6.02E + 23	9.70E + 22	1.62E + 23	7.84E + 22	−1.20E + 23	−9.12E + 23
25/02/2007	01:18:33.7	26	−110.46	10	7.14E + 16	−3.65E + 23	1.54E + 23	1.15E + 23	−6.90E + 22	−5.39E + 22	−6.42E + 23
25/02/2007	15:00:41.2	26.15	−110.42	6	6.21E + 17	−4.05E + 24	1.40E + 24	2.00E + 24	4.98E + 24	−1.99E + 24	−1.86E + 24
27/02/2007	20:08:09.0	25.98	−110.46	10	4.28E + 16	−3.17E + 23	9.08E + 22	9.20E + 22	1.12E + 23	−7.23E + 22	−2.30E + 23
13/03/2007	02:59:04.4	26.26	−110.54	6	1.06E + 18	−6.70E + 24	−7.89E + 23	4.76E + 24	6.34E + 24	−2.51E + 24	−4.65E + 24
28/03/2007	13:21:37.1	25.42	−109.71	14	9.32E + 16	−3.73E + 23	2.08E + 23	1.82E + 23	7.07E + 23	−2.62E + 23	3.47E + 23
28/03/2007	14:28:54.7	25.5	−109.63	16	1.75E + 17	−2.05E + 23	2.89E + 23	2.02E + 23	1.36E + 24	−6.43E + 23	6.30E + 23
17/07/2007	04:13:48.3	24.55	−109.09	16	1.15E + 17	−1.01E + 24	−2.72E + 23	3.03E + 23	6.88E + 23	3.47E + 23	2.60E + 23
01/09/2007	19:14:22.8	24.91	−109.59	8	1.79E + 18	−9.22E + 24	6.88E + 24	−1.28E + 24	9.54E + 24	−2.80E + 24	−8.83E + 24
02/09/2007	11:04:46.5	24.61	−109.82	8	6.56E + 16	−5.27E + 23	1.91E + 23	1.33E + 23	4.02E + 23	1.32E + 23	−1.39E + 23
05/09/2007	03:07:37.3	24.35	−109.95	18	2.04E + 17	2.35E + 23	1.08E + 24	−1.96E + 22	−4.48E + 23	5.17E + 23	−1.25E + 24
24/09/2007	01:50:00.3	24.37	−109.91	10	3.30E + 16	−1.11E + 23	2.09E + 23	2.95E + 22	−9.63E + 21	−2.16E + 21	−1.42E + 23
05/05/2008	16:28:06.2	24.8	−109.58	10	3.88E + 16	−3.56E + 23	−1.15E + 23	8.05E + 21	6.15E + 22	6.55E + 22	−6.91E + 22

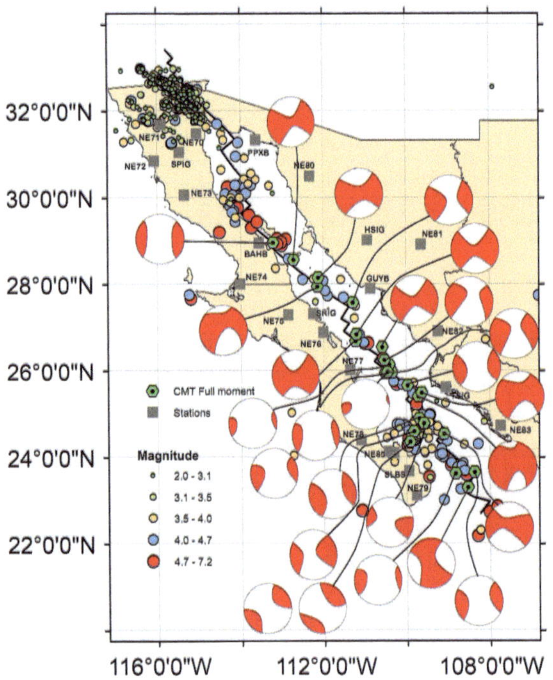

Figure 5
Seismicity map of the studied area. The events that were analyzed are plotted with a lower hemisphere stereographic projection, stations are represented as *solid squares*

as well as the double couple plus a volumetric component, thus the full moment tensor represents the overall force acting at the source. Also, it has been widely recognized that the parameter T characterizes complex sources (Frohlich 1995). Regular determinations of the parameter T gave rise to the hope that the plane-shear source model can be replaced by more sophisticated and supposedly more realistic models, for example, models in which the motion in the source occurs on a curved surface or in the form of echelon cracks that can also be approximated by a curved surface. Previous authors (Frohlich *et al.* 1989; Dufumier and Rivera 1997) described the non-double couple parts as non-plane faults and non-tectonic isotropic components, respectively.

Any symmetric moment tensor can be described by this model; however, the rheological and geometrical properties of each fault segment requires independent studies since they cannot be described uniquely by the six moment tensor parameters. For example, two faults would require eight direction values (four for each pair) and a six-component tensor could not fully describe them. However, $\hat{n}$ and $\hat{\upsilon}$, the slip displacement, D, and the source area, S represent the overall result. The contributions of each subarea and slip displacement are not needed to obtain the full moment tensor. Therefore, the $\hat{n}$ and $\hat{\upsilon}$. vectors are obtained by the total contribution of $\hat{n}S$ and $\hat{\upsilon}D$. In the next section we will discuss an example for the extensional province of the Gulf of California.

3. Application to the Extensional Province of the Gulf of California

We analyzed 27 earthquakes that occurred in the Gulf of California (Table 1); the full moment tensor inversion was carried out using the methodology described by Minson and Dreger (2008) and (Ford *et al.* 2009, Fig. 5). We used the Global Centroid Moment Tensor (CMT) catalog from 1992 to 1998 as our initial database. During this period, a cooperative effort between Utrecht University (the Netherlands), Centro de Investigación Científica y de Educación Superior de Ensenada (CICESE, Mexico), and the California Institute of Technology established the NARS-Baja array (Trampert *et al.* 2003), which consisted of 17 broad band stations that were operated autonomously, recording 20 samples per second.

Even though the Global Moment Tensor Project is capable, in theory, of inverting the six elements of the moment tensor, on a routine basis they only report the deviatoric case. Since the early 1980s, the Global Moment Tensor Project has computed globally the moment tensor of earthquakes >5 Mw. The original algorithm was designed to solve the six moment tensor elements and the centroid location. However, in practice, the trace of the moment tensor is constrained to zero in their standard processing; therefore, it was necessary to analyze the earthquakes with six moment tensor elements.

We observed that few earthquakes occurred in a very short time in doublets or multiples events. In the recorded wavefield there was a clear interference among those earthquakes that occurred almost simultaneously. We avoided all the events that have those multiple events in the waveform.

Table 2

Parameters obtained from the full moment tensor (see "Methods")

Date mm/dd/yyyy hh:mm	Θ (°)	T	k	n_{trend} (°)	n_{plunge} (°)	v_{trend} (°)	v_{plunge} (°)	VR (%)
03/10/2002 16:08	64.1	0.51	0.14	275.3	56.8	251.6	−12.8	82.5
12/03/2003 23:41	81.3	0.11	0.09	76.1	4.7	98.5	0.4	76.1
12/11/2003 04:54	123.4	−0.38	−0.26	121.5	−2.6	249.7	0.6	73.5
18/02/2004 10:59	138.7	−0.29	−0.38	43.6	−11	16	−12	81.6
24/09/2004 14:43	82.7	0.05	0.1	120.2	−17	61.8	−1.6	81.5
22/02/2005 19:15	136.1	−0.64	−0.26	188.2	−5.8	311.1	7.3	70.1
05/06/2005 08:28	142.3	−0.67	−0.28	59.8	8.9	183.8	−10.6	80.6
04/01/2006 01:05	130.4	−0.91	−0.09	316.5	9.4	40.3	−19.8	75.8
04/01/2006 08:32	69.5	0.41	0.12	102.5	−15.5	77.3	−11.1	79.2
28/05/2006 14:18	73.3	0.05	0.22	193.1	−6.4	313.3	−3.4	82.8
06/12/2006 14:09	100.9	−0.29	−0.05	283.4	17.2	260.4	13	74.7
30/07/2006 01:20	80.7	0.03	0.14	51.5	6.7	110.4	2.4	80
23/02/2007 04:16	152.5	−0.13	−0.47	351.6	14.3	252.7	1.3	72.6
23/02/2007 07:12	128	−0.07	−0.39	83.4	1.3	311.6	−6.5	75.9
23/02/2007 07:30	151.9	−0.2	−0.45	57	−1.2	351.5	−8.3	68
25/02/2007 01:18	165.1	−0.07	−0.5	73.7	17.3	336.3	−17.9	62.4
25/02/2007 15:00	97.6	−0.17	−0.05	219.7	−4	29.4	6.5	88.3
27/02/2007 20:08	125.5	−0.19	−0.34	296.7	7.6	308.5	9.6	79.4
13/03/2007 02:59	99.2	0.03	−0.16	90.8	−17.1	170.2	−6.4	86.2
28/03/2007 13:21	65.4	0.22	0.24	5.4	11.9	215.1	2.1	84.2
28/03/2007 14:28	69.1	−0.31	0.34	114.4	−29.2	113.9	−29.1	75.8
17/07/2007 04:13	79.9	0.39	−0.02	233.2	−14.5	341.9	−19.5	74.5
01/09/2007 19:14	134.9	−0.85	−0.16	167.9	−18.8	0.1	17.1	81.4
02/09/2007 11:04	105.1	−0.23	−0.13	52	18.8	152.6	−0.2	78.7
05/09/2007 03:07	123.6	−0.45	−0.24	171.5	−47.9	265	37.8	54.2
24/09/2007 01:50	125.3	−0.42	−0.26	272.9	−8.8	287.3	2.5	73
05/05/2008 16:28	108.7	0.25	−0.31	333.3	−21.5	222.1	−0.4	65.3

Table 3

Velocity model

No	Depth (km)	Velocity P (km/s)	Velocity S (km/s)	Density (g/cm^3)
0	2	4.00	2.28	2.30
1	5	6.00	3.42	2.40
2	8	6.40	3.64	2.67
3	55	7.50	4.20	3.10
4	22	7.52	4.21	3.20
5	23	7.60	4.25	3.20

We filtered the synthetic and observed seismograms in the 25–50 s passband range after instrument removal, and then we computed all the parameters described in the previous section. In Table 2 we present the full moment tensor parameters ($\theta, T, k\ \hat{n}$ and $\hat{v}$) as described in the method section. The velocity model is represented in Table 3. The values of λ and μ were obtained using the Vp and Vs values from the CRUST2.0 (Bassian *et al.* 2000) project and the depth was estimated by choosing the maximum variance reduction from systematic moment tensor inversions with depths ranging from 2 to 26 km at 2-km steps.

4. Discussion and Results

As described in the previous section, the moment tensor inversion was carried out in order to estimate the six components of the moment tensor. Also, the scalar moment M_0, as defined by Bowers and Hudson (1999), was estimated. The scalar moment of a full moment tensor includes the isotropic moment and the deviatoric moment. In addition, we observe that the waveform moment tensor inversion is better adjusted with the six components moment tensor if it is compared with the five component deviatoric case. In Fig. 6 we present three examples of the moment tensor inversion. We observe that the azimuthal

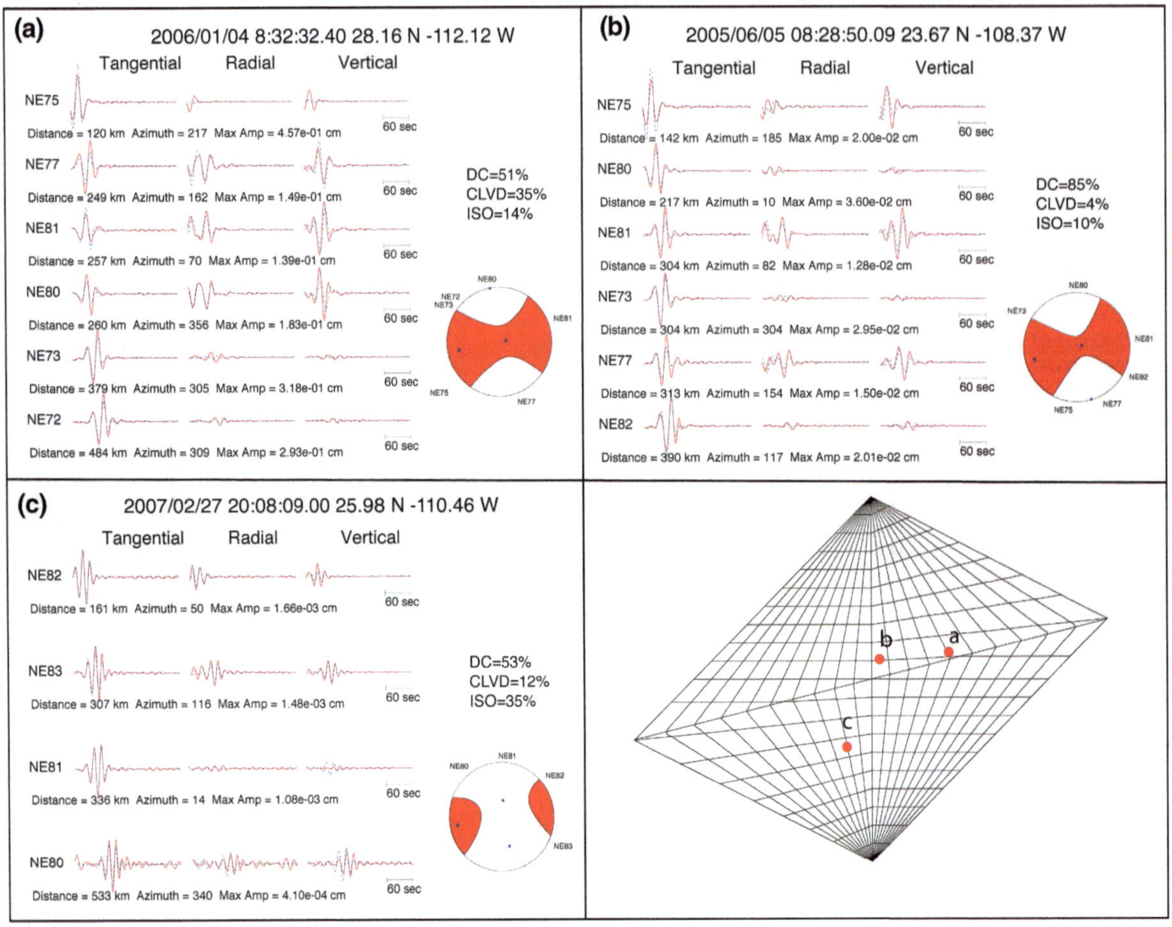

Figure 6

Examples of the waveform modeling **a** Event that occurred on 5 June 2005, **b** 4 January 2006 and **c** 27 February 2007. Synthetic seismograms are represented in *solid red* and observed seismograms are depicted in *dotted blue*. At the *bottom right* we depict the source type plot of the events

coverage is satisfactory and the moment tensor estimation is robust. In Fig. 6a we present a moment tensor that is close to a double couple (DC = 85 %, CLVD = 4 % and ISO = 10 %), in contrast, in the examples given in Fig. 6b, c, the double couple component is about 50 % and the contribution of CLVD and ISO is higher. The good observed-synthetic seismogram fit and the high variance reduction VR give us great confidence that the assumed velocity model and depth are correct and the estimated moment tensor solutions are robust.

We do not find that the *k* and *T* parameters of the source type plot have preference to any specific sign nor do they show any tendency as increasing scalar moment (Fig. 7). Previous authors (e.g. Kawakatsu 1991) observed that T tends to zero as increasing size; however, he only used deviatoric solutions from the systematic analysis of the Global CMT solutions. Therefore, his analysis was projected to a space solution of five moment tensor elements as pointed out by Dufumier and Rivera (1997) and the analysis is different if we use instead the full moment tensor inversion. More recently, some authors (Aptekman and Tatevossian 2007) analyzed the Global CMT solutions in order to detect complex sources and they observed that, at the entire set of observations, T does not depend on M_0. In Fig. 7 we observe that, at different earthquake sizes, the T value varies from roughly −0.8 to 0.8 suggesting that earthquakes of different magnitude are equally complex. In our case M_0 was computed following the definition of Bowers and Hudson (1999).

Note that it is not possible to distinguish between $\hat{n}$ and $\hat{\upsilon}$ since both parameters are defined mathematically as the algebraic addition of the eigenvectors s_1 and s_3 (Eq. 5), thus it is necessary to identify $\hat{n}$ and $\hat{\upsilon}$ based on geological and field observations. This is similar to the ambiguity problem of the principal and auxiliary planes of the double couple focal mechanisms. Therefore, we decided to represent all the normal plane surfaces of $\hat{n}$ and $\hat{\upsilon}$ in a rose diagram with steps of 10° and plot them with the general trend of the transform fault system of the Gulf of California (Fig. 8a) and we also computed $\hat{n}$ and $\hat{\upsilon}$ for the deviatoric case (Fig. 8b). The trend was estimated using a least squares regression of the Cartesian coordinates of the transform fault system of the Gulf of California. In Fig. 8a we observe that the general trend of the fault system (dotted line) follows one of the principal directions of the rose diagram; however, there is another population of the normal surface trends that is slightly higher than that which coincides with the fault system trend. On the other hand, we observe that the deviatoric solution fails to predict the general trend of the fault system and we believe that this discrepancy can be caused by the deviatoric restriction of projecting focal mechanisms into CLVD and DC. This deviation to the principal direction is probably caused by the contribution of two or more faults of different orientations that construct the $\hat{n}$ vector. The simple case of one single planar surface requires that the normal surface trend

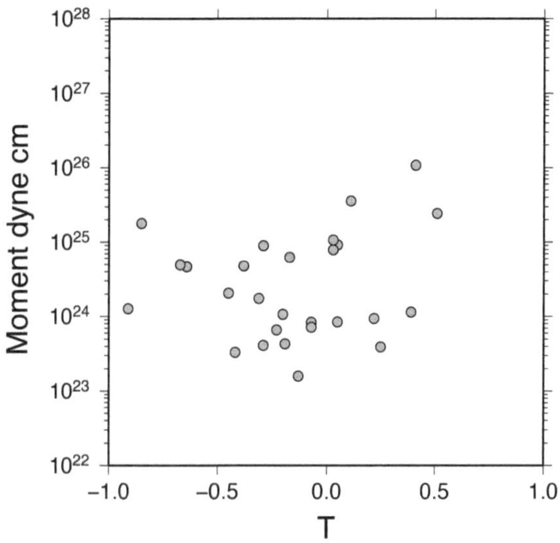

Figure 7
Constant volume (shear) component (T) versus scalar moment. We observe that, at the entire set of observations, T does not depend on M_0

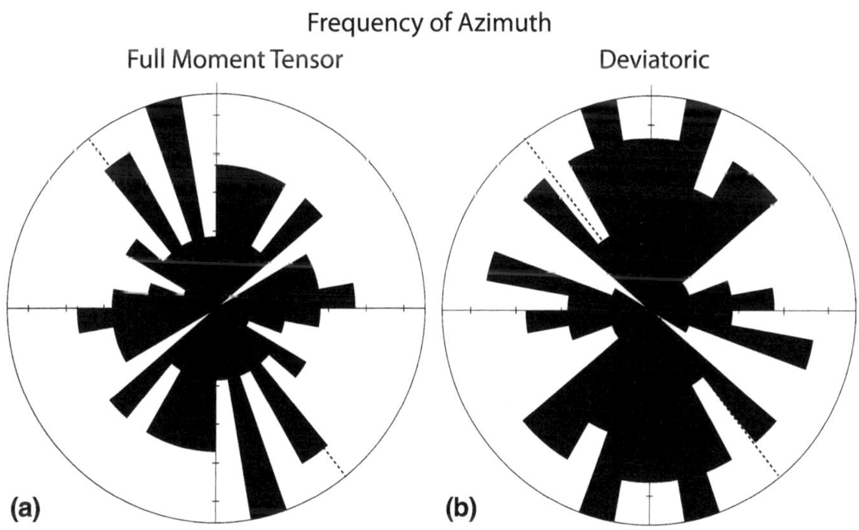

Figure 8
Rose diagram of the trend of all data orthogonal to $\hat{n}$ and $\hat{\nu}$ used in this article, also the general trend of the transform fault obtained using a least squares regression is plotted with a discontinuous line. At the *left* we show the full moment tensor direction and at the right the standard deviatoric case. We observe that the general trend of the fault system (*dotted line*) follows one of the principal directions of the rose diagram for the full moment tensor solution; however, in the deviatoric solution the direction fails to predict the general trend of the transform fault (see text for details)

coincide with the actual fault trend. This work is useful, since we have added more quantitative information about the complexity of the source.

Recently, there have been many studies that confirm the complexity of strike-slip earthquakes at different sizes. For example, DUPUTEL *et al.* (2012) show that the great Sumatra earthquake of 2012 ($Mw = 8.6$) is better explained with two point sources separated by about 209 km; moreover, they do not exclude more complicated models. On the other hand, WEI *et al.* (2011) suggest that the Cucapah—El Mayor earthquake ($Mw = 7.1$) on the Mexico–US border near the Gulf of California, is better explained by four segments that accommodate at depth and this distribution of sources also explains the complex pattern of the radiated seismic waves. In addition, we believe that better instrumentation will help us to add independent observations to those that are obtained by regional data; for example, SUMY *et al.* (2013), published a comprehensive catalog of the earthquakes in the Southern Gulf of California using eight four-component ocean-bottom seismographs (OBS) from October 2005 to 2006. During this period, there were some earthquakes that were clearly identified as doublets and they were able to locate them in detail; however, in the regional waveforms we were able to observe only some complex short-period body-wave waveforms. Using a visual inspection we decided not to analyze those events.

Future work is needed to obtain the full moment tensor systematically. We have shown that the six elements of the moment tensor provide important information about the earthquake source. This information is useful for further analysis, such as source complexity at different sizes. The overall result suggests that moderate earthquakes are equally complex as large earthquakes in the Gulf of California.

5. Conclusions

We can summarize the major conclusions as following:

- Full moment tensor inversions are useful for retrieving parameters that provide information about slip and average fault direction. Slip, fault direction and isotropic parameters are in turn used to analyze fault complexity.
- In tectonic environments such as the Extensional Province of Baja California where transform faults control the principal earthquakes, the full tensor inversion resolves better the general trend of the fault orientation compared to the standard deviatoric inversion.
- Small and moderate size earthquakes are equally complex in the Gulf of California Transform Fault system.

Acknowledgments

This work was funded by CONACYT Mexico, under Grant Number 133910. Additionally, we would like to thank Sean Ford and an anonymous reviewer for their insightful comments.

REFERENCES

ABERCROMBIE, R.E., ANTOLIK, M., EKSTROM G. (2003), *The June 2000 Mw 7.9 earthquakes south of Sumatra: deformation in the India–Australia plate*, J. Geophys. Res. *108*, B1, 2018. doi:2010.1029/2001JB000674, 1–16.

ADAMOVA, P., and SILENY, J. (2010), *Non-double-couple earthquake mechanism as an artifact of the point-source approach applied to a finite-extent focus,* Bull. Seismol. Soc. Am. *100*, doi:410.1785/0120090097, 447–457.

APTEKMAN, ZH. YA., and TATEVOSSIAN, R. E. (2007), *On the possibility of detecting complex sources of earthquakes from data of the centroid moment tensor catalog*, Izvestiya, Physics of the Solid Earth, *43*, 5, 354.

AKI, K. (1984), *Evidence for magma intrusion during the Mammoth Lakes Earthquakes of May 1980 and implications of the absence of volcanic (Harmonic) tremor,* J. Geophys. Res. *89*, B9, 7689–7696.

AKI, K., and RICHARDS, P., *Quantitative Seismology* (University Science Books, Sausalito, 2002).

BASSIAN, C., LASKE, G., and MASTERS, G. (2000), *The current limits of resolution for surface wave tomography in North America,* Eos Trans. AGU 81.

BEN-MENAHEN, A., and SINGH, S.J., *Seismic waves and sources* (Springer Verlag New York, 1981).

BEN-ZION, Y., and AMPUERO, J.P. (2009), *Seismic radiation from regions sustaining laterial damage*, Geophys. J. Int. *178*, doi:10.1111/j.1365-246X.2009.04285, 1351–1356.

BOWERS, D., and HUDSON, J. A. (1999), *Defining the Scalar Moment of a Seismic Source with a General Moment Tensor*, Bull. Seism. Soc. Am. *89*, (5), 1390–1394.

BRUNE, J. N., BROWN, N. S., and JOHNSON, P. A. (1993), *Rupture mechanism and interface separation in foam rubber models of earthquakes: a possible solution to the heat flow paradox and the paradox of large overthrusts*, Tectonophysics *218*, 59–67.

DREGER, D.S., TKALCIC, H., and JOHNSTON, P. A. (2000), *Dilatational processes accompanying earthquakes in the Long Valley Caldera*, Science *288*, 122–125.

DUFUMIER, H. and RIVERA, L. (1997), *On the resolution of the isotropic component in moment tensor inversión*, Geophys. J. Int. *131*, doi:10.1111/j.1365-246X.1997.tb06601.x, 595–606.

DUPUTEL, Z., KANAMORI, H., TSAI, V. C., RIVERA, L., MENG, L., AMPUERO, J. P. and STOCK, J. M. (2012), *The 2012 Sumatra great earthquake sequence*, Earth Planet. Sci. Lett. 351-352, 247–257.

ESTABROOK, C. H. (1999), *Body wave inversion of the 1970 and 1963 South American large deep-focus earthquakes*, J. Geophys. Res. Solid Earth (1978-2012), *104*, B12, 28751–28767.

FORD, S. R., DREGER, D. S., and WALTER, W. R. (2009), *Identifying isotropic events using a regional moment tensor inversion*, J. Geophys. Res. *114*, B01306, doi:10.1029/2008JB005743, 1–12.

FOULGER, G., and LONG, R.E. (1984), *Anomalous focal mechanisms: tensile crack formation on an accreting plate boundary*, Nature *310*, 43–45.

FROHLICH, C., RIEDESEL, M. A., and APPERSON, K. D. (1989), *Note concerning posible mechanisms for non doublé couple earthquake sources*, Geophys. Res. Lett. *16*, 523–526.

FROHLICH, C. (1995), *Characteristics of well-determined non-double-couple earthquakes in the Harvard CMT catalog*, Phys. Earth Planet. Int. *91*, 213–228.

HARA, T. (1996), *Determination of the isotropic component of deep focus earthquakes by inversion of normal-mode data*, Geophys. J. Int. *127*, (2), 515–528.

HAUKSSON, E., STOCK, J., HUTTON, K., YANG, W., VIDAL – VILLEGAS, J. A., and KANAMORI, H. (2011), The 2010 Mw 7.2 *El Mayor-Cucapah Earthquake Sequence, Baja California, Mexico and Southernmost California, USA: Active Seismotectonics along the Mexican Pacific Margin*, Pure Appl. Geophys. *168*, (8–9), 1255–1277.

HUDSON, J. A., PEARCE, R. G., and ROGERS, R. M. (1989), *Source type plot for inversion of the moment tensor*, J. Geophys. Res. *94*, (B1), 765–774.

JOST, M. L., and HERRMANN, R.B. (1989), *A students guide to and review of moment tensors*, Seism. Res. Lett. *60*, 37–57.

JULIAN, B.R., and SIPKIN, S.A. (1985), *Earthquake processes in the Long Valley caldera area, California*, J. Geophys. Res. *90*, 11155–11169.

KAWAKATSU, H. (1991), *Enigma of earthquakes at ridge-transform-fault plate boundaries distribution of non-double couple parameter of Harvard CMT Solutions*, Geophys. Res. Lett. *18*, (6), 1103–1106.

KAWAKATSU, H. (1996), *Observability of the isotropic component of a moment tensor*, Geophys. J. Int. *126*, (2), doi:10.1111/j.1365-246X.1996.tb05308.x, 525–544.

KIRBY, S. H., STEIN, S., OKAL, E. A., and RUBIE, D. C. (1996), *Metastable mantle phase transformations and deep earthquakes in subducting oceanic lithosphere*, Rev. Geophys. *34*, (2), 261–306.

KNOPOFF, L., and RANDALL, M. J. (1970), *The compensated linear vector dipole: a possible mechanism for deep earthquakes*, J. Geophys. Res. *75*, 4957–4963.

KŘÍŽOVÁ, D., J. ZAHRADNÍK and A. KRATZI (2013), *Resolvability of isotropic component in regional seismic moment tensor inversion*, Bull. Seis. Soc. Am. *103*, 2460–2473 doi:10.1785/0120120097.

LAY, T., AMMON, C. J., KANAMORI, H., RIVERA, L., KOPER, K. D., and HUTKO, A. R. (2010), *The 2009 Samoa-Tonga great earthquake triggered doublet*, Nature, *466*, doi:10.1038/nature09214, 964–968.

MINSON, S., and DREGER, D. (2008), *Stable Inversions for Complete Moment Tensors*, Geophys. Journ. Int., *174*, 585–592.

OKAL, E. A. (1996), *Radial modes from the great 1994 Bolivian earthquake: No evidence for an isotropic component to the source*, Geophys. Res. Lett. *23*, (5), 431–434.

ORTEGA, R., and QUINTANAR, L. (2010), *Seismic evidence of a ridge-parallel strike slip fault off the transform system in the Gulf of California*, Geoph, Res. Lett. *37*, L06301, doi:10.1029/2009GL042208.

RUSSAKOFF, D., EKSTROM, G., and TROMP, J. (1997), *A new analysis of the great 1970 Colombia earthquake and its isotropic component*, J. Geophys. Res. *102*, (B9), 20,423–20,434.

SHI, Z., and BEN-ZION, Y. (2009), *Seismic radiation from tensile and shear point dislocations between similar and dissimilar solids,* Geophys. J. Int. *179*, 444–458.

SUMY, D. F., GAHERTY, J. B., WON-YOUNG, K., DIEHL T., and COLLINS, J. A. (2013), *The mechanisms of earthquakes and faulting in the Southern Gulf of California*, Bull. Seism. Soc. Am. *103*, 487–506. doi:10.1785/0120120080.

TRAMPERT, J., PAULSSEN, H., VAN WETTUM, A., RITSEMA, J., CLAYTON, R., CASTRO, R., REBOLLAR, C., and PEREZ-VERTTI, A. (2003), *New array monitors seismic activity near the Gulf of California in Mexico*, Eos Trans. AGU *84*, 29–32.

VAVRYČUK, V. (2004), *Inversion for anisotropy from non-double-couple components of moment tensors*, J. Geophys. Res. Solid Earth *109*, (B7), doi:10.1029/2003JB002926

WEI, S., FIELDING, E., LEPRINCE, S., SLADEN, A., AVOUAC, J. P., HELMBERGER, D., HAUKSSON, E., CHU, R., SIMONS, M., HUDNUT, K., HERRING, T., and BRIGGS, R. (2011), *Superficial simplicity of the 2010 El Mayor–Cucapah earthquake of Baja California in Mexico*, Nature Geoscience *4*, doi:10.1038/ngeo1213, 615–618.

(Received May 1, 2013, revised October 29, 2013, accepted December 3, 2013, Published online December 14, 2013)

Pure Appl. Geophys. 171 (2014), 2819–2840
© 2014 Springer Basel
DOI 10.1007/s00024-013-0769-4

Pure and Applied Geophysics

Diagnosing Source Geometrical Complexity of Large Earthquakes

L. Rivera[1,2] and H. Kanamori[2]

Abstract—We investigated the possible frequency dependence of the moment tensor of large earthquakes by performing W phase inversions using teleseismic data and equally-spaced narrow, overlapping frequency bands. We investigated frequencies from 0.6 to 3.8 mHz. Our focus was on the variation with frequency of the scalar moment, the amount of non-double-couple, and the focal mechanism. We applied this technique to 30 major events in the period 1994–2013 and used the results to detect source complexity. Based on the results, we classed them into three groups according to the variability of the source parameters with frequency: simple, complex and intermediate. Twelve of these events fell into the simple category: Bolivia-1994, Kuril-1994, Sanriku-1994, Antofagasta-1995, Andreanoff-1996, Peru-2001, Sumatra-2004, Sumatra-2005, Tonga-2006, Sumatra-2007, Japan-2011, and the recent Sea of Okhotsk-2013. Seven exhibited significant complexity: Balleny-1998, Sumatra-2000, Indian Ocean-2000, Macquarie Island-2004, Sichuan-2008, and Samoa-2009. The remaining 11 events showed a moderate degree of complexity. Here, we discuss the results of this study in light of independent observations of source complexity, made by various investigators.

1. Introduction

Since high-quality global digital seismic data became available in the 1980s, long-period seismic waves have been successfully used to obtain rapid point-source representations of earthquakes. Some early attempts include the works of Dziewonski *et al.* (1981), Dziewonski and Woodhouse (1983), and Kanamori and Given (1982). The method of Dziewonski *et al.* (1981) has matured as the Centroid Moment Tensor (CMT) method (Dziewonski *et al.* 1981, 1984; Ekström *et al.* 2012, and references therein) and has provided an invaluable catalog of global earthquake focal mechanisms. Global CMT solutions, currently generally called the GCMT solutions, are available through several sources (e.g., Gcmt 2013, National Earthquake Information Center (NEIC) of the U.S. Geological Survey). More recently, another point-source inversion method using a very-long-period W phase (Kanamori and Rivera 2008) has been implemented at NEIC and is providing a global database of long-period point-source solutions of large earthquakes (Hayes *et al.* 2009; Duputel *et al.* 2011, 2012a).

The GCMT solutions are estimated from body, surface, and/or mantle waves with the lower cut-off frequencies ranging from 5 to 10 mHz, except for the 2004 Sumatra–Andaman islands earthquake, for which it is lowered to 3 mHz. The W-phase source inversion operates nominally on a frequency band from 1 to 5 mHz (Kanamori and Rivera 2008), but with the 1 Hz sampling and with the distance-dependent W-phase time window (P, $P + 15\Delta$ s/°) we use for global studies, the effective response is somewhat narrower and distance dependent. Figure 1 represents the effective amplitude response for some typical epicentral distances. Both GCMT and W-phase CMT (WCMT) solutions provide robust source mechanisms for very large earthquakes (i.e. Mw > 8.0), which are usually difficult to achieve with other methods. In the GCMT method, the effects of the first-order lateral heterogeneities of the earth are corrected for. In the case of the W phase algorithm, the longer period and the particular W-phase time window minimize the influence of shallow heterogeneities, and no correction is applied (e.g. Duputel *et al.* 2012a, their figure 12). These solutions are generally considered to represent the source characteristics up to 200–300 s.

With the commonly used centroid moment tensor inversions, just one number represents the long period size of the event, the seismic moment M_0 (or corresponding Mw). However, several recent studies

[1] Institut de Physique du Globe de Strasbourg, Université de Strasbourg/CNRS, Strasbourg, France. E-mail: luis.rivera@unistra.fr
[2] Seismological Laboratory, California Institute of Technology, Pasadena, CA 91125, USA.

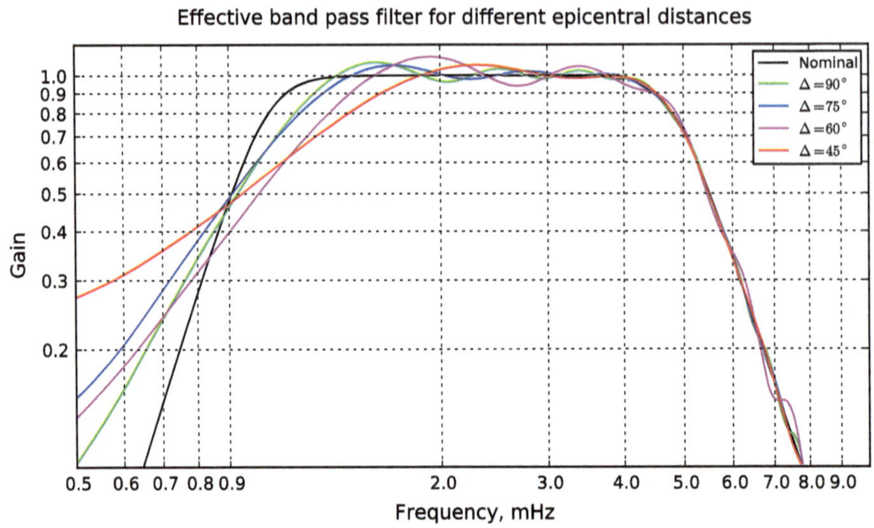

Figure 1
Effective bandpass filter used for large earthquakes (Mw > 8.0) in the standard W-phase source inversion algorithm (DUPUTEL *et al.* 2012a). The *black line* represents the amplitude response of a fourth-order, ideal Butterworth bandpass filter. With the 1 Hz sampling and the distance-dependent W-phase time window (P, $P + 15\Delta$ s/°) we use for global studies, the effective response is distance dependent. The figure represents the effective amplitude response for 45° (*red*), 60° (*magenta*), 75° (*blue*), and 90° (*green*) of epicentral distance

have clearly demonstrated that this is not satisfactory, at least for some earthquakes. For example, for the 2004 Sumatra–Andaman Is. earthquake, the amplitudes of long-period normal modes indicated that the effective seismic moment increases from 4.0×10^{22} to 7.1×10^{22} N-m as the period increases from 300 to 1,000 s (OKAL and STEIN 2009; PARK *et al.* 2005; LAMBOTTE *et al.* 2006).

For the 2009 Samoa Is. earthquake (Mw = 8.1), the moment tensor was found to be strongly dependent on frequency because the source of this earthquake consisted of at least two distinct events with very different mechanisms (LAY *et al.* 2010a). It is possible that other large and great earthquakes may have similar complex characteristics, but with the standard moment tensor inversion with a single frequency band, we may not notice this easily. In this study, we investigated the possible frequency dependence of the moment tensor of large earthquakes by performing W-phase inversions using multiple frequency bands from 1.1 to 3.3 mHz.

2. Method

Central to the WCMT inversion is time-domain causal deconvolution of instrument response from the data followed by bandpass filtering; also, the synthetic seismograms are convolved with a source moment-rate function and bandpass filtering. Once these operations have been performed, data and synthetics can be directly compared. When applying the W-phase inversion algorithm to large events (Mw > 8), we normally use the frequency band 1–5 mHz (KANAMORI and RIVERA 2008). Here, instead, we perform several W-phase inversions using equally-spaced narrower, overlapping frequency bands. After some testing of the overall frequency band, the width of the individual bands, and the separation between adjacent bands, we settled on the following scheme. We explore the spectrum from 0.6 to 3.8 mHz with a bandwidth of 1 mHz separated by 0.2 mHz steps. This choice is guided by the S/N ratio, which deteriorates at longer periods, and is consistent with the maximum W-phase time window length that we use for global studies (i.e., 1,300 s). The filtering is made in the time domain and is implemented as an infinite impulse response (IIR) filter derived by bilinear-transformation from a fourth-order, causal bandpass Butterworth filter.

For each frequency band, we obtain a moment tensor, which for simplicity we tag with the corresponding central frequency. The inversion includes a scheme of data screening. Two criteria are used for this purpose. First, data channels with a root mean square (rms) value too far from the median of the rms values of all the available channels are rejected. Second, data channels fitting too poorly with respect to previous solution synthetics are successively rejected. More precisely, we compute a misfit per-channel as follows:

$$\mathrm{nrms} = \sqrt{\frac{\sum_i (d_i - s_i)^2}{\sum_i d_i^2}}, \qquad (1)$$

where d_i stands for the sample "i" of the data and s_i for the sample "i" of the corresponding synthetics. The value of nrms ("normalized" rms), although not strictly normalized, can be used to quantify the quality of the data fit. We perform three successive steps of cumulative screening. On each step, we perform an inversion and compare the nrms of each channel with a given threshold to decide if it should be kept or rejected in the next iteration. In this paper, we use the thresholds 5.0, 3.0, and 0.6. The collection of channels passing the complete set of screenings is then fixed and used for the final W-phase inversion, including a grid search on the centroid timing and location.

As a result, for each frequency band, we obtain several source parameters: scalar moment, best double-couple mechanism, non-double couple (NDC), nrms, etc. Because we are working here at a very long period with mainly shallow earthquakes and because we are using frequency bands that are significantly narrower than the one used in the standard W-phase application, we are potentially exposed to the well-known trade-off between the seismic moment (M_0) and the dip angle (δ), (Kanamori and Given 1982). This is related to the relative proximity of the source to the free surface and the resulting lack of constraint on $M_{r\theta}$ and $M_{r\phi}$. This trade-off is such that for low-angle dip-slip events, M_0 and δ can vary quite freely, while the product $M_0 \sin(2\delta)$ remains well-constrained (Kanamori and Given 1982). Since we are interested in the possible variation of M_0 with frequency for a given event, this simple relationship allows us to alleviate the trade-off effect by proceeding as follows. For each event, we choose a reference focal mechanism and fix from it a reference dip (δ_r). As reference here, we use a standard W-phase solution from the catalog in Duputel *et al* (2012a) (also accessible at WCMT 2013). These solutions are presented in Table 1. Then, for all the solutions related to that event, we update the moment value to $M_0{}'$ as follows:

$$M_0{}'(f) = M_0(f) = \frac{\sin\, 2\delta(f)}{\sin\, 2\delta_r}, \qquad (2)$$

Table 1

List of events

GCMT event ID	Geographical region	Date mm/dd/yy	Mw	WCMT
060994A	Bolivia	06/09/94	8.2	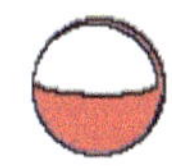
100494B	Kuril Islands	10/04/94	8.3	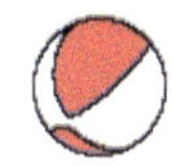
122894C	Sanriku	12/28/94	7.7	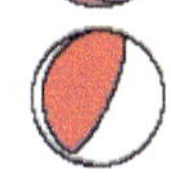
073095A	Antofagasta, Chile	07/30/95	8.0	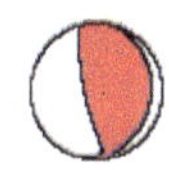
100995C	Jalisco, Mexico	10/09/95	8.0	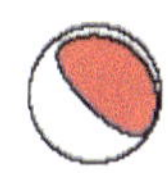
120395E	Kuril Islands	12/03/95	7.9	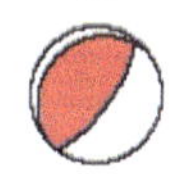
010196C	Minahasa, Indonesia	01/01/96	7.9	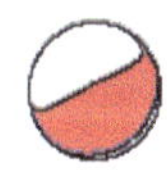
021796B	Irian	02/17/96	8.2	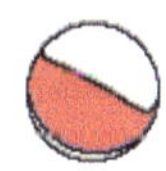
061096B	Andreanof Islands	06/10/96	7.9	
032598B	Balleny Islands	03/25/98	8.1	
060400D	Sumatra	06/04/00	8.1	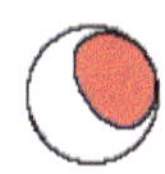
061800A	South Indian Ocean	06/18/00	7.9	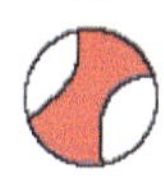
111600B	New Ireland Region	11/16/00	8.0	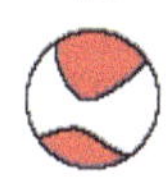

Table 1 *continued*

GCMT event ID	Geographical region	Date mm/dd/yy	Mw	WCMT
062301E	Peru	06/23/01	8.4	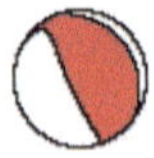
092503C	Hokkaido, Japan	09/25/03	8.3	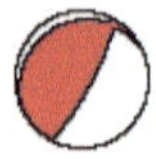
122304A	Macquaire Island	12/23/04	8.1	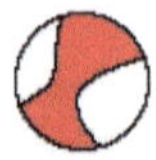
122604A	Sumatra	12/26/04	9.2	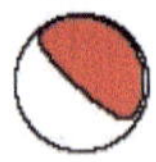
200503281609A	Sumatra	03/28/05	8.6	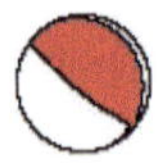
200605031527A	Tonga Islands	05/03/06	8.0	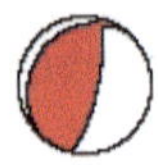
200611151114A	Kuril Islands	11/15/06	8.3	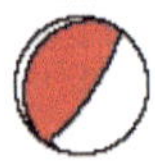
200701130423A	Kuril Islands	01/13/07	8.1	
200704012039A	Solomon Islands	04/01/07	8.1	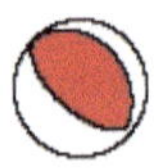
200708152340A	Peru	08/15/07	8.0	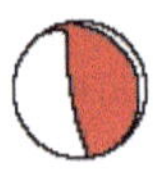
200709121110A	Sumatra	09/12/07	8.5	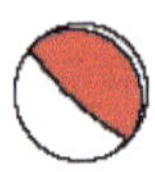
200805120628A	Sichuan, China	05/12/08	7.9	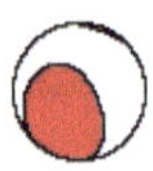
200909291748A	Samoa	09/29/09	8.1	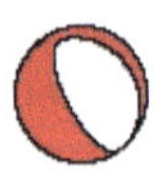
201002270634A	Maule, Chile	02/27/10	8.8	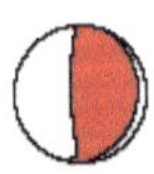
201103110546A	Tohoku-oki, Japan	03/11/11	9.0	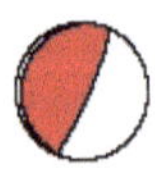
201204110838A	Off-Sumatra	04/11/12	8.6	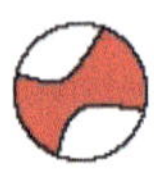
201305240545A	Sea of Okhotsk	05/24/13	8.3	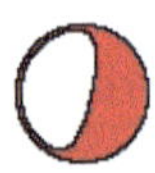

where δ is the dip angle of the shallowest dipping plane from the best double couple. In a way, since δ_r is kept constant, this is equivalent to tracking the variation of the well-constrained parameter $M_0 \sin(2\delta)$. Hereafter, we apply this update recipe to all the events having $\delta_r < 30°$ and NDC < 40 %; for other events, we use the non-modified value $M_o(f)$. Whenever the term "scalar moment" is used, we are tacitly referring to the updated value of M_0, updated with this ruler while dropping the prime notation.

A usual parameter to quantify the amount of non-double couple in a deviatoric moment tensor is

$$\mathrm{NDC}(f) = 200\frac{\lambda_2}{\lambda_1 - \lambda_3}(\%), \qquad (3)$$

where $\lambda_1 \geq \lambda_2 \geq \lambda_3$ $(\lambda_1 > 0 > \lambda_3)$ are the three eigenvalues of the deviatoric moment tensor. In the context of this study on large earthquakes, this parameter does not necessarily mean involvement of an exotic, non-fault component (e.g., explosion, dyke injection, etc.). More often, it simply represents a finite source for which more than one fault is involved. In this case, even if the rupture on each fault is a double couple, the total moment tensor can have a large NDC.

Another aspect we would like to keep track of is the similarity/variability of the focal mechanism across the frequency band. Hereafter, we define a parameter for such a purpose. Let $\boldsymbol{M}_1$ and $\boldsymbol{M}_2$ be two moment tensors (symmetric and traceless), and let's suppose we want to define a scalar parameter measuring their geometric difference. By geometric difference we mean to measure their difference regardless of their size (scalar moment). A first step is of course to normalize each one of the two moment tensors:

$$\widehat{\boldsymbol{M}}_1 = \frac{\boldsymbol{M}_1}{\sqrt{\boldsymbol{M}_1 : \boldsymbol{M}_1}} \text{ and } \widehat{\boldsymbol{M}}_2 = \frac{\boldsymbol{M}_2}{\sqrt{\boldsymbol{M}_2 : \boldsymbol{M}_2}},$$

where the symbol ":" represents double tensor contraction $\mathbf{M} : \boldsymbol{M} = M_{ij}M_{ij}$.

If the two normalized moment tensors have similar eigenvalues (e.g., if they correspond to two double couples), then a very natural measure of their difference is the minimal rotation angle necessary to bring one into coincidence with the other (Kagan 1991). However, this parameter can lead to very counterintuitive results when the amount of non-double couple is significant. An alternative parameter can be defined by the difference in the radiation patterns. Let's define Δ_p as the RMS of the difference of the P radiation patterns:

$$\begin{aligned}\Delta_p(\boldsymbol{M}_1,\boldsymbol{M}_2) &= \frac{1}{2\langle P\rangle}\left(\frac{1}{4\pi}\int_{\Omega}\left(\gamma^{\mathrm{T}}\widehat{\boldsymbol{M}}_1\gamma-\gamma^{\mathrm{T}}\widehat{\boldsymbol{M}}_2\gamma\right)^2\mathrm{d}S\right)^{1/2}\\ &= \frac{1}{2\langle P\rangle}\left(\frac{1}{4\pi}\int_{\Omega}\left(\gamma^{\mathrm{T}}\left(\widehat{\boldsymbol{M}}_1-\widehat{\boldsymbol{M}}_2\right)\gamma\right)^2\mathrm{d}S\right)^{1/2}\\ &= \frac{1}{2\langle P\rangle}\left(D_{ij}D_{kl}\frac{1}{4\pi}\int_{\Omega}\gamma_i\gamma_j\gamma_k\gamma_l dS\right)^{1/2}\\ &= \frac{1}{2\sqrt{2}}(\boldsymbol{D}:\boldsymbol{D})^{1/2}\end{aligned}.$$

In this expression, $\gamma^{\mathrm{T}}\widehat{\boldsymbol{M}}\gamma$ is the P radiation pattern (Aki and Richards 1980, p. 118), $\boldsymbol{D}$ is the difference between the two normalized moment tensors, $\widehat{\boldsymbol{M}}_1-\widehat{\boldsymbol{M}}_2$, $\langle P\rangle=\sqrt{4/15}$ is the RMS value of the P radiation pattern, Ω is the unit sphere, and γ is the unit radial vector. Alternatively, if we use the S wave instead of the P wave, we can define:

$$\begin{aligned}\Delta_S(\boldsymbol{M}_1,\boldsymbol{M}_2) &= \frac{1}{2\langle S\rangle}\left(\frac{1}{4\pi}\int_{\Omega}\left\|\left[\widehat{\boldsymbol{M}}_1\gamma-\left(\gamma^{\mathrm{T}}\widehat{\boldsymbol{M}}_1\gamma\right)\gamma\right]\right.\right.\\ &\qquad\left.\left.-\left[\widehat{\boldsymbol{M}}_2\gamma-\left(\gamma^{\mathrm{T}}\widehat{\boldsymbol{M}}_2\gamma\right)\gamma\right]\right\|^2\mathrm{d}S\right)^{1/2}\\ &= \frac{1}{2\langle S\rangle}\left(\frac{1}{4\pi}\int_{\Omega}\|\boldsymbol{D}\gamma-(\gamma^{\mathrm{T}}\boldsymbol{D}\gamma)\gamma\|^2\mathrm{d}S\right)^{1/2}\\ &= \frac{1}{2\langle S\rangle}\left(D_{ij}D_{ik}\frac{1}{4\pi}\int_{\Omega}\gamma_j\gamma_k dS\right.\\ &\qquad\left.-D_{ij}D_{kl}\frac{1}{4\pi}\int_{\Omega}\gamma_i\gamma_j\gamma_k\gamma_l\mathrm{d}S\right)^{1/2}\\ &= \frac{1}{2\sqrt{2}}(\boldsymbol{D}:\boldsymbol{D})^{1/2}\end{aligned}.$$

Here, $\widehat{M}\gamma-\left(\gamma^{\mathrm{T}}\widehat{M}\gamma\right)\gamma$ is a concise way of writing the S radiation pattern and particle motion, and $\langle S\rangle=\sqrt{2/5}$ is the RMS value of the S radiation pattern. Δ_P is in fact identical to Δ_S, and we can then drop the subscripts. Explicitly written in terms of the components of $\boldsymbol{D}$, the above expression leads to:

$$\Delta=\frac{1}{2\sqrt{2}}\left(D_{11}^2+D_{22}^2+D_{33}^2+2(D_{12}^2+D_{23}^2+D_{31}^2)\right)^{1/2}. \tag{4}$$

This is a well-defined scalar quantity in the sense that it is independent of the orientation of the reference frame. Expression (4) for Δ is particularly simple and well-suited for practical applications. Written in terms of the eigenvalues of $\boldsymbol{D}$: d_1, d_2 and d_3, it reduces to:

$$\Delta=\frac{1}{2\sqrt{2}}\left(d_1^2+d_2^2+d_3^2\right)^{1/2}.$$

$\Delta(\boldsymbol{M}_1,\boldsymbol{M}_2)$, so defined, has the following properties:

$0\le\Delta(\boldsymbol{M}_1,\boldsymbol{M}_2)\le 1$.

$\Delta(\boldsymbol{M}_1,\boldsymbol{M}_2)=0$ if and only if $\widehat{\boldsymbol{M}}_1=\widehat{\boldsymbol{M}}_2$.

$\Delta(\boldsymbol{M}_1,\boldsymbol{M}_2)=1$ if and only if $\widehat{\boldsymbol{M}}_1=-\widehat{\boldsymbol{M}}_2$, namely, if $\widehat{\boldsymbol{M}}_1$ and $\widehat{\boldsymbol{M}}_2$ correspond to opposite focal mechanisms.

In the following, we use the quantity $\alpha=1-\Delta$ as a measure of the geometrical similarity of two moment tensors.

In deriving the above results, we used the two identities:

$$\frac{1}{4\pi}\int_{\Omega}\gamma_i\gamma_j\gamma_k\gamma_l dS=\frac{\delta_{ij}\delta_{kl}+\delta_{ik}\delta_{jl}+\delta_{il}\delta_{jk}}{15} \tag{5}$$

$$\frac{1}{4\pi}\int_{\Omega}\gamma_i\gamma_j\mathrm{d}S=\frac{\delta_{ij}}{3}. \tag{6}$$

Note that the left-hand side of (2) is an isotropic fourth-order tensor, and it is well-known (Jeffreys 1931, page 70) that there are only three such linearly independent tensors: $\delta_{ij}\delta_{kl}$, $\delta_{ik}\delta_{jl}$, and $\delta_{il}\delta_{jk}$. Note furthermore that it is invariant under arbitrary permutation of its four indices, leading to a common coefficient. Finally, by evaluating any particular non-null component (e.g., "1111"), the coefficient is found to be 1/15.

An alternative version of this parameter, α', is obtained by using only the sign of the radiation pattern (first-motion polarity) instead of the radiation pattern itself in the above expression. In such case, α' represents the fraction of the surface of the focal sphere on which the sign of the P radiation patterns of the two focal mechanisms coincide. In the figures below, we show both quantities. For each event, we use the standard W-phase solution (Duputel *et al.* 2012a; WCMT 2013) as a reference. We then measure the similarity $\alpha(f)$, between the focal mechanism we obtain at each frequency f, and the reference solution. We also keep track of two more parameters: the number of channels, $N(f)$, that passed all the screening processes and the corresponding azimuthal gap $\gamma(f)$. These are quality control parameters that we use to detect and flag poorly-constrained solutions.

Lundgren and Giardini (1995) also made a series of narrow-frequency band CMT inversions on several hours of long-period teleseismic signals to study the source of the Bolivia 1994 deep-focus earthquake. Barth *et al.* (2007), on the other hand, applied a similar technique at shorter periods on regional data to study moderate magnitude earthquakes in eastern Africa.

3. Data and Analysis

We selected the events that occurred since 1990 with Mw (GCMT) ≥7.9. These events had enough signal strength to clearly overcome the background noise at periods longer than 500 s. There are, however, two exceptions. We included in the dataset the 1994 Sanriku earthquake (Mw 7.7) and the June 4th 2000 Sumatra earthquake (Mw 7.8), considering their potential source complexity (Heki and Tamura 1997; Abercrombie *et al.* 2003). On the other hand, we removed from the dataset two large aftershocks: 200709122348A (Sumatra, Mw 7.9) and 201103110615A (Tohoku-oki, Mw 7.9). Both of them occurred soon after a much larger event, and their signals are heavily perturbed by the long-period seismic waves excited by the main shock. The dataset so defined contains 30 events. These are listed in Table 1 and shown in Fig. 2. To study these earthquakes, we used three component records archived at IRIS (Incorporated Research Institutions for Seismology) for epicentral distances between 5° and 90°, sampled at 1 sps (LH). Data from the following networks were used: II, IU, G, IC, CI, TS, GE, CN, BK, MN, GT (and a few others from CU, US, CZ, PS, MS). These include all the available stations equipped with the STS-1 (Streckeisen 1) sensors and also several other broadband sensors.

Since the S/N ratio generally improves with the size of the events, we first present the results for events with Mw ≥ 8.5 (2004 Sumatra–Andaman Is., 2005 Nias, 2007 Sumatra, 2010 Maule, Chile, 2011

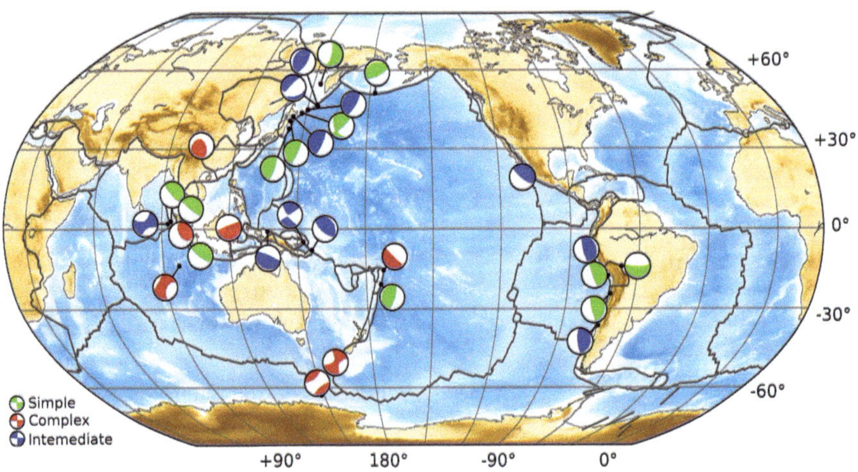

Figure 2
Geographic distribution and focal mechanisms of the events studied in the present work. The *beach balls* represent the stereographic projection of the lower hemisphere of the W-phase focal mechanism (Duputel *et al.* 2012b). The *color of the focal spheres* corresponds to the classification according to the source geometric complexity, as described in Sect. 4. *Green, blue* and *red colors* are used to represent "simple", "intermediate" and "complex" events, respectively

Tohoku-Oki, and 2012 Indian Ocean). In order to avoid repeating some of the figures, we refer hereafter directly to Figs. 4, 5, 6, 7, 8, 9 where the complete dataset has been split into three groups according to their geometric complexity (c.f. Sect. 4). For these events, Figs. 4g, h, j, k, 6j, and k show the focal mechanism as a function of the frequency band. Figures 7g, h, j, k, 9j, and k show the variation of the parameters: moment, M_0; NDC, similarity index, number of channels, azimuthal gap, and normalized misfit (nrms). In spite of being very large events, the focal mechanisms are remarkably simple and stable for the frequency bands from 1.1 to 3.3 mHz, indicating that the source process is relatively simple without involving sub-events with drastically different mechanisms. We briefly comment on the result for these events.

3.1. 2004 Sumatra–Andaman Is. Earthquake, Mw 9.2 (Figs. 4g, 7g)

As shown in Figs. 4g and 7g, the NDC component is small, <8 %, and the similarity index is larger than 0.9 over the frequency band from 1.1 to 3.3 mHz. Thus, despite the extremely long rupture length, the 2004 Sumatra–Andaman Is. earthquake remains essentially a double couple with about the same geometry (Tsai *et al.* 2005). However, the moment, M_0, increases by a factor of 1.35 as the frequency decreases from 3.3 to 1.1 mHz. This increase is subtle, but the trend is consistent with the source duration of about 10 min observed by Lambotte *et al.* (2007), and with the result of Park *et al.* (2005), who showed (their Fig. 5) that the moment rate spectral amplitude of this earthquake at 1 mHz estimated from the normal-mode amplitudes is about 1.6 times

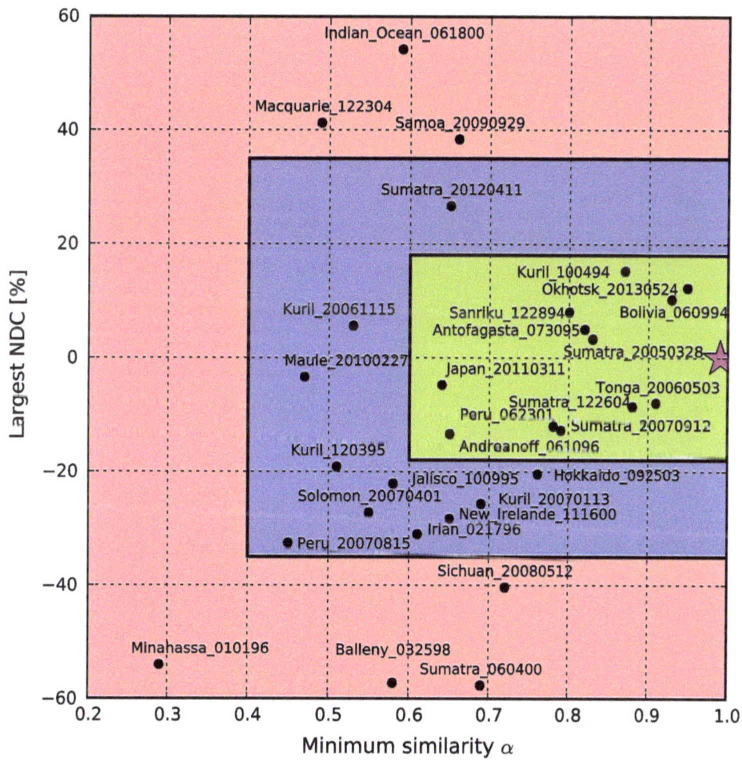

Figure 3

Summary of the classification of the 30 events studied in this work according to their geometrical complexity. The *horizontal axis* represents the minimum α value obtained for each event and the *vertical axis* signifies the largest NDC value. Events appearing on the right side, (α ∼ 1), and at mid-height, (NDC ∼ 0), near the *star* symbol on this diagram, are simple events showing a stable, pure double-couple focal mechanism over the whole frequency band. The three regions: *inner*, *middle* and *outer* correspond to "simple", "intermediate" and "complex" events, respectively

Figure 4
Focal mechanism variation with frequency for events classified as "simple": **a** 1994-Bolivia, **b** 1994-Kuril, **c** 1994-Sanriku, **d** 1995-Antofagasta, **e** 1996-Andreanoff, **f** 2001-Peru, **g** 2004-Sumatra, **h** 2005-Sumatra, **i** 2006-Tonga, **j** 2007-Sumatra, **k** 2011-Japan, and **l** 2013-Okhotsk. Each focal mechanism corresponds to that obtained over a 1-mHz-wide frequency band. The number at the *bottom* of each focal mechanism is the central frequency in mHz. The last focal sphere (*red color*) is the "reference solution"; it is the result of a broadband 1–5-mHz W-phase source inversion (Duputel *et al.* 2012a). See Fig. 7 for details

larger than the moment of the GCMT solution. It is unclear exactly at what period the GCMT moment was effectively determined, but it is probably somewhat shorter than 300 s. Thus, the increase in M_0 shown in Fig. 7g is consistent with the result shown in Fig. 5 of Park *et al.*(2005). Park *et al.* (2005) showed that at 0.3 mHz (i.e., ${}_0S_2$ frequency), the moment-rate spectral amplitude is about 2.6 times the GCMT moment. With our method, we cannot decrease the central frequency to 0.3 mHz while maintaining the 1 mHz bandwidth. Thus, we cannot confirm Park *et al.*'s (2005) result with our method.

Figure 5
Same as Fig. 4, for events classified as "complex": **a** 1996-Minahassa, **b** 1998-Balleny, **c** 2000-Sumatra, **d** 2000-Indian Ocean, **e** 2004-Macquarie Island, **f** 2008-Sichuan and **g** 2009-Samoa. See Fig. 8 for details

3.2. 2005 Nias, Sumatra, Earthquake, Mw 8.6 (Figs. 4h, 7h)

Figures 4h and 7h show the result for the 2005 Nias, Sumatra, earthquake. The moment, M_0, stays essentially constant over the frequency band from 1 to 3 mHz. The NDC component is <4 %, and the similarity index remains high. The variation of the moment tensor is generally small, and no significant complexity of the source is detected.

3.3. 2007 Sumatra Earthquake, Mw 8.5 (Figs. 4j, 7j)

Figures 4j and 7j show the results for the 2007 Sumatra earthquake. The general behavior is essentially the same as that for the 2005 Nias earthquake, and no significant complexity of the source is detected.

3.4. 2010 Maule, Chile, Earthquake, Mw 8.8 (Figs. 6j, 9j)

The moment, M_0, is essentially constant and the NDC component remains very small. However, at frequencies lower than 2.4 mHz, the similarity index decreases to 0.6, indicating some mechanism changes at a long period. The directions of the coseismic displacements reported by Vigny *et al.* (2011) vary considerably from place to place (e.g., Fig. 1 of Vigny *et al.* 2011), and further detailed studies are warranted.

3.5. 2011 Tohoku-Oki, Japan Earthquake, Mw 9.0 (Figs. 4k, 7k)

The behavior is similar to that of the 2005 Nias and 2007 Sumatra earthquakes. Although the similarity index decreases to 0.7 at the long period, its variation with frequency is gradual and no significant complexity of the source is detected.

3.6. 2012 Indian Ocean (Sumatra) Earthquake, Mw 8.6 (Figs. 6k, 9k)

Although the mechanisms are relatively similar (similarity index >0.75) over the entire frequency band, the moment, M_0, grows by a factor of 1.4 when

Figure 6

Same as Fig. 4, for events classified as "intermediate": **a** 1995-Jalisco, **b** 1995-Kuril, **c** 1996-Irian, **d** 2000-New Ireland, **e** 2003-Hokkaido, **f** 2006-Kuril, **g** 2007-Kuril, **h** 2007-Solomon, **i** 2007-Peru, **j** 2010-Maule and **k** 2012-Sumatra. See Fig. 9 for details

passing from 3.3 to 1.1 mHz. Duputel *et al.* (2012b) applied a multiple-event W-phase inversion method to this event and found that this quake actually consisted of two distinct events (Mw = 8.5, 8.3) about 70 s apart with approximately the same mechanism.

Clearly, most of these very large events present stable, pure double-couple focal mechanisms. Only the 2010 Maule and the 2012 Indian Ocean events slightly departed from such a simple model. Next, we look at the earthquakes for which the source is known to be complex in time and space.

3.7. 1998 Balleny Is. Earthquake, Mw 8.1 (Figs. 5b, 8b)

This earthquake has been studied in great detail by Henry *et al.* (2000) and Hjörleifsdóttir *et al.* (2009). Although the mechanism is primarily strike-

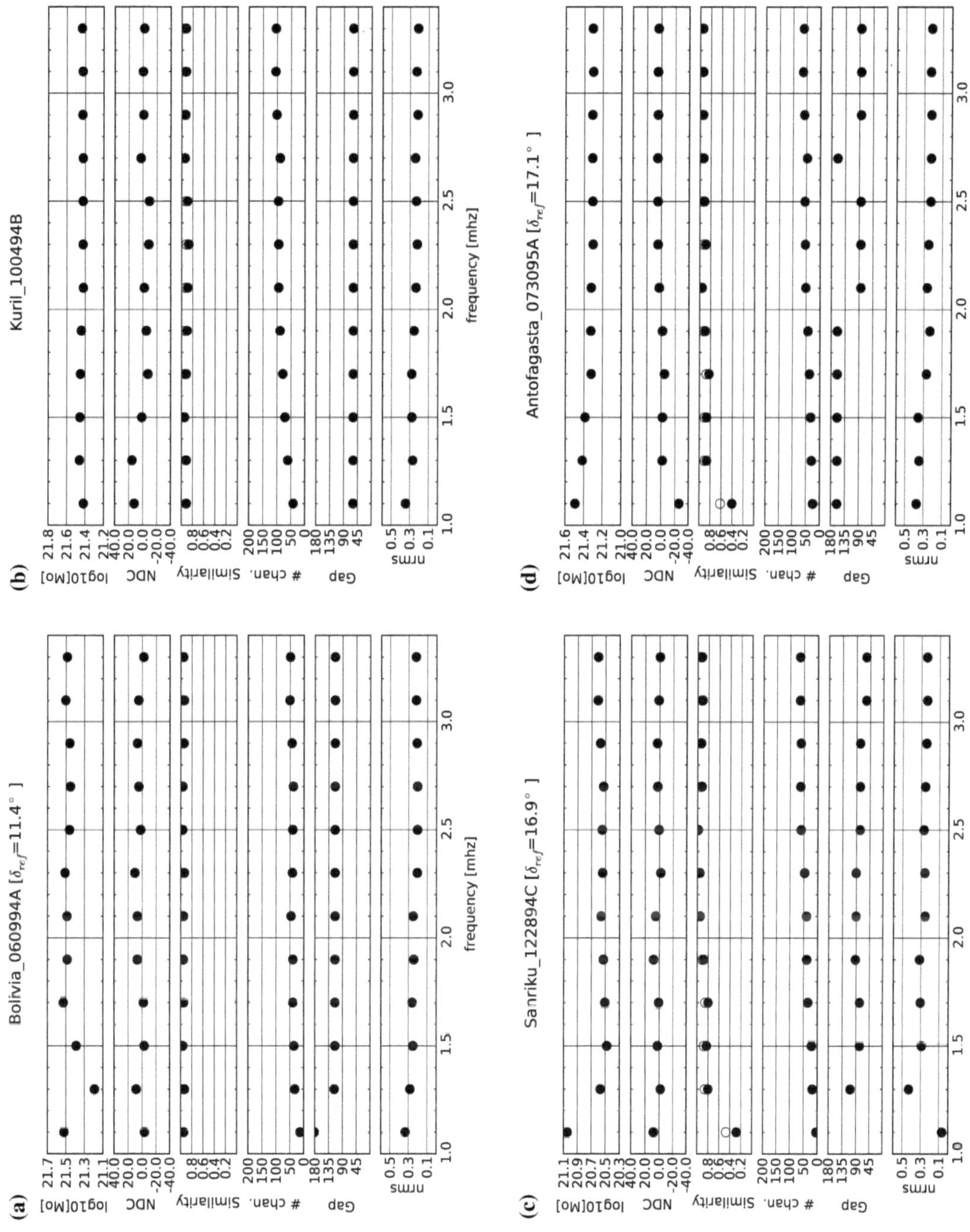

Figure 7
Variations with frequency of the source parameters for some major earthquakes. From *top to bottom*, each *panel* presents: scalar moment, $\log_{10}M_o$ (Nm); non-double-couple component, NDC, (%), similarity index with respect to the reference solution (the filled symbol corresponds to α and the open symbol to α' (c.f. text after Eq. 3); number of channels after data screening, azimuthal gap, and normalized misfit (nrms). For events with dip <30° and NDC < 40 %, the quantity plotted on the top is the updated scalar moment (Eq. 2). In such cases, the dip of the reference solution is indicated besides the name of the event, on top of the panel. The events included in this figure are the same as in Fig. 4: **a** 1994-Bolivia, **b** 1994-Kuril, **c** 1994-Sanriku, **d** 1995-Antofagasta, **e** 1996-Andreanoff, **f** 2001-Peru, **g** 2004-Sumatra, **h** 2005-Sumatra, **i** 2006-Tonga, **j** 2007-Sumatra, **k** 2011-Japan, and **l** 2013-Okhotsk

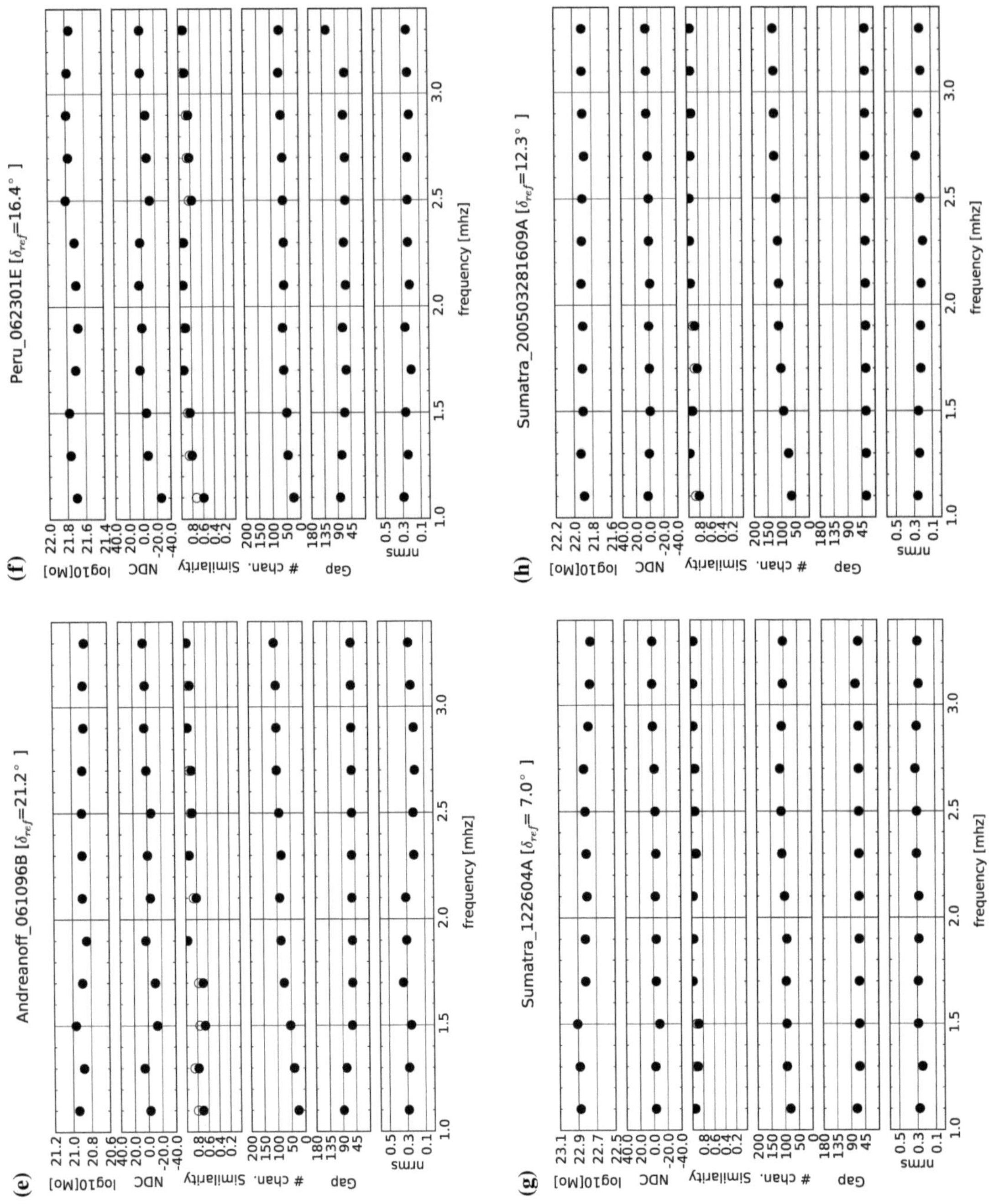

Figure 7
continued

slip, the orientation of the strike significantly rotated during rupture propagation, which is reflected in a large NDC component (−52 % in GCMT solution, −49 % in WCMT solution). As shown in Figs. 5b and 8b, our result shows that the moment varies by a factor of 1.7 and the non-DC component varies over a range (−46 %, 57 %), reflecting the complexity of this earthquake.

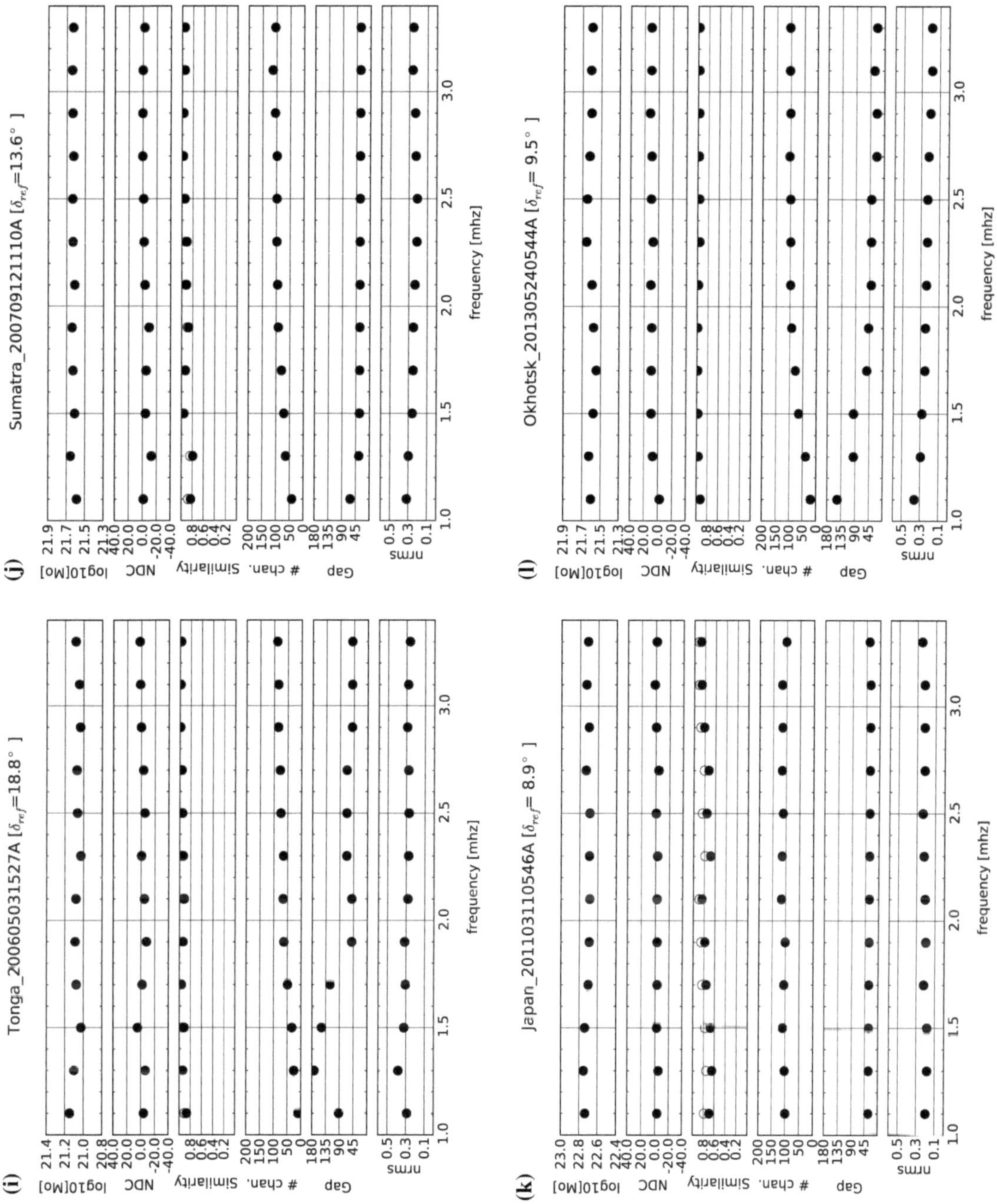

Figure 7
continued

3.8. 2009 Samoa Is. Earthquake, Mw 8.1 (Figs. 5g, 8g)

BEAVAN *et al.* (2010) and LAY *et al.* (2010a) demonstrated that this earthquake consists of an Mw = 8.1 outer-rise earthquake and one or two Mw = 7.8 thrust earthquakes on the mega-thrust boundary. The temporal and spatial separations of the normal-fault event and the thrust-fault events are about 100 s and 80 km, respectively. DUPUTEL *et al.* (2012b) confirmed the earlier results using their application of a multiple-source W-phase inversion method to this sequence (Figure S10 of DUPUTEL

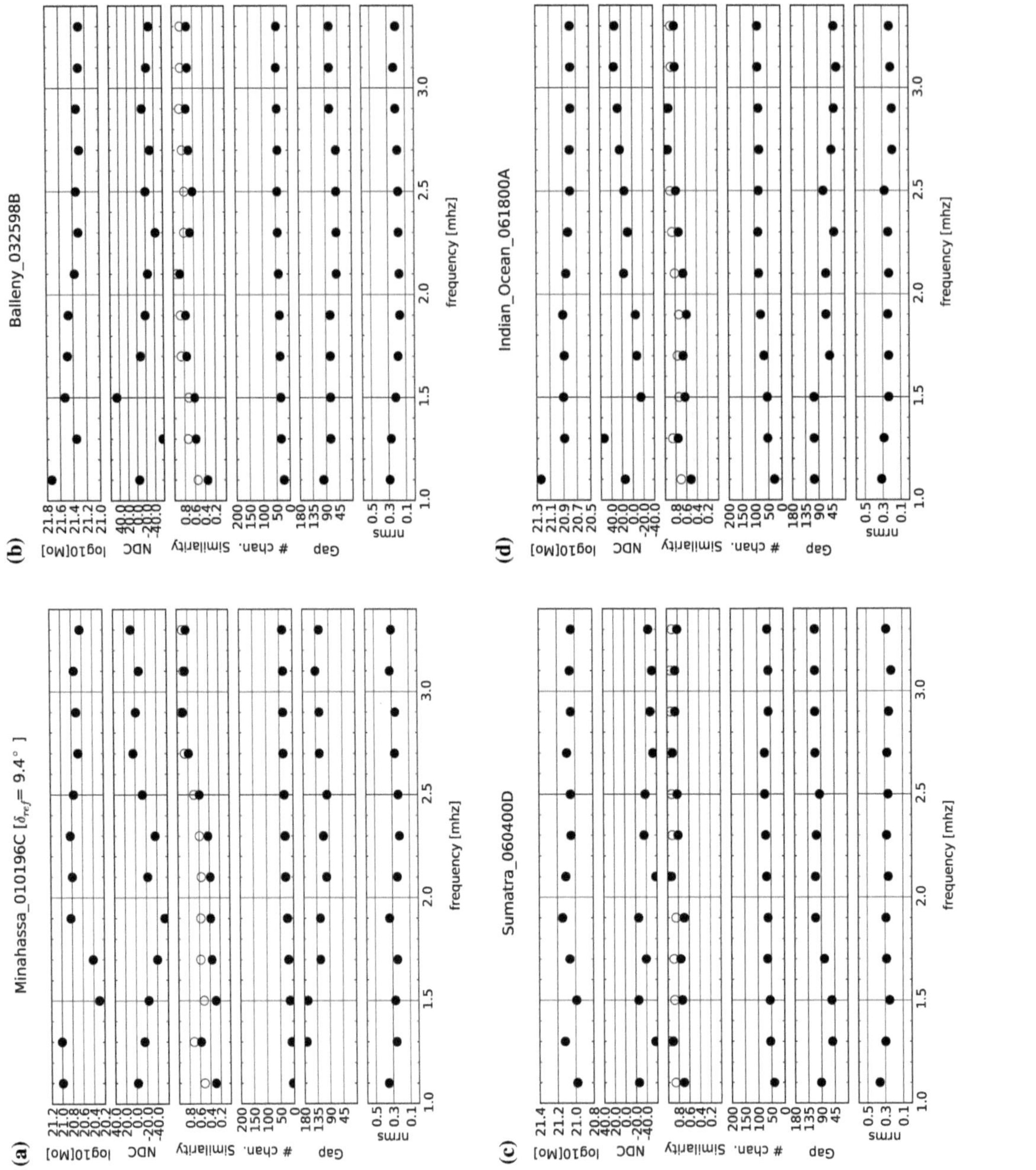

Figure 8
Same as Fig. 7, for events classified as "complex": **a** 1996-Minahassa, **b** 1998-Balleny, **c** 2000-Sumatra, **d** 2000-Indian Ocean, **e** 2004-Macquarie Island, **f** 2008-Sichuan and **g** 2009-Samoa

et al. 2012b). LAY *et al.* (2010a) showed that if two distinct events with different mechanisms occurred separated in time by about 100 s, the composite source can no longer be represented by a single moment tensor. Thus, if inversion is performed for a single moment tensor, as is done in the routine practice, the solution becomes frequency-dependent. Also, the solutions usually exhibit a large NDC component. As shown in Figs. 5g and 8g, the mechanism diagrams exhibit significant variations over the entire frequency band from 1.1 to 3.3 mHz, and the NDC component presents a large fluctuation

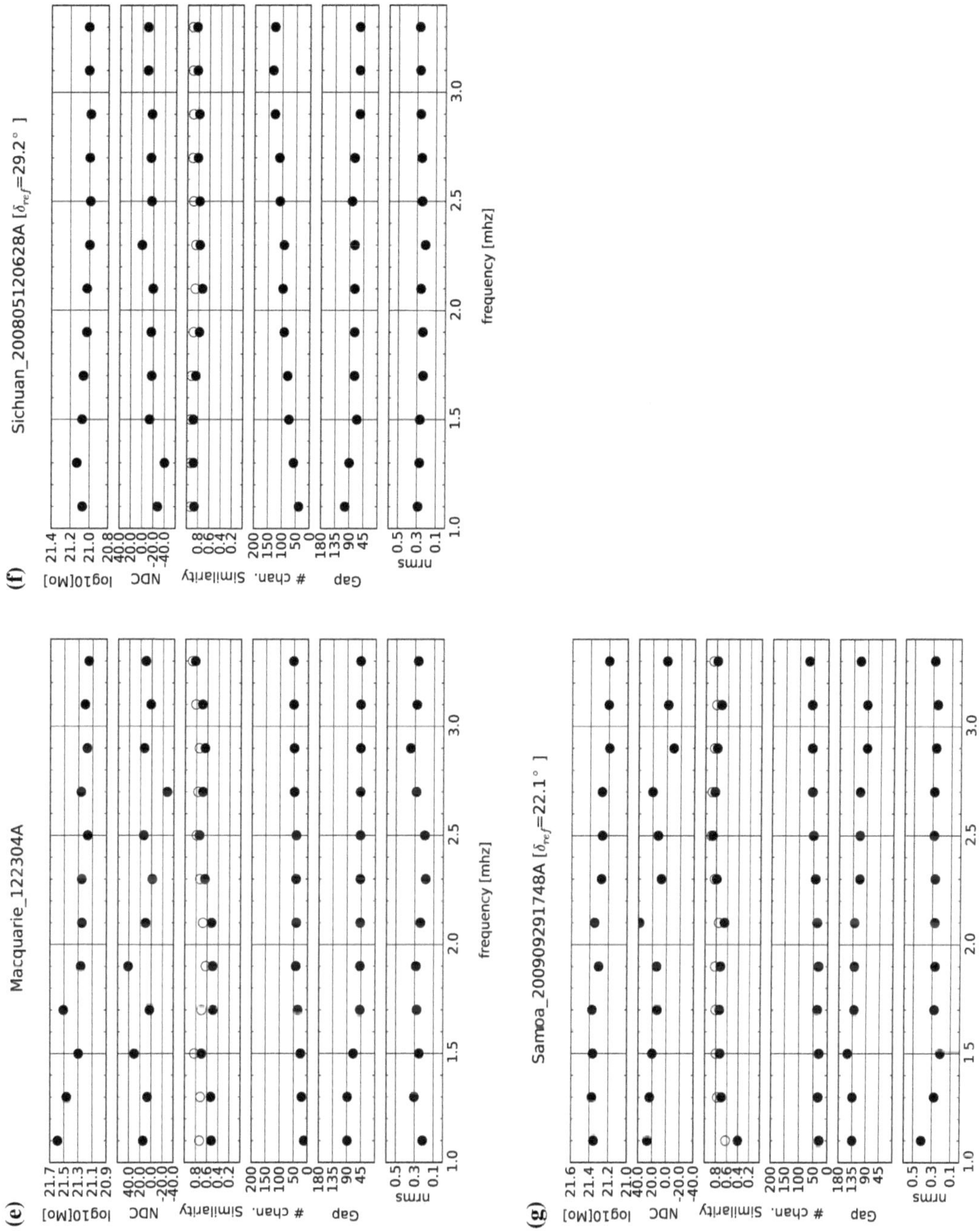

Figure 8
continued

from −10 to 38 % over the same frequency band. This behavior reflects the complexity of the source. Our method cannot determine the details of the complexity, but can diagnose it. To determine the details, we need more elaborate methods such as the one used by ABERCROMBIE *et al.* (2003) and DUPUTEL *et al.* (2012b), but at present the application of such methods cannot be done routinely.

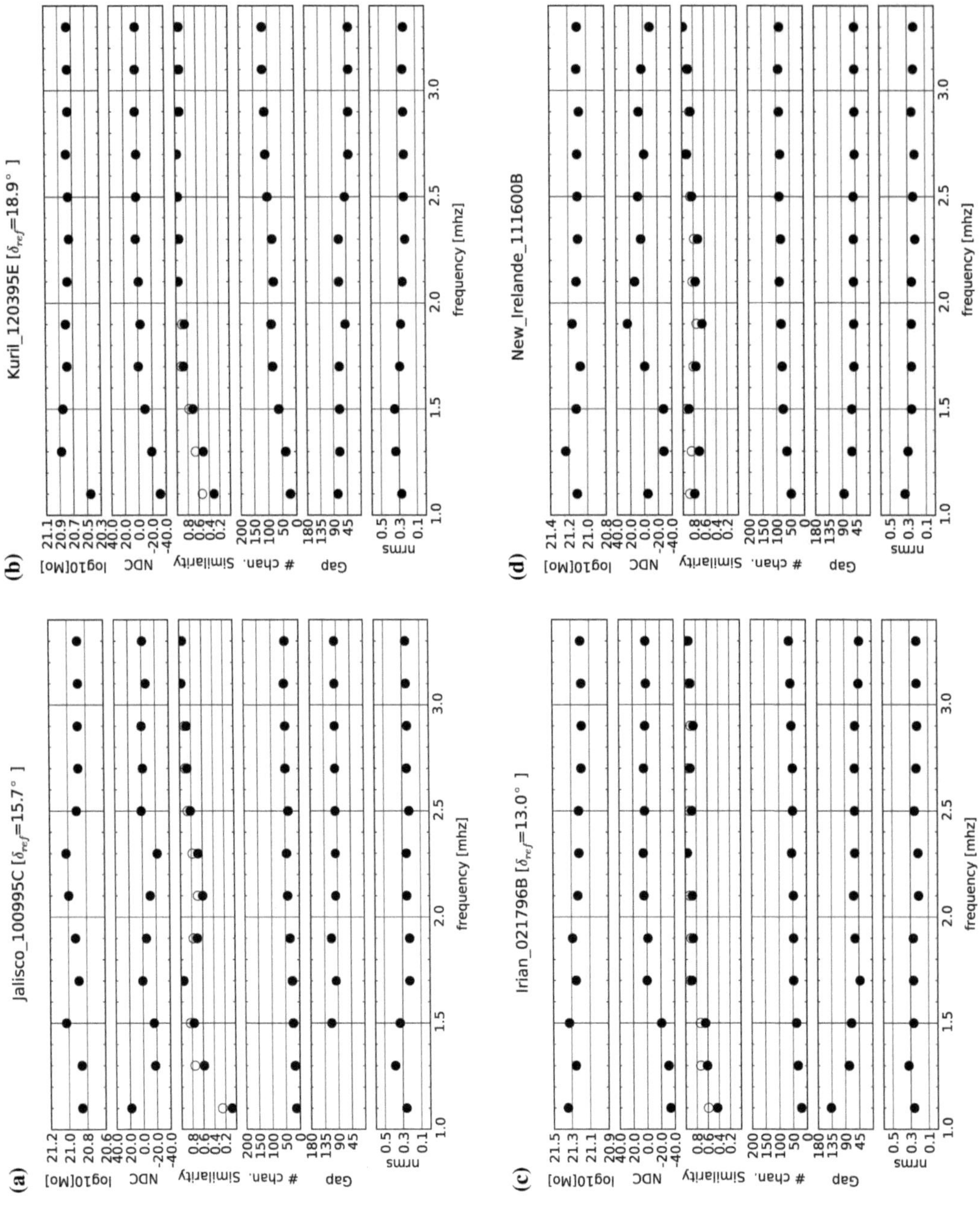

Figure 9

Same as Fig. 7, for events classified as "intermediate": **a** 1995-Jalisco, **b** 1995-Kuril, **c** 1996-Irian, **d** 2000-New Ireland, **e** 2003-Hokkaido, **f** 2006-Kuril, **g** 2007-Kuril, **h** 2007-Solomon, **i** 2007-Peru, **j** 2010-Maule and **k** 2012-Sumatra

4. *Classification*

As we have demonstrated above, the amount of non-double-couple (NDC) and the similarity parameter, α, allow us to diagnose source geometrical complexity. Figure 3 plots the events on a diagram with the minimum alpha value on the abscissa and the largest NDC value on the ordinate. Events appearing on the right side, (α ∼ 1), and at mid-height, (NDC ∼ 0), near the star symbol on this diagram, are simple events showing a stable, pure double-couple focal mechanism over the whole frequency

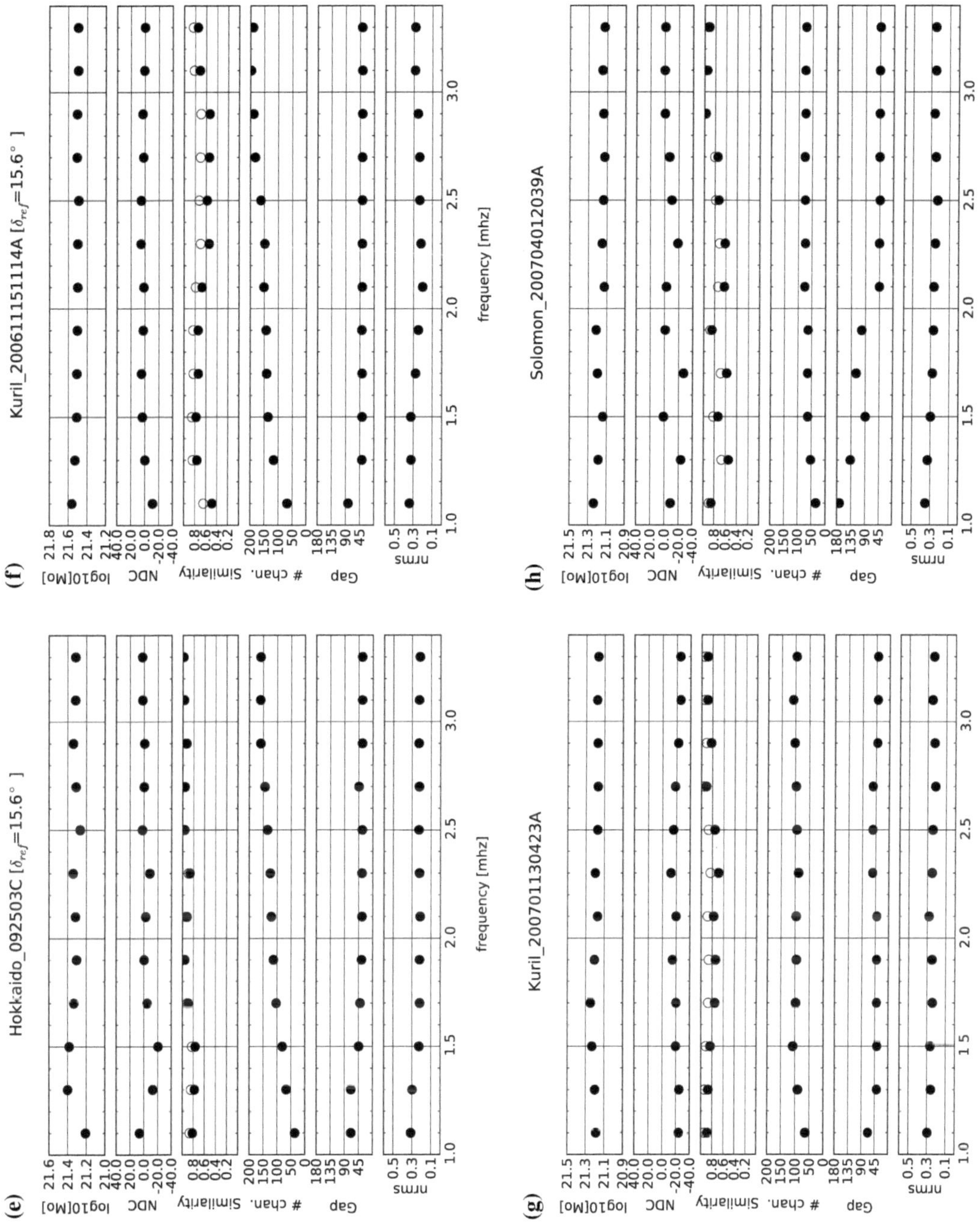

Figure 9
continued

band. Events far from that point present either significant non-double-couple or focal mechanisms that vary with frequency. Along with the complete set of solutions for all the 30 events in the frequency band 1.1–3.3 mHz, we use this diagram as a guide for classification into three groups: simple, complex and intermediate. Figures 4, 5, and 6 show the variation of the focal mechanism for these three groups. Figures 7, 8, and 9 show the detail of the variation of the different parameters.

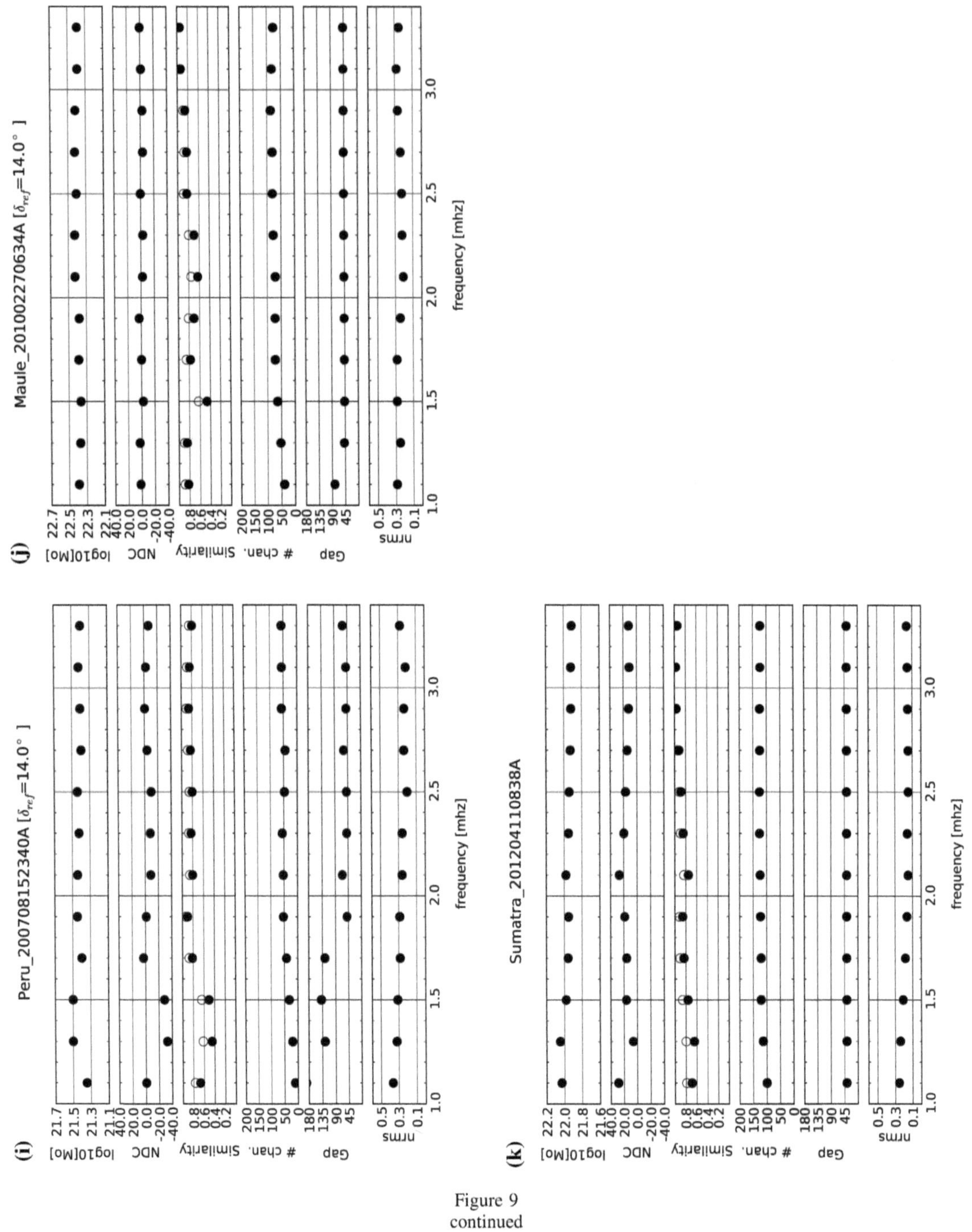

Figure 9
continued

4.1. *Simple Events (Figs. 4, 7)*

Twelve out of 30 events that we studied belong to this category and do not present any significant geometrical complexity: Bolivia-1994, Kuril-1994, Sanriku-1994, Antofagasta-1995, Andreanoff-1996, Peru-2001, Sumatra-2004, Sumatra-2005, Tonga-2006, Sumatra-2007, Japan-2011, and the recent Sea of Okhotsk-2013. The focal mechanism, the scalar moment, and the amount of NDC are very stable over the frequency band 1.3 to 3.3 mHz. As a general trend, at very long period ($\sim$1.1–1.3 mHz) the number of channels decreases sharply and the solution starts to be less stable. This is particularly true for the smallest events: Sanriku-1994 (Mw 7.7), Antofagasta-1995 (Mw 8.0), and Tonga-2006 (Mw 8.0), which exhibit some variability at very long period. It is noteworthy that the two deep-focus events, Bolivia-1994 (Mw 8.2) and Sea of Okhotsk-2013 (Mw 8.3), belong to this group. The source characteristics are remarkably stable over the entire frequency band and the amount of NDC is relatively modest ($\sim$5 and $\sim$10 %, respectively) for such large earthquakes. Remarkably, as mentioned above, most of the largest events (Mw > 8.5) belong to this group.

4.1.1 1994 Sanriku, Earthquake, Mw 7.7 (Figs. 4c, 7c)

As is mentioned above, we included the Sanriku 1994 earthquake in the dataset because it is one of the first events with a well-documented, significant after-slip (Heki *et al.* 1997; Heki and Tamura 1997). These authors report a 1-year after-slip comparable to the co-seismic slip. On the time scale we are looking at in this work, 300–1,000 s, we do not see substantial geometrical complexity. The scalar moment we observe is about 4.0×10^{20} Nm and it remains stable down to 1.5 mHz, which is our lower-limit frequency for this relatively small event. This value is similar to the one for GCMT (4.9×10^{20} Nm). The amount of NDC is also stable and mostly smaller than 5 %. There is a slight rotation of the focal sphere at frequencies lower than 2 mHz (Figs. 4c, 7c), but it remains marginal. The similarity index is larger than 0.8. Our result indicates that there was no significant slip on the scale of hundreds of seconds. This conclusion is not in contradiction with the works quoted above. Given the relatively low magnitude of this earthquake (Mw 7.7), we cannot determine the behavior at longer periods with the narrow band approach that we use here.

4.1.2 2001 Peru Earthquake, Mw 8.4 (Figs. 4f, 7f)

Robinson *et al.* (2006) found a complex rupture distribution for this event. They argued that the rupture was stalled by a subducting fracture zone, which caused a complex spatial and temporal rupture pattern. Back projection images found by Lay *et al.* (2010b) also suggest a spatially complex rupture distribution. However, neither of these two studies find significant variability of the focal mechanism, either on time or space.

4.2. Complex Events

Seven events exhibited considerable variations of the focal parameters with frequency: Minahassa-1996 (Mw 7.9), Balleny-1998 (Mw 8.1), Sumatra-2000 (Mw 8.1), Indian Ocean-2000 (Mw 7.9), Macquarie Island-2004 (Mw 8.1), Sichuan-2008 (Mw 7.9) and Samoa-2009 (Mw 8.1). We classify them as "complex" events. Not surprisingly, we find in this group the Balleny-1998 and the Samoa-2009 earthquakes, which we already discussed above as examples of earthquakes with known source complexity. Besides these, two other events from this group have known evidence for source complexity: the Indian Ocean-2000 and Macquarie Island-2004 earthquakes. Their complexity is successfully detected by significant variation of the source parameters with frequency.

4.2.1 1998 Minahassa Earthquake, Mw 7.9 (Figs. 5a, 8a)

This event shows a relatively simple behavior at periods shorter than 350 s. In this frequency band, the NDC parameter is lower than 8 %, the similarity index larger than 0.8, the scalar moment is nearly constant (between 4×10^{20} and 5×10^{20} Nm), and the focal mechanism presents only minor variations. At longer periods, however, both the scalar moment and the focal mechanism exhibit large variations and the NDC attains more than 50 %. Little is known about this event. Gómez *et al.* (2000) used teleseismic data (5–125 s) and performed body-wave inversion to retrieve a double-couple focal mechanism. Their result is in very good agreement with the GCMT and with our shortest-period solutions. More detailed analysis at long periods is warranted.

4.2.2 2000 Sumatra Earthquake, Mw 8.1 (Figs. 5c, 8c)

This is another event known to present some source complexity. Abercrombie *et al.* (2003) found that it is composed of two distinct sub-events. The first has a strike-slip focal mechanism and the second, slightly delayed in time (15 s), is a thrust. The seismic moment is relatively stable with some variation for frequencies below 1.5 mHz. The focal mechanisms are also quite regular, but they present systematically a very significant amount of NDC (~40 %). This is certainly related to the multiple events with different focal mechanisms. The similarity index is larger than 0.75 over the complete frequency band we used.

4.2.3 2000 Indian Ocean Earthquake, Mw 7.9 (Figs. 5d, 8d)

This earthquake is known as a complex multiple event (Abercrombie *et al.* 2003). The scalar moment increases by a factor of 1.3 when passing from 3.3 to 1.3 mHz. The amount of NDC changes gradually, but significantly between −20 % and +20 %. This is a typical case in which the NDC parameter reflects the existence of multiple events having different fault planes. The similarity index shows large variations (0.6–1.0), reflecting the visible variation of the focal mechanism (Fig. 5d).

4.2.4 2004 Macquarie Island Earthquake, Mw 8.1 (Figs. 5e, 8e)

Robinson (2011) studied this earthquake using both body waves and mantle waves and found that it consisted of two distinct events on a curved fossil fracture zone. The first event is a strike-slip event (s/d/r = 160°/86°/5°) and the second event involves an oblique mechanism (s/d/r = 178°/54°/65°). Figure 5f shows that both the scalar moment and NDC vary quite substantially and irregularly. The scalar moment varies by a factor of 2.0 and the NDC from −28 to 41 %, while the similarity index decreases to 0.6. Since the similarity is defined with respect to the broadband solution, this indicates that the solutions determined with the 1-mHz bandwidth vary significantly from the broadband solution, as shown in Fig. 5f.

4.2.5 2008 Sichuan Earthquake, Mw 8.1 (Figs. 6f, 9f)

As shown by many investigators, this earthquake involved several faults. It started out as a thrust event, but as the rupture propagated northeast, it turned into a strike-slip event (Hao *et al.* 2009; Pei *et al.* 2009; Hashimoto *et al.* 2009; Liu-Zeng *et al.* 2010; Kurahashi *et al.* 2010; De Michele *et al.* 2010). Figures 6h and 9h show our results. The moment, M_0, varies by a factor of 1.2 from 3.3 mHz to 1.5 mHz, and the NDC component varies from −20 to 0 %. The variation is not as wild as for the 2009 Samoa earthquake, but this result reflects a significant change in the mechanism during rupture propagation, and the possible simultaneous activation of more than a single fault plane.

4.3. Intermediate Case (Figs. 6, 9)

The remaining 11 events exhibited moderate variations of the source parameters over the frequency band: Jalisco-1995, Kuril-1995 (Mw 8.0), Irian-1996 (Mw 8.2), New Ireland-2000 (Mw 8.0), Hokkaido-2003 (Mw 8.3), Kuril-2006 (Mw 8.3), Kuril-2007 (Mw 8.1), Solomon-2007 (Mw 8.1), Peru-2007 (Mw 8.0), Maule-2010 (Mw 8.8) and Sumatra-2012 (Mw 8.6).

4.3.1 2003 Hokkaido (Tokachi-oki) Earthquake, Mw 8.3 (Figs. 6e, 9e)

This event has a well-documented after-slip expanding away from the co-seismic rupture zone (Miyazaki *et al.* 2004; Sato *et al.* 2010). Our results exhibit moderate variations of the scalar moment (a factor of 1.3 from 3.3 to 1.5 mHz). The NDC parameter is largest at low frequency, −20 % at 1.5 mHz. This is clearly visible on Fig. 6e. These observations could be a symptom related to the after-slip mentioned above, but more detailed studies are required for confirmation.

4.3.2 2007 Solomon Island Earthquake, Mw 8.1 (Figs. 6h, 9h)

This event involved simultaneous subduction of two plates, the Woodlark Basin plate and the Australian plate, beneath the Pacific plate (Taylor *et al.* 2008; Furlong *et al.* 2009; Chen *et al.* 2009). The slip inversion by Furlong *et al.* (2009) indicates different rake angles for the Australia-Pacific boundary and the Woodlark Basin-Pacific boundary, which may be the cause of the large variation of similarity indices for this event (Fig. 9f).

4.3.3 2007 Peru Earthquake, Mw 8.0, (Figs. 6i, 9i)

Back projection results by LAY *et al.* (2010b) suggest some spatial complexity for the rupture patterns of this event, but do not require substantial changes in the mechanism. DUPUTEL *et al.* (2013) found an anomalously long source-time duration that may be a manifestation of the complexities seen by LAY *et al* (2010b).

5. Conclusions

As the quality of the global broadband seismic network has improved, we can now determine the moment tensor of large earthquakes with a relatively narrow frequency band, 1 mHz in our case. This allows us to investigate the frequency-dependent behavior of large earthquakes over the frequency band from 1.1 to 3.3 mHz.

We have demonstrated that the variation of the moment tensor as a function of frequency from 1.1 to 3.3 mHz can be used to quickly diagnose the complexity of large earthquakes. For about 50 % of the earthquakes with Mw $\geq$ 8 for the period from 1990 to 2013, all the attributes of the moment tensor, seismic moment, the amount of NDC component, and similarity index are very stable as a function of frequency. However, about 25 % of the events exhibited significant variations of the moment tensor as a function of frequency, which suggests significant changes in the mechanism, multiple events, or the combination of both. Other events exhibited moderate complexity. Our method alone cannot determine the details of the complexity, but provides impetus for more detailed body-wave modeling or long-period, multiple source inversion.

Acknowledgments

This work uses seismic time series from the Federation of Digital Seismic Network (FDSN) retrieved through the Incorporated Research Institutions for Seismology (IRIS) Data Management System (DMS) and from Geoscope. We acknowledge two anonymous reviewers for their comments that helped to improve the original manuscript. L.R. was partially supported by the Caltech Seismological Laboratory and by the Tectonic Observatory.

REFERENCES

ABERCROMBIE, R., M. ANTOLIK and G. EKSTRÖM, 2003, *The June 2000 Mw 7.9 earthquakes south of Sumatra: Deformation in the India–Australia Plate*, J. Geophys. Res. *108*, B1, 2018, doi:10.1029/2001JB000674.

AKI, K. and P.G. RICHARDS, 1980. Quantitative Seismology, 1st edition, W. H. Freeman and company, San Francisco, United States.

BARTH, A., F. WENTZEL and D. GIARDINI, 2007, *Frequency sensitive moment tensor inversion for light to moderate magnitude earthquakes in eastern Africa*, Geophys. Res. Lett., *34*, L15302. doi:10.1029/2007GL030359.

BEAVAN, J., X. WANG, C. HOLDEN, K. WILSON, W. POWER, G. PRASETYA, M. BEVIS, R. KAUTOKE; 2010, *Near-simultaneous great earthquakes at Tongan megathrust and outer rise in September 2009*, Nature, *466*, 7309; 959–963.

CHEN, T., A. V. NEWMAN, L. FENG, and H. M. FRITZ, 2009, *Slip distribution from the 1 April 2007 Solomon Islands earthquake: A unique image of near-trench rupture*, Geophys. Res. Lett., *36*, L16307, doi:10.1029/2009GL039496.

DE MICHELE, M., D. RAUCOULES, J. DE SIGOYER, M. PUBELLIER and N. CHAMOT-ROOKE, 2010, *Three-dimensional surface displacement of the 2008 May 12 Sichuan earthquake (China) derived from Synthetic Aperture Radar: evidence for rupture on a blind thrust*, Geophys. J. Int. doi:10.1111/j.1365-246X.2010.04807.x.

DUPUTEL, Z., L. RIVERA, H. KANAMORI, G.P. HAYES, B. HIRSORN and S. WEINSTEIN, 2011. *Real-time W Phase inversions during the 2011 Tohoku-oki earthquake*, Earth, Planets and Space, *63*, 7, 535–539.

DUPUTEL, Z., L. RIVERA, H. KANAMORI, and G. P. HAYES. 2012a, *W phase source inversion for moderate to large earthquakes*. Geophys. J. Int., *189*, 1125–1147, 2012.

DUPUTEL, Z., H. KANAMORI, V.C. TSAI, L. RIVERA, L. MENG, J.-P. AMPUERO and J. STOCK, 2012b. *The 2012 Sumatra great earthquake sequence, Earth Planet Sci. Lett.*, v. 351–352, p. 247–257.

DUPUTEL, Z, V. C. TSAI, L. RIVERA, and H. KANAMORI, 2013. *Using centroid time-delays to characterize source durations and identify earthquakes with unique characteristics*, Earth Planet Sci. Lett., in press.

DZIEWONSKI, A. M. and J. H. WOODHOSE, 1983, *An experiment in systematic study of global seismicity: centroid-moment tensor solutions for 201 moderate and large earthquakes of 1981. J. of Geophys. Res.*, *88*, B4, 3247–3271.

DZIEWONSKI, A. M., T.-A. CHOU and J. H. WOODHOUSE, 1981, *Determination of earthquake source parameters from waveform data for studies of global and regional seismicity*, J. Geophys. Res., *86*, 2825–2852. doi:10.1029/JB086iB04p02825.

DZIEWONSKI, A.M., FRANZEN, J.E., WOODHOUSE, J.H., 1984. *Centroid-moment tensor solutions for January–March 1984*. Phys. Earth Planet. Int. *34*, 209–219.

EKSTRÖM, G., M. NETTLES, A.M. DZIEWONSKI, 2012. *The global CMT project 2004–2010: Centroid-moment tensors for 13,017 earthquakes*. Physics of the Earth and Planetary Interiors, 200–201, 1–9.

FURLONG, K., T. LAY and C. J. AMMON' 2009. *A Great Earthquake Rupture Across a Rapidly Evolving Three-Plate Boundary*. Science 10 April 2009: 226–229. doi:10.1126/science.1167476.

GCMT, 2013, Global CMT Project, www.globalcmt.org.

GÓMEZ, J. M., R. MADARIAGA, A. WALPERSDORF and e. CHALARD, 2000, *The 1996 Earthquake in Sulawesi, Indonesia*. Bull. Seism. Soc. Am., *90*(3), 739–751.

Hao, K. X., H. Si, H. Fujiwara, and T. Ozawa, 2009, *Coseismic surface-ruptures and crustal deformations of the 2008 Wenchuan earthquake Mw7.9, China*, Geophys. Res. Lett., *36*, L11303, doi:10.1029/2009GL037971.

Hashimoto, M., M. Enomoto and Y. Fukushima., 2009; *Coseismic Deformation from the 2008 Wenchuan, China, Earthquake Derived from ALOS/PALSAR Images*, Tectonophysics, doi:10.1016/j.tecto.2009.08.034.

Hayes, G., L. Rivera, and H. Kanamori, 2009. *Source Inversion of the W-Phase : Real-time Implementation and Extension to Low Magnitudes*. Seismological Research Letters, *80*(5), 817.

Heki, K., S. Miyazaki and H. Tsuji, S, 1997, *Fault slip following an interplate thrust earthquake at the Japan Trench*, Nature, *386*, 595–597, doi:10.1038/386595a0.

Heki, K. and Y. Tamura, 1997, *Short term afterslip in the 1994 Sanriku-Haruka-oki earthquake*, Geophys. Res. Lett., *24* (24), 3285–3288.

Henry, C., Das, S., Woodhouse, J.H., 2000. *The great March 25, 1998, Antarctic Plate earthquake: moment tensor and rupture history*. J. Geophys. Res. *105*, 16097–16118.

Hjörleifsdóttir, V., H. Kanamori, and J. Tromp, 2009. *Modeling 3-D wave propagation and finite slip for the 1998 Balleny Islands earthquake*, J. Geophys. Res., *114*, B03301, doi:10.1029/2008JB005975.

Jeffreys, H., 1931, Cartesian Tensors. Cambridge University Press. 93 pages., 1st edition, Cambridge, Great Britain.

Kagan, Y. Y., 1991. *3-D rotation of double-couple earthquake sources*, Geophys. J. Int., *106*(3), 709–716.

Kanamori, H., and J.W. Given, 1982, *Use of long-period surface waves for rapid determination of earthquake source parameters. Preliminary determination of source mechanisms of large earthquakes (MS > = 6.5) in 1980*. Phys. Earth Planet. Int., *30*, 260–268.

Kanamori, H. and L. Rivera, 2008, *Source inversion of W phase speeding up seismic tsunami warning*. Geophys. J. Int., *175*, 222–238.

Kurahashi S. and K. Irikura, 2010, *Characterized Source Model for Simulating Strong Ground Motions during the 2008 Wenchuan Earthquake*, Bull. Seism. Soc. Am. *100*, 2450–2475; doi:10.1785/0120090308.

Lambotte, S., L. Rivera, and J. Hinderer, 2006. *Rupture length and duration of the 2004 Aceh-Sumatra earthquake from the phases of the Earth's gravest free oscillations*. Geophys. Res. Lett., *33*, L03307, doi:10.1029/2005GL024090.

Lambotte, S., L. Rivera, and J. Hinderer, 2007, *Constraining the Overall Kinematics of the 2004 Sumatra and the 2005 Nias Earthquakes Using the Earth's Gravest Free Oscillations*. Bull. Seism. Soc. Am., *97*,1A, S128–S138, doi: 10.1785/0120050621.

Lay, T., C.J. Ammon, H. Kanamori, L. Rivera, K. D. Koper and A.R. Hutko, 2010a. *The 2009 Samoa–Tonga great earthquake triggered doublet*. Nature, doi:10.1038/466931a.

Lay, T., C. J. Ammon, A. R. Hutko, and H. Kanamori, 2010b. *Effects of kinematic constraints on teleseismic finite-source rupture inversions: Great Peruvian earthquakes of 23 June 2001 and 15 August 2007*, Bull. Seism. Soc. Am., *100*, 969–994, doi:10.1785/0120090274.

Liu-Zeng, J., L. Wen, J. Sun, Z. Zhang, G. Hu, X. Xing, L. Zeng, and Q. Xu, 2010; *Surficial Slip and Rupture Geometry on the Beichuan Fault near Hongkou during the M_w 7.9 Wenchuan Earthquake, China*, Bull. Seism. Soc. Am., *100*, 2615–2650; doi:10.1785/0120090316.

Lundgren, P and D. Giardini, 1995, *The June 9 Bolivia and March 9 Fidji deep earthquakes of 1994 : I. Source process*. Geophys. Res. Lett., *22* (16), 2241–2244.

Miyazaki, S., P. Segall, J. Fukuda, and T. Kato, 2004. *Space time distribution of afterslip following the 2003 Tokachi-oki earthquake: Implications for variations in fault zone frictional properties*, Geophys. Res. Lett., *31*, L06623, doi:10.1029/2003GL019410.

Okal, E.A., and S. Stein, 2009. *Observations of ultra-long period normal modes from the 2004 Sumatra-Andaman earthquake*, Phys. Earth Planet. Inter., *175*, 53-62.

Park, J., T. R. A. Song, J. Tromp, E. Okal, S. Stein, G. Roult, E. Clevede, G. Laske, H. Kanamori, P. Davis, J. Berger, C. Braitenberg, M. Van Camp, X. Lei, H. Sun, H. Xu, and S. Rosat, 2005. *Earth's free oscillations excited by the 26 December 2004 Sumatra-Andaman earthquake*, Science *308*, 1139–1144.

Pei, S., et al., 2009; *Three-dimensional seismic velocity structure across the 2008 Wenchuan Ms 8.0 earthquake, Sichuan, China*, Tectonophysics, doi:10.1016/j.tecto.2009.08.039.

Robinson, D.P, Das, S, Watts, AB, 2006. *Earthquake rupture stalled by a subducting fracture zone*, Science. 1203–1205 doi:10.1126/science.1125771.

Robinson, D.P., 2011. *A rare great earthquake on an oceanic fossil fracture zone*, Geophys. J. Int. *186*, 1121–1134 doi:10.1111/j.1365-246X.2011.05092.x.

Sato K., T. Baba, T. Hori, M. Hyodo' and Y. Kaneda; 2010, *Afterslip distribution following the 2003 Tokachi-oki earthquake: An estimation based on the Green's functions for an inhomogeneous elastic space with subsurface structure*; Earth Planets Space, *62* (12), 923–932, doi:10.5047/eps.2010.11.007.

Taylor, F., R. W. Briggs, C. Frohlich, A. Brown, M. Hornbach, A. K. Papabatu, A. J. Meltzner and D. Billy, 2008. *Rupture across arc segment and plate boundaries in the 1 April 2007 Solomons earthquake*. Nature Geoscience *1*, 253–257.

Tsai, Victor C., M. Nettles, G. Ekström, and A. M. Dziewonski; 2005; *Multiple CMT source analysis of the 2004 Sumatra earthquake*; Geophys. Res. Lett., *32*, 17; doi:10.1029/2005GL023813; L17304.

Vigny, C., A. Socquet, S. Peyrat, J.-C. Ruegg, M. Métois, R. Madariaga, S. Morvan, M. Lancieri, R. Lacassin, J. Campos, D. Carrizo, M. Bejar-Pizarro, S. Barrientos, R. Armijo, C. Aranda, M.-C. Valderas-Bermejo, I. Ortega, F. Bondoux, S. Baize, H. Lyon-Caen, A. Pavez, J. P. Vilotte, M. Bevis, B. Brooks, R. Smalley, H. Parra, J.-C. Baez, M. Blanco, S. Cimbaro, E. Kendrick' 2011, *The 2010 M_w 8.8 Maule Megathrust Earthquake of Central Chile, Monitored by GPS*. Science, *332* (6036), 1417–1421, doi:10.1126/science.1204132.

WCMT, 2013. Wphase source inversion. http://wphase.unistra.fr.

(Received June 25, 2013, revised December 24, 2013, accepted December 28, 2013, Published online January 31, 2014)

Pure Appl. Geophys. 171 (2014), 2841–2862

DOI 10.1007/s00024-014-0804-0

Pure and Applied Geophysics

Radiated Energy of Great Earthquakes from Teleseismic Empirical Green's Function Deconvolution

Annemarie S. Baltay,[1,3] Gregory C. Beroza,[1] and Satoshi Ide[2]

Abstract—We expand on the empirical Green's function deconvolution method of Ide *et al.* (2011) to estimate radiated energy for the six largest earthquakes worldwide over the last 10 years: 2011 M_w 9.0 Tohoku-Oki, 2004 M_w 9.1 Sumatra, 2010 M_w 8.8 Maule, 2005 M_w 8.7 Nias, 2007 M_w 8.5 Bengkulu, and 2012 M_w 8.6 off-Sumatra. Deconvolution of *P*, *SV* and *SH* components gives consistent energy results that are comparable to estimates found independently by other researchers. Apparent stress for the five great thrust earthquakes is between 0.4 and 0.8 MPa, while the 2012 off-Sumatra strike-slip earthquake has a higher apparent stress of 3 MPa, which is consistent with other studies that find a tendency for strike-slip events to be more energetic. Our results are within the spread of apparent stress from the wider global earthquake population over a large magnitude range. The azimuthal distribution of energy in each case shows signs of directivity, and in some cases, shows less energy radiated in the trench-ward direction, which may suggest enhanced tsunami potential. We find that eGfs as small as ~M 6.5 can be used for teleseismic deconvolution, and that an eGf-mainshock magnitude difference of 1.5 units yields stable results. This implies that M 8 is the minimum mainshock size for which teleseismic eGf deconvolution will work well. We propose that a database of eGf events could be used to calculate radiated energy and apparent stress of great, hazardous events in near real time, i.e., promptly enough that it could contribute to rapid response measures.

1. Introduction

Great earthquakes, $M_w \geq 8.5$, are rare and potentially devastating events. Until the occurrence of the 2004 M_w 9.1 Sumatra earthquake, no broadband digital recordings existed for these very large events, and precise source parameters were unavailable. The 2004 Sumatra event came with surprises: an extended rupture zone of almost 1,300 km in length, and a large tsunami that caught much of the Indian Ocean Basin unprepared (Ishii *et al.* 2005). The 2011 M_w 9.0 Tohoku-Oki earthquake had a much more compact source zone, and in contrast to the 2004 Sumatra event, generated high frequencies at the deep edge of the rupture with slower, higher amplitude slip at the trench, with the latter making an unfortunately large contribution to the devastating tsunami that followed. Although the site of the Tohoku-Oki event was deemed the most likely location for a future magnitude (M) 7.5–8 earthquake, the extent of rupture and magnitude of slip was unanticipated [Lay *et al.* 2013; National Seismic Hazard Maps for Japan (2005), http://www.jishin.go.jp/main/index-e.html]. The 2012 M_w 8.6 off-Sumatra event is the largest strike-slip earthquake ever recorded, and while it caused little damage, it raises the prospect that the maximum possible size of continental strike-slip earthquakes may be underestimated. Because each of these earthquakes is uncommon in the instrumental record, they merit intense study.

Radiated seismic energy is a scalar measure of the strength of an earthquake that can be related to other physical parameters, such as fracture energy, rupture velocity and stress drop. By studying the relationship between radiated seismic energy, a dynamic parameter, and seismic moment, a static measure of the earthquake's size, we can gain insight into the dynamic processes of an earthquake. If all earthquakes follow self-similar physics, then simple power-law scaling will explain earthquake size-

---

For submission to the special issue of Pure and Applied Geophysics (PAGEOPH), entitled "Earthquake Source Physics on Various Scales."

Electronic supplementary material The online version of this article (doi:10.1007/s00024-014-0804-0) contains supplementary material, which is available to authorized users.

[1] Department of Geophysics, Stanford University, Stanford, CA, USA. E-mail: abaltay@usgs.gov
[2] Department of Earth and Planetary Science, University of Tokyo, Tokyo, Japan.
[3] *Present Address*: United States Geological Survey, Menlo Park, CA, USA.

dependence. If not, then it would point to an intrinsic length scale related to some physical factor controlling the earthquake faulting process. Whether all earthquakes have the same ratio of seismic energy to moment, E_R/M_o, remains an open question (e.g. Boatwright and Choy 1986; Abercrombie 1995; Ide and Beroza 2001; Choy *et al.* 2006; Bormann and Di Giacomo 2011), and how the very largest earthquakes radiate energy is not well understood, due to the fact that great earthquakes are uncommon and it is challenging to measure their energy accurately.

Strike-slip events have been shown to have larger scaled energy than dip-slip events. Choy and Boatwright (1995) estimate an order of magnitude higher scaled energy for strike-slip vs. reverse-faulting events, and Pérez-Campos and Beroza (2001) found this difference to be a factor of five, using a revised method to account for potential biases. Convers and Newman (2011) determine strike-slip events to be twice as energetic as thrust events. The largest strike-slip earthquakes occurring in oceanic lithosphere may be associated with emergent or diffuse plate boundaries, rupturing cold, intact crustal material. The associated earthquakes may have high stress drop and large scaled energy. In contrast, the largest thrust events tend to occur on mature plate boundaries that repeatedly rupture with lower stress drop and scaled energy.

A second class of earthquakes that may display abnormal ratios of E_R/M_o are "tsunami earthquakes", which excite a much larger tsunami than expected for their magnitude (Kanamori 1972). Newman and Okal (1998) found that many of these tsunami earthquakes have anomalously low ratios of radiated energy to seismic moment, which provides a motivation for rapid determination of this source parameter. On the other hand, great earthquakes may be tsunamigenic without being deficient in radiated energy. In either situation, radiated energy provides important information that is highly relevant to understanding both tsunamigenesis and earthquake dynamics.

Estimates of energy for large earthquakes are typically computed by making path corrections to account for attenuation and geometrical spreading, as well as site and surface corrections, using teleseismically recorded data (Boatwright and Choy 1986; Perez-Campos *et al.* 2003; Convers and Newman 2011). Empirical Green's functions have been employed as an effective means for making these corrections for local earthquakes, and avoid the need for explicit determination of path corrections or radiation pattern coefficients. Use of eGfs assumes that a smaller, collocated event, usually at least one magnitude unit smaller, can approximate the point-source response to a double-couple source (e.g. Hartzell 1978; Courboulex *et al.* 1996; Hough 2001; Kane *et al.* 2011). Then, the source time function of the larger mainshock event can be determined by deconvolving the smaller Green's function event from the larger event.

Empirical Green's functions have been used profusely on a local scale (within a few hundred km) to correct to source spectra of earthquakes up to ~M 7 (Venkataraman *et al.* 2002; Ide *et al.* 2003; Baltay *et al.* 2010, 2011); however, for great earthquakes in which the rupture dimension can no longer be considered negligible compared to the source-station distance, the local empirical Green's function assumptions are not met. Ammon *et al.* (1993) showed that teleseismic surface wave deconvolution could work between earthquakes of magnitude 7.2 and 7.4 with eGf events of magnitude ~6 to resolve fault finiteness. When considered at teleseismic distances ($\Delta > 30°$), even very large ruptures can be considered point sources, which leads us to consider large earthquakes (~M 6.5–7.5) as empirical Green's functions, to correct for path and site effects of great earthquakes.

In this study, we expand the teleseismic eGf deconvolution approach (Ide *et al.* 2011) to model source spectra of the six recent great earthquakes. We can accurately estimate the radiated seismic energy using either P or S wave groups, and find the scaled energy, E_R/M_o, to be fairly consistent for the five thrust events, and about a factor of five larger for the 2012 off-Sumatra strike slip event. Our estimates of E_R are consistent with those from previous teleseismic studies of great earthquakes, and the higher energy associated with the 2012 off-Sumatra strike slip event is in agreement with the mechanism dependence shown in previous studies (Choy and Boatwright 1995; Pérez-Campos and Beroza 2001; ConversNewman 2011). With the dense azimuthal coverage of the global seismic network (GSN), we find that some of the events, especially the 2011 Tohoku-Oki earthquake, have strong directivity, consistent with the known direction of rupture. Others show less clear azimuthal dependence, such as the 2007

Bengkulu earthquake, or no variation with azimuth, as in the case of the 2012 off-Sumatra event, which ruptured multiple conjugate fault planes. We find at least a factor of two, and up to an order of magnitude, difference in energy from maximum to minimum with some events, implying that directivity effects should be considered when estimating energy teleseismically. Finally, we suggest that this teleseismic eGf deconvolution could be adapted for use in real time, for rapid characterization of the dynamic properties of an earthquake, as also pointed out by Di Giacomo *et al.* (2010) and Convers and Newman (2011). An eGf catalog could be built of preprocessed eGf events for all regions in the world that are available, covering most of the regions that are most likely to host a great earthquake in the future. Since so little is known about the great, hazardous events, any information discerned quickly after the occurrence of an event has the potential to aid in rapid response.

2. *Teleseismic eGf Deconvolution*

We consider six great earthquakes recorded worldwide over the past 10 years: 2011 M_w 9.0

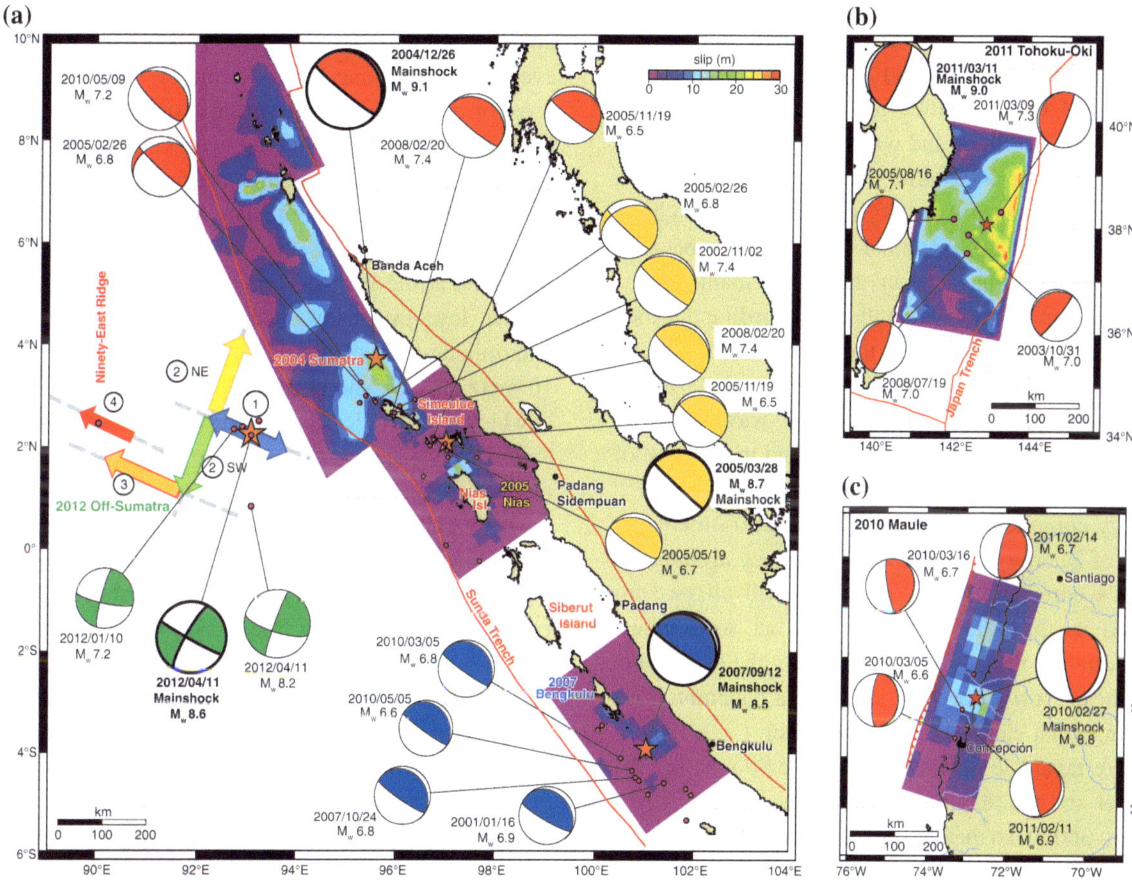

Figure 1

Maps of great earthquake locations, rupture history and gCMT locations and mechanisms of candidate eGf earthquakes. **a** Greater Sumatra map, showing: 2004 Sumatra earthquake and eGf events (*red* focal mechanisms), slip distribution from Ammon *et al.* (2005); 2005 Nias earthquake and eGf events (*yellow* focal mechanisms), slip distribution from Konca *et al.* (2007); 2007 Bengkulu earthquake and eGf events (blue focal mechanisms), slip distribution from Konca *et al.* (2008); and 2012 off-Sumatra earthquake and eGf events (*green* focal mechanisms), schematic of rupture propagation from Meng *et al.* (2012), who model the rupture as starting at (*1*), rupturing bi-laterally along the *blue arrow* before moving to a conjugate fault and rupturing again bilaterally to the northeast (*yellow arrow*) and southwest (*green arrow*) at the (*2*)s, then moving again to a conjugate fault at (*3*), *orange arrow*, before jumping to a parallel plane shown with the *red arrow* at (*4*). **b** Map of 2011 Tohoku-Oki rupture area and candidate eGf events. Slip distribution from Ide *et al.* (2011). **c** Map of 2010 Maule rupture area and candidate eGf events. Slip distribution from Hayes (2010) USGS Teleseismic model, as displayed by Moreno *et al.* (2010)

Tohoku-Oki, 2004 M_w 9.1 Sumatra, 2010 M_w 8.8 Maule, 2005 M_w 8.7 Nias, 2007 M_w 8.5 Bengkulu, and 2012 M_w 8.6 off-Sumatra, and nearby large earthquakes from M ~6.0 to ~7.9 as potential empirical Green's functions (Fig. 1). We estimate the energy of each mainshock by deconvolution of several candidate eGfs located near the mainshock, and with similar mechanism to that of the mainshock. The eGfs considered occur both before and after the mainshock, except for the 2004 Sumatra and 2010 Maule eGfs, which are all afterwards.

We use both *P* and *S* waves recorded by broadband global seismic stations in the GSN. For both the mainshock and eGf events, we analyze vertical (BHZ) records, windowed from 90 s before the *P* arrival to 360 s after, and *S*-horizontal components (BHE and BHN), rotated to radial (SV) and transverse (SH) directions and windowed from 90 s before the *S* arrival to 360 s after. We use recordings from stations of distance $\Delta = 30°$ to $\Delta = 90°$, which have relatively simple propagation characteristics because their turning points are in the lower mantle. This combination of window length and recording distance were chosen to maximize the signal of the *P* and *S* waves separately, without contamination of too many other phases. Closer than 30°, *P*-wave windows will contain *S*-waves within 270 s, and triplication of arrivals from upper mantle structure and reflected *PP* energy complicate the record; Beyond 90°, there is a loss of energy as body waves diffract around the earth's core. *P*, *pP*, and *sP* waves will all be contained in the vertical component, but for shallow earthquakes, these phases have nearly the same path, and hence arrive together to create a coherent wave package that can be analyzed together.

We taper the time series over the 90 s before the arrival and the 90 s of the end of the window (from 270 to 360 s), so that 270 s of record with no tapering are used, and calculate the spectrum of each component at each station. We smooth each spectrum into 90 bins that are evenly spaced in log frequency from 1 mHz to 10 Hz. Figure 2 shows example windowed *P* and *S* waveforms at one global station, KEV, for the 2011 Tohoku-Oki earthquake, and the four candidate eGf events.

We correct both the mainshock and eGf spectra together, such that the corrected spectrum of the eGf is an ideal Brune ω-squared model, as in Baltay *et al.* (2010), using the USGS centroid moment given in Tables 1, 2, 3, 4, 5, and 6 and a reference stress drop of 3 MPa to define the corner frequency. We follow Hanks and Thatcher (1972) and assume $\beta = 3{,}600$ m/s, so that corner frequency is,

$$f_c = \left(\frac{\Delta\sigma}{8.47M_o}\right)^{1/3}\beta. \quad (1)$$

In log–log frequency space, deconvolution of the time series is equivalent to subtraction of the eGf spectra from the mainshock spectra. In the deconvolution (subtraction), the correction is the same for both the eGf and the mainshock at each frequency, so that the relative spectral levels between the two events are maintained. We set the absolute moment level of all spectra based on the moment of the eGf from the USGS Centroid Moment Tensor (Fig. 4a). We then compare all of the mainshock spectra to investigate their low-frequency behavior. The frequency at which all the mainshock spectra, as measured at the various stations, converge is chosen as the low frequency cutoff for that eGf event, maximizing the coherency. This point is usually near 10 mHz (100 s), and represents the lowest frequency at which the eGf recording is above the noise, and the lowest frequency at which we use direct data for the mainshock spectrum.

Although we can measure the eGf spectrum down to the low-frequency cutoff near 10 mHz (100 s), the long-period moment level of the mainshock is not completely captured at these periods, due to the exceptionally long rupture durations of great earthquakes. The duration of the events we have studied range from 100 s for the 2007 M_w 8.5 Bengkulu earthquake (Konca *et al.* 2008) to over 500 s for the 2004 M_w 9.1 Sumatra event (Lay *et al.* 2005). For this reason, we extrapolate the mainshock spectrum to lower frequencies following an ω^{-2} trend, until the independent moment of both the eGf and mainshock are recovered, at a frequency of about 1–4 mHz (depending on the event, eGf and component). Figure 4 (and Figs. S2a–b, S4a–b, S6a–b, S8a–b and S10a–b) shows the extrapolated part of the spectra in thin lines. Because the eGf spectra are unstable below about 100 s, we are unable to adequately correct the mainshock for these very low frequencies. It is also

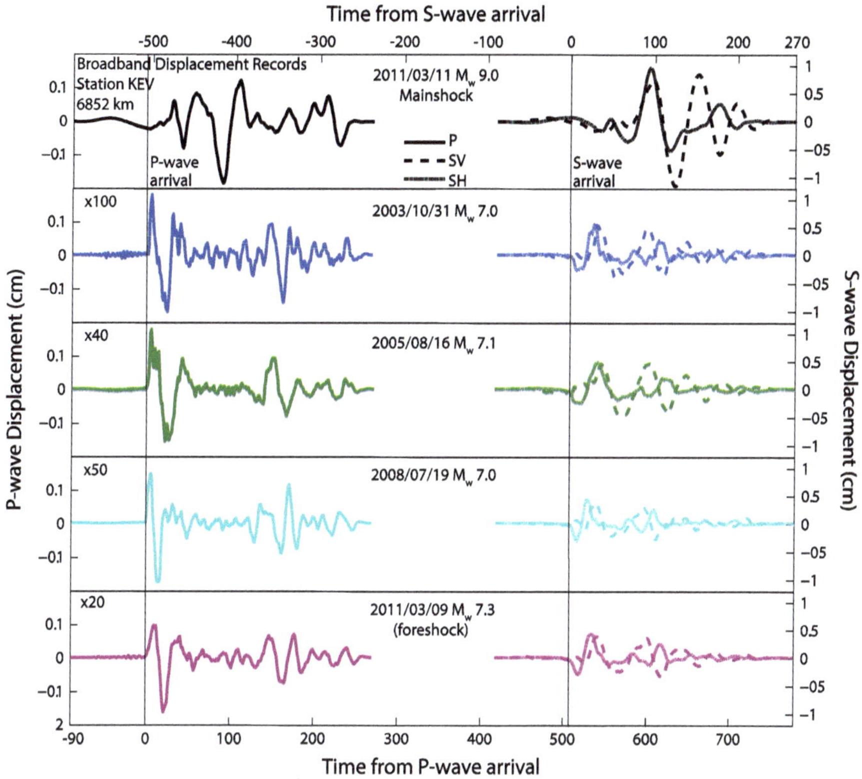

Figure 2
Broadband displacement waveforms from 2011 Tohoku-Oki mainshock (*top panel*) and four candidate eGf events, recorded at station KEV with a distance of 6,852 km. *Left* seismograms are *P*-wave vertical recordings, *right* are the SV and SH components, each windowed from 90 s before the first arrival to 270 s afterwards, and tapered. The eGf recordings are scaled by a factor ranging from 20× to 100×, shown in the *upper left corner* of each panel

Table 1

Radiated seismic energy of the 2011 M_w 9.0 Tohoku-Oki earthquake, corrected by each eGf, with given eGf moment and magnitude

Catalog information				Radiated energy (J $\times 10^{17}$)						
#	Date (Y/M/D)	M_o (Nm, $\times 10^{19}$)	M_w	*P*	SV	SH	Average	min	max	*N*
4	2003/10/31	3.4	7.0	12.19	9.58	10.58	10.73	9.14	12.59	28
6	2005/08/16	4.7	7.1	6.56	5.25	5.84	5.86	5.07	6.77	27
9	2008/07/09	3.8	7.0	7.90	6.50	6.59	6.97	6.03	8.06	32
10	2011/03/09	9.9	7.3	7.58	4.88	4.89	5.66	5.19	6.16	41
Average of each component				8.32	6.32	6.68				
Average of S-components					6.50					
Overall average							7.06	6.58	7.57	368

The $\log_{10}$ average is weighted azimuthally. *N* is the number of stations recording each eGf event (note that not all components at each station are always used, so that the sum of all components at all stations may be less than 3**N*). The min and max are the aleatory 95 % confidence intervals based on the $\log_{10}$ distribution of the energy estimates at individual stations. The 95 % confidence intervals are equal to ±2*$\log_{10}$ standard error (the data sample standard deviation over all three components and all *N* station divided by the square root of *N**3). They are symmetric in $\log_{10}$ space, and hence asymmetric in linear space, as presented here

possible that teleseismic spectra are deficient in low-frequency energy in the time window considered, due to possible interference effects from free surface reflections (Kanamori and Given 1981). This extrapolation accounts for only about 10–20 % of the total energy of the mainshock, depending on the

Table 2

Radiated seismic energy of the 2004 M_w 9.1 Sumatra earthquake

Catalog information				Radiated energy (J $\times 10^{17}$)						
#	Date (Y/M/D)	M_o (Nm, $\times 10^{19}$)	M_w	P	SV	SH	Average	min	max	N
2	2005/02/26	1.7	6.8	8.93	12.91	9.37	10.26	9.39	11.21	45
4	2005/11/19	0.62	6.5	6.52	9.48	6.67	7.45	6.73	8.23	41
5	2008/02/20	11	7.4	5.08	7.37	8.02	6.70	6.08	7.37	36
7	2010/05/09	8.5	7.2	4.46	6.38	5.56	5.41	4.72	6.20	36
Average of each component				6.03	8.71	7.27				
Average of S-components					7.96					
Overall average							7.25	6.84	7.69	463

Columns as described for Table 1

Table 3

Radiated seismic energy of the 2010 M_w 8.8 Maule earthquake

Catalog information				Radiated energy (J $\times 10^{17}$)						
#	Date (Y/M/D)	M_o (Nm, $\times 10^{19}$)	M_w	P	SV	SH	Average	min	max	N
1	2010/03/05	0.73	6.6	1.39	0.92	1.12	1.13	1.01	1.27	35
2	2010/03/16	1.20	6.7	2.28	1.80	1.77	1.93	1.72	2.17	35
4	2011/02/11	5.30	6.9	3.92	2.48	2.06	2.72	2.41	3.06	31
5	2011/02/14	1.70	6.7	2.92	1.78	1.57	2.01	1.75	2.32	30
Average of each component				2.45	1.65	1.59				
Average of S-components					1.62					
Overall average							1.86	1.74	1.99	383

Columns as described for Table 1

Table 4

Radiated seismic energy of the 2005 M_w 8.7 Nias earthquake

Catalog information				Radiated energy (J $\times 10^{17}$)						
#	Date (Y/M/D)	M_o (Nm, $\times 10^{19}$)	M_w	P	SV	SH	Average	min	max	N
2	2002/11/02 01:26:10	13	7.4	2.56	2.69	2.26	2.49	2.32	2.68	52
4	2005/02/26	1.7	6.8	3.28	4.08	3.46	3.59	3.42	3.77	64
8	2005/05/19	1.2	6.7	2.07	1.62	1.52	1.72	1.63	1.82	62
10	2005/11/19	0.62	6.5	1.98	1.85	1.58	1.80	1.70	1.89	57
13	2008/02/20	11	7.3	2.15	1.77	1.63	1.84	1.66	2.03	44
Average of each component				2.36	2.25	1.98				
Average of S-components					2.11					
Overall average							2.19	2.11	2.28	819

Columns as described for Table 1

station and eGf used (Fig. 4b), which is less than the variability of ±0.2 magnitude units in energy magnitude quoted in other studies (i.e. Di Giacomo *et al.* 2010). The close correspondence of the mainshock spectra to each other, when corrected by different eGf events, indicates that the corrected source spectra and

Table 5

Radiated seismic energy of the 2007 M_w 8.5 Bengkulu earthquake

Catalog information				Radiated energy (J $\times 10^{17}$)						
#	Date (Y/M/D)	M_o (Nm, $\times 10^{19}$)	M_w	P	SV	SH	Average	min	max	N
4	2001/01/16	2.0	6.8	1.66	1.05	1.28	1.31	1.18	1.45	46
6	2007/10/24	2.3	6.9	1.56	0.88	0.73	1.00	0.89	1.12	57
9	2010/03/05	1.6	6.8	2.00	1.45	1.12	1.48	1.36	1.62	55
10	2010/05/05	0.8	6.6	2.20	1.41	1.32	1.60	1.44	1.78	50
Average of each component				1.84	1.17	1.08				
Average of S-components					1.13					
Overall average							1.33	1.26	1.40	608

Columns as described for Table 1

Table 6

Radiated seismic energy of the 2012 M_w 8.6 off-Sumatra earthquake, corrected by the Mw 7.2 eGf

Catalog information				Radiated energy (J $\times 10^{17}$)						
#	Date (Y/M/D)	M_o (Nm, $\times 10^{19}$)	M_w	P	SV	SH	Average	min	max	N
4	2012/01/10	7.2	7.2	8.03	6.17	6.09	6.71	5.24	8.59	57
Average of S-components					6.13					
Overall average							6.71	5.24	8.59	164

Columns as described for Table 1

extrapolation to low frequencies are robust with respect to the eGf choice.

Once we calculate corrected displacement source spectra from eGf deconvolution, we further extrapolate the spectra to the high and low frequencies following the Brune ω-squared model (similar to illustration in BALTAY *et al.* 2010; Fig. 1d). The low frequency point near 1–4 mHz at which the mainshock spectrum reaches the moment level is taken as an effective corner frequency, below which the spectrum is extrapolated along the flat, moment, part of the spectrum, to frequencies less than 0.1 mHz. In this high frequency range, the spectrum is extrapolated as ω-squared to frequencies greater than 100 Hz. Figure 4 fully illustrates these limits, with the thick lines representing the range of the eGf events, the thin lines in Fig. 4a the extrapolation to the moment level, and finally, the cumulative curve in Fig. 4b showing the large range over which the integration is performed.

Radiated energy is proportional to the integral of the corrected velocity spectra squared,

$$E_R = \frac{I}{4\pi^2 \rho \beta^5} \int_0^\infty \left| \omega \cdot \dot{M}(\omega) \right|^2 \mathrm{d}\omega, \tag{2}$$

where $\dot{M}(\omega)$ is the moment rate (displacement) spectrum, $\beta = 3.6$ km/s, $I = 2/5$ and $\rho = 3{,}000$ kg/m^3. We estimate E_R for the mainshock as corrected by each eGf event, and for each of three components of motion, at each GSN station. Although we use both P and S wave, the relative eGf deconvolution yields similar source spectra (i.e., Fig. 4a) and hence the single Eq. (2) is valid to calculate the energy, regardless of the component. We also calculate scaled energy and apparent stress after WYSS and BRUNE (1968),

$$\tau_a = \mu \frac{E_R}{M_o}, \tag{3}$$

with the shear modulus $\mu = 39$ GPa, using $\mu = \rho\beta^2$.

3. *Choice of Empirical Green's Function*

While our use of large, teleseismic empirical Green's functions in this study allows us to calculate radiated seismic energy from the largest earthquakes, the method is not without some of the limitations that also affect local and regional eGf deconvolution. Typically, it is considered that the eGf event should be (1) small enough in comparison to the main event to approximate a point-source; (2) close enough to the main event so that path effects are similar; and (3) of similar focal mechanism so that the radiation patterns of the eGf and main event are similar. The choice of eGf events can be a source of apparent variability in the spectral signature of the main event, and deviations from the conditions above can introduce errors in the parameters estimated for the main event (e.g., KANE *et al.* 2013; MORI and FRANKEL 1990; ABERCROMBIE 2013). KANE *et al.* (2013) determined that the most important criteria for an eGf event is co-location to the mainshock. They suggest an optimal eGf-mainshock separation distance of about 1 km for events with radius ~0.1 km when observing at local distances (~10 km); extending these ratios to teleseismic recording distances (3,000–10,000 km) and great earthquake dimensions (tens of km radius) implies an eGf-mainshock separation of ~100 km is reasonable. KANE *et al.* (2013) also suggest that eGf size in relation to the mainshock and similarity of mechanism are important. Because teleseismic eGf deconvolution to estimate seismic energy has not been undertaken before, we test some of these assumptions.

In the 2005 M_w 8.7 Nias and 2007 M_w 8.5 Bengkulu data sets, we consider up to 14 possible eGf events each, ranging from M_w 6.0 to M_w 7.8, with normal and strike-slip mechanisms, as well as thrust events that match the mechanism of the mainshock. The locations are spread throughout the fault plane. We use these events to test the limitations on the choice of eGf. For the events considered, the $\log_{10}$ standard deviation of the mainshock energy determined using all potential eGf events is 0.25 (2005 Nias) and 0.40 (2007 Bengkulu). With just the chosen eGf events, shown in Tables 4 and 5, the $\log_{10}$ std drops to 0.14 and 0.09, showing that the correct choice of eGf events gives much more stable and reliable mainshock energy estimates.

We find that anything less than ~M 6.5 is too small to use as an eGf, at least for teleseismic distances, and that energy results determined with these eGfs are systematically smaller than the average (Fig. 3; Events 5, 7 and 8). Events of this size have peak amplitudes near 10^{-6} m/s (vertical) at teleseismic distances, and corner frequencies near 0.1 Hz. Microseismic noise has a peak at 8 s, with a median amplitude of about 3×10^{-7} m/s (ASTER *et al.* 2008), and hence the eGfs of M < ~6.5 are too close to the microseismic background noise, and may display artificially elevated amplitudes. When these events are used as eGfs, the corrected mainshock spectrum is biased towards low amplitude in the microseismic band, and leads to lower estimates of radiated energy.

We also consider events as large as M_w 7.8–7.9 as potential eGfs. A rule of thumb for choosing an appropriate eGf is a magnitude separation of 1.5 units between the mainshock and eGf (e.g. COURBOULEX *et al.* 1996; HOUGH 2001; KANE *et al.* 2011). In the case of the 2005 M_w 8.7 Nias, we find that the largest potential eGf (M_w 7.8 2010/04/06) is too large to be used as an eGf for this event, with a magnitude difference of only 0.9. Our teleseismic eGf deconvolution method is very sensitive to relative spectral amplitudes, and it is possible for the large eGfs to have amplitudes as high or higher than the mainshock at some frequencies, which leads to unstable, unreliable radiated energy estimates. Event 11 shown in Fig. 5 from the Bengkulu sequence is actually the October 2010 M_w 7.8 tsunami earthquake, with depleted radiated energy when compared with a typical earthquake of its magnitude (LAY *et al.* 2011; NEWMAN *et al.* 2011). Therefore, we would expect its spectra to have lower amplitudes than the ideal Brune spectra to which it was corrected. This gives rise to corrected mainshock spectra that are enhanced in energy, and thus an overestimated value of radiated energy. Because this event is known to be an enervated tsunami earthquake, it is not ideal to use as an eGf, and emphasizes the need for careful eGf selection.

To test the assumption of like mechanisms and locations, we consider potential eGfs with differing mechanisms and locations. We find that mainshock energies estimated using eGfs of similar location and mechanism (shallow thrust) have much smaller inter-station standard error, shown by error bars in Fig. 3. These are the results used in the final analysis. While eGfs with a normal sense of motion yield mainshock

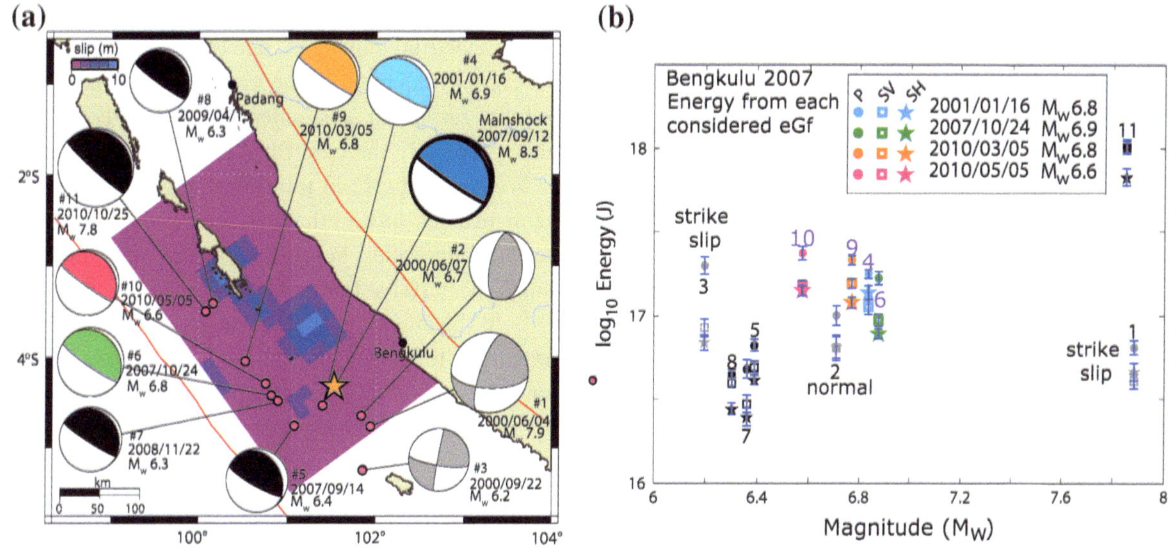

Figure 3

a Map showing location, size and mechanism of the 11 potential eGf events and **b** three components of radiated energy of the M_w 8.5 Bengkulu mainshock, as measured by the 11 different considered eGfs. Shown in *colors* are the four eGfs used in the analysis (#4, #6, #9, #10). Estimates from each of three components are shown in different *symbols* (*P*-wave *vertical* in *circles*; SV in *squares*; and SH in *stars*). *Blue bars* show standard error of the estimate using inter-station standard deviation. Event 1 is strike slip and gives greater variability between station measurements (*larger error bars*); Event 11, ~M 7.9, is the October 2010 tsunami earthquake, and is hence depleted in energy, so that its use as an eGf gives artificially elevated mainshock energy. Event 2 is discarded due to its opposite (normal) mechanism. Events 3, 5, 7 and 8 are too small, and hence underestimate the mainshock energy. In most cases, the *P* wave gives slightly higher radiated energy estimates, but all three components yield similar results

energies consistent with the average, the azimuthal distribution of energy determined by a normal eGf is more strongly variable (larger error bars in Fig. 3; Event #2). Strike-slip eGfs yield highly variable estimates of mainshock energy compared to the other events because the corrected mainshock spectra are less unstable, while eGFs with mechanisms consistent with that of the mainshock yield robust and stable results.

In the 2005 Nias sequence, we test a M_w 6.8 with a very similar mechanism to that of the mainshock, but located much closer to the trench and nearly 150 km from the area of high slip. Mainshock energy estimated from this eGf has the highest inter-station standard error of any considered thrust eGf. We find that eGfs close to the epicenter (within ~100 km) yield energy estimates with less variation between *P* and *S* components, as well as smaller inter-station standard error.

Also of interest is the effect of depth of the eGf event. However, the events considered here as potential eGfs are all located at similar depths between 21 and 36 km, on the subducting plate interface. Most of the events are located not far from the areas that ruptured in the main event, as can be seen in Fig. 3a. Due to the lack of diversity in depth, we don't see any correlation between the eGf quality and depth. We may expect to see an effect if eGf events were significantly deeper or below the subduction plate interface, or if they were very shallow in the wedge. It is also possible that at teleseismic distances, the depth is a less important factor, as compared to the mechanism, magnitude and location. Hence, we conclude that a radius of 100 km is the upper limit for mainshock-eGf distances, within which eGfs provide stable radiated energy estimates for the range of magnitudes we are working with.

4. *Radiated Energy from Six Recent Great Earthquakes*

For each of the six recent great earthquakes, we consider as many as 14 different empirical Green's function events, and use as many as five for the final analysis (Tables 1, 2, 3, 4, 5, 6). In each case, there is

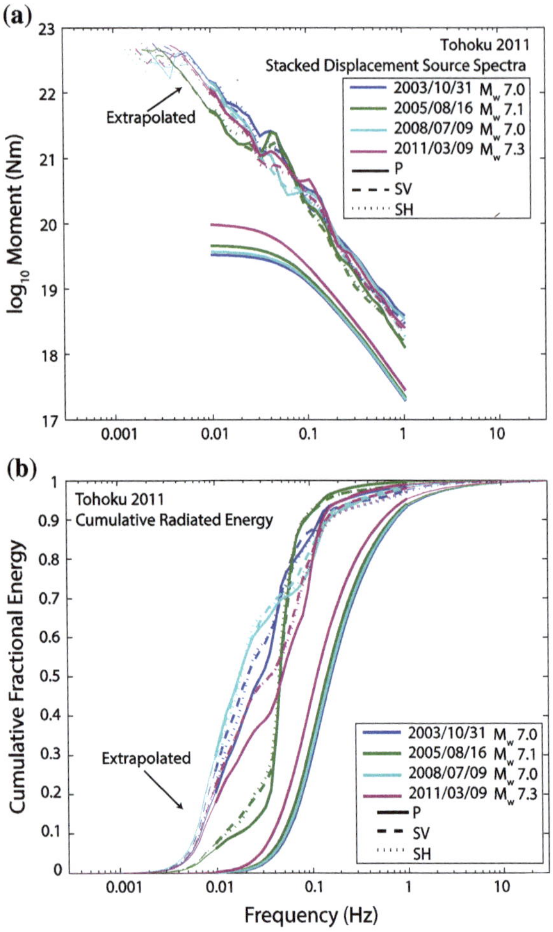

Figure 4

a 2011 Tohoku-Oki displacement source spectra for each eGf, and for the mainshock after deconvolving the different eGf events, stacked over all stations. **b** Cumulative fractional energy of the mainshock corrected individually by each eGf, and stacked over all stations. *Light lines* in the low frequencies of the mainshock in (**a**) and (**b**) show the missing, extrapolated energy, which amounts to ~10–20 % on average

close correspondence between the mainshock spectra when corrected by different eGfs, indicating that the corrected source spectra are robust with respect to the assumed eGf (Fig. 4a, S2a, S4a, S6a, S8a, S10a). By examining the cumulative fractional energy of the mainshock at each frequency, we find that the extrapolated energy is very small in some cases, and at most accounts for ~20 % of the total energy (Figs. 4b, S2b, S4b, S6b, S8b, S10b). In studies of smaller earthquakes, the energy integral is routinely extrapolated to higher frequencies (e.g. BALTAY *et al.* 2010). In the case of great earthquakes, however, we are able to measure the high frequency end of the energy spectrum quite well, due to their exceptionally low corner frequencies. The modest extrapolation we perform results in a smaller percentage correction than is typically the case for smaller earthquakes.

We also indicate in thin lines (Fig. 4) the part of the mainshock spectra that has been extrapolated into the low frequencies, below the low cut-off of the eGf, usually near 100 s. For an ideal Brune ω^{-2} spectrum, 80 % of the energy is at frequencies up to 6.25 times the corner frequency (e.g., SINGH and ORDAZ 1994). For these very large events with corner frequencies of several hundred seconds, that value is well below our 1 Hz high frequency cutoff, so we are measuring most of the radiated energy.

We find that analysis of *P*, *SV* and *SH* waves give very similar mainshock corrected spectra and energy values, and that the two *S*-wave energies are even closer in value, emphasizing the precision of the energy estimates. In many cases, the energy we estimate from the *P* wave is slightly higher. BOATWRIGHT and CHOY (1986) use primarily *P*-wave spectra for energy determination because *S* waves are strongly attenuated at teleseismic distances. Due to the extremely low frequencies of great earthquakes, the effects of attenuation are manageable, and allow us to make use of *S* waves, essentially tripling the available data.

For all the events, we study the azimuthal dependence of the radiated energy. Empirical Green's functions correct the mainshock for radiation patterns, but not for azimuthal variation due to the effect of source-finiteness. The eGf events are much smaller and do not display strong directivity at the periods we consider. The azimuthal distribution of mainshock energy is thus a source-finiteness effect. To avoid a bias in the station-averaged total radiated energy due to uneven station distribution azimuthally, we bin the station measurements into groups every 30° of azimuth. The final energy estimates for each event represent this azimuthally binned average, so that no cluster of nearby stations (in Europe or North America, for example) has too large an effect on the final radiated energy value.

Lastly, we quantify the aleatory variability of each final energy estimate. For each eGf correction, the energy estimate is made for three components at between 25 and 60 stations, and these estimates are then averaged to find the overall estimate of the radiated energy for the great earthquake. The minimum and maximum values quoted in Tables 1, 2, 3, 4, 5, 6 are 95 % confidence intervals on the average energy value, determined as ± twice the $\log_{10}$ standard error of the estimate, where the standard error is the ratio of the standard deviation of all the estimates (for all components and stations) and the square root of the number of records. We translate this $\log_{10}$ value into linear space in the table for ease of reading. While this 95 % confidence interval gives us valuable information about how stable each estimate is, it represents only the confidence within our method of the mean value dependent on different observations, and can yield no information about the epistemic uncertainty that is associated with our lack of knowledge about the true model or underlying processes of earthquakes. In this sense, some of the other tests we have considered, i.e., for choice of eGf, comparison to previous studies, contrasting the azimuthal variations or different energy values from the three components, or understanding the shortcomings of the spectral extrapolation of the mainshock into the low frequencies, may give a more enlightened view of the uncertainty associated with our estimates.

4.1. 2011 M_w 9.0 Tohoku-Oki

The M_w 9.0 Tohoku-Oki earthquake of 2011 occurred on March 11 offshore of Honshu, Japan, on the Japan Trench. The earthquake lasted about 160 s, on a rupture plane of about 240 km down dip × 440 km along strike of 200°, with prodigious shallow slip at the trench that generated a large tsunami (Fig. 1) (Ide *et al.* 2011). Waveforms from the four eGf events show very similar patterns in both the *P* and *S* waves. It is apparent from the source time function that the mainshock initially developed slowly, before rapidly increasing in size about a minute after the origin time (Fig. 2). The event was unusual in that it occurred on a compact rupture plane for an event of this size, but also had large, slow shallow slip coupled with deep slip that generated strong high frequency radiation (e.g., Ide *et al.* 2011; Suzuki *et al.* 2011; Koper *et al.* 2011).

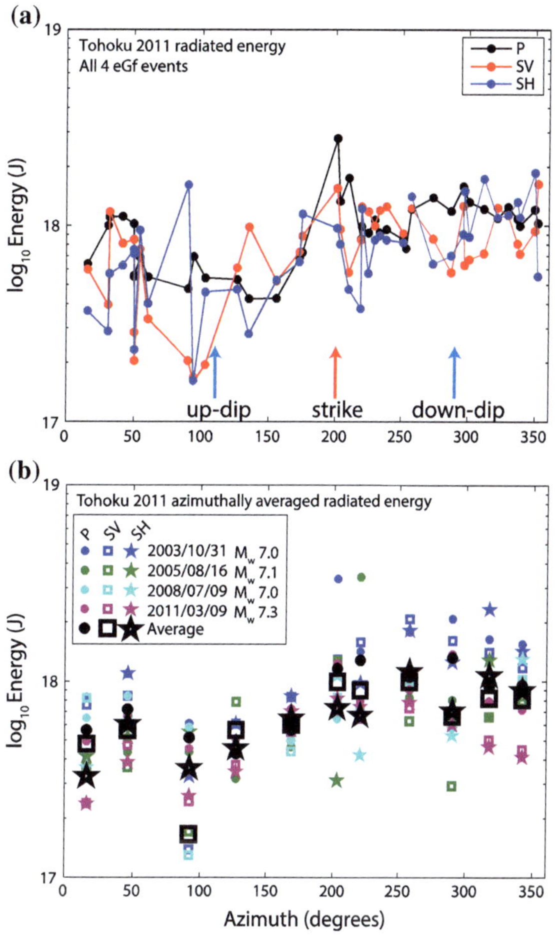

Figure 5
Mainshock radiated energy as a function of station azimuth. **a** $\log_{10}$ average over all eGfs, for each component, shown at all stations. **b** Energy shown as corrected from each eGf event and each component, azimuthally binned every 30°. Average of each bin shown in *black*, used for the final reported energy estimate. Different components shown with different *symbol shapes*

We find a radiated energy of 7.06×10^{17} J, a scaled energy of $E_R/M_o = 1.57 \times 10^{-5}$, and an apparent stress of 0.61 MPa (Table 1). This event shows strong directivity in radiated energy, with a minimum near 100° and a maximum near 200° (Fig. 5). This corresponds to energy directivity found by Convers and Newman (2011), Newman (2011): http://geophysics.eas.gatech.edu/anewman/research/RTerg/. Our cumulative energy shows that the extrapolated energy into the low frequencies is around 10–20 % on average, and that about 90 % of the energy is radiated below 0.1 Hz (Fig. 4).

4.2. 2004 M_w 9.1 Sumatra

The 2004 M_w 9.1 Sumatra earthquake was the first great earthquake to test the GSN. It generated a devastating tsunami, which was made worse by the lack of a tsunami warning system for the Indian Ocean Basin. The earthquake lasted for nearly 500 s, almost three times as long as the Tohoku-Oki event, and its rupture length was the greatest ever recorded—nearly 1,300 km along strike, with a variable slip distribution from the epicenter in the south to the northern tip of the rupture (Ammon *et al.* 2005; Lay *et al.* 2005; Chlieh *et al.* 2007). This event may have had a large slow slip component, releasing energy for up to 1 h after the initiation (Ammon *et al.* 2005; Bilham 2005; Banerjee *et al.* 2005; Lay *et al.* 2005).

Even at teleseismic distances, the 2004 Sumatra earthquake, with its unusually long rupture plane, may not be able to be modeled using a single eGf event; however, Ammon *et al.* (2005) demonstrated that much of the slip occurred in the first 230 s, on the southern-most 420 km of the fault plane (Fig. 1). Since we use the first 270 s of *P* and *S* arrivals, we should capture this initial intense rupture, and be able to model that contribution to the source spectrum from single event eGf deconvolution, provided that the eGfs are located near the patches of high slip. This part of the rupture was along a plane approximately 420 × 240 km, with a strike of about 330°, and average slip of 7 m, compared to 2–5 m average slip on the northern parts of the fault. There are four events with similar mechanism to the mainshock, occurring very near the location of high slip. While the four eGf events have consistent, impulsive arrivals, the mainshock, like the Tohoku-Oki earthquake, is less impulsive and has an obviously longer duration overall (Fig. S1).

We find the energy of the 2004 Sumatra event to be 7.19×10^{17} J, with a scaled energy of 1.12×10^{-5}, and apparent stress of 0.43 MPa, which is slightly less than that of the Tohoku-Oki event (Table 2). The *S* components of the corrected mainshock have more energy in the higher frequencies than do the *P* components, which can also be seen in the cumulative energy (Figure S2b). It is possible that the 270 s window we use around the *S*-wave arrival is also capturing other *P* arrivals due to the long rupture length, which would affect the frequency content. About 80 % of the mainshock energy is below 0.1 Hz.

The 2004 Sumatra event also shows strong directivity in radiated energy, which is expected due to the long and narrow fault plane that ruptured nearly unilaterally to the north (Fig. S2c–d). This directivity is consistent with that found for Rayleigh wave amplitudes, with a large peak towards ~270°–320°, a smaller peak near 20°, and a minimum around 135° (Ammon *et al.* 2005). Although Kanamori (2006) states that there is little directivity effect in his estimation of radiated energy from the 2004 Sumatra earthquake, the stations that he uses show similar energy results. By including more stations, we can better resolve the signature of azimuthal variability in radiated energy.

4.3. 2010 M_w 8.8 Maule

The 2010 M_w 8.8 Maule earthquake occurred along the Chilean subduction zone, just to the north of the 1960 Chilean M_w 9.5 megathrust event (Moreno *et al.* 2010). The rupture occurred on a plane about 150 km along-dip, up to the trench, by 400 km along strike, in NNW direction, and lasted 150 s (Hayes 2010). This event caused a tsunami, but of smaller amplitude than either the 2004 Sumatra or 2011 Tohoku-Oki tsunami. We identify four potential empirical Green's functions near the area of high slip, and discard one due to its distance from the epicenter. The 2010 Maule mainshock shows somewhat more impulsive arrivals, compared to the 2004 Sumatra and 2011 Tohoku-Oki earthquakes, yet the eGf events are even more impulsive (Fig. S3).

We find a radiated energy of 1.86×10^{17} Joules, $E_R/M_o = 1.03 \times 10^{-5}$ and $\tau_a = 0.40$ MPa, averaged over all four eGf events and three components (Table 3). The mainshock source spectra contain a consistent bump near 0.1 Hz that we interpret as a source effect (Fig. S4a). About 20 % of the total energy is from extrapolation to low frequencies, below 10 mHz. The three components of motion show a similar pattern as a function of azimuth, with a minimum of radiated energy towards the southwest, at about 250°, but the azimuthal variation is not as strong as that for the two larger events (Fig. S4c–d).

4.4. 2005 M_w 8.7 Nias

The 2005 M_w 8.7 Nias earthquake occurred south of the 2004 M_w 9.1 Sumatra, along the Sunda trench. The rupture plane is about 150 km wide and 300 km along strike, to the NNW. The event lasted about 120–150 s, and ruptured bilaterally, to the north and south, in two distinct patches (Konca *et al.* 2007). The rupture did not come as close to the trench as the 2004 Sumatra event, however, and generated a much smaller, local tsunami (Dean *et al.* 2010). In this case, we consider 14 different eGf events, but discard most of them to use five earthquakes, occurring both before and after the mainshock (see discussion above). The five chosen eGf events show very similar waveforms to each other, especially in the *P* components (Fig. S5).

The final estimate of radiated energy for all eGf events is 2.19×10^{17} J, with a scaled energy of 2.09×10^{-5} and apparent stress of 0.81 MPa (Table 4). On average, <20 % of the energy is extrapolated into the low frequencies, and in the cumulative fractional energy, the mainshock spectra look fairly Brune-like (Fig. S6a-b). The azimuthal dependence of radiated energy is not as strong as that of some of the other great events, but we do see a local maximum near 150°, along strike to the south (Fig. S6c–d).

4.5. 2007 M_w 8.5 Bengkulu

The 2007 M_w 8.5 Bengkulu event occurred south of the 2005 Nias along the trench, and just off shore of the town of Bengkulu. The rupture occurred in two distinct patches along strike, over 100 s (Konca *et al.* 2008). The rupture plane was 100×200 km, striking about 330°, similar to the 2005 Nias and 2004 Sumatra events. For this mainshock, we considered 12 potential empirical Green's function events, ranging from M_w 6.3 to M_w 7.9, but choose to use four, and discard those with different mechanism or magnitude too large or small (see discussion above). The four eGf events and the mainshock have very similar waveforms, as shown at station MDJ (Fig. S7), comparable to the Nias mainshock and eGf events.

The corrected source spectra show a large hole at about 100 s (Fig. S8a). This is likely due to the rupture occurring in two distinct bursts, one represented by a spectrum with a higher corner frequency, and one with a lower corner frequency, so that the interference of the two yields the spectral hole we observe. This peculiar feature shows up in the cumulative fractional energy, so that the amount extrapolated is very low, <10 % on average.

We find an average radiated energy of 1.33×10^{17} J, with $E_R/M_o = 1.98 \times 10^{-5}$, and $\tau_a = 0.77$ MPa (Table 5). The azimuthal distribution of energy is not as pronounced as for the other great events, but there is a slight peak to the north, between 300° and 50° (Fig. S8c–d). We also note that the *P* wave spectra show consistently higher energy values than the *S* waves.

4.6. 2012 M_w 8.6 Off-Sumatra

The 2012 M_w 8.6 off-Sumatra earthquake is the largest strike slip event ever recorded, occurring deep (~40 km) in an intraplate region off of the north coast of Sumatra (McGuire and Beroza 2012; http://www.globalcmt.org/) (Fig. 1a). The region between the Sunda trench, the location of three of the thrust events discussed here, and the Ninety-East ridge is a zone of distributed seismicity that may be associated with a wide diffuse plate boundary bisecting the Indo-Australian Plate (Royer and Gordon 1997). Back projection results indicate that the M_w 8.6 ruptured along several NW–SE and NE–SW trending orthogonal faults, deep in the oceanic lithosphere, spanning a fault surface of about 400×300 km total (Meng *et al.* 2012). The following M_w 8.2 aftershock occurred along the southern portion of these faults. Whether this oceanic strike-slip event has implications for the maximum size of continental faulting is an interesting topic of conversation.

We use a magnitude 7.2 event from January 2012 for an eGf, as the aftershocks of ~M 6.2 are too close to the noise level to use teleseismically, yielding a lower energy estimate. From the M_w 7.2 eGf, we estimate an azimuthally averaged radiated energy of 6.71×10^{17} J, and apparent stress of 3.07 MPa, about five times more energetic than the five thrust events (Tables 6 and 7). There is no discernable azimuthal pattern to the radiated energy (Fig. S10c–d), but given that the rupture occurred bilaterally on multiple, conjugate fault planes, we should not expect much directivity overall.

Table 7

Radiated energy for all five great earthquakes, compared to estimates from other studies. The $\log_{10}$ 95% confidence intervals in energy from previous tables are translated into apparent stresses, τ_a.

Event	Radiated energy, E_R (J $\times 10^{17}$)					E_R/M_o ($\times 10^{-5}$)	τ_a (MPa)		
	Newman[a, b]	Lay[c]	USGS[d]	This study	This study[e]		Mean	Min	Max
2011 Tohoku-Oki	4.2[a]	4.18	5.1	7.06	4.22	1.57	0.61	0.57	0.65
2004 Sumatra	8.2[b]	2.98	1.1	7.25	4.33	1.12	0.43	0.40	0.46
2010 Maule	2.6[b]	2.10	0.47	1.86	1.11	1.03	0.40	0.38	0.43
2005 Nias	1.1[b]	0.82	0.37	2.19	1.31	2.09	0.81	0.78	0.84
2007 Bengkulu	0.69[b]	0.44	0.26	1.33	0.79	1.98	0.77	0.73	0.81
2012 Off-Sumatra	2.1[a]	n/a	2.7	6.71	4.01	7.89	3.07	2.40	3.93

[a] http://geophysics.eas.gatech.edu/anewman/research/RTerg/

[b] Convers and Newman (2011)

[c] Lay *et al.* (2012)

[d] As reported on http://earthquake.usgs.gov, following Boatwright and Choy (1986)

[e] Based on differences in material parameters, estimates from this study would be 1.67 times larger that those of Newman, so we provide the comparison of our results/1.67

5. Constant Scaled Energy and Apparent Stress

We find that the scaled energy of the six great earthquakes is similar in size to the scaled energy for smaller earthquakes. The scaled energy of all five events is between ~ 1 and 2×10^{-5}, corresponding to an apparent stress of 0.4–0.8 MPa, and the strike-slip off-Sumatra earthquake has apparent stress of ~ 3 MPa. These apparent stresses are well within the range of global values, and the mechanism dependence is consistent with the notion of an immature strike-slip event in cold lithosphere vs. thrust faulting on mature subduction boundaries.

We compare our results to other studies (Convers and Newman 2011; Lay *et al.* 2012; NEIC reported radiated energy, http://earthquake.usgs.gov; Boatwright and Choy 1986) in Table 7. These studies make attenuation, source, and station corrections to teleseismic *P* waves to estimate source spectra and radiated energy. Results from our study compare well with those of Convers and Newman (2011) and Lay *et al.* (2012). With the exception of the 2011 Tohoku-Oki event, the USGS estimates are systematically much lower than the others, as also noted by Newman and Okal (1998) and Di Giacomo *et al.* (2010). This discrepancy is potentially caused by too short a time window in the NIEC analysis. Although we use the same value of ρ as Convers and Newman (2011), their use of $\alpha = 7{,}000$ m/s and a *S*-to-*P* wave partitioning factor of 15.6 actually implies that their values should be 1.67 times smaller than ours. To clarify this comparison, Table 7 also shows the results of this study divided by 1.67. Then, the radiated energies are even more similar from our results. However, there are still sources of uncertainty between the different methods, such as how different station energies are averaged, that could account for the differences between the studies. That our teleseismic eGf deconvolution yields radiated energy results in agreement with other studies validates this method, and demonstrates that it can be used to estimate robust source spectra and other source parameters derived from it. An inherent advantage of eGf techniques is that specific source and attenuation information is not required, and the result is only dependent on the eGf assumptions. We also show that *S* waves, in addition to *P* waves, can be used in teleseismic energy calculation for great earthquakes, which greatly increases the number of records that can be analyzed. For smaller events, the increased attenuation of *S* waves may not allow the same calculation at teleseismic distances.

Convers and Newman (2011) study 342 global earthquakes with $M > 6.5$, and find an average scaled energy, $E_R/M_o = 1.82 \times 10^{-5}$ for thrust events, and 3.63×10^{-5} for strike-slip earthquakes. For our

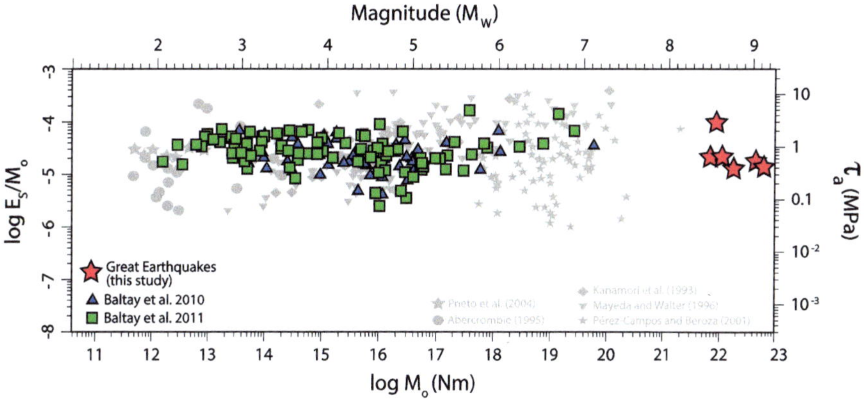

Figure 6
Scaled energy and apparent stress of the six great earthquakes (*red stars*) overlain on the figure of Ide and Beroza (2001) compiling many other studies. Also shown are the scaled energy results from Baltay *et al.* (2010) from the western US, and Baltay *et al.* (2011) from Honshu, Japan. The energy of great earthquakes is consistent with that of other earthquakes, from ~M 2 to M 7. Apparent stress is similar and between 0.4 and 0.8 MPa for the five thrust events, and higher for the 2012 off-Sumatra strike-slip event

sample size of only six earthquakes, it is problematic to determine a difference between mechanisms, yet the factor of ~5 difference between the thrust events and the 2012 Sumatra strike slip event is consistent with Pérez-Campos and Beroza (2001) between those mechanisms for a much larger data set of somewhat smaller earthquakes.

Figure 6 compares our results for scaled energies to previous studies of Baltay *et al.* (2010) and (2011), who used a local empirical Green's function technique to find scaled energy of ~200 events in the western United States and eastern Honshu, Japan, ranging from magnitude ~2 to ~7. Inclusion of the six recent great earthquakes emphasizes the lack of dependence of scaled energy on moment, as these observations show no systematic variation in apparent stress with magnitude.

6. *Azimuthal Dependence of Radiated Energy*

We investigate the azimuthal variations of radiated energy from the six great earthquakes. Because the eGf events are much smaller compared to the mainshock, directivity in the mainshock at long periods should be preserved in measurements of the radiated energy. We see the strongest azimuthal dependence and clearest trends for the largest events, 2004 M_w 9.1 Sumatra and 2011 M_w 9.0 Tohoku-Oki. The other three thrust events also show azimuthal variation of radiated energy, but the patterns are not as well resolved. The 2012 M_w 8.6 off-Sumatra earthquake shows little azimuthal variation, likely due to the fact that it includes complex rupture on multiple perpendicular fault planes. In all cases, although the initial waveforms are quite different; the *P* and *S* components show similar trends, which implies that the relative differences between mainshock and eGf events is the same regardless of the component of motion (Fig. 7).

In the Tohoku-Oki case, we find an energy peak in the strike direction at ~200°, a small peak in the anti-strike direction (along-strike in the opposite direction) at ~20°, and a low near ~100°. Based on the spatio-temporal slip inversion of Ide *et al.* (2011), we interpret the peaks at 200° and 20° to be generated by high apparent velocity along and parallel to the trench as the rupture reaches the trench. The minimum in radiated energy at 100° is in the up-dip, trench direction, which is explained by the observation of high-frequency energy, >0.1 Hz, generated from the down dip (~270°) area, so that the low-frequency energy in the trench-ward direction doesn't contribute much to the overall broadband radiated energy estimated here.

Because of the extended, 1,300 km long rupture of the 2004 Sumatra earthquake, and the strong unilateral rupture propagation for that event, we expect

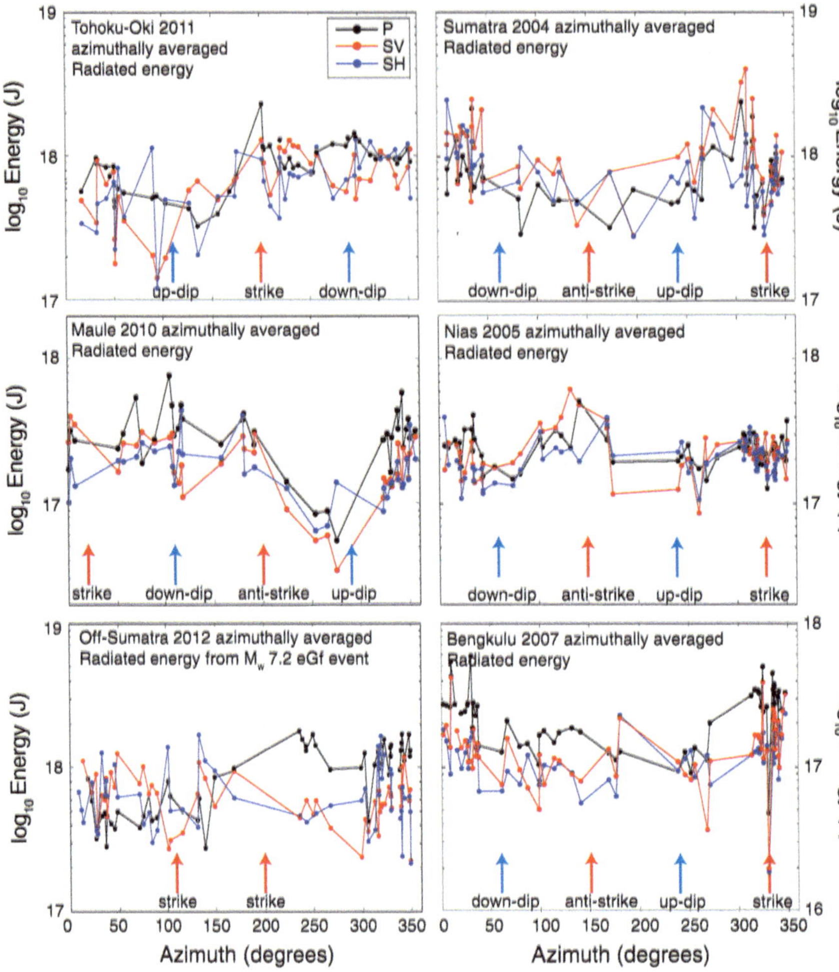

Figure 7
Azimuthal dependence of radiated seismic energy all six great earthquakes, shown as the $\log_{10}$ mean overall eGf corrections, for each component. "Strike" indicates direction or rupture propagation, while "anti-strike" is along the strike in the opposite direction

directivity along strike, towards the north (Ammon *et al.* 2005). We see this clearly in the radiated energy, with a strong peak just west of the 330° strike. In addition, there is a minimum of radiated energy in the anti-strike direction, at ~150°, which is also explained by the northward rupture directivity. In contrast to the Tohoku-Oki event, which occurred on a compact fault plane with rupture moving up- and down-dip, the 2004 Sumatra event started in the south and moved unilaterally to the north, along strike.

The 2010 M_w 8.8 Maule earthquake shows a strong minimum of radiated energy to the west, at about 250°, and a slight peak to the east, near ~90°. The other azimuths show fairly constant radiated energy. The rupture started in the south, with a concentration of slip near the epicenter, and then moved to the north with another concentration of high slip (Kiser and Ishii 2012). The minimum in radiated energy to the west is in the trench-ward direction, and in that sense our observations are similar to the minimum in energy seen in the up-dip direction for the Tohoku-Oki earthquake.

Both the 2005 M_w 8.7 Nias and 2007 M_w 8.5 Bengkulu earthquakes show weaker azimuthal variation in radiated energy, perhaps due to their smaller size. The Nias earthquake consisted of bilateral rupture, to the north at 330° and to the south at 150° (Konca *et al.* 2007). We find peaks in radiated energy in these two directions, with a stronger peak to the south, consistent with the larger slip patch in the southern portion. The Bengkulu rupture also occurred in two distinct south and north patches, but with rupture direction to the north only. In the radiated energy, we resolve only a broad peak in the northerly, along-strike direction.

The 2012 off-Sumatra strike-slip event shows no distinct azimuthal variation in radiated energy; however, this lack of directivity is attributable to rupture on multiple, orthogonal fault planes (Meng *et al.* 2012).

We can discern general source rupture characteristics from the radiated energy. The larger events show stronger azimuthal variation. If radiated energy estimates are to be used for rapid response (see discussion below), then the possibility of variations in single-station estimates due directivity must be taken into account. When paired with a time dependent slip inversion, we can interpret the maxima and minima in energy as particular rupture processes. Venkataraman and Kanamori (2004) conclude that directivity for dip-slip events with rupture propagating along strike is less than a factor of two at teleseismic distances. However, even for the 2005 Nias and 2007 Bengkulu events, which show the smallest directivity, we see at least a factor of two difference in the radiated energy between stations. The 2004 Sumatra earthquake, which does propagate along strike, shows nearly an order of magnitude difference of energy from station to station. Favreau and Archuleta (2003) forward modeled the energy radiated by the 1979 Imperial Valley earthquake, and found extreme directivity from the super-shear rupture, with 86 % of the energy radiated in the along-strike direction. Therefore, it is essential to take variability in energy due to directivity into account when estimated radiated energy teleseismically. With our method, we calculate energy at many stations, and take an azimuthally weighted average to account for the directivity. Expressing the uncertainty in estimates of radiated energy becomes imperative for rapid response; in cases of extreme directivity, the measure of radiated energy could be off by an order of magnitude depending on the stations considered. In an unlucky case where the early data is from the direction of negative directivity, the underestimation of energy would present an unrealistically enervated view of a potentially devastating earthquake.

7. Real-Time Determination of Radiated Energy

Radiated seismic energy is a fundamental earthquake source parameter, and reflects both the amount of material displaced and the speed at which that occurs. Earthquakes radiating greater amounts of energy per unit fault area are expected to produce larger ground motion, and the high frequency energy content of earthquakes can greatly affect the built environment. On the other hand, earthquakes that have a low ratio of radiated energy to seismic moment release energy more slowly and have the potential to create anomalously large tsunami for their size (Newman and Okal 1998). Rapid determination of radiated energy as a dynamic measure of the earthquake size could aid in hazard mitigation efforts or rapid emergency response, and could quickly indicate if an earthquake will generate unusually large tsunami amplitudes, or will likely cause large ground shaking.

During the 2011 Tohoku-Oki earthquake, the Japanese early warning and information systems worked well, as many Japanese were warned of ground shaking and tsunami waves before their arrival. However, Sagiya *et al.* (2011) noted that although the system worked well, more accurate, quick determination of source parameters would increase the efficacy of all of the systems, and in any case redundancy from independent real-time hazard assessment would have great value. Tsunami are generated most strongly at the shallowest part of rupture (i.e., at the trench), which is typically far enough offshore, that even local tsunami will take up to 30 min to arrive, so a rapid assessment of the radiated energy from an earthquake made within ~10 min would still allow 20 min of warning.

We propose that our method could be used for rapid energy estimation to support other ongoing approaches. A global eGf database could be set up so that the largest global earthquakes can quickly be analyzed after their occurrence. Given a magnitude difference of 1.5 magnitude units between the eGf and the mainshock and our eGf magnitude limit of ~6.5, we surmise that the smallest earthquake that could be studied teleseismically with our method is about M 8. Figure 8 shows the distribution of magnitude 6.5 and greater earthquakes globally for the

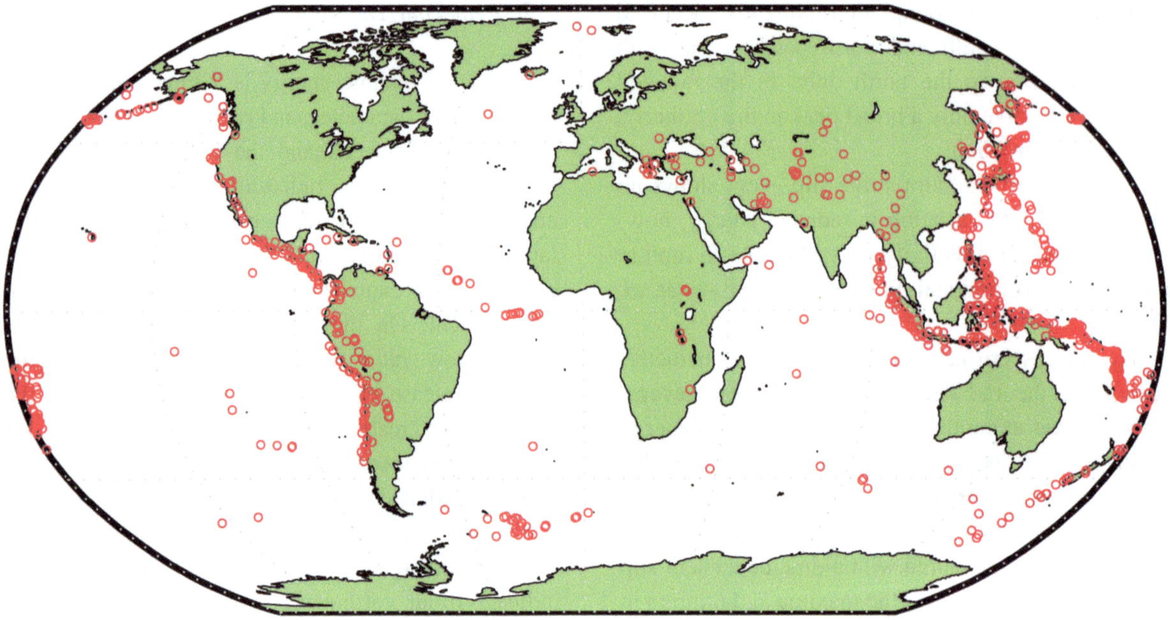

Figure 8
Global earthquakes > M6.5 from 20 years (1990–2010). Each earthquake is shown as a *circle* of radius 100 km, which we find to be the approximate radius of influence of eGf events. Mainshocks occurring within a *circle* can be corrected acceptably well by eGf deconvolution to estimate radiated energy

20-year period (1990–2010). Each earthquake is shown as a circle of radius 100 km, which we found to be the most effective range of eGf events for mainshock deconvolution. Hence, any $M \geq 8$ mainshock located within any 100 km radius circle could be studied with our method. There is dense coverage of $M_w > 6.5$ earthquakes along most of the world's subduction zones, where we expect the largest future hazardous earthquakes are most likely to occur. Regions such as the Mediterranean, San Andreas Fault system, and the Himalayan system have sparser coverage, so energy estimation using eGfs would not be as rapid. The radius over which the eGf deconvolution is robust and stable, however, would be frequency dependent. Because smaller events have proportionally more of their energy at shorter periods, where there is more anelastic attenuation, the correction may only work at shorter distances.

Potential eGf earthquakes, as shown in Fig. 8, recorded at reliable GSN stations would be prescreened and pre-processed (quality checked, windowed and tapered), and their three-component spectra stored in the database. Once a large earthquake occurs, it could quickly be located and matched with appropriate eGf events. Our energy estimation method is not computationally intensive, so the mainshock energy could be determined rapidly. The time consuming steps are in screening the potential eGf events and determining their parameters, which would be accomplished before the occurrence of a mainshock event, cutting back significantly on the processing time. Then, the radiated energy could be calculated within a few minutes of the arrival of waveforms at a station.

For the closest GSN stations considered here at 30° distance, *P* waves first arrive 6 min after rupture begins, and *S* waves begin to arrive at about 12 min. We have shown that there is no dependence of estimated energy on station distance, so that the closest stations could reliably be used, but for redundancy, at least six stations should be used for the estimation. We have shown that a directivity correction through averaging is important for an accurate result if stations are well distributed in azimuth. Since energy estimated from P and S waves yields consistent results, the estimation could be performed in real time using only the P arrivals, and then updated as the S waves arrive. Thus, an estimate of radiated seismic energy could be calculated within $\sim$10 min. That number could be reduced if the method is shown to be stable for earthquakes recorded at shorter distances.

Given that most tsunami are generated at the trench, and the propagation time from there to shore is on the order of 30 min, use of this method for quick tsunami potential determination would allow approximately 20 min for evacuation or early warning. Furthermore, the azimuthal energy dependence may quickly inform the rupture model and directivity, which could also be used to improve rapid hazard characterization.

Convers and Newman (2011) developed a method for fast determination of the radiated energy, which was recently implemented as an IRIS Data Management Systems product (http://www.iris.edu/dms/products/eqenergy/). Their method estimates radiated energy teleseimically from averaged attenuation models for the whole earth to account for energy loss, which is dependent on radiation pattern, and hence specific knowledge of the mainshock depth, location and mechanism. Di Giacomo *et al.* (2010) also proposed a similar system for rapid energy determination, using pre-computed numerically simulated Green's functions. They also pointed out the necessity of determining the radiated energy as a dynamic measure of the earthquake's size for rapid earthquake assessment, rather than simply the static moment magnitude.

The limitations of these two methods, as well as our own, include the fact that the data is teleseismic. In all cases, the size of the smallest event that can be analyzed is limited, although the method of Convers and Newman (2011) and Di Giacomo *et al.* (2010) are able to analyze smaller events than our proposed method. Secondly, the waves must travel a far distance from the source to the station before the energy can be estimated, which automatically delays the rapidity of the estimate; however, teleseismic methods are very valuable when there are few local stations, as is the case in some subduction zones, or when local stations go off-scale or off-line, which happened during the 2011 Tohoku-Oki event.

Our method has its own limitations, in that we need a priori moment, and the eGf events have to be computed from real data. However, rapid moment magnitude determinations are now routinely made, so they could be implemented into our method. The use of real eGf events implies that any heterogeneity in the earth structure near the source or site, or along the travel path, will automatically be accounted for, unlike in the other methods.

On the other hand, because our method is empirical and not specifically dependent on the mechanism, we could simultaneously run deconvolution with many potential eGf events, and find a converging best answer as more information arrives about the large earthquake. As well, although we have not tested the method for deep events, it is likely that they will not pose a large problem if appropriate eGf events of similar depth are used, while the approach of Convers and Newman (2011) cannot handle deep events without corrections to the assumed attenuation structure.

In any case, redundancy from independent real-time hazard assessment would have great value as we work towards mitigating the effects of great earthquakes. There is still work to be done to get these rapid energy estimates to be implemented for tsunami warning, and hence all possible approaches and information should be taken into account.

8. *Conclusions*

We expand on the teleseismic empirical Green's function deconvolution approach, using eGf events as large as M_w 7.4. We show that both P and S waves can be used in the eGf deconvolution teleseismically to give consistent energy estimates, whereas before, primarily only P waves were used at teleseismic distances (e.g. Boatwright and Choy 1986).

We find that radiated energy of six great earthquakes is consistent with the self-similar source physics seen for smaller events. The scaled energy of each of the thrust events is in the range 1–2 $\times$ 10^{-5}, and apparent stress ranges from 0.4 to 0.8 MPa. The great strike-slip 2012 off-Sumatra event has a higher apparent stress of 3 MPa. These values are very consistent with those found for a wide range of events, ranging from micro-earthquakes to other subduction zone megathrust earthquakes. Our results support a mechanism dependence on radiated energy, as has been found in previous studies, between energetic strike-slip (particularly oceanic lithospheric faulting) and enervated subduction zone thrust faulting. Our estimates of radiated energy for each event are similar to those of other researchers.

We demonstrated that teleseismic eGf deconvolution is robust and stable method for removing path and site effects using both *P* and *S* waves, to measure radiated seismic energy. EGF deconvolution could thus be used in other teleseismic applications to isolate source spectra or time series. IDE *et al.* (2011) used teleseismic eGf deconvolution to invert for a spatio-temporal slip distribution of the 2011 Tohoku-Oki earthquake, a method that could be applied to the great earthquakes we have analyzed here as well.

The azimuthal distribution of radiated energy is consistent with the sense of rupture propagation for each earthquake. The larger events show stronger and more varied directivity in energy, which can be interpreted as high apparent velocity along the trench, and slow rupture in the direction towards the trench. The smaller events show azimuthal dependence of energy as well, so that directivity effects should be considered when estimating radiated energy from teleseismic data for great earthquakes. Further work to couple time-dependent rupture propagation with the energy radiation may further illuminate source characteristics of megathrust earthquakes.

Acknowledgments

IRIS Standing Order of Data (SOD) (OWENS *et al.* 2004) was used to acquire Global Seismic Network (GSN) data. The authors thank Jesse Lawrence for help with acquiring GSN data and use of SOD. The authors thank Adrien Oth, Kevin Mayeda and Lusi Rivera for organizing and convening the workshop "Earthquake source physics on various scales," and ECGS for hosting. We also thank two anonymous reviewers and the guest editor, Adrien Oth, for their constructive comments and reviews which improved the manuscript. A. Baltay was partially supported at Stanford by a Gabilan Stanford Graduate Fellowship, as well as through financial support from Pacific Gas and Electric.

REFERENCES

ABERCROMBIE, R. E. (1995), *Earthquake source scaling relationships from −1 to 5* M_L *using seismograms recorded at 2.5-km depth*, J. Geophys. Res., *100*(B12), 24015–24036, doi:10.1029/95JB02397.

ABERCROMBIE, R. E. (2013), *Comparison of direct and coda wave stress drop measurements for the Wells, Nevada, earthquake sequence*, J. Geophys. Res., *118*, doi:10.1029/2012JB009638.

AMMON, C. J., A. A. VELASCO and T. LAY (1993), *Rapid estimation of rupture directivity: application to the 1992 Landers (MS = 7.4) and Cape Mendocino (Ms = 7.2), California earthquakes*, Geophy. Res. Lett., *20*(2), 97–100.

AMMON, C. J., C. JI, H.-K. THIO, D. ROBINSON, S. NI, V. HJORLEIFSDOTTIR, H. KANAMORI, T. LAY, S. DAS, D. HELMBERGER, G. ICHINOSE, J. POLET and D. WALD (2005), *Rupture Process of the 2004 Sumatra-Andaman Earthquake*, Science, *308*(5725), 1133–1139. doi:10.1126/science.1112260.

ASTER, R. C., D. E. MCNAMARA, and P. D. BROMIRSKI (2008), *Multidecadal climate-induced variability in microseisms*, Seismo. Res. Lett., *79*(2):194–202.

BALTAY, A., G. PRIETO and G. C. BEROZA (2010), *Radiated seismic energy from coda measurements and no scaling in apparent stress with seismic moment.* J. Geophys. Res. 115, B08314.

BALTAY, A., S. IDE, G. PRIETO, and G.BEROZA (2011), *Variability in earthquake stress drop and apparent stress*, Geophys. Res. Lett., *38*, L06303, doi:10.1029/2011GL046698.

BANERJEE, P., F. F. POLLITZ, and R. BURGMANN (2005), *The size and duration of the Sumatra-Andaman earthquake from far-field offsets*, Science, *308*, 1769–1772.

BILLHAM, R. (2005), *A flying start, then a slow slip*, Science, *308*, 1126–1127.

BOATWRIGHT, J. and G. CHOY (1986), *Teleseismic estimates of the energy radiated by shallow earthquakes*, J. Geophys. Res., *91*, 2095–2112.

BORMANN, P., and D. DI GIACOMO, (2011), *The moment magnitude* M_w *and the energy magnitude* M_e*: common roots and differences*, Journ. of Seismol., *15*(2), 411–427.

CHLIEH, M., J-P. AVOUAC, V. HJORLEIFSDOTTIR, T-R A. SONG, C. JI, K. SIEH, A. SLADEN, H. HEBERT, L. PRAWIRODIRDJO, Y. BOCK, and J. GALETZKA (2007), *Coseismic Slip and Afterslip of the Great Mw 9.15 Sumatra-Andaman Earthquake of 2004*, Bull. Seismol. Soc. Am., *97*, 152–173.

CHOY, G. L., and J. L. BOATWRIGHT (1995), *Global patterns of radiated seismic energy and apparent stress*, J. Geophys. Res., *100*, 18,205–18,228.

CHOY, G. L., A. MCGARR, S. H. KIRBY and J. BOATWRIGHT (2006), An overview of the global variability in radiated energy and apparent stress, in *Earthquakes: Radiated energy and the physics of faulting*, Geophysical Monograph Series, *170*, 43–57, doi:10.1029/170GM01.

CONVERS, J. A., and A. V. NEWMAN (2011), *Global evaluation of large earthquake energy from 1997 through mid-2010*, J. Geophys. Res., *116*, B08304, doi:10.1029/2010JB007928.

Courboulex, F., Virieux, J., Deschamps, A., Gibert, D. and Zollo, A. (1996), *Source investigation of a small event using empirical Green's functions and simulated annealing*. Geophys. J. Int., *125*:768–780. doi:10.1111/j.1365-246X.1996.tb06022.x.

Dean, S.M., L. C. McNeill, T. J. Henstock, J. M. Bull, S. P. S. Gulick, J. A. Austin, Jr., N. L. B. Bans, Y. S. Djajadihardja and H. Permana (2010), *Contrasting decollement and prism properties over the Sumatra 2004–2005 earthquake rupture boundary*, Science, *329*, 207–210.

Di Giacomo, D., S. Parolai, P. Bormann, H. Grosser, J. Saul, R. Wang, and J. Zschau (2010), *Suitability of rapid energy magnitude determinations for emergency response purposes*, Geophys. J. Int., *180*(1): 361–374, doi:10.1111/j.1365-246X.2009.04416.x.

Favreau, P. and R. J. Archuleta (2003), *Direct seismic energy modeling and application to the 1979 Imperial Valley earthquake*, Geophy. Res. Lett., *30*(5), 1198, doi:10.1029/2002GL015968.

Hanks, T. C., and W. Thatcher (1972), *A graphical representation of seismic source parameters*, J. Geophys. Res. *77*, 4292–4405.

Hartzell, S. H. (1978), *Earthquake aftershocks as Green's functions*, Geophys. Res. Lett. *5*(1), doi:10.1029/GL005i001p00001.

Hayes, G. (2010), *Finite Fault Model. Updated Result of the Feb 27, 2010 Mw 8.8 Maule, Chile Earthquake*. http://earthquake.usgs.gov/earthquakes/eqinthenews/2010/us2010tfan/finite_fault.php, (US Geological Survey/NEIC, 2010).

Hough, S. E. (2001), *Empirical Green's function analysis of recent moderate events in California*, Bull. Seismol. Soc. Am., *91*, 456–467.

Ide, S., and G. C. Beroza (2001), *Does apparent stress vary with earthquake size?*, Geophys. Res. Lett., *28*(17), 3349.

Ide, S., G. C. Beroza, S. G. Prejean, and W. L. Ellsworth (2003), *Apparent break in earthquake scaling due to path and site effects on deep borehole recordings*, J. Geophys. Res., *108*(B5), 2271, doi:10.1029/2001JB001617.

Ide, S., A. Baltay and G. C. Beroza (2011), *Shallow dynamic overshoot and energetic deep rupture in the 2001 M_w 9.0 Tohoku-Oki Earthquake*, Science, *332*, 6036, 1426–1429. doi:10.1126/science.1207020.

Ishii, M., P. M. Shearer, H. Houston and J. E. Vidale (2005), *Extent, duration and speed of the 2004 Sumatra-Andaman earthquake imaged by the Hi-Net array*, Nature, *435*, doi:10.1038/nature03675.

Kanamori, H. (1972), *Mechanism of tsunami earthquakes*, Phys. Earth Planet. Inter. *6*, 346–359.

Kanamori, H. (2006), The radiated energy of the 2004 Sumatra-Andaman earthquake, in *Earthquakes: radiated energy and the physics of faulting*, Geophysical Monograph Series, *170*, 59–68, doi:10.1029/170GM10.

Kanamori, H. and J. Given (1981), *Use of long-period surface waves for rapid determination of earthquake-source parameters*. Phys. Earth Planet. Inter. *27*, 8.

Kane, D. L., G. A. Prieto, F. L. Vernon, and P. M. Shearer (2011), *Quantifying seismic source parameter uncertainties*, Bull. Seism. Soc. Am., *101*, 535–543.

Kane, D. L., D. L. Kilb and F. L. Vernon (2013), *Selecting empirical Green's functions in regions of fault complexity; a study of data from the San Jacinto fault zone, Southern California*, Bull. Seismol. Soc. Am., *103*(2A), 641–650, doi:10.1785/0120120189.

Kiser, E. and Ishii, M. (2012), *Combining seismic arrays to image the high-frequency characteristics of large earthquakes*, Geophys. Jour. Inter., *188*:1117–1128. doi:10.1111/j.1365-246X.2011.05299.x.

Konca, A.O., V. Hjorleifsdottir, T-R. A. Song, J.-P. Avouac, D. V. Helmberger, C. Ji, K. Sieh, R. Briggs, and A. Meltzner (2007), *Rupture kinematics of the 2005 Mw 8.6 Nias-Simeulue earthquake from the joint inversion of seismic and geodetic data (in The 2004 Sumatra-Andaman earthquake and the Indian Ocean tsunami)*, Bull. Seismol. Soc. Am., *97*(1A):S307–S322.

Konca, A. O., J-P Avouac, A. Sladen, A. J. Meltzner, K. Sieh, P. Fang, Z. Li, J. Galetzka, J. Genrich, M. Chlieh, D. H. Natawidjaja, Y. Bock, E. J. Fielding, C. Ji and D. Helmberger (2008), *Partial rupture of a locked patch of the Sumatra megathrust during the 2007 earthquake sequence*, Nature, *456*. doi:10.1038/nature07572.

Koper, K. D., Hutko, A. R., Lay, T., Ammon, C. J., and Kanamori, H. (2011). *Frequency-dependent rupture process of the 2011 M_w 9.0 Tohoku Earthquake: Comparison of short-period P wave backprojection images and broadband seismic rupture models*. Earth, planets and space, *63*(7), 599–602.

Lay, T., H. Kanamori, C. J. Ammon, M. Nettles, S. N. Ward, R. C. Aster, S. L. Beck, S. L. Bilek, M. R. Brudzinski, R. Butler, H. R. DeShon, G. Ekstrom, K. Satake, and S. Sipkin (2005). *The great Sumatra- Andaman earthquake of 26 December 2004*, Science *308*, 1127–1133.

Lay, T., C. J. Ammon, H. Kanamori, Y. Yamazaki, K. F. Cheung, and A. R. Hutko (2011), *The 25 October 2010 Mentawai tsunami earthquake (M_w 7.8) and the tsunami hazard presented by shallow megathrust ruptures*, Geophys. Res. Lett., *38*, L06302, doi:10.1029/2010GL046552.

Lay, T., H. Kanamori, C. J. Ammon, K. D. Koper, A. R. Hutko, L. Ye, H. Yue, and T. M. Rushing (2012), *Depth-varying rupture properties of subduction zone megathrust faults*, J. Geophys. Res., *117*, B04311, doi:10.1029/2011JB009133.

Lay, T., Y. Fujii, E. Geist, K. Koketsu, J. Rubinstein, T. Sagiya, and M. Simons (2013), *Introduction to the Special Issue on the 2011 Tohoku Earthquake and Tsunami*, Bull. Seism. Soc. Am., *103*:1165–1170, doi:10.1785/0120130001.

McGuire, J. J. and G. C. Beroza (2012), *A rogue earthquakes off Sumatra*, Science, *336*, 6085, 1118–1119, doi:10.1126/science.1223983.

Meng, L., J.-P. Ampuero, J. Stock, Z. Duputel, Y. Luo and V. C. Tsai (2012), *An earthquake in a maze: compressional rupture branching in a weakened oceanic lithosphere during the April 11 2012 M8.6 Sumatra earthquake*, Science, *337*:6095. doi:10.1126/science.1224030.

Moreno, M., M. Rosenau and O. Oncken (2010), 2010 *Maule earthquake slip correlates with pre-seismic locking of Andean subduction zone*, Nature *467*, 198–202 doi:10.1038/nature09349.

Mori, J. Frankel, A. (1990). *Source parameters for small events associated with the 1986 North Palm Springs, California, earthquake determined using empirical Green functions*, Bull. Seism. Soc. Am. *80*, 278–295.

Newman, A. V., and E. A. Okal (1998), *Teleseismic estimates of radiated seismic energy: The E/M $_0$ discriminant for tsunami earthquakes*, J. Geophys. Res., *103*(B11), 26,885–26,898, doi:10.1029/98JB02236.

Newman, A. V., G. Hayes, Y. Wei, and J. Convers (2011), *The 25 October 2010 Mentawai tsunami earthquake, from real-time discriminants, finite-fault rupture, and tsunami excitation*, Geophys. Res. Lett., *38*, L05302, doi:10.1029/2010GL046498.

NEWMAN, A. (2011) Real-Time Earthquake Energy and Rupture Duration Determinations, http://geophysics.eas.gatech.edu/anewman/research/RTerg/2011/11031100/, Accessed November 6, 2011.

OWENS, T.J., H. P. CROTWELL, C. GROVES, and P. OLIVER-PAUL (2004), *SOD: Standing Order for Data*. Seismol. Res. Lett., *75*:515–520.

PÉREZ-CAMPOS, X., and G. C. BEROZA (2001), *An apparent mechanism dependence of radiated seismic energy*, J. Geophys. Res.; *106*, 11,127–11,136.

PEREZ-CAMPOS, X., S. K. SINGH and G. C. BEROZA (2003) *Reconciling teleseismic and regional estimates of seismic energy*, Bull. Seismol. Soc. Am. *95*(5), 2123–2130.

ROYER, J. Y., and R. G. GORDON (1997). *The motion and boundary between the Capricorn and Australian plates*. Science, *277*(5330), 1268–1274.

SAGIYA, T., H. KANAMORI, Y. YAGI, M. YAMADA, and J. MORI (2011), *Rebuilding seismology*, Nature, *473*, 146–148, doi:10.1038/473146a.

SINGH, S. K., and ORDAZ, M. (1994). *Seismic energy release in Mexican subduction zone earthquakes*. Bull. Seismol. Soc. Am, *84*(5), 1533–1550.

SUZUKI, W., S. AOI, H. SEKIGUCHI, and T. KUNUGI (2011), *Rupture process of the 2011 Tohoku-Oki mega-thrust earthquake (M9.0) inverted from strong-motion data*, Geophys. Res. Lett., *38*, L00G16, doi:10.1029/2011GL049136.

VENKATARAMAN, A., L. RIVERA, and H. KANAMORI (2002), *Radiated energy from the 16 October 1999 Hector Mine earthquake: Regional and teleseismic estimates*, Bull. Seismol. Soc. Am., *92*, 1256–1265, doi:10.1785/0120000929.

VENKATARAMAN, A. and H. KANAMORI (2004), *Effect of directivity on estimates of radiated seismic energy*, J. Geophys. Res., *109*, B04301, doi:10.1029/2003JB002548.

WYSS, M. and J. BRUNE (1968), *Seismic moment, stress, and source dimensions for earthquakes in the California-Nevada region*, J. Geophys. Res., *73*(14), 4681–4694, doi:10.1029/JB073i014p0468.

(Received May 13, 2013, revised February 17, 2014, accepted February 19, 2014, Published online April 4, 2014)

MIX
Papier aus verantwortungsvollen Quellen
Paper from responsible sources
FSC® C105338

If you have any concerns about our products,
you can contact us on
ProductSafety@springernature.com

In case Publisher is established outside the EU,
the EU authorized representative is:
Springer Nature Customer Service Center GmbH
Europaplatz 3, 69115 Heidelberg, Germany

Printed by Libri Plureos GmbH
in Hamburg, Germany